Categories for the Working Philosopher

Categories for the Working Philosopher

EDITED BY
Elaine Landry

"Trickery + treachery are the practices of fools who have not wits enuf to be honest."
— Benjamin Franklin

"One cannot continually disappoint a continent." James Wistler

OXFORD
UNIVERSITY PRESS

Great Clarendon Street, Oxford, OX2 6DP,
United Kingdom

Oxford University Press is a department of the University of Oxford.
It furthers the University's objective of excellence in research, scholarship,
and education by publishing worldwide. Oxford is a registered trade mark of
Oxford University Press in the UK and in certain other countries

© the several contributors 2017

The moral rights of the authors have been asserted

First Edition published in 2017

Impression: 1

All rights reserved. No part of this publication may be reproduced, stored in
a retrieval system, or transmitted, in any form or by any means, without the
prior permission in writing of Oxford University Press, or as expressly permitted
by law, by licence or under terms agreed with the appropriate reprographics
rights organization. Enquiries concerning reproduction outside the scope of the
above should be sent to the Rights Department, Oxford University Press, at the
address above

You must not circulate this work in any other form
and you must impose this same condition on any acquirer

Published in the United States of America by Oxford University Press
198 Madison Avenue, New York, NY 10016, United States of America

British Library Cataloguing in Publication Data

Data available

Library of Congress Control Number: 2017940285

ISBN 978-0-19-874899-1

Printed and bound by
CPI Group (UK) Ltd, Croydon, CR0 4YY

Links to third party websites are provided by Oxford in good faith and
for information only. Oxford disclaims any responsibility for the materials
contained in any third party website referenced in this work.

For Aldo, for everthing.
$(40 + n) - 8$

Preface

Often people have wondered why there is no introductory text on category theory aimed at philosophers. The answer is simple: what makes categories interesting and significant is their specific use for specific purposes. These uses and purposes, however, vary over many areas, both "pure", e.g., mathematical, foundational, and logical, and "applied", e.g., applied to physics, biology, and the nature and structure of mathematical models.

Borrowing from the title of Saunders Mac Lane's seminal work *Categories for the Working Mathematician*, this book aims to bring the concepts of category theory to philosophers working in areas ranging from mathematics to proof theory to computer science to ontology, from physics to biology to cognition, from mathematical modeling to the structure of scientific theories to the structure of the world.

Moreover, it aims to do this in a way that is accessible to a general audience. Each chapter is written by either a category-theorist or a philosopher working in one of the represented areas, and in a way that is accessible and is intended to build on the concepts already familiar to philosophers working in these areas.

As a rough and ready characterization, the "pure" chapters (Chapters 1–11) consider the use of category theory for mathematical, foundational, and logical purposes. I say "rough and ready" because, along the way, these chapters also investigate the application of category theory so considered to geometry, arithmetic, physics, and mathematical knowledge itself. The "applied" chapters (Chapters 12–18) consider the use of category theory for representational purposes; that is, they investigate using category theory as a framework for theories of physics and biology, for mathematical modeling more generally, and for the structure of scientific theories.

Chapter 1, by Colin McLarty, shows the sense in which categorical set theory, or the Elementary Theory of the Category of Sets (ETCS), grew to meet the varying needs of mathematical practice, including the analysis of specific problems in both physics and logic. His aim is to show that what mathematicians know and find useful about set theory is better captured by ETCS than by philosophers' typically preferred Zermelo Fraenkel (ZF) set theory. In Chapter 2, David Corfield argues that the best way to understand modern geometry, and its historical roots as found in the writings of Weyl and Cassirer, is via the notion of a homotopy topos and its internal language, Homotopy Type Theory (HoTT). Along the way, he demonstrates how this framework leads to new conceptions of space useful for our understanding of both arithmetic and mathematical physics. In Chapter 3, Michael Shulman provides an argument for taking Homotopy Type Theory and Univalent Foundations (HoTT/UF) as a new foundation for both mathematics and logic. Specifically, he uses homotopy theory, including Voevodsky's Univalence Axiom, to give the notion of equality a central role, without appealing to the often confused set-theoretic notion of isomorphism. In Chapter 4, Steve Awodey continues this thread by showing the geometric and logical significance of the Univalence Axiom, viz., that it can be used to capture the mathematical practice of identifying isomorphic objects. Thus, in so far as philosophically motivated structural foundations aim to capture objects "up to isomorphism", this work also aims to capture the foundational goals of the category-theoretic mathematical structuralist. Armed with the considerations of technical adequacy and autonomy, in Chapter 5 Michael Ernst takes up the philosophical debate of whether set theory or category theory is the best candidate for a mathematical foundation. He further develops an often-neglected aspect of the debates by considering the naturalness and usability of categorical foundations and further investigates how this might impact our accounts of both mathematical thinking and mathematical practice. Likewise driven by an attempt to understand the nature of mathematical knowledge and its relation to the systematic architecture of mathematics, in Chapter 6, Jean-Pierre Marquis underscores the foundational role of category-theoretic canonical maps.

Turning to logical considerations, Chapter 7, by John Bell, provides a history of the development of categorical logic from a topos-theoretic perspective. It covers both the semantics and syntax of logically formalized systems, including models of first-order languages, and ends by using the notion of a classifying topos to present the generic model of a geometric theory. The following Chapter 8, by Jean-Pierre Marquis, considers the philosophical motivation of Michael Makkai's First-Order Logic with Dependent Sorts (FOLDS) and shows the sense in which this can be used as a foundational framework for presenting abstract mathematical concepts. Kohei Kishida, in Chapter 9, uses category theory to develop a model theory for modal logic by focusing on the familiar Stone duality. Specifically, he aims to bring together Kripke semantics, topological semantics, quantified modal logic, and Lewis' counterpart theory by taking categorical principles as both mathematically and philosophically unifying. The presentation of categorical proof theory is the focus of Robin Cockett and Robert Seely's chapter. Starting with the cut rule as a basic component, they develop two logical calculi and show how the standard logical features can be added. Moreover, they consider both the symbolic power and the intuitive appeal of the use and significance of graphical representations. Samson Abramsky, in Chapter 11, uses category theory to analyze and frame the notion of contextuality, from within both a logical and a quantum mechanical domain, with the aim of dealing with those phenomena "at the borders of paradox". He shows how sheaf-theoretic notions arise naturally in this analysis, and how they can be used to develop a general structural theory for contextuality, with extensive applications to quantum information.

Shifting our focus to applications, and continuing to use both category theory and categorical diagrams for quantum mechanical purposes, in Chapter 12, Bob Coecke and Aleks Kissinger begin with a process ontology and frame quantum mechanics as a category-theoretic theory of systems, processes, and their interactions. They then use this formalism to argue for a compatibility between quantum and relativity theory, thus allowing them to recover a notion of causality that can be used to prove the no-signaling theorem. Concentrating on theories of spacetime, James Weatherall, in Chapter 13, provides an overview of recent applications of category theory to both general considerations of the theoretical structure of spacetime theories and particular claims of the theoretical equivalence of classical field theories, general relativity, and Yang–Mills theories. Ending the investigation into the application of category theory to physics, in Chapter 14, Joachim Lambek argues for the use of a six-dimensional Lorentz category, wherein spacetime, including basic physical quantities, are represented by six vectors which form the objects of an additive category in which Lorentz transformations appear as arrows, thus allowing for a unifying account of the structure of both known and unknown particles. In contrast to typical accounts of biological and cognitive processes that simply make use of mathematical models borrowed from physics, in Chapter 15, Andrée Ehresmann uses category theory to present global dynamical models for living systems. Specifically, she uses category theory to frame a Memory Evolution System (MES), defined by a hierarchy of "configuration" categories with partial "transition" functors between them.

Narrowing our investigation to a more general account of the application of category theory to the structure of both mathematical models and scientific theories, David Spivak, in Chapter 16, uses category theory to present an account of mathematical modeling that highlights the relationship between objects, over talk of objects themselves, so that mathematical objects themselves are considered as categorical models. Moving to a more philosophical consideration of the structure of scientific theories, in Chapter 17, Hans Halvorson and Dimitris Tsementzis argue that a category-theoretic account of the structure of scientific theories allows us to transcend the syntax-semantics debate that has encumbered philosophy of science. Finally, in Chapter 18, I argue that while the structural realist can use category theory as a tool to answer how we can conceptually speak of relations without relata or structures without objects, it cannot be used to underpin any ontic structural realist claim that runs from the structure of a scientific theory to the structure of the world.

Finally, I end this Preface with a short story that might assist the weary, or even fearful, reader. Years ago, as a graduate student, I attended a category theory conference in Montreal and was sitting beside Saunders Mac Lane. During one of the talks, Saunders looked to me and said: "Are you following all of this?" I replied, rather embarrassed: "No." There was a slight pause (for effect, I'm sure), and Saunders then turned with a grin and said: "Neither am I (another pause) and I invented it!" I share this if only to remind the reader that it's ok not to follow all of it!

Elaine Landry

Contents

Notes on Contributors   xiii

1. The Roles of Set Theories in Mathematics   1
   Colin McLarty

2. Reviving the Philosophy of Geometry   18
   David Corfield

3. Homotopy Type Theory: A Synthetic Approach to Higher Equalities   36
   Michael Shulman

4. Structuralism, Invariance, and Univalence   58
   Steve Awodey

5. Category Theory and Foundations   69
   Michael Ernst

6. Canonical Maps   90
   Jean-Pierre Marquis

7. Categorical Logic and Model Theory   113
   John L. Bell

8. Unfolding FOLDS: A Foundational Framework for Abstract Mathematical Concepts   136
   Jean-Pierre Marquis

9. Categories and Modalities   163
   Kohei Kishida

10. Proof Theory of the Cut Rule   223
    J. R. B. Cockett and R. A. G. Seely

11. Contextuality: At the Borders of Paradox   262
    Samson Abramsky

12. Categorical Quantum Mechanics I: Causal Quantum Processes   286
    Bob Coecke and Aleks Kissinger

13. Category Theory and the Foundations of Classical Space–Time Theories   329
    James Owen Weatherall

14. Six-Dimensional Lorentz Category   349
    Joachim Lambek

15. Applications of Categories to Biology and Cognition 358
    Andrée Ehresmann

16. Categories as Mathematical Models 381
    David I. Spivak

17. Categories of Scientific Theories 402
    Hans Halvorson and Dimitris Tsementzis

18. Structural Realism and Category Mistakes 430
    Elaine Landry

Name Index 451

Subject Index 457

Notes on Contributors

SAMSON ABRAMSKY is Christopher Strachey Professor of Computing at the Department of Computer Science, University of Oxford. His speciality areas are concurrency, domain theory, lambda calculus, semantics of programming languages, abstract interpretation, and program analysis.

STEVE AWODEY is a professor at the Department of Philosophy, Carnegie Mellon University. His specialty areas are category theory, logic, philosophy of mathematics, and early analytic philosophy.

JOHN L. BELL is a professor in the Department of Philosophy, University of Western Ontario. His specialty areas are mathematical logic, philosophy of mathematics, set theory, boolean algebras, lattice theory, and category theory.

J. R. B. COCKETT is a professor in the Department of Computer Science, University of Calgary. His specialty areas are distributive categories, restriction categories, linearly distributive categories, differential categories, categorical proof theory, semantics of computation, semantics of concurrency, categorical programming, and quantum programming.

BOB COECKE is Professor of Quantum Foundations, Logics, and Structures at the Department of Computer Science, University of Oxford. His specialty areas are foundations of physics, logic, order and category theory, and compositional distributional models of natural language meaning.

DAVID CORFIELD is a senior lecturer in philosophy at the University of Kent. His specialty areas are philosophy of science and philosophy of mathematics.

ANDRÉE EHRESMANN is Professeur Emérite at the Department of Mathematics, Université de Picardie Jules Verne. His specialty areas are analysis, category theory, self-organized multiscale systems with applications to biology and cognition.

MICHAEL ERNST is a graduate student in the Department of Logic and Philosophy of Science, University of California, Irvine. His specialty area is philosophy of mathematics.

HANS HALVORSON is a professor in the Department of Philosophy, Princeton University. His specialty areas are category theory, logic, philosophy of science, and philosophy of physics, science, and religion.

KOHEI KISHIDA is a post-doctoral researcher in the Department of Computer Science, University of Oxford. His specialty areas are category theory, philosophical logic, and modal logic.

ALEKS KISSINGER is a post-doctoral research assistant in the Department of Computer Science, University of Oxford. His specialty areas are category theory, quantum information, graphical calculi, and graph rewriting.

JOACHIM (JIM) LAMBEK was an emeritus professor, Department of Mathematics and Statistics, McGill University. His specialty areas were categorical logic and mathematical linguistics.

ELAINE LANDRY is a professor at the Department of Philosophy, University of California, Davis. Her specialty areas are philosophy of mathematics, philosophy of science, analytic philosophy, philosophy of language, and logic.

JEAN-PIERRE MARQUIS is Professeur titulaire, Department of Philosophy, Université de Montréal. His specialty areas are logic, philosophy of mathematics, philosophy of science, foundations of mathematics, and epistemology.

COLIN MCLARTY is Truman P. Handy Professor of Philosophy & Professor of Mathematics, Department of Philosophy, Case Western Reserve University. His specialty areas are logic, philosophy of logic, philosophy of mathematics, philosophy of science, and contemporary French philosophy.

R. A. G. SEELY is an adjunct professor at the Department of Mathematics, McGill University. His specialty areas are categorical logic, linear logic, computational logic, and proof theory.

MICHAEL SHULMAN is an assistant professor at the Department of Mathematics and Computer Science, University of San Diego. His specialty areas are category theory and algebraic topology.

DAVID I. SPIVAK is a research scientist at the Department of Mathematics, Massachusetts Institute of Technology. His specialty areas are category theory, categorical informatics, and derived manifolds.

DIMITRIS TSEMENTZIS is with the Department of Philosophy, Princeton University. His specialty areas are logic, foundations of mathematics, and category theory.

JAMES OWEN WEATHERALL is an assistant professor in the Department of Logic and Philosophy of Science, University of California, Irvine. His specialty areas are philosophy of physics, philosophy of science, and mathematical physics.

1

The Roles of Set Theories in Mathematics

Colin McLarty

> This examination consists in specifying topics within mathematics for which the appropriate branches of logical foundations [l.f.] do or do not contribute to effective knowledge.
>
> Correspondingly, the demand, accepted (uncritically) in the early days of l.f., for foundations of all mathematics, by logical means to boot, is replaced below by a question: In which areas, if any, of mathematics do such foundations contribute to effective knowledge?
>
> (Kreisel, 1987, 19)

Like Kreisel (1987), we here do not argue about logical foundations for all mathematics. We look at how specific set theories in fact advance mathematics. However, Georg Kreisel's great concern was *effective* knowledge in the logician's sense of finding explicit numerical solutions or at least explicit numerical bounds on solutions to arithmetic problems. Here I use "effective" in the colloquial sense of widely successful in producing a desired or intended result, whether or not it is specifically a numerical solution to an arithmetic problem.

Let me explain because philosophers sometimes miss this topic. I do not ask here what kind of foundations are necessary in principle, nor what all ideas of sets have been used for something at some time, nor what might lead to progress in the future. Those fine questions are not the topic here. This paper addresses set theories in widespread, currently productive use. Excursuses 1.2.1 and 1.2.2 on the Continuum Hypothesis and Grothendieck Universes discuss two interesting gray areas where it matters just how widespread and productive you want it to be

1.1 Overview

No one can be surprised that the role of set theory in mathematics varies with the kind of mathematics. Topologists and analysts face set-theoretic issues, notably the

Continuum Hypothesis, which number theorist do not. Some people will be surprised, and may even object, on hearing different set theories are typical in different parts of mathematics. Linnebo and Pettigrew (2011) apparently make a counterclaim:

Many textbooks that introduce elementary areas of mathematics, such as algebra, analysis, and number theory, include an elementary section surveying the elements of set theory, and this is explicitly orthodox set theory. (249)

But they do not say what they mean by "orthodox". It is true that textbook set theory is rarely intuitionistic, predicative, modal, or non-classical in other ways familiar to philosophers. And in logic texts it is almost always Zermelo Fraenkel set theory, which philosophers tend to take as orthodox. But texts outside of logic rarely come close to Zermelo–Fraenkel.

Section 1.2 gives a few reasons why "orthodox set theory" does not per se mean Zermelo–Fraenkel set theory. Section 1.3 looks at set theory in two standard first year graduate textbooks: James Munkres' *Topology* (2000) and Serge Lang's *Algebra* (2005).[1] It also looks at the recent article by mathematician Tom Leinster (2014). All support the claim that mathematicians know and use the concepts and axioms of the Elementary Theory of the Category of Sets (ETCS), often without knowing or caring that they are the ETCS axioms.

Section 1.4 uses a concept of "mathematical gauge invariance" to show why the category of sets described in ETCS is a closer fit to the practical needs of most mathematicians than is the cumulative hierarchy of sets described in ZFC. For example, philosophers of science may suspect that nothing in mathematical practice depends on solving the multiple reduction problem of whether the number 2 *is* the ZFC set $\{\{\phi\}\}$ or the set $\{\phi, \{\phi\}\}$, or some other set. Few mathematicians have ever heard of this alleged problem.

Mathematicians do constantly meet a similar problem: take the tangent bundle $T(M)$ of a manifold M as a geometric example. Spivak (1999, ch. 3) gives three quite different geometric constructions starting from M. The three give naturally isomorphic results, any one of which will be used as "the" tangent bundle $T(M)$ of M for some purposes. Let me stress: the results are not just different sets when formalized in ZFC. They rely on different aspects of geometry. So the difference between them matters in geometry. Geometers daily rely on nontrivial theorems showing the results are isomorphic, which is why Spivak spends a chapter on the definitions and the proofs.

For this and many other reasons geometers have developed agile, rigorous techniques for handling spaces that are only defined up to isomorphism. Most fields of

---

[1] Some may object that I only consider two cases. They are influential books, and McLarty (2008a) and (2012) discuss several more. But really that objection gets things backwards because no matter how many books I cite not using Zermelo–Fraenkel theory with choice (hereafter ZFC) one could suspect I left out scores more that do. Rather, philosophers who believe many mathematics texts use ZFC should specify at least a few and show how those use ZFC any more than Munkres (2000) does.

mathematics rely on such techniques . These are not the techniques of philosophical structuralism! They are categorical and functorial techniques. ETCS was created from these same techniques.

Section 1.5 describes the positive uses of the extra structure of *an iterated hierarchy* which is assumed for ZFC sets, and compares it to the yet further structure of *constructibility*. Constructibility in this sense was introduced by Gödel (1939, 577) as "a new axiom [which] seems to give a natural completion of the axioms of set theory" and which implies the Continuum Hypothesis. Today constructibility is assumed in most uses of set theories adapted to arithmetic. But it is incompatible with most kinds of large cardinals, so ZFC set theorists treat it as a technical tool and not a property of all sets, as did Gödel himself as he soon rejected the view he had taken in 1938.

1.2 Why ZFC is not Synonymous with "Set Theory"

Zermelo–Fraenkel set theory with choice has indeed been orthodox in set-theoretic research especially since Cohen (1966) used it for the method of *forcing*. But other parts of logic use other set theories. For example, Kreisel's work on effective knowledge led to an array of set theories at the far extreme from ZFC, and closely related to arithmetic, as expounded by Hájek and Pudlák (1993, ch. 1) and Simpson (2010, ch. 1). These theories are provably too weak for some standard mathematics but that is exactly what adapts them to elucidating effective knowledge in the logical sense, that is in Kreisel's sense.

Outside of research logic most mathematicians succeed in the practice of mathematics without ever seeing the ZFC axioms or the set theories close to arithmetic. Few mathematicians could state axioms for any set theory, and more than a few insist this is as it should be. When Alexander Grothendieck began creating the now standard tools of algebraic geometry he used a large cardinal axiom added to a set theory similar to ZFC, and he commissioned a 40-page exposition of it by N. Bourbaki (Artin et al., 1972, 185ff.).[2] A number of number theorists have deplored, not the new axiom itself which is modest by set theorists' standards for large cardinals, but the very idea of mixing axiomatic set theory with number theory. Excursus 1.2.2 returns to this.

Could it be that most mathematicians use ZFC the way most people drive a car without knowing how the engine works? That is, are most mathematicians content to sketch how proofs of their theorems should go, while letting others, trained in logic, worry about actual proofs? Certainly mathematicians often use results in mathematics that they have never seen fully proved.

The extent of this varies with the field. When one algebraist said he would never cite a theorem in a paper unless he knew the proof, my teacher Charles Wells responded, "We can do that in abstract algebra. Differential geometers have to use more theorems

---

[2] Bourbaki is the pseudonym of a group of mathematicians. Pierre Cartier, who was in the group at the time, says Pierre Samuel wrote this appendix (conversation February 19, 2015).

than they have personally checked". But on one hand neither of those algebraists took proof to mean a proof in ZFC. And more to our point the results a mathematician will cite in research without learning the proofs are generally not basic facts used throughout the work. They are more special theorems. Mathematicians normally can prove the facts they use daily. And I will maintain this includes the facts of set theory that they actually use. An analyst, algebraist, or topologist sometimes mentions an independence result in set theory without having worked through its proof. But they rarely *use* independence results.

A philosopher of mathematics might take a hard stand, saying most mathematicians do not and cannot justify their theorems: because justification requires proofs in ZFC which few mathematicians know. The philosopher could say nonetheless the theorems are justifiable because philosophers and logicians can recast the proofs in ZFC. I cannot argue against such epistemology here. But I prefer the hard-won insight of Russell (1924, reprint 326) that far from foundations justifying mathematics, foundations themselves must be "believed chiefly because of their consequences" in mathematics.

Alternatively, could it be like speaking in prose without knowing it is called "prose"? Perhaps mathematicians generally know the ZFC axioms without knowing they are the ZFC axioms? That is more in line with Russell's mature thinking. And indeed most mathematics texts say a lot about sets without calling anything an axiom of set theory. But the things they say are not distinctive of ZFC. Far from tacitly knowing the ZFC Axiom of Foundation, or the Axiom Schemes of Separation and Replacement, most mathematicians are unfamiliar even with the notions of *global membership* and *first-order axiom scheme*. Those notions are basic to stating these axioms which are in turn basic to ZFC. We will see that mathematicians do know the concepts and axioms of the *Elementary Theory of the Category of Sets*, which, after all, Lawvere (1964) took from the practice of mathematics around him.

1.2.1 The Continuum Hypothesis

Among Cantor's first decisive achievements was his proof that the real numbers $\mathbb{R}$ are *uncountable*: they cannot be put into one simply infinite sequence $r_0, r_1, \ldots$. He then conjectured that $\mathbb{R}$ has the smallest uncountable cardinality: every infinite subset $S \subseteq \mathbb{R}$ either is countable or can be put in bijection $S \cong \mathbb{R}$ with the set of all reals. This is Cantor's Continuum Hypothesis (CH). It remains a leading topic of set theory and arguably the leading topic ever since Cantor, because no generally accepted set theory can either prove it or refute it. It is not just that no one knows a proof or refutation in any accepted set theory. Masses of research descended from Gödel (1939) and Cohen (1966) prove all generally accepted set theories are consistent with both the truth and the falsity of CH. Nearly all of this set theoretic work is conducted in terms of ZFC and extensions of ZFC.

This is obviously relevant to analysis, and while it might be surprising how rarely it matters, it does come up sometimes. For example, it matters exactly once in Munkres'

Topology. An exercise, mentions a question about *box topologies* whose answer is not known, though it is known that "the answer is affirmative if one assumes the Continuum Hypothesis" (Munkres 2000, 205). He cites a proof by Mary Ellen Rudin (1972), which is also inexplicit about its set theory. Rudin cites Bourbaki and also Kelley (1955), who each give their own set theories (Kelley's is stronger than ZFC). But she says nothing about either one's set theory. Her topological argument is insensitive to any difference among ZFC, ETCS, and those others.

The long and short of it for Munkres was that this question on box topologies could only be settled by making a controversial assumption, the CH. He felt topology students should know this happens sometimes.

1.2.2 Grothendieck universes

Grothendieck found penetrating new ways to calculate with small (often finite) structures, by organizing those small structures into very large categories and functors, e.g., *Abelian categories* and *derived functors*. He was not very interested in set theory, yet thought it was worth getting right, and he saw that his large categories and functors were far too big for ZFC to prove they exist. He did not only use collections such as the class of all sets, which is too big to be a set. He used collections of those too-large collections, and larger-yet collections of those, and more. McLarty (2010) gives details for logicians.

Grothendieck cared enough to give a rigorous set-theoretic foundation. But he did not linger on minimizing the foundation in any way. He gave a quick solution using *Grothendieck universes*, which are sets so large that ZFC does not prove they exist.

The impact of this set theory on mainstream mathematics is attenuated by a mix of factors:

1. While Grothendieck's cohomology is entirely standard in research number theory and geometry, it remains a rather advanced specialty.
2. As Grothendieck intended, the large structure tools are not the focus of attention but merely a framework organizing calculations.
3. As Grothendieck knew, those calculations can be done without the large structures, at the loss of conceptual unity and general theorems.
4. Texts such as Milne (1980), Freitag and Kiehl (1988), and Tamme (1994) use large structure theorems informally without discussing foundations.
5. Others, like (Fantechi et al., 2005, 10) and Lurie (2009, 50f.), invoke Grothendieck universes only to dismiss a technical problem.
6. The large structures can be founded on a conservative extension of ETCS, far weaker than ZFC, let alone a Grothendieck universe (McLarty, 2011).

In sum, this has not made axiomatic set theory a standard topic in geometry or even in specifically Grothendieck-inspired geometry.

1.3 Sets for the Working Mathematician

André Weil, whose research in number theory and geometry put him among the leading mathematicians of his time even apart from his role as the initiating member of Bourbaki, launched the meme "X for the working mathematician" with a talk titled "Foundations of Mathematics for the Working Mathematician", delivered to the Association for Symbolic Logic (1949). He aimed to bring mathematical practice closer to logical principles. However much people complain about Bourbaki's abstractness, or neglect of geometry or of advanced logic, it remains that Bourbaki was a leader in making mathematical writing radically more accessible. And they did this using set theory. Mathematics textbooks and published research became more uniform in style, and more rigorous, than before World War II. The spread of explicit set theory was a great part of this, going hand in hand with the demand for readily readable textbooks and more uniform terminology in the burgeoning new fields of mathematics (McLarty, 2008b). This has become the norm in graduate mathematics teaching worldwide, because experience showed it made the subject easier to learn.

It is valuable to match a modern introduction to differential geometry with classical readings from Riemann, for example, as Spivak (1999) does. But few people today could read Riemann without the help of modern texts. Few ever did succeed at reading Riemann in the nineteenth century, or in the first half of the twentieth century. It is easier today because of the set-theorization of mathematics associated with Bourbaki among others. Nostalgia for the good old days of easy, intuitive mathematics is misplaced.

Bourbaki (1958) created their own set theory similar to ZFC. But their volumes on various fields of mathematics rarely referred to it and few mathematicians, or even logicians, ever learned it.

To see how the set-theorization actually happened in widespread practice let us look at set theory in two currently influential graduate textbooks. Neither of them gives precise axioms. Rather they discuss more or less basic facts about sets which they go on to use. We focus on two questions bearing on the distinction of ZFC from ETCS set theory:[3]

1. How does the book handle the elements of sets?
2. Does the book define functions as a kind of set?

Munkres (2000) opens with seventy leisurely pages:

---

[3] As a more technical issue: every textbook I looked at freely forms sets of sets without discussing the Axiom Scheme of Replacement. One explanation would be that these avowedly naive treatments accept unlimited comprehension though that is actually inconsistent. Or one might notice these apparent uses of replacement can be reduced to the more innocuous Bounded Separation Axiom Scheme, which in turn can be stated as a single axiom. There is usually a suitable ambient superset ready at hand. The point here, though, is that none of these textbooks is sufficiently explicit about logic even to state Replacement or Separation axiom schemes, or to explain the difference between an axiom and an axiom scheme. Those schemes are not part of ETCS, though they are available there as add-ons if wanted.

THE ROLES OF SET THEORIES IN MATHEMATICS 7

I begin with a fairly thorough chapter on set theory and logic. It starts at an elementary level and works up to a level that might be described as "semisophisticated". It treats those topics (and only those) that will be needed later in the book. (2000, xi)

In order to "introduce the ideas of set theory", Munkres says, "Commonly we shall use capital letters $A, B, \ldots$ to denote sets, and lowercase letters $a, b, \ldots$ to denote the objects or elements belonging to these sets" (2000, 4). Of course in ZFC everything is a set, including that the elements of sets are sets. Munkres hardly denies that all objects are sets. But, just as he says he will do, he commonly writes as if elements are of a different type than sets, which they indeed are in ETCS.

Munkres comes close to ZFC practice on cartesian products. He says $A \times B$ is the set of all ordered pairs $\langle a, b \rangle$ with $a \in A$ and $b \in B$. He treats it as optional to say that $\langle a, b \rangle$ is a set itself. He notes it can be defined as the set $\{\{a\}, \{a, b\}\}$ and he shows that this set uniquely determines a and b. But he concludes, "it is fair to say that most mathematicians think of an ordered pair as a primitive concept rather than thinking of it as a collection of sets" (2000, 13).

Munkres says mathematicians think of a function $f : A \to B$ as a rule assigning values $f(a) \in B$ to arguments $a \in A$, but they also need a more precise definition (2000, 15). For this he follows the ZFC practice of defining a function $f : A \to B$ as a set. Namely, f is a subset of the product $f \subseteq A \times B$ with the property that for each $a \in A$ there exists a unique $b \in B$ with $\langle a, b \rangle \in f$.

He comes to "what we might call the mathematical foundations for our study—the integers and the real number system" (2000, 36). He notes there are two approaches to this. One is to construct these sets by building them up from the empty set. That is the standard approach of ZFC, though Munkres does not say so. He does say that approach "takes a good deal of time and effort and is of greater logical than mathematical interest" (2000, 36). He will rather "assume a set of axioms for the real numbers and work from these axioms" without proving from any more basic axioms that any such set exists.

He never entertains the question of what set is 2. For him 2 is an integer, the sum of 1 and 1, it is an element of the set $\mathbb{R}$ of real numbers. He never affirms or denies that 2 is itself a set. We have just seen he declines to give any set theoretic specification of the elements of the set $\mathbb{R}$. Nor does he ever specify what sets are the elements of the sets $\mathbb{Q}, \mathbb{C}$ of rational and complex numbers, respectively. This has a practical advantage as it allows Munkres to assume actual subset inclusions from the natural numbers through the integers, the rational, and the real numbers, up to the complex numbers:

$$\mathbb{N} \subset \mathbb{Z} \subset \mathbb{Q} \subset \mathbb{R} \subset \mathbb{C}.$$

Most ZFC definitions make these not actually subset inclusions. We return to this at the start of Section 1.4.

This is the second edition of a book Munkres first published in 1975, when he had probably not heard of ETCS. There is no evidence he ever heard of any axiomatic set theory but ZFC. On the other hand, in this book he never actually gives the central concepts or axioms of ZFC. When he does bring up the occasional idea from ZFC he always says mathematicians do not usually think that way.

Lang (2005) is greatly expanded from the first edition (Lang, 1965). The treatment of sets did not change significantly. Lang has none of Munkres's switching between how he believes mathematicians think and what he believes rigor requires. He simply gives the principles he uses in the rest of his book. It is possible that by 1965 Lang had heard of ETCS in conversation with Peter Freyd but also possible he had not. And his book had to be well underway before Lawvere had shown ETCS to anyone at all. In any case I do not claim Lang cared about ETCS. His account simply says nothing that would distinguish between ZFC and ETCS, because nothing in his book depends on those differences.

The discussion of sets begins: "We assume that the reader is familiar with sets" (Lang, 2005, xi). Yet Lang proves very simple facts. An appendix proves, for example, that the product of a non-empty finite set with a countably infinite one is countably infinite (878). Without ever saying what a function *is*, indeed without saying functions are sets at all, Lang says what he needs about them: a function $f: A \to B$ may also be called a *mapping*, it has a domain A and a codomain B, and it takes a well-defined value for each element of the domain. His entire explicit discussion of the term is the following:

If $f: A \to B$ is a mapping of one set into another, we write

$$x \mapsto f(x)$$

to denote the effect of f on an element x of A.   (Lang, 1993, ix)

Either ZFC or ETCS would spell this out in just a few steps:

1. ZFC formalizes function evaluation by taking elementhood as primitive and defining an ordered pair as a set by some means such as $\langle x, y \rangle = \{\{x\}, \{x, y\}\}$. Then prove that indeed on this definition $\langle x, y \rangle = \langle w, z \rangle$ implies that $x = w$ and $y = z$. Then define a function f as a suitable set of ordered pairs $f \subseteq A \times B$, and define $f(x) = y$ to mean $\langle x, y \rangle \in f$.
2. Categorical set theory takes function composition as primitive, posits a set 1 such that every set has exactly one function $A \to 1$, and defines elements $x \in A$ as functions $x: 1 \to A$. This implies that 1 has exactly one element, namely the sole function $1 \to 1$. Then for any function $f: A \to B$ and element x of A, define $f(x) \in B$ as the composite $fx: 1 \to B$.

Either way works formally and mathematicians generally learn set theory without ever hearing a formal logical treatment of either one. I will mention that in some parts of geometry a point p of a space M is defined as a function $p: 1 \to M$ from a one point space, so that a function $f: M \to N$ acts on points $p: 1 \to M$ by composition, giving $f(p): 1 \to N$.

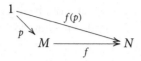

Lang never defines cartesian products $A \times B$ in a ZFC way. Rather, he defines $A \times B$ up to isomorphism in his introduction to "categories and functors" (2005, 53–65). He handles products of sets along with products of groups and other structures by the standard commutative diagram definition (2005, 55). The product is not just a set but a set $A \times B$ plus two functions $p_1\colon A \times B \to A$ and $p_2\colon A \times B \to B$. And given any set T and functions $f\colon T \to A$ and $g\colon T \to B$ there is a unique function u making all the composites in this diagram equal:

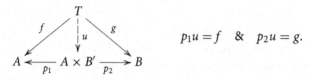

$$p_1 u = f \quad \& \quad p_2 u = g.$$

A convenient notation has $u = \langle f, g \rangle$. In that notation, the special case $T = 1$ of the definition says the elements of $A \times B$ are precisely the pairs $\langle x, y \rangle$ with $x \in A$ and $y \in B$.

Again let us be clear about the issue: every mathematician must know cartesian products have this property. It is implicit in all use of products since, say, Descartes, long before the notion of a function or of a cartesian product could be clearly stated. In fact, this property determines the product triple $A \times B, p_1, p_2$ uniquely up to isomorphism, as proved by Lang and by any introduction to category theory. And, in fact, this is the property of products that Lang uses throughout his book. That is why he gives it.

So, can these properties that Lang and Munkres and others actually use be stated precisely in just the terms that Lang and Munkres use? Or must they remain a rough guide which actually requires ZFC for a precise statement? That is exactly what Lawvere asked himself in the years leading up to his Lawvere (1964). For this history, see McLarty (1990).

It is not that Lang, or Munkres, or many other authors chose to use this set theory. Rather this set theory is based on the very techniques that mathematicians use in algebra, topology, and so on, every day.

We need not look at ETCS in all formality. That is done in Lawvere (1964) and many other places. We will just quote the less formal summary given by Leinster (2014, 404):

1. Composition of functions is associative and has identities.
2. There is a set with exactly one element.
3. There is a set with no elements.
4. A function is determined by its effect on elements.
5. Given sets X and Y, one can form their cartesian product $X \times Y$.
6. Given sets X and Y, one can form the set of functions from X to Y.

7. Given $f: X \to Y$ and $y \in Y$, one can form the inverse image $f-1(y)$.
8. The subsets of a set X correspond to the functions from X to $\{0, 1\}$.
9. The natural numbers form a set.
10. Every surjection has a right inverse.

These are Leinster's informal summary of Lawvere's ETCS axioms and Leinster spells them all out in categorical terms, just as we spelled several of them out already.

All these facts are familiar to every mathematician and are used routinely in textbooks. Munkres is fairly typical in suggesting that ZFC is the way to make them precise—though typical too in that he never actually names ZFC let alone states its axioms. And he repeatedly says the "precise" versions are not how mathematicians think. Lang is more artful in stating all and only the properties of sets that he actually uses, and never suggesting any others, are needed for rigor. All the properties Lang describes are easy consequences of these ETCS axioms.

The precise formal relationship between ETCS and ZFC has been known since Osius (1974). But, for a longer more practical/intuitive proof that ETCS suffices for standard mathematics, just read ordinary textbooks like those described here. Their proofs fit easily and naturally into the ETCS formalism.

Anyone who believes the ZFC axioms necessarily believes those of ETCS, but not conversely. The axioms of ETCS are all theorems of ZFC, when "function" is defined in the standard way for ZFC. The converse is not true.

The ZFC axioms say much more about sets than ETCS. These further claims are rarely noted, let alone used, in mathematics textbooks outside of set theory. They are pervasive in some kinds of advanced set theory. The ZFC axioms say everything is a set; notably functions are a special kind of set; and every set is built up from the empty set by transfinitely iterated set formation. We have seen how Munkres hints at these ZFC devices without going much into them and Lang avoids them. Section 1.5 will describe some reasons set theorists took them up, but first Section 1.4 describes why most textbooks don't.

1.4 Gauge Invariant Set Theory

Weatherall (2015) argues that the word "gauge" is ubiquitous in modern physics, and ambiguous. Without offering a thorough survey he describes two meanings, and we can use one of those:

On the first strand, a "gauge theory" is a theory that exhibits excess structure... in such a way that (perhaps) one could remove some structure from the theory without affecting its descriptive or representational power. (Weatherall, 2015, 1-12)

As an example he gives electromagnetic theory using an electromagnetic potential A_a. Distinct potentials A_a, A'_a produce exactly the same observable consequences if their difference is the derivative of some scalar χ:

$$A_a - A'_a = \nabla_a \chi$$

In other words, given even ideal observations of the total behavior of some electromagnetic system, we have a choice of infinitely many different potentials A_a to describe that system.

So the specific potential is somehow more than we need. And in fact the potential is eliminated from versions of electromagnetic theory using electromagnetic force F_{ab} instead. Any two different forces $F_{ab} \neq F'_{ab}$ do produce, in principle, observably different consequences.

While mathematics has no good analogue of observable consequences, it certainly deals with descriptive or representational power. We will argue that the global membership relation of ZFC is a gauge on sets as it does not contribute to the descriptions and representations of structures in most of mathematics. This is not meant as any very close analogue to the situation in electromagnetic theory! But then, neither is the situation electromagnetic theory exactly the same as in General Relativity (Weatherall, 2015, passim).

We must distinguish between *global* and *local* membership. Throughout mathematics it is crucial to know which elements $x \in A$ of a set A are members of which subsets $S \subseteq A$. We say this relation is *local* to elements and subsets of the ambient set A. For example, arithmeticians need to know which natural numbers $n \in \mathbb{N}$ are in the subset of primes $Pr \subset \mathbb{N}$. On the other hand, nearly no one ever asks whether the imaginary unit $i \in \mathbb{C}$ is also a member of the unit sphere $S^2 \subset \mathbb{R}^3$, because they do not lie inside of any one natural ambient.

In ZFC the elements of sets are sets and it always makes sense to ask of sets X and Y whether $X \in Y$. This is a *global membership relation* since it does not rest on any sense of an ambient set but is meaningful for every two sets.

As a prominent example, it makes sense in ZFC to ask whether each rational number $q \in \mathbb{Q}$ is also a real number $q \in \mathbb{R}$, and the answer is "yes" or "no", depending on how we define real numbers. It is no if we use Dedekind cuts on the rational numbers, or equivalence classes of Cauchy sequences of rational numbers. It is yes if we use the definitions by Quine (1969, 136ff).

In ETCS membership is local, and it makes no sense to ask whether each rational number $q \in \mathbb{Q}$ is also a real number $q \in \mathbb{R}$ until both $\mathbb{Q}$ and $\mathbb{R}$ are taken as embedded in some common ambient set—and the most common ambient in textbook practice is $\mathbb{R}$ itself. In other words, ETCS takes it as true by stipulation that all rational numbers are real. It makes no sense in ETCS to ask for any answer except a stipulation. And most math texts that address the issue at all, including the two we looked at in Section 1.3, do solve it by stipulation. They say that every natural number is also a rational number, and every rational number is also a real number, without ever saying how this is to be achieved by a ZFC set-theoretic definition of those numbers. See McLarty (1993, 2008a) for more examples of textbook treatments of related issues and comparison with ETCS.

The global membership relation in ZFC is a gauge in the sense that it is rarely used in the "descriptive or representational" practices of mathematics, to recall the words of Weatherall (2015, 1–12) quoted earlier. Like the gauges in physical gauge theories it has important uses but the uses are of a technical kind which we will get to in the next section. Normal practice in most parts of mathematics describes and represents sets just up to isomorphism.[4]

Philosophers sometimes suppose definition up to isomorphism is a recent, abstract idea, which would not be found in classical mathematics. But this is backwards. Newton and Leibniz did understand the real numbers in terms of their algebraic and analytical relations to one another—which today are called isomorphism invariant— they did not understand real numbers as sets built up by transfinite accumulation from the empty set! They understood the arithmetic of whole numbers as dealing with the laws of addition and multiplication—which we today say define the integers only up to isomorphism—they never imagined one could explain what 2 *is* by using the set $\{\{\emptyset\}\}$ or the set $\{\emptyset, \{\emptyset\}\}$.

The novelty that arose in the nineteenth century, grew in the twentieth, and is growing faster today, was the scope, agility, and fecundity of structural methods.

For example, the idea of a tangent space to a manifold was reasonably clear, and extremely useful in expert hands, in the nineteenth century. But it became more useful and more widely accessible as the twentieth century gave clearer articulation to several distinct geometric constructions of it expressing different geometric aspects. One approach constructs it using coordinate systems placed on patches of the manifold. Another uses (equivalence classes of) curves through points of the manifold. Another uses (equivalence classes of) real-valued functions defined on neighborhoods of points of the manifold. Each of the different realizations relates most directly to some problems. All these constructions yield "the same structure" in some sense, which mathematicians at first described vaguely by saying all the realizations are "naturally isomorphic". That idea became more or less clear, to the more modern-minded mathematicians, through the early twentieth century. It was first addressed explicitly, and first made into a rigorous means of proving theorems, by Eilenberg and Mac Lane (1945) in the first general paper on categories and functors.

---

[4] Readers who compare Weatherall (2015) will note he associates gauges with a paucity of isomorphisms (so that some observationally identical models are not isomorphic), while we associate them with abundance of isomorphisms (so that gauges are not preserved by isomorphisms). That is because he takes 'isomorphism' in the way common throughout most of mathematics, where an isomorphism preserves the structures at hand. But isomorphisms in ZFC set theory do not preserve the membership structure. In ZFC or in ETCS, and indeed in Cantor's work long before either of those theories existed, an isomorphism of sets $f: A \to B$ is a function with an inverse, $f^{-1}: B \to A$, or in other words a one-to-one onto function. Such a function need not preserve the membership structure of ZFC sets (and almost never does, in fact). Given an isomorphism $f: A \to B$ and $z \in x \in A$ we can hardly conclude that $f(z) \in f(x)$. We cannot even conclude that $z \in A$, so we cannot conclude that $f(z)$ is defined at all. And if, in some particular case, $f(z)$ is defined, there is no reason for it to be an element of $f(x)$. This situation in ZFC is like the situation Weatherall (2015, 12) considers, where gauge transformations are taken as isomorphisms.

Philosophers often underestimate the practical need for rigorous proof in mathematics. There may be some quick, elementary proof of Fermat's Last Theorem that would avoid long difficult considerations—but honestly most people looking for proofs of Fermat's Last Theorem through the centuries since Fermat have looked for just that kind of proof. Most people working on Fermat's Last Theorem today continue to seek such a proof. Most have never learned the apparatus used by Wiles (1995). But up to now everyone who has found new and simpler proofs has used essentially that apparatus. All the yet-known proofs remain long and rather difficult, notably Kisin (2009).

Proofs like this require that each step be as clear and concise as possible while keeping extreme rigor. Otherwise, one could hardly credit the conclusion of such long reasoning. For many other examples, see McLarty (2008a, c).

Today mathematicians use extensive, efficient functorial means to deal with structures defined only up to isomorphism. And naturally, as we have seen in Lang and Munkres, they tend to treat sets also by isomorphism-invariant means. Why handle sets by one set of conceptual tools, and groups, rings, differential manifolds, and other algebaic or geometric structures by another, when the same tools will work for both? Those are the tools used in ETCS, and not in ZFC. But of course each part of mathematics also has its own special character which can call for specially adapted devices.

1.5 The Use of a Gauge

By and large mathematicians will agree with Munkres (2000, 36) that it is uninteresting to analyze a natural number $n \in \mathbb{N}$ or a real number $r \in \mathbb{R}$ as itself a set. But advanced investigations in set theory require much more detailed analysis of sets than most of mathematics does. Zermelo's approach to this was to give each set itself much more structure than Cantor had.[5] This attitude was canonized when the Axiom of Foundation and the Axiom Scheme of Replacement were added to Zermelo's early axiomatization to give today's Zermelo–Fraenkel Axioms. These axioms imply that every set is built up from the empty set by (possibly transfinitely) iterated set formation.

In other words every ZFC set has a well-founded downward-growing *membership tree* structure where the top node indicates the set, and each node has as many nodes below it as elements. For example, the set $\mathbb{N} = \{0, 1, 2, 3, \ldots\}$ can be represented by the set of finite von Neumann ordinals:

$$\{\emptyset, \{\emptyset\}, \{\emptyset, \{\emptyset\}\}, \{\emptyset, \{\emptyset\}, \{\emptyset, \{\emptyset\}\}\}, \ldots\}.$$

---

[5] Zermelo declared his preference for Frege on this question, over Cantor, in his comments on Cantor's collected works. See especially Cantor (1932, 351, 353, 441f.) and Lawvere (1994).

So it has this tree:

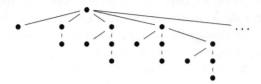

Each bottom node stands for the empty set ∅, each node with nothing below it but a single bottom node stands for the singleton set {∅}, and so on. Each branch is finite but they get longer and longer without bound as they go to the right.

Now be clear, both ETCS and ZFC can describe tree structures like this. The difference is that ZFC says each set S has such a tree intrinsically given by the membership relation on its elements, and their elements, and so on. Because the elements of S are ZFC sets, so are their elements, and so on. So ZFC sets come intrinsically arranged in an *cumulative hierarchy* also called the *iterative hierarchy* where each set is located at the lowest stage higher than any of its elements.

This tree structure is extremely useful in set theory. The *Mostowski embedding theorem*, found in any advanced ZFC textbook, tells exactly which trees correspond to ZFC sets. And categorical set theory also uses it for logical investigations (Osius, 1974, 88). A *tree interpretation* lets ETCS interpret ZFC with no loss of information at all, by interpreting a ZFC set as a suitable tree as discussed in McLarty (2004). So there is a translation routine taking any theorem or proof in ZFC to an equivalent theorem or proof in ETCS.[6]

In short, what ZFC takes as a set S, ETCS takes as a set S plus a suitable (extensional, well-founded) tree with the elements of S as the first level of nodes below the top. The tree structure in ETCS captures the iterative hierarchy structure in ZFC. From the viewpoint of ETCS the iterative hierarchy is a gauge, very useful in set theory research.

Then there is a yet further structure, called *constructibility*, forming the *constructible hierarchy*, with similarly pervasive research uses. A constructible set is not only accumulated level by level from earlier constructible sets, but each set is formed with an explicitly stated definition of which earlier sets are to be collected into it.[7] This gives an extremely tight handle on each set and it is useful at every level of research set theory. The set theories closest to arithmetic generally assume every set is constructible to gain enough control on the sets, as explained by Simpson (2010, 282).

---

[6] The ETCS version may need to specify further assumptions corresponding to the fact that ZFC assumes stronger axioms than ETCS. The corresponding stronger assumptions are always available in ETCS, if only by translating the ZFC version, though in most natural cases there is also a natural correspondent in ETCS.

[7] Actually, on conventional accounts, each constructible set S is formed infinitely many times, each time with a different explicitly stated definition. But the definitions are well ordered so each constructible S has a unique *first* definition forming it.

However, nearly the whole theory of large cardinals disappears if all sets are constructible. *Measurable cardinals* cannot exist if all sets are constructible, and so of course no larger cardinals can either. So ZFC set theorists treat constructibility as a valuable feature of *some* sets, not *all* sets. But note this is not a distinction at the level of set isomorphism. The ZFC axioms themselves trivially prove every set is *isomorphic to* some constructible set: choice proves every set is isomorphic to some ordinal. All ordinals are constructible by definition.

In short, ZFC and ETCS both treat membership trees and constructibility as gauges. Both traits are extremely useful at times (as are potentials in electromagnetism). Neither trait is preserved by set isomorphisms. Neither trait is much mentioned in math outside of research set theory. These gauges are indispensable to the kind of questions handled in set theory texts such as Kunen (1983) or Kanamori (1994). These texts find it natural to use ZFC, which has the first gauge (membership trees) built in, and which does not imply that every set admits the second gauge at all. But far the greatest part of mathematics in algebra, geometry, or analysis is gauge invariant—or, as logicians prefer to say, isomorphism invariant—both theoretically and in daily practice.

References

Artin, M., Grothendieck, A., and Verdier, J.-L. (1972). *Théorie des Topos et Cohomologie Etale des Schémas*. Séminaire de géométrie algébrique du Bois-Marie, 4. Springer-Verlag, Paris. Three volumes, usually cited as SGA 4.

Bourbaki, N. (1949). Foundations of mathematics for the working mathematician. *Journal of Symbolic Logic* 14, 1–8.

Bourbaki, N. (1958). *Théorie des Ensembles*, 3rd edition. Hermann, Paris.

Cantor, G. (1932). *Gesammelte Abhandlungen mathematischen und philosophischen Inhalts*, ed. E. Zermelo. Springer, Berlin.

Cohen, P. (1966). *Set Theory and the Continuum Hypothesis*. W. A. Benjamin, New York.

Eilenberg, S., and Mac Lane, S. (1945). General theory of natural equivalences. *Transactions of the American Mathematical Society* 58, 231–94.

Fantechi, B., Vistoli, A., Gottsche, L., Kleiman, S. L., Illusie, L., and Nitsure, N. (2005). *Fundamental Algebraic Geometry: Grothendieck's FGA Explained*, vol. 123 of *Mathematical Surveys and Monographs*. American Mathematical Society, Providence, RI.

Freitag, E., and Kiehl, R. (1988). *Étale cohomology and the Weil conjecture*. Springer-Verlag, New York.

Gödel, K. (1939). The consistency of the axiom of choice and of the generalized continuum-hypothesis. *Proceedings of the National Academy of Sciences of the United States of America* 24, 556–7.

Hájek, P., and Pudlák, P. (1993). *Metamathemtics of First Order Arithmetic*. Springer-Verlag, New York.

Kanamori, A. (1994). *The Higher Infinite*. Springer-Verlag, New York.

Kelley, J. (1955). *General Topology*. Van Nostrand, New York.

Kisin, M. (2009). Moduli of finite flat group schemes, and modularity. *Annals of Mathematics* 170(3), 1085–180.

Kreisel, G. (1987). So-called formal reasoning and the foundational ideal, in Paul Weingartner and Gerhard Schurz (eds), *Logik, Wissenschaftstheorie und Erkenntnistheorie*, 19–42. Hölder-Pichler-Tempsky, Wien.

Kunen, K. (1983). *Set Theory: An Introduction to Independence Proofs*. North-Holland, Amsterdam.

Lang, S. (1965). *Algebra*, 1st edn. Addison-Wesley, Reading, MA.

Lang, S. (1993). *Algebra*, 3rd edn. Addison-Wesley, Reading, MA.

Lang, S. (2005). *Algebra*. Addison-Wesley, Reading, MA.

Lawvere, F. W. (1964). An elementary theory of the category of sets. *Proceedings of the National Academy of Science of the United States of America* 52, 1506–11.

Lawvere, F. W. (1994). Cohesive toposes and Cantor's "lauter Einsen". *Philosophia Mathematica* 2, 5–15.

Leinster, T. (2014). Rethinking set theory. *American Mathematical Monthly* 121(5), 403–15.

Linnebo, Ø., and Pettigrew, R. (2011). Category theory as an autonomous foundation. *Philosophia Mathematica* 19, 227–54.

Lurie, J. (2009). *Higher Topos Theory*. Annals of Mathematics Studies 170. Princeton University Press, Princeton, NJ.

McLarty, C. (1990). The uses and abuses of the history of topos theory. *British Journal for the Philosophy of Science* 41, 351–75.

McLarty, C. (1993). Numbers can be just what they have to. *Noûs* 27, 487–98.

McLarty, C. (2004). Exploring categorical structuralism. *Philosophia Mathematica*, 37–53.

McLarty, C. (2008a). What structuralism achieves, in Paolo Mancosu (ed.), *The Philosophy of Mathematical Practice*, 354–69. Oxford University Press, Oxford.

McLarty, C. (2008b). Articles on Claude Chevalley, Jean Dieudonné, and André Weil, in Noretta Koertge (eds), *New Dictionary of Scientific Biography*, vol. II, 116–20 and 289–93, 254–8. Scribner's, Detroit.

McLarty, C. (2008c). "There is no ontology here": visual and structural geometry in arithmetic, in Paolo Mancosu (ed.), *The Philosophy of Mathematical Practice*, 370–406. Oxford University Press, Oxford.

McLarty, C. (2010). What does it take to prove Fermat's Last Theorem? *Bulletin of Symbolic Logic* 16, 359–77.

McLarty, C. (2011). A finite order arithmetic foundation for cohomology. Preprint on the mathematics arXiv, arXiv:1102.1773v3, 2011.

McLarty, C. (2012). Categorical foundations and mathematical practice. *Philosophia Mathematica* 20, 111–13.

Milne, J. (1980). *Étale Cohomology*. Princeton University Press, Princeton, NJ.

Munkres, J. (2000). *Topology*, 2nd edn. Prentice Hall, Upper Saddle River, NJ.

Osius, G. (1974). Categorical set theory: A characterization of the category of sets. *Journal of Pure and Applied Algebra* 4, 79–119.

Quine, W. V. O. (1969). *Set Theory and Its Logic*. Harvard University Press, Cambridge, MA.

Rudin, M. E. (1972). The box product of countably many compact metric spaces. *General Topology and its Applications* 2, 293–8.

Russell, B. (1924). Logical atomism, in J. Muirhead (ed.), *Contemporary British Philosophers*, 356–83. Allen and Unwin, London. Reprinted in Russell, B. (1956) *Logic and Knowledge*, 323–43. Allen and Unwin, London.

Simpson, S. (2010). *Subsystems of Second Order Arithmetic*. Cambridge University Press, Cambridge, UK.

Spivak, M. (1999). *A Comprehensive Introduction to Differential Geometry*, 3rd edn. Publish or Perish, Houston.

Tamme, G. (1994). *Introduction to Etale Cohomology*. Springer-Verlag, New York.

Weatherall, J. (2015). Understanding gauge. Preprint on the mathematics arXiv, arXiv: 1505.02229.

Wiles, A. (1995). Modular elliptic curves and Fermat's Last Theorem. *Annals of Mathematics* 141, 443–551.

2
Reviving the Philosophy of Geometry

David Corfield

2.1 Introduction

Leafing through Robert Torretti's book *Philosophy of Geometry from Riemann to Poincaré* (1978), it is natural to wonder why, at least in the Anglophone community, we have little activity meriting this name today. Broadly speaking, we can say that any philosophical interest in geometry shown here is directed at the appearance of geometric constructions in physics, without any thought being given to the conceptual development of the subject within mathematics itself. This is in part a result of a conception we owe to the Vienna Circle and their Berlin colleagues that one should sharply distinguish between *mathematical* geometry and *physical* geometry. Inspired by Einstein's relativity theory, this account, due to Schlick and Reichenbach, takes mathematical geometry to be the study of the logical consequences of certain Hilbertian axiomatizations. For its application in physics, in addition to a mathematical geometric theory, one needs laws of physics and then 'coordinating principles' which relate these laws to empirical observations. From this viewpoint, the mathematics itself fades from view as a more or less convenient choice of language in which to express a physical theory. No interest is taken in which axiomatic theories deserve the epithet 'geometric'.

However, in the 1920s this view of geometry did not go unchallenged as Hermann Weyl, similarly inspired by relativity theory, was led to very different conclusions. His attempted unification of electromagnetism with relativity theory of 1918 was the product of a coherent geometric, physical, and philosophical vision, inspired by his knowledge of the works of Fichte and Husserl. While this unification was not directly successful, it gave rise to modern gauge field theory. Weyl, of course, also went on to make a considerable contribution to quantum theory. And while Einstein gave initial support to Moritz Schlick's account of his theory, he later became an advocate of the idea that mathematics provides important conceptual frameworks in which to do physics:

Experience can of course guide us in our choice of serviceable mathematical concepts; it cannot possibly be the source from which they are derived; experience of course remains the sole criterion of the serviceability of a mathematical construction for physics, but the truly creative principle resides in mathematics. (Einstein, 1934, 167)

We may imagine then that an important chapter in any sequel to Torretti's book would describe both Reichenbach's and Weyl's views on geometry. This is done in Thomas Ryckman's excellent *The Reign of Relativity* (2005), where the author also discusses further overlooked German-language philosophical writings on geometry from the 1920s, this time by Ernst Cassirer. Cassirer's extraordinary ability to assimilate the findings of a wide range of disciplines sees him discuss the work of important mathematicians such as Felix Klein, Steiner, Dedekind, and Hilbert.

Ryckman ends his book with a call for philosophical inquiry into what sense a 'geometrized physics' can have, to emulate the work of these thinkers from the interwar period. And he is not alone in thinking that this was a golden age. There is now an impressive concentration on this era. Today, in Ryckman and similar-minded thinkers, such as Michael Friedman and Alan Richardson (see the contributions of all three in Domski and Dixon (2010)), we find fascinating discussions about these themes. However, these discussions by themselves are unlikely to give rise today to the kind of primary work that they are studying from the past. It is one thing to make a careful, detailed study of the interweaving of philosophy, mathematics, and physics of a period, quite another to begin to take the steps necessary for a revival of such activity.

Michael Friedman (2001) points out two connected, yet somewhat distinct, activities dealing with mathematical physics which might be called 'philosophical'. One termed 'meta-scientific' is much as Weyl does, reconceiving the idea of space and thereby generating foundational advances. Meta-science is typically done by philosophically informed scientists, such as Riemann, Helmholtz, Poincaré, and Einstein. By contrast, the other activity is much as Cassirer and the Vienna Circle did, reflecting on the broader questions of the place of mathematics and science in our body of knowledge in light of important events in the histories of those practices. While there spontaneously arises work of the first kind in any era, work of the second kind requires a philosophical orientation which may be lost. One very obvious difference is that today we have so few philosophers emulating Cassirer by keeping abreast of the mathematics of the recent past. This simply must change if we are to generate the forms of discussion to parallel those of the 1920s. For too long, philosophy has thought to constrain its interest in any current mathematical research largely to set theory, when it has long been evident that it offers little or nothing as far as many core areas of mathematics are concerned, and especially the mathematics needed for physics. Casting the differential cohomology of modern quantum gauge theory in set-theoretic clothing would do no favours to anyone. So, with some notable exceptions, such as

Marquis (2008) and McLarty (2008), we let the bulk of mainstream mathematical research pass us by.

However, there are reasons to be hopeful. I shall argue in this chapter that our best hope in reviving a 1920s-style philosophy of geometry lies in following what has been happening at the cutting edge of mathematical geometry over the past few decades, and that while this may appear a daunting prospect, we do now have ready to hand a means to catch up rapidly. These means are provided by what is known as *cohesive homotopy type theory*.

Univalent foundations, or plain homotopy type theory (cf. Awodey's and Shulman's chapters), provide the syntax for theories which can be interpreted within $(\infty, 1)$-toposes, a generalization of the ordinary notion of toposes. The basic shapes of mathematics are now taken to be the so-called 'homotopy n-types'. However, these are not sufficient to do what needs to be done in modern geometry, and especially in the geometry necessary for modern physics, since we need to add further structures to express continuity, smoothness, and so on. As we add extra properties and structures to $(\infty, 1)$-toposes, characterized by qualifiers—local, ∞-connected, cohesive, differentially cohesive—increasing amounts of mathematical structure are made possible internally. The work of Urs Schreiber (2013) has shown that cohesive $(\infty, 1)$-toposes provide an excellent environment to approach Hilbert's sixth problem on axiomatizing physics, allowing the formulation of relativity theory and all quantum gauge theories, including the higher-dimensional ones occurring in string theory.

Cohesiveness in this sense arose from earlier formulations of the notion in the case of ordinary toposes by William Lawvere (2007), motivated in turn by philosophical reflection on geometry and physics. Schreiber's claim, however, is that for these concepts to take on their full power they must be extended to the context of higher-topos theory, that is the theory of $(\infty, 1)$-toposes, where differential cohomology finds its natural setting. Now, rather than the mathematics necessary for physics being viewed, as it often is at present from a set-theoretic foundation, as elaborate and unprincipled, we can see the simplicity of the necessary constructions through the universal constructions of higher-category theory.

In a single article it will only be possible to outline the kind of work necessary to fill in the spaces we have left ourselves. There is important interpretative work to do already in making sense of plain homotopy type theory, so here I can only indicate further work to be done. At the same time, in that mathematics finds itself once again undergoing enormous transformations in its basic self-understanding, it is important as philosophers to take this opportunity to remind ourselves that we should provide an account of mathematical enquiry where such changes are to be expected. It is striking that Hegel should be found informing both those wishing to characterize the dynamic growth of mathematics and those striving to refashion the very concepts of modern geometry itself.

2.2 Current Geometry

If any reassurance is needed that geometry is alive and well today, one need only look at the variety of branches of mathematics bearing that name which are actively being explored:

algebraic, differential, metric, symplectic, contact, parabolic, convex, Diophantine, tropical, conformal, Riemannian, Kähler, Arakelov, analytic, rigid analytic, global analytic, ...

Continuing our search, we find noncommutative versions of some of these items, also 'derived' versions, and so on. Indeed, there has never been so much 'geometric' research being carried out as there is today, from constructions that Gauss or Riemann might have recognized to ones which would seem quite foreign. So the question arises of whether this list provides just a motley of topics which happen to bear the same name, or whether there is something substantial that is common to all of them, or at least many of them.

Evidence for the latter option comes from people still making unqualified use of 'geometry' and its cognates to mean something. Some such uses are informal, as in the following example:

The fundamental aims of geometric representation theory are to uncover the deeper geometric and categorical structures underlying the familiar objects of representation theory and harmonic analysis, and to apply the resulting insights to the resolution of classical problems.
(MRSI, 2014)

Other uses are technical, such as where Jacob Lurie (2009) uses the term 'geometry' to name a certain kind of mathematical entity, here a small $(\infty, 1)$-category with certain additional data. The question then arises as to what features of these structures make Lurie single them out as geometries. At first glance this seems a rather technical matter. Let us return to it once we have some motivation from the past.

Something that would have seemed novel to Gauss and Riemann, and which might give rise to doubts concerning the unity of geometry, is the thorough injection of spatial ideas brought into algebraic geometry by Alexandre Grothendieck in the 1960s. By the late 1800s it was already known of the collection of complex polynomials, $\mathbb{C}[z]$, that the space on which these functions are defined could be recaptured from the algebraic structure of the collection itself. $\mathbb{C}[z]$ forms a ring, and it is possible to construct an associated space from points corresponding to its maximal ideals. These are the ideals generated by $(z-a)$, for each $a \in \mathbb{C}$. Picking up on the complex function field/algebraic number field analogy of Dedekind and Weber, as developed by Weil in his Rosetta Stone account (see Corfield, 2003, ch. 4), it was then shown that even apparently nonfunctional rings, such as the ring of integers and others encountered in arithmetic, might be treated likewise. Grothendieck's scheme theory (see McLarty, 2008) provided such a space, in the case of the integers denoted $Spec(\mathbb{Z})$, again constructed out of prime ideals. Now an integer is considered as a function defined at each prime, a point in $Spec(\mathbb{Z})$, as the function $n(p) \equiv n \pmod{p}$. Where this differs from the complex function case is that while the values of 'integer-as-function' still land in a field, here the field varies according to the point where the function is evaluated, $\mathbb{F}_p$ as p varies. This suggests a space whose points are not identical.

This attempt to geometrize arithmetic is not an empty game. It feeds through to mathematical practice as can readily be discovered thanks to the growth of online informal discussion.

I like to picture $Spec\,\mathbb{Q}$ as something like a 2-manifold which has had all its points deleted. The extra complication is that what we think of as the points are actually very small circles. So it's really a three manifold with all of the loops inside it deleted.

For example, let's look first at function fields. $Spec\,\mathbb{C}[z]$ is just the complex line $\mathbb{C}$. As we start inverting elements of $Spec\,\mathbb{C}[z]$, as we must do to make $Spec\,\mathbb{C}(z)$, the effect on the spectrum is to remove bigger and bigger finite sets of points. The limit is where we remove all the points and we're just left with some kind of mesh.

If we had started with a Riemann surface of genus g, then we'd be left with a mesh of genus g, a surface sewn out of the cloth from which fly screens for windows are made. If we want to recover the original surface from the surface mesh, we just put it out back in the shed for a while and let the mesh fill up with dirt. This is just the familiar fact that a (smooth compact, say) Riemann surface can be recovered from the field of meromorphic functions on it.

If we replace $\mathbb{C}$ by a finite field $\mathbb{F}$, then everything is the same but what we thought of as the point is now a very small circle, and so our original surface reveals itself to be a 3-manifold fibered over a very small circle when we zoom in. And when we delete points, we're really deleting not just single-valued sections of this fibration but also multivalued sections. So $Spec\,\mathbb{F}[z]$ is some kind of 3-manifold fibered over the circle with all the loops over the base circle deleted.

For the passage from $\mathbb{Z}$ to $\mathbb{Q}$, I don't have anything better to say than that it's sort of the same but there's no base circle. We're just removing lots of loops from a 3-manifold. Maybe some should be seen as bigger than others, corresponding to the fact that there are prime numbers of different magnitudes. (Borger, 2009)

Here we see Borger passing across each of the three columns of Weil's Rosetta Stone.

Now, not only do we find a geometrized arithmetic, but these ideas and constructions are the structural cousins of those appearing in cutting-edge physics, as we see in this comment by David Ben-Zvi:

the geometric analog of a number field or function field in finite characteristic should not be a Riemann surface, but roughly a surface bundle over the circle. This explains the "categorification" (need for a function-sheaf dictionary, which is the weak part of the analogy) that takes place in passing from classical to geometric Langlands—if you study the corresponding QFT on such three-manifolds, you get structures much closer to those of the classical Langlands correspondence. (Ben-Zvi, 2014)

So we have *both* widespread current interest in classically geometric areas of mathematics and geometric approaches to other areas, including arithmetic, *and* ways of thinking about the subject matter expressed, at least informally, in a very visual language. Geometry as a whole is something larger that that which has application in mathematical physics, and applied mathematics more generally. Similar structures are now found to lie at the heart of number theory. But how to go about saying something satisfactorily general about geometry?

One might throw up one's hands at the task of bringing this wealth of subject matter under the umbrella of a straightforward description. To the extent that people try to do this it is largely left to the doyens of mathematics. For example, Sir Michael Atiyah writes

> Broadly speaking, I want to suggest that geometry is that part of mathematics in which visual thought is dominant whereas algebra is that part in which sequential thought is dominant.
> (Atiyah, 2003, 29)

Such a distinction is reminiscent of Kant, for whom space and time were considered to be forms of sensibility, and yet Atiyah continues:

> This dichotomy is perhaps better conveyed by the words "insight" versus "rigour" and both play an essential role in real mathematical problems.   (Atiyah, 2003, 29)

A more careful treatment is required here, since there seems nothing to object to in the idea of 'rigorous geometry' or 'algebraic insight'. We need to turn back the clock to when philosophical research directed itself towards then current geometry.

2.3 Regaining the Philosophy of Geometry

What led to the demise of the philosophy of geometry in the English-speaking world? I think this can be attributed largely to the success of logical empiricism. Many of those dispersed from Germany and Austria in the 1930s were accepted into the universities of the USA, welcomed by existing empiricists such as Ernst Nagel. In a long paper published in 1939, Nagel uses the history of projective geometry to explain the new understanding of then modern axiomatic mathematics:

> It is a fair if somewhat crude summary of the history of geometry since 1800 to say that it has led from the view that geometry is the apodeictic science of space to the conception that geometry, in so far as it is part of natural science, is a system of "conventions" or "definitions" for ordering and measuring bodies.   (Nagel, 1939, 143)

> The distinction between a pure and an applied mathematics and logic has become essential for any adequate understanding of the procedures and conclusions of the natural sciences.
> (Nagel, 1939, 217)

So axiom systems are proposals for stipulations. As pure mathematics they are to be studied for their logical properties. Some of them may be found to be well adapted to allow the expression of scientific laws, which may then be used in applied sciences. This is made possible by coordinating principles which tie the scientific laws to empirical measurements. For example, Riemannian geometry allows for the expression of Einstein's field equations, which can be coordinated to observation by stipulating that light follows null geodesics.

While now the ideational content of mathematics is left to one side, other tasks fall to the philosopher of mathematics:

> the concepts of structure, isomorphism, and invariance, which have been fashioned out of the materials to which the principle of duality is relevant, dominate research in mathematics, logic, and the sciences of nature. (Nagel, 1939, 217)

Had philosophers at least heeded this, more attention might have been paid to category theory, the language par excellence of structure, isomorphism, and invariance, which emerged shortly after Nagel's paper (Corfield, 2015). As it was, a uniform treatment of mathematics as the logical consequences of definitions, or of the set-theoretic axioms, came to prevail.

In the process, as Heis (2011) argues, two lines of thought from earlier in the century were being ruled out, each responding to other nineteenth-century developments in geometry:

1. What could be saved of Kantian philosophy given the appearance of non-Euclidean geometry, and then Riemannian geometry? What are the conditions for spatial experience?
2. How should we understand the ever-changing field of geometry given the introduction of ideal elements, imaginary points, and so on?

Let us take each of these in turn.

2.3.1 Weyl: the essence of space

With an ever-expanding variety of geometries emerging through the nineteenth century, it became implausible to maintain with Kant that our knowledge of Euclidean geometry is a priori. Helmholtz had argued that because empirical measurement requires that objects undergo only 'rigid motions', we can work out which geometries are presupposed by our physics. He concluded that only those spaces which possessed the property of constant curvature were permissible. With the contribution of the technical expertise of Sophus Lie, this line of research resulted in the Helmholtz–Lie theorem, characterizing Euclidean, elliptic, and hyperbolic geometries.

Research such as this was certainly discussed by philosophers. Indeed, Russell, and later Schlick and Reichenbach, responded to Helmholtz's work, but perhaps the most profound response came from Hermann Weyl. After the success of Einstein's general theory of relativity, Helmholtz's results were evidently far too limited. Weyl, inspired by Husserl and perhaps more profoundly by Fichte (see Scholz, 2005), sought to discern what he termed "the essence of space". In a letter to Husserl, he wrote

> Recently, I have occupied myself with grasping the essence of space [das Wesen des Raumes] upon the ultimate grounds susceptible to mathematical analysis. The problem accordingly

concerns a similar group theoretical investigation, as carried out by Helmholtz in his time   (Ryckman, 2005, 113)

Weyl conceived of spaces in which it was only possible to compare the lengths of two rods if they were situated at the same point.

> Only the spatio-temporally coinciding and the immediate spatial-temporal neighborhood have a directly clear meaning exhibited in intuition.
> (Weyl, 'Geometrie und Physik', quoted in Ryckman (2005, 148))

Along the lines of Helmholtz and Lie, this led him to prove a group theoretic result: the only groups satisfying certain desiderata (involving the "widest conceivable range of possible congruence transfers" and a demand for a single affine connection) are the special orthogonal groups of any signature with similarities, $G \simeq SO(p,q) \times \mathbb{R}^+$ (Scholz, 2011, 230).

There is an interesting story to be told here of how Einstein and other physicists found implausible the possibility allowed by this geometry that rods of identical lengths, as measured at one point, if transported along different paths to a distant point might have different resulting lengths. One usually tells the story of how the beauty of the mathematics got the better of Weyl, and how physicists eventually uncovered what was good about the idea whilst modifying his original idea to allow a $U(1)$ gauge group. This story needs to be told in a much more nuanced way (see Giovanelli, 2013), and in any case is complicated by the survival of Weyl's original idea in forms of conformal gauge theory.

In any case, Weyl himself later became sceptical of this kind of mathematical speculation about the geometry required for physics that had so consumed him in his earlier years. With the demise of other philosophical attempts to study our a priori geometric intuition, for example, Carnap's doctoral thesis on how our intuitive concept of space was required to be n-dimensional topological space, such attempts largely came to an end. We will take a look at more recent 'meta-scientific' kinds of work, but first let us turn to Friedman's other form of philosophical research.

2.3.2 Cassirer: beyond intuition

In an unusual paper, published the year before his death, Ernst Cassirer (1944) argued for an important connection to be seen between Felix Klein's Erlanger programme and our everyday perception. Where Klein had given a presentation of many forms of geometry as the study of invariants of space under the action of groups of symmetry, Cassirer saw the seeds for this idea in our abilities to perceive the invariant size, colour, and shape of objects under varying viewing conditions. Now evidently these abilities are rooted in our distant evolutionary past, and yet the full-blown mathematical idea had only crystallized in modern mathematical thinking around 1870.

Klein's ideas on geometry marked an important stage in the course of a revolutionary century for geometric thought. Not only had the range of geometries been extended from the single Euclidean geometry to hyperbolic and elliptical forms, but there had been many kinds of extension of the notion of space by the introduction of elements that seemed to lead us away from the intuitively familiar. For example, the nondegenerate conic sections had been unified as curves of degree 2 in complex projective space, brought about by the addition of 'points at infinity' and complex coordinates. Now all circles were seen to pass through two imaginary points at infinity, and so 'intersect' there.

Such forays beyond the intuitive led those wishing to retain what they took to be valuable in Kant to take a different tack. As Heis (2011) convincingly shows, the neo-Kantian Cassirer had to come to terms with just such developments. This is evident in his later work:

> It is hence obvious that mathematical theories have developed in spite of the limits within which a certain psychological theory of the concept tried to confine them. Mathematical theory ascended higher and higher in order to look farther and farther. Again and again it ventured the Icarian flight which carried it into the realm of mere "abstraction" beyond whatever may be given and represented in intuition. (Cassirer, 1944, 24)

But then without any firm rootedness in intuition, what provides us with guidance that our "Icarian flights" are heading in the best direction? This problematic runs through Cassirer's career, and is answered by the unity of the history of the discipline.

> Though a properly Neo-Kantian philosophy of mathematics will appreciate that mathematics itself has undergone fundamental conceptual changes throughout its history, such a philosophy will also have to substantiate the claim that the various stages in the historical development of mathematics constitute *one history*... we can say that they [mathematicians] were studying the same objects only because we can say that they are parts of the same history. (Heis, 2011, 768)

It is worth quoting Cassirer at length on this point:

> it is not enough that the new elements should prove equally justified with the old, in the sense that the two can enter into a connection that is free from contradiction—it is not enough that the new should take their place beside the old and assert themselves in juxtaposition. This merely formal combinability would not in itself provide a guarantee for a true inner conjunction, for a *homogeneous logical structure of mathematics*. Such a structure is secured only if we show that the new elements are not simply adjoined to the old ones as elements of a different kind and origin, but the new are *a systematically necessary unfolding of the old*. And this requires that we demonstrate a primary logical kinship between the two. Then the new elements will bring nothing to the old, other than *what was implicit in their original meaning*. If this is so, we may expect that the new elements, instead of fundamentally changing this meaning and *replacing* it, will first bring it to *its full development* and *clarification*. (Cassirer, 1957, 392)

If one can hear an overtone of Hegelian thought here, this is not surprising. In the introduction to this third volume of *The Philosophy of Symbolic Forms* Cassirer explained the debt to Hegel as shown by the subtitle of the book—*The Phenomenology of Knowledge*:

The truth is the whole—yet this whole cannot be presented all at once but must be unfolded progressively by thought in its own autonomous movement and rhythm. It is this unfolding which constitutes the being and essence of science. The element of thought, in which science is and lives, is consequently fulfilled and made intelligible only through the movement of its becoming.   (Cassirer, 1957, xiv)

This line of thought sits very happily with the idea that important developments in a discipline allow its history to be written in such a way that it makes best sense of what was only obscurely seen in the past or of what became the means to overcome perceived obstacles or limitations, in other words, a history of rational unfolding out of an older stage. As I argue in Corfield (2012), we find this position very well expressed by the moral philosopher Alasdair MacIntyre. A similar idea is expressed by Friedman's 'retrospective' rationality (2001).

In the Anglophone revival (Friedman, Ryckman, Richardson, Heis, and Everett) of interest in Cassirer of recent years, there has been particular focus on the place of the 'constitutive' and the 'regulative' in his account of the progress of science. This amounts to rival interpretations of the relative importance for Cassirer of the prospective overcoming of limitations within a discipline and the retrospective rationalization of its course, eventually as seen from an ideal future point. However these debates turn out, it is intriguing then to see what we might call a further strand added in the 1944 paper, that in mathematics we may devise concepts which owe their origin to unnoticed cognitive structures. In the case of the Erlanger Program

the mathematical concepts are only the full actualisation of an achievement that, in a rudimentary form, appears also in perception. Perception too involves a certain invariance and depends upon it for its inner constitution.   (Cassirer, 1944, 17)

Taken together, we can see in the work of Weyl and Cassirer just how far we are here in attitude towards mathematical geometry from what was bequeathed to us by the logical empiricists. Heis quotes Hans Reichenbach:

It has become customary to reduce a controversy about the logical status of mathematics to a controversy about the logical status of the axioms. Nowadays one can hardly speak of a controversy any longer. The problem of the axioms of mathematics was solved by the discovery that they are definitions, that is, arbitrary stipulations which are neither true nor false, and that only the logical properties of a system—its consistency, independence, uniqueness, and completeness—can be subjects of critical investigation.   (Heis, 2011, 790)

He very aptly writes: "One could hardly find a point of view further from Cassirer's own" (Heis, 2011, 790). Indeed so, but now to be true to Cassirer's spirit we should try to work out our own position on the rationality of mathematical enquiry in the process of coming to frame what has been happening in the mathematics of the recent past. That this has typically not been felt to be a requirement of philosophy makes this no easy task, but we should try to make a start anyway.

2.4 Capturing Modern Geometry

Even to summarize one particular line of development here will not be easy. Along with Grothendieck's invention of scheme theory, mentioned in Section 2.2, we would also need to talk about topos theory. This could take the form of a story of natural unfolding. Indeed, we could report the originator's own words.

> one can say that the notion of a topos arose naturally from the perspective of sheaves in topology, and constitutes a substantial broadening of the notion of a topological space, encompassing many concepts that were once not seen as part of topological intuition... As the term "topos" itself is specifically intended to suggest, it seems reasonable and legitimate to the authors of this seminar to consider the aim of topology to be the study of topoi (not only topological spaces).
> (Grothendieck and Verdier, 1972, 302)

However, for the purposes of this chapter, we need to race forward to much more recent work. Jacob Lurie motivates his 'Structured Spaces' paper (2009) by means of an account of the passage to less restricted forms of Bézout's theorem. This is a result that goes back to the eighteenth century, involving just the kind of achievement of unity through addition of ideal elements that interested Cassirer. While it was known to Newton that the number of real solutions to the intersection of a pair of plane curves was bounded by the product of their degrees, by the nineteenth century we find a form of the result that states that two complex projective plane curves of respective degrees m and n which share no common component have $m \cdot n$ points of intersection, counted with multiplicity. Any two non identical conics meet four times, including at those imaginary points at infinity in the case of two circles that were mentioned in the previous section.

Lurie takes this result up, looking to understand it in terms of cohomology and the cup product of fundamental classes of the curves, which corresponds to the class of their intersection. Since this method does not work for non-transverse intersections, using Grothendieck's constructions we then turn to 'nonreduced' schemes. Further, according to Lurie, we should look at a Euler characteristic involving the dimension of the local ring of the scheme-theoretic intersection plus various corrections.

Now an interesting thing happens when we attempt to retain the fundamental result $[C] \cup [C'] = [C \cap C']$ in the very general setting where there may even be coinciding components. Here we need *derived* algebraic geometry.

> To obtain the theory we are looking for, we need a notion of generalized ring which remembers not only whether or not x is equal to 0, but how many different ways x is equal to 0. One way to obtain such a formalism is by categorifying the notion of a commutative ring. That is, in place of ordinary commutative rings, we consider categories equipped with addition and multiplication operations (which are encoded by functors, rather than ordinary functions). (Lurie, 2009, 3)

Lurie is drawing attention here to the passage from the proposition 'x is equal to 0' to the set of ways in which it is equal. To do so is to take the first step up an infinitely tall ladder of weakenings of identity. In Corfield (2003, ch. 10) I give an account of categorification, the replacement of set by category by 2-category. In the dozen or so years since my book, the emphasis has swung round to the groupoid version of this ladder, where sets become groupoids become 2-groupoids, etc. Lurie, in particular, was instrumental in this change in showing that most constructions of ordinary category theory have their analogues in the $(\infty, 1)$ setting, where instead of hom-sets between objects, we have ∞-groupoids.

Now, since when dealing with a pair of coinciding lines, we need to make identifications in the form of isomorphisms, we find the following:

These isomorphisms are (in general) distinct from one another, so that the categorical ring C "knows" how many times x and y have been identified.   (Lurie, 2009, 4)

Of course, we never stop with a single step up this ladder, and eventually we seek further generalized forms of ring, such as E_∞-ring spectra and simplicial commutative rings. The key lesson here is that to retain a simple formulation, we must change our framework, for one thing here to allow homotopic weakening. In Cassirerian terms, this is forced upon us by the natural unfolding of the discipline. And it is not just algebraic geometry that demands this richer notion of space, so does physics. The moduli spaces of today's gauge field theories are often stacks, such as the moduli stack of flat connections for some gauge group. Higher gauge theory requires similar homotopic weakening to higher stacks (Schreiber, 2013).

Naturally, Lurie is not a lone voice in calling for this change of outlook. Bertrand Toën likewise gives an account of *derived algebraic geometry*:

Derived algebraic geometry is an extension of algebraic geometry whose main purpose is to propose a setting to treat geometrically special situations (typically bad intersections, quotients by bad actions, ...), as opposed to generic situations (transversal intersections, quotients by free and proper actions, ...).   (Toën, 2014, 1)

He explains the need for 'homotopical perturbation' in Kuhnian terms,

the expression *homotopical mathematics* reflects a shift of paradigm in which the relation of equality relation is weakened to that of homotopy.   (Toën, 2014, 3)

At the same time he points the reader to the Homotopy Type Theory and Univalent Foundation (HoTT/UF) programme as the new foundational language for this homotopical mathematics.

Now, despite this shift to what appears to be the more complex *derived* setting, familiar features are retained:

Just as an ordinary scheme is defined to be "something which looks locally like *SpecA* where A is a commutative ring", a derived scheme can be described as "something which looks locally like *SpecA* where A is a simplicial commutative ring".   (Lurie, 2009, 5)

So, an apparently complicated space is being stuck together from pieces. This theme is taken up by Carchedi in a recent paper:

we will make precise what is means to glue structured ∞-topoi along local homeomorphisms (i.e. étale maps) starting from a collection of local models. This parallels the way one builds manifolds out of Euclidean spaces, or schemes out of affine schemes. Since we are allowing our "spaces" to be ∞-topoi however, in these two instances we get much richer theories than just the theory of smooth manifolds, or the theory of schemes, but rather get a theory of higher generalized orbifolds and a theory of higher Deligne-Mumford stacks respectively. This same framework extends to the setting of derived and spectral geometry as well.

(Carchedi, 2013, 43)

The obvious point to be made is that all of this is just simply unthinkable without category theory. No category theory, no modern geometry of this kind. On the other hand, it may strike the reader as rather daunting that we may need to get a good handle on what Lurie, Toën, and Carchedi are doing with $(\infty, 1)$-toposes. However, we are in luck since just the right kind of foundational language is at hand to help. As Toën noted, in recent years there has emerged homotopy type theory (see Shulman's and Awodey's chapters, this volume), which is expected to play the role of the internal language of $(\infty, 1)$-toposes. Now, this language can be extended to describe large tracts of the constructions of geometry. Schreiber found the ingredients for such an extension in the writings of Lawvere, but needed to transplant them from the original topos setting to the setting of $(\infty, 1)$-toposes. Schreiber and Shulman (2014) worked out how this can be done synthetically by adding 'modalities' to homotopy type theory.

I say it is fortunate for us that this is so, but we should not underestimate the work that is still required. If we recall Friedman's scheme of meta-scientific work leading up to a revolution followed by philosophical interpretative work to make sense of it, we might say that the cycle was largely broken through the twentieth century. Even the lessons of the seventy-year-old category theory are still very far from having been absorbed within philosophy. There have been many contributions made over the decades, but not the kind of sustained work that would make it matter of course for someone entering on a career in philosophy of mathematics to know the basics of category theory. At the very least adjunctions and monads are needed to make any headway.

There will not be space to go into much detail here, but let us begin at the ordinary 1-category level with Lawvere's notion of *cohesion* (Lawvere, 2007) expressed as a chain of adjunctions between a category of spaces and the category of sets. If we take the former to be topological spaces, then one basic mapping takes such a space and gives its underlying set of points. All the cohesive 'glue' has been removed. Now there are two ways to generate a space from a set: one is to form the space with the discrete topology, where no point sticks to another; the other is to form the space with the codiscrete topology, where the points are all glued together into a single blob so that no part is separable, in the sense that there are only constant maps from a codiscrete space to the discrete space with two points. Finally, we need a second map from spaces to sets, one which 'reinforces' the glue by reducing each connected part to an element of a set, the connected components functor, π_0:

$$(\pi_0 \dashv Disc \dashv U \dashv coDisc) : Top \rightarrow Set$$

These four functors form an adjoint chain, where any of the three compositions of two adjacent functors ($U \circ coDisc$, $U \circ Disc$, $\pi_0 \circ Disc$) from the category of sets to itself is the identity, whereas, in the other direction, composing adjacent functors to produce endofunctors on *Top* ($coDisc \circ U$, $Disc \circ U$, $Disc \circ \pi_0$) yields two idempotent monads and one idempotent comonad.

Adjoint modalities where the monad is the right adjoint, $\Box \dashv \bigcirc$, can be thought of as two different opposite 'pure moments', such as codiscreteness and discreteness in this case, or in another example by Lawvere (2000), oddness and evenness of integers. There is an equivalence between types which are pure according to one of the moments and those pure according to the other, but these pure collections inject into the whole differently, as with odd and even integers into all integers, and discrete and codiscrete spaces into all spaces.

On the other hand, in adjoint modalities where the monad is on the left, $\bigcirc \dashv \Box$, there is a single moment, but the full collection of types projects onto those pure according to this moment in two different ways. Here, cohesive spaces project in two opposite ways to discrete spaces, either by the complete removal of the cohesion or by the identication of any cohering points. Another simple example has the real numbers project to the integers (entities which are purely integral) in two ways, via the floor and ceiling functions.

What Schreiber does is to find analogous modalities generated by an adjoint quadruple between an $(\infty, 1)$-topos, H, and the base $(\infty, 1)$-topos of ∞-groupoids, $\infty Grpd$:

$$(\Pi \dashv Disc \dashv \Gamma \dashv coDisc) : H \to \infty Grpd.$$

The three induced 'adjoint modalities' are called shape modality $\dashv$ flat modality $\dashv$ sharp modality and denoted $\int \dashv \flat \dashv \sharp$. In a sense, this H can be seen as spaces modelled on a 'thickened' point.

Now a very similar pattern repeats itself in the form of a further string of four adjunctions, this time between H and another $(\infty, 1)$-topos, corresponding to extending the thickened point infinitesimally. The three resulting adjoint modalities now comprise two comonads and one monad.

The existence of these two related sets of three adjoint modalities is extraordinarily powerful, allowing the expression of a rich internal higher geometry, including Galois theory, Lie theory, differential cohomology, and Chern–Weil theory, and allows for the synthetic development of higher gauge theory (Schreiber, 2013). There remains plenty more interpretative work to be done in making these ideas more accessible, but for our purposes here let us just retain the monad of the second adjoint modality triple, denoted $\Im$ or sometimes $\int_{inf}$. It is the important one for us to continue the story from Lurie and Carchedi.

So now, despite the apparently intimidating complexity of modern geometry, it is possible to maintain, as Schreiber does, that there remains a simplicity.

It would seem to me that the old intuition, seemingly falling out of use as the theory becomes more sophisticated, re-emerges strengthened within higher topos theory... Notably all those "generalized schemes", "étale infinity-groupoids" and so forth are nothing but the implementation of the old intuition of "big spaces glued from small model spaces" implemented in homotopy theory... I think it's a general pattern, in the wake of homotopy type theory we find that much of what looks super-sophisticated in modern mathematics is pretty close to the naive idea, but implemented internally in an ∞-topos. (Schreiber, 2014b)

With homotopy type theory and the six modalities briefly mentioned earlier, and in particular the infinitesimal shape modality, it is possible to describe *synthetically* what it is to be a 'formally étale morphism'. Now choosing types, $\{U_i\}$, as 'model spaces', then a general geometric space is a type X equipped with a map of the form

$$\coprod_j U_j \longrightarrow X,$$

such that this map is a 1-epimorphism and formally étale.[1] We have arrived thus at a synthetic formulation of one of the very basic ideas of geometry.

Of course, there are many such basic ideas for us to consider. In this section, I have sketched some ideas of an extraordinarily ambitious body of scientific and *meta-scientific* work. It may appear that by proposing that we understand cutting-edge geometry, I risk being caught up with the changeable fashions of research, but let us not forget that these projects are rooted in the ideas of Grothendieck from many decades ago, and that later developments were foreseen to some considerable extent by him (see, e.g., Grothendieck, 1983). Current ideas thus emerge out of a vast body of work. Indeed, Toën motivates a section where he constructs "a brief, and thus incomplete, history of the mathematical ideas that have led to the modern developments of derived algebraic geometry" as follows:

As we will see the subject has been influenced by ideas from various origins, such as intersection theory in algebraic geometry, deformation theory, abstract homotopy theory, moduli and stacks theory, stable homotopy theory, and so on. Derived algebraic geometry incorporates all these origins, and therefore possesses different facets and can be comprehended from different angles. We think that knowledge of some of the key ideas that we describe below can help to understand the subject from a philosophical as well as from technical point of view. (Toën, 2014, 6)

If some details will inevitably change, that $(\infty, 1)$-categories lie at the heart of modern geometry will very likely not.

---

[1] In the case of schemes, one needs to modify slightly to *pro-étale* morphisms, in some sense a reflection of the less homogeneous nature of the spaces.

2.5 Conclusion

I have sketched a broad canvas in this chapter. This is to some degree forced upon us by the state we are in where philosophy has drifted from its task. Had the course of philosophy after the famous Davos meeting (Friedman, 2000) favoured Cassirer, we might have had a generation of philosophers keen to search for the emergence of new self-understandings in mathematics. Surely in that case category theory, and its higher forms, would have been absorbed much more fully into philosophical consciousness. With the emergence of homotopy type theory, which is already generating considerable philosophical interest, we may see this happen at last. What I have described in this chapter should suggest that there is a great deal of further work to be done in coming to understand extensions of homotopy type theory, certainly the cohesive variety so far as geometry goes. It should also be noted that with a linear logic variant of homotopy type theory it is possible to express synthetically many aspects of the quantization of higher gauge theory (Schreiber, 2014a).

We have seen Weyl-like meta-scientific work in the formulation of cohesive homotopy type theory, requiring a range of modalities to be added to the basic type theory. Unlike Weyl with Fichte, Schreiber follows Lawvere (1970, 1991) in finding inspiration in Hegel. One can even tell a 'Hegelian' story starting from the opposition between $\emptyset$ and 1, rising through a process of 'Aufhebung' to the six modalities (Schreiber, 2014a, sect. 2.4), and even beyond to a further set of three modalities which may be interpreted as capturing the supergeometry needed for dealing with fermions.

Contrast this with a different kind of use of Hegel by Cassirer and also Lakatos, a philosopher more familiar to the Anglophone community. With the new framework for geometry in place, we should be able to tell the Cassirerian story of the unfolding of the past in mathematics and physics, as mathematicians such as Toën are inclined to do by themselves. Mathematics is to be understood by the fact that it constitutes a single tradition of intellectual enquiry. Ideas found at particular stages possess the seeds of later formulations, which retrospectively allow us to understand them better.

We can use this opportunity to gain a grip on some real mathematical content, offering the opportunity for a more interesting dialogue between philosophy of physics and philosophy of mathematics. For one thing, the duality between geometry and algebra that we saw between rings and affine schemes, and which lies behind the relation between the Heisenberg and Schrödinger pictures, continues to higher geometry and higher algebra, where it manifests itself in different formulations of higher gauge theory. Fundamentally, this duality relates to the operation of taking opposites of $(\infty, 1)$-categories (Corfield, 2015).

Finally, as with Cassirer's observation about the seeds of the Erlanger Programme lying within our perception, it is sometimes revealed during and after moments of synthesis in mathematics that there is a reliance on aspects of cognition, perception, and language, which had possibly gone unnoticed. I think at the very least a form of dependent type theory is present in our cognition as manifested in ordinary language

(Ranta, 1995). Likewise, the idea of big spaces glued from small model spaces seems very basic. It is surely no accident that mathematicians speak of an 'atlas' to define a manifold, since an ordinary atlas provides a collection of maps which overlap. It seems likely we employ something like this in the cognitive maps by which we navigate our domain. Perhaps one of the invariants of geometry has been found here.

References

Atiyah, M. (2003). What is geometry? in *The Changing Shape of Geometry: Celebrating a Century of Geometry and Geometry Teaching*, 24–30. Cambridge University Press, Cambridge, UK.

Ben-Zvi, D. (2014). Blog comment. https://www.math.columbia.edu/ woit/wordpress/?p=7114& cpage=1# comment-214353

Borger, J. (2009). Blog comment. https://golem.ph.utexas.edu/category/2009/02/lakatos_as_ dialectical_realist.html#c022225

Carchedi, D. (2013). Higher orbifolds and Deligne-Mumford stacks as structured infinity topoi. Arxiv preprint: http://arxiv.org/abs/1312.2204

Cassirer, E. (1944). The concept of group and the theory of perception. *Philosophy and Phenomenological Research* 5(1), 1–36.

Cassirer, E. (1957). *The Philosophy of Symbolic Forms: The Phenomenology of Knowledge*, Vol. 3. Yale University Press, New Haven, CT.

Corfield, D. (2003). *Towards a Philosophy of Real Mathematics*. Cambridge University Press, Cambridge, UK.

Corfield, D. (2012). Narrative and the rationality of mathematical practice, in A. Doxiadis and B. Mazur (eds) *Circles Disturbed: The Interplay of Mathematics and Narrative*, 244–80. Princeton University Press, Princeton, NJ.

Corfield, D. (2015). Duality as a category-theoretic concept. *Studies in History and Philosophy of Modern Physics*, doi:10.1016/j.shpsb.2015.07.004.

Domski, M., and Dickson, M. (eds) (2010). *Discourse on a New Method: Reinvigorating the Marriage of History and Philosophy of Science*. Open Court Publishing Company.

Einstein, A. (1934). On the method of theoretical physics. *Philosophy of Science* 1(2), 163–9.

Friedman, M. (2000). *The Parting of the Ways: Carnap, Cassirer, and Heidegger*. Open Court Publishing Company.

Friedman, M. (2001). *The Dynamics of Reason*. University of Chicago Press, Chicago.

Giovanelli, M. (2013). Talking at cross-purposes: how Einstein and the logical empiricists never agreed on what they were disagreeing about. *Synthese* 190(17), 3819–63.

Grothendieck, A. (1983). Pursuing stacks, Letter to D. Quillen, in G. Maltsiniotis, M. Künzer, and B. Toen (eds), *Documents Mathématiques*. Society Mathematics Paris, France.

Grothendieck, A., and Verdier, J. (1972). *Théorie des topos et cohomologie étale des schémas*, Tome 1: Théorie des topos. *Lecture Notes in Mathematics*, Vol. 269. Springer-Verlag, Berlin.

Heis, J. (2011). Ernst Cassirer's neo-Kantian philosophy of geometry. *British Journal for the History of Philosophy* 19(4), 759–94.

Lawvere, W. (1970). Quantifiers and sheaves. *Actes du congres international des mathematiciens*, Nice 1, 329–34.

Lawvere, W. (1991). Some thoughts on the future of category theory, in A. Carboni et. al. (eds), *Category Theory*, 1–13 Springer, New York.

Lawvere, W. (2000). Adjoint cylinders. Categories mailing list comment, http://permalink.gmane.org/gmane.science.mathematics.categories/1683

Lawvere, W. (2007). Axiomatic cohesion. *Theory and Applications of Categories* 19(3), 41–9.

Lurie, J. (2009). Derived algebraic geometry V: structured spaces. Arxiv preprint: http://arxiv.org/abs/0905.0459

Marquis, J.-P. (2008). *From a Geometric Point of View: A Study of the History and Philosophy of Category Theory*. Springer, New York.

McLarty, C. (2008). There is no ontology here: visual and structural geometry in arithmetic, in P. Mancosu (ed.), *The Philosophy of Mathematical Practice*, 370–406. Oxford University Press, Oxford.

MRSI (2014). Geometric Representation Theory, Programme announcement, https://www.msri.org/programs/276

Nagel, E. (1939). The formation of modern conceptions of formal logic in the development of geometry. *Osiris* 7, 142–223.

Ranta, A. (1995). *Type-Theoretical Grammar*. Oxford University Press, Oxford.

Ryckman, T. (2005). *The Reign of Relativity*. Oxford University Press, Oxford.

Scholz, E. (2005). Philosophy as a cultural resource and medium of reflection for Hermann Weyl. *Revue de Synthèse* 126, 331–351.

Scholz, E. (2011). H. Weyl's and E. Cartan's proposals for infinitesimal geometry in the early 1920s. *Boletim da Sociedada Portuguesa de Matematica* Numero Especial A, 225–45.

Schreiber, U. (2013). Differential cohomology in a cohesive infinity-topos. Arxiv preprint: http://arxiv.org/abs/1310.7930

Schreiber, U. (2014a). Quantization via Linear homotopy types. Arxiv preprint: http://arxiv.org/abs/1402.7041

Schreiber, U. (2014b). nForum discussion comment. http://nforum.ncatlab.org/discussion/2084/higher-geometry/?Focus=49931# Comment_49931

Schreiber, U., and Shulman, M. (2014). Quantum gauge field theory in cohesive homotopy type theory. Arxiv preprint: http://arxiv.org/abs/1408.0054

Toën, B. (2014). Derived algebraic geometry. Arxiv preprint: http://arxiv.org/abs/1401.1044

Torretti, R. (1978). *Philosophy of Geometry from Riemann to Poincaré*. Springer, New York.

3

Homotopy Type Theory: A Synthetic Approach to Higher Equalities

Michael Shulman

3.1 Introduction

Ask an average mathematician or philosopher today about the foundations of mathematics, and you are likely to receive an answer involving set theory: an apparent consensus in marked contrast to the foundational debates of the early twentieth century. Now, at the turn of the twenty-first century, a new theory has emerged to challenge the foundational ascendancy of sets. Arising from a surprising synthesis of constructive intensional type theory and abstract homotopy theory, Homotopy Type Theory and Univalent Foundations (HoTT/UF) purports to represent more faithfully the everyday practice of mathematics, but also provides powerful new tools and a new paradigm. So far, its concrete influence has been small, but its potential implications for mathematics and philosophy are profound.

There are many different aspects to HoTT/UF,[1] but in this chapter I will focus on its use as a foundation for mathematics. Like set theory, it proposes to found mathematics on a notion of *collection*, but its collections (called *types*) behave somewhat differently. The most important difference is that in addition to having elements as sets do, the types of HoTT/UF come with further collections of *identifications* between these elements (i.e. ways or reasons that they are equal). These identifications form a structure that modern mathematicians call an ∞-*groupoid* or *homotopy type*, which is a basic object of study in homotopy theory and higher category theory; thus, HoTT/UF offers mathematicians a new approach to the latter subjects.

Of greater importance philosophically, however, is HoTT/UF's proposal that such types can be the fundamental objects out of which mathematics and logic are built.

---

[1] Though HoTT and UF are not identical, the researchers working on both form a single community, and the boundary between them is fluid. Thus, I will not attempt to distinguish between them, even if it results in some technically incorrect statements.

In other words, HoTT/UF suggests that whenever we mentally form a collection of things, we must *simultaneously* entertain a notion of what it means for two of those things to be the same (in contrast to the position of Zermelo-Fraenkel theory with choice (ZFC) that all things have an identity criterion *prior* to their being collected into a set). As stated, this is closely related to the conception of "set" promulgated by Bishop, but HoTT/UF generalizes it by allowing two things to "be the same" in *more than one way*. This is perhaps not a common everyday occurrence, but it is a fundamental part of category theory and thus an integral part of mathematics, including many modern theories of physics. Thus, like other initially unintuitive ideas such as relativistic time dilation and quantum entanglement, it can be argued to be basic to the nature of reality. The innovation of HoTT/UF is that this idea can be made basic to the foundational logical structure of mathematics as well, and that doing so actually *simplifies* the theory.

In this chapter, I will attempt to convey some of the flavor and advantages of HoTT/UF; we will see that in addition to expanding the discourse of mathematics, it also represents certain aspects of *current* mathematical practice more faithfully than set theory does. In Sections 3.2 and 3.3, I will describe HoTT/UF very informally; in Sections 3.4–3.6, I will discuss some of its features in a bit more detail; and in Section 3.7, I will attempt to pull together all the threads with an example. For space reasons, I will not be very precise, nor will I discuss the history of the subject in any depth; for more details, see Univalent Foundations Program (2013). Other recent survey articles on HoTT/UF include Awodey (2012); Awodey et al. (2013); and Pelayo and Warren (2014).

For helpful conversations and feedback, I would like to thank (in random order) Emily Riehl, David Corfield, Dimitris Tsementzis, James Ladyman, Richard Williamson, Martín Escardó, Andrei Rodin, Urs Schreiber, John Baez, and Steve Awodey, as well as numerous other contributors at the n-Category Café and the HoTT email list, and the referees.

3.2 ∞-groupoids

The word "∞-groupoid" looks complicated, but the underlying idea is extremely simple, arising naturally from a careful consideration of what it means for two things to be "the same". Specifically, it happens frequently in mathematics that we want to define a collection of objects that are determined by some kind of "presentation", but where "the same" object may have more than one presentation. As a simple example, if we try to define a *real number* to be an infinite decimal expansion[2] such as $\pi = 3.14159\cdots$, we encounter the problem that (for instance)

---

[2] Like any mathematical object, there are many equivalent ways to define the real numbers. This specific definition is rarely used in mathematics for technical reasons, but it serves as a good illustration, and the common definition of real numbers using Cauchy sequences has exactly the same issues.

$$0.5 = 0.50000\cdots \quad \text{and} \quad 0.4\overline{9} = 0.49999\cdots$$

are distinct decimal expansions but ought to represent the same real number. Therefore, "the collection of infinite decimal expansions" is not a correct way to define "the collection of real numbers".

If by "collection" we mean "set" in the sense of ZFC, then we can handle this by defining a real number to be a *set* of decimal expansions that all "define the same number", and which is "maximal" in that there are no *other* expansions that define the same number. Thus, one such set is $\{0.5, 0.4\overline{9}\}$, and another is $\{0.\overline{3}\}$. These sets are *equivalence classes*, and the information about which expansions define the same number is an *equivalence relation* (a binary relation $\sim$ such that $x \sim x$, if $x \sim y$ then $y \sim x$, and if $x \sim y$ and $y \sim z$ then $x \sim z$). The set of equivalence classes is the *quotient* of the equivalence relation.

Similarly, Frege (1884, sect. 68) defined the *cardinality* of a set X to be (roughly, in modern language) the set of all sets related to X by a bijection. Thus, for instance, 0 is the set of all sets with no elements, 1 is the set of all singleton sets, and so on. These are exactly the equivalence classes for the equivalence relation of bijectiveness. That is, we consider a cardinal number to be "presented" by a set having that cardinality, with two sets presenting the same cardinal number just when they are bijective.

An example outside of pure mathematics involves Einstein's theory of general relativity, in which the universe is represented by a differentiable manifold with a metric structure. In this theory, if two manifolds are *isomorphic* respecting their metric structure, then they represent the same physical reality. (An isomorphism of manifolds is often called a "diffeomorphism", and if it respects the metric it is called an "isometry".) Thus we find, for instance, in Sachs and Wu (1977, sect. 1.3) that

> A general relativistic *gravitational field* [$(M, \mathbf{g})$] is an equivalence class of spacetimes [manifolds M with metrics $\mathbf{g}$] where the equivalence is defined by ... isometries.

This sort of situation, where multiple mathematical objects represent the same physical reality, is common in modern physics, and the mathematical objects (here, the manifolds) are often called *gauges*.[3]

Definitions by equivalence classes are thus very common in mathematics and its applications, but they are not the only game in town. A different approach to the problem of "presentations" was proposed by Bishop (1967, sect. 1.1):

> A set is defined by describing exactly what must be done in order to construct an element of the set and what must be done in order to show that two elements are equal.

In other words, according to Bishop, a *set* is a collection of things *together with* the information of when two of those things are equal (which must be an equivalence

---

[3] Whether general relativity should be technically considered a "gauge theory" is a matter of some debate, but all that matters for us is that it exhibits the same general phenomenon of multiple models.

relation).[4] Thus, the real numbers would *be* infinite decimal expansions, but "the set of real numbers" would include the information that (for instance) 0.5 and 0.4$\overline{9}$ are the same real number. One advantage of this is that if we are given "a real number", we never need to worry about *choosing* a decimal expansion to represent it. (Of course, for decimal expansions there are canonical ways to make such a choice, but in other examples there are not.)

As a much older example of this style of definition, in Euclid's *Elements* we find the following:

> **Definition 4.**   Magnitudes are said to *have a ratio* to one another which can, when multiplied, exceed one another.
>
> **Definition 5.**   Magnitudes are said to be *in the same ratio*, the first to the second and the third to the fourth, when, if any equimultiples whatever are taken of the first and third, and any equimultiples whatever of the second and fourth, the former equimultiples alike exceed, are alike equal to, or alike fall short of, the latter equimultiples respectively taken in corresponding order.

That is, Euclid first defined how to *construct* a ratio, and then second he defined when two ratios are *equal*, exactly as Bishop says he ought.

On its own, Bishop's conception of set is not a very radical change. But it paves the way for our crucial next step, which is to recognize that frequently there may be more than one "reason" why two "presentations" define the same object. For example, there are two bijections between $\{a, b\}$ and $\{c, d\}$: one that sends a to c and b to d, and another that sends a to d and b to c. Likewise, a pair of manifolds may be isometric in more than one way.

This should not be confused with the question of whether there is more than one *proof* that two things are the same. Rather, the question is whether substituting one for the other in a mathematical statement or construction can yield multiple inequivalent results. For instance, there is a predicate P on $\{a, b\}$ such that $P(a)$ is true and $P(b)$ is false. We can "transport" P along a bijection from $\{a, b\}$ to $\{c, d\}$ to obtain a predicate Q on $\{c, d\}$, but the resulting Q will depend on which bijection we use. If we use the bijection that sends a to c and b to d, then $Q(c)$ will be true and $Q(d)$ will be false, but if we use the other bijection, then $Q(c)$ will be false and $Q(d)$ will be true. Thus, $\{a, b\}$ and $\{c, d\}$ "are the same" in more than one way.

If a predicate or construction is left literally unchanged by this sort of substitution, it is called *invariant*. Thus, physicists speak of *gauge invariance* when talking about theories with multiple mathematical models of the same reality. More generally, a construction that "varies appropriately" under such substitutions (but in a way potentially dependent on the "reason" for sameness, as explained earlier) is called

---

[4] Although Bishop's goal was to give a constructive treatment of mathematics, this notion of "set" is meaningful independently of whether one's logic is constructive or classical.

covariant. In particular, general relativity is said to be *generally covariant*, meaning that a mathematical model of reality can be replaced by any isometric one—but in a way dependent on the particular isometry chosen.

This behavior lies at the root of Einstein's famous *hole argument*, which can be explained most clearly as follows. Suppose M and N are manifolds with spacetime metrics **g** and **h**, respectively, and ϕ is an isometry between them. Then any point $x \in M$ corresponds to a unique point $\phi(x) \in N$, both of which represent the same "event" in spacetime. Since ϕ is an isometry, the gravitational field around x in M is identical to that around $\phi(x)$ in N. However, if ψ is a *different* isomorphism from M to N which does *not* respect the metrics, then the gravitational field around x in M may be quite different from that around $\psi(x)$ in N.

So far, this should seem fairly obvious. But Einstein originally considered only the special case where M and N happened to be the same manifold (though not with the same metric), where ψ was the identity map id_M, and where ϕ was the identity outside of a small "hole". In this case, it seemed wrong that two metrics could be the same outside the hole but different inside of it. The solution is clear from the more general situation in the previous paragraph: the fact that the two metrics "represent the same reality" is witnessed by the isomorphism ϕ, not ψ. Thus, even for a point x inside the hole, we should be comparing **g** at x with **h** at $\phi(x)$, not with **h** at $\text{id}_M(x) = x$.[5]

This and other examples show that it is often essential to *remember which* isomorphism we are using to treat two objects as the same. The set-theoretic notion of equivalence classes is unable to do this, but Bishop's approach can be generalized to handle it. Indeed, such a generalization is arguably already latent in Bishop's constructive phrasing: both the construction of elements and the proofs of equality are described in terms of *what must be done*, so it seems evident that just as there may be more than one way to construct an element of a set, there may be more than one way to show that two elements are equal. Bishop made no use of this possibility, but HoTT/UF takes it seriously. The laws of an equivalence relation then become algebraic structure on these "reasons for equality": given a way in which $x = y$ and a way in which $y = z$, we must have an induced way in which $x = z$, and so on, satisfying natural axioms. The resulting structure is called a *groupoid*. Thus, for instance, spacetime manifolds form a groupoid, in which the ways that $M = N$ are the isometries from M to N (if any exist).

If it should happen that for every x and y in some groupoid, there is *at most one* reason why $x = y$, then our groupoid is essentially just a set in Bishop's sense; thus, the universe of sets is properly included in that of groupoids. This is what happens with decimal expansions: there is only one way in which 0.5 and $0.4\overline{9}$ represent the same real number (i.e. in any statement or construction involving 0.5, there is only one way to replace 0.5 by $0.4\overline{9}$). This is in contrast to the situation with manifolds, where using

---

[5] While this description in modern language makes it clear why there is no paradox, it does obscure the reasons why for many years people *thought* there was a paradox! I will return to this in Section 3.7.

a different isomorphism ϕ or ψ from M to N can result in different statements, e.g. one which speaks about $\phi(x) \in N$ and another about $\psi(x) \in N$.

The final step of generalization is to notice that we introduced sets (and generalized them to groupoids) to formalize the idea of "collection", but we have now introduced, for each pair of things x and y in a groupoid, an *additional* collection, namely the ways in which x and y are equal. Thus, it seems natural that this collection should itself be a set, or more generally a groupoid, so that two ways in which $x = y$ could themselves be equal or not, and perhaps in more than one way. Taken to its logical conclusion, this observation demands an infinite tower consisting of elements, ways in which they are equal, ways in which those are equal, ways in which *those* are equal, and so on. Together with all the necessary operations that generalize the laws of an equivalence relation, this structure is what we call an *∞-groupoid*.

This notion may seem very abstruse, but over the past few decades ∞-groupoids have risen to a central role in mathematics and even physics, starting from algebraic topology and metastasizing outwards into commutative algebra, algebraic geometry, differential geometry, gauge field theory, computer science, logic, and even combinatorics. It turns out to be very common that two things can be equal in more than one way.

3.3 Foundations for Mathematics

In Section 3.2, I introduced the notion of ∞-groupoid informally. At this point a modern mathematician would probably try to give a *definition* of ∞-groupoid, such as "an ∞-groupoid consists of a collection of elements, together with for any two elements x, y a collection of ways in which $x = y$, and for any two such ways f, g a collection of ways in which $f = g$, and so on, plus operations …". Clearly, any such definition must refer to a *prior* notion of "collection", which a modern mathematician would probably interpret as "set". Such definitions of ∞-groupoids are commonly used, although they are quite combinatorially complicated.

However, in Section 3.2, we considered ∞-groupoids not as *defined in terms of* sets, but as *substitutes* or rather *generalizations* of them. Thus, we should instead seek a theory at roughly the same ontological level as ZFC, whose basic objects are ∞-groupoids. This is exactly what HoTT/UF is: a *synthetic theory of ∞-groupoids*.[6]

The word "synthetic" here is, as usual, used in opposition to "analytic". In modern mathematics, an analytic theory is one whose basic objects are defined in some other theory, whereas a synthetic theory is one whose basic objects are undefined terms given meaning by rules and axioms. For example, *analytic geometry* defines points

---

[6] Since ∞-groupoids are a formalization of the idea that things can be equal in more than one way, that these ways can themselves be equal in more than one way, and so on, we may equivalently (but more informally) call HoTT/UF a *synthetic theory of higher equalities*, as in the chapter title.

and lines in terms of numbers, whereas *synthetic geometry* is like Euclid's with "point" and "line" essentially undefined.[7]

Thus, our first step to understanding HoTT/UF is that it is an axiomatic system in which "∞-groupoid" is essentially an undefined term. One advantage of this can already be appreciated: it allows us to say simply that for any two elements x and y of an ∞-groupoid, the "ways in which $x = y$" form another ∞-groupoid, so that ∞-groupoids are really the only notion of "collection" that we need consider. As part of a *definition* of ∞-groupoid, this would appear circular, but as an *axiom*, it is unobjectionable.

So far, this description of HoTT/UF could also be applied (with different terminology) to the field of mathematics called "abstract homotopy theory". However, although HoTT/UF is strongly influenced by homotopy theory, there is more to it: as suggested earlier, its ∞-groupoids can substitute for sets as a foundation for mathematics.

When I say that a synthetic theory can be a *foundation for mathematics*, I mean simply that we can encode the rest of mathematics into it somehow.[8] This definition of "foundation" is reasonably precise and objective, and agrees with its common usage by most mathematicians. A computer scientist might describe such a theory as "mathematics-complete", by analogy with Turing-complete programming languages (that can simulate all other languages) and NP-complete problems (that can solve all other NP problems). For example, it is commonly accepted that ZFC set theory has this property. On the other hand, category theory in its role as an organizing principle for mathematics, though of undoubted philosophical interest, is not foundational in this sense (although a synthetic form of category theory like that of Lawvere (1966) could be).

In particular, a synthetic theory cannot fail to be foundational because some analytic theory describes similar objects. The fact that we *can* define and study ∞-groupoids inside of set theory says nothing about whether a *synthetic* theory of ∞-groupoids can be foundational. To the contrary, in fact, it is highly *desirable* of a new foundational theory that we can translate back and forth to previously existing foundations; among other things it ensures the relative consistency of the new theory. Similarly, we cannot dismiss a new synthetic foundational theory by claiming that it "requires some pre-existing notions": the simple fact of being synthetic means that it does not. Of course, humans always try first to *understand* new ideas in terms of old ones, but that doesn't make the new ideas *intrinsically* dependent on the old. A student may learn that dinosaurs are like "big lizards", but that doesn't make lizards logically, historically, or genetically prior to dinosaurs.

---

[7] Euclid's *Elements* as they have come down to us do contain "definitions" of "point" and "line", but these are not definitions in a modern mathematical sense, and more modern versions of Euclidean geometry such as that of Hilbert (1899) do leave these words undefined.

[8] Or into some natural variant or extension of it, such as by making the logic intuitionistic or adding stronger axioms.

In addition, we should beware of judging a theory to be more intuitive or fundamental merely because we are familiar with it: intuition is not fixed, but can be (and is) trained and developed. At present, most mathematicians think of ∞-groupoids in terms of sets because they learned about sets early in their mathematical education; but even in its short existence the HoTT/UF community has already observed that graduate students who are "brought up" thinking in HoTT/UF form a direct understanding and intuition for it that sometimes outstrips that of those who "came to it late". Moreover, the ZFC-like intuitions about set theory now possessed by most mathematicians and philosophers also had to be developed over time: Lawvere (1994) has pointed out that Cantor's original "sets" seem more like those of Lawvere's alternative set theory, the Elementary Theory of the Category of Sets (ETCS) (see Lawvere (2005) and McLarty's chapter in the present volume).

The point being made, therefore, is that HoTT/UF, the synthetic theory of ∞-groupoids, can be a foundation for mathematics in this sense. There is quite an easy proof of this: we have already seen that the universe of ∞-groupoids properly contains a universe of sets. More precisely, there is a subclass of the ∞-groupoids of HoTT/UF which together satisfy the axioms of ETCS.[9] A model of ZFC can then be constructed using trees as described in McLarty's chapter, or directly as in Univalent Foundations Program (2013, sect. 10.5). Thus, any mathematics that can be encoded into set theory can also be encoded into HoTT/UF. (Of course, if we intended to encode *all* of mathematics into HoTT/UF via set theory this way, there would be no benefit to choosing HoTT/UF as a foundation over set theory. The point is that *some* parts of mathematics can be also encoded into HoTT/UF in *other*, perhaps more natural, ways.)

In sum, if we so desire, *we may regard the basic objects of mathematics to be ∞-groupoids rather than sets*. Our discussion in Section 3.2 suggests some reasons why we might want to do this; I will mention some further advantages as they arise. But it is now time to say something about what HoTT/UF actually looks like.

3.4 Type Theory and Logic

The basic objects of HoTT/UF behave like ∞-groupoids; but we generally call them *types* instead, and from now on I will switch to this usage. This particular word is due to the theory's origins in Martin-Löf type theory (Martin-Löf, 1975); but (in addition to being five syllables shorter) it also fortuitously evokes the terminology "homotopy type" from algebraic topology, which is essentially another word for "∞-groupoid" (see e.g. Baez, 2007).

---

[9] In fact, HoTT/UF is not (yet) a single precisely specified theory like ZFC and ETCS: as befits a young field, there are many variant theories in use and new ones under development. In particular, when I say "HoTT/UF" I mean to encompass both "classical" versions that have the Axiom of Choice and Law of Excluded Middle and also "intuitionistic" or "constructive" ones that do not. In the latter cases, the universe of sets satisfies not ETCS (which is classical) but an "intuitionistic" version thereof.

Like sets, the types of HoTT/UF have *elements*, also called *points*. We write $x : A$ when x is a point of A; the most salient difference between this and ZFC's "$x \in A$" is that (like in ETCS) we cannot compare elements of different types: a point is always *a point of some type*, that type being part of its nature. Whenever we introduce a variable, we must specify its type: whereas in ZFC "for every integer x, $x^2 \geq 0$" is shorthand for "for every thing x, if x happens to be an integer then $x^2 \geq 0$", in HoTT/UF the phrase "for every integer x" is atomic. This arguably matches mathematical practice more closely, although the difference is small.

The basic theory of HoTT/UF is a collection of *rules* stipulating operations we can perform on types and their points. For instance, if A and B are types, there is another type called their cartesian product and denoted $A \times B$. Any such rule for making new types comes with some number of rules for making points of these types: in the case of products, this rule is that given $a : A$ and $b : B$, we have an induced point of $A \times B$ denoted (a, b). We also have dual rules for extracting information from points of types, e.g. from any $x : A \times B$ we can extract $\pi_1(x) : A$ and $\pi_2(x) : B$. Of course, $\pi_1(a, b)$ is a and $\pi_2(a, b)$ is b.

It is important to understand that these *rules* are not the same sort of thing as the *axioms* of a theory like ZFC or ETCS. Axioms are statements *inside* an ambient superstructure of (usually first-order) logic, whereas the rules of type theory exist at the same level as the deductive system of the logic itself. In a logic-based theory like ZFC, the "basic act of mathematics" is to deduce a conclusion from known facts using one of the rules of logic, with axioms providing the initial "known facts" to get started. By contrast, in a type theory like HoTT/UF, the "basic acts of mathematics" are specified directly by the rules of the theory, such as the rule for cartesian products which permits us to construct (x, y) once we have x and y. Put differently, choosing the axioms of ZFC is like choosing the starting position of a board game whose rules are known in advance, whereas choosing the rules of HoTT/UF is like choosing the rules of the game itself.

To understand the effect this distinction has on mathematical practice, we observe that the everyday practice of mathematics can already be separated into two basic activities: constructing (a.k.a. defining or specifying) and proving. For instance, an analyst may first construct a particular function, then prove that it is continuous. This distinction can be found as far back as Euclid, whose Postulates and Propositions are phrased as things to be *done* ("to draw a circle with any center and radius") rather than statements of existence, and which are "demonstrated" by making a *construction* and then *proving* that it has the desired properties. Rodin (2017) has recently argued that this distinction is closely related to Hilbert's contrast between *genetic* and *axiomatic* methods.[10]

---

[10] At least in Hilbert and Bernays (1934–1939); in Hilbert (1900) the same words seem to refer instead to analytic and synthetic theories, respectively.

When encoding mathematics into ZFC, however, the "construction" aspect of mathematics gets short shrift, because in fully formal ZFC the only thing we *can* do is prove theorems. Thus, the encoding process must translate constructions into proofs of existence. By contrast, in HoTT/UF and other type theories like it, it appears that the pendulum has swung the other way: the *only* thing we can do is perform constructions. How, then, do we encode proofs?

The answer begins with an idea called *propositions as types*: we interpret every *statement* that we might want to prove as a *type*, in such a way that it makes sense to interpret *constructing an element* of that type as *proving the original statement*. In this way we obtain a form of logic *inside of* type theory, rather than starting with a background logic as is done in set theory. Thus, as a foundation for mathematics, type theory is "closer to the bottom" than set theory: rather than building on the same "sub-foundations" (first-order logic), we "re-excavate" the sub-foundations and incorporate them into the foundational theory itself. In the words of Pieter Hofstra, type theory is "the engine and the fuel all in one".

One reason this idea is so useful is an observation called the *Curry–Howard correspondence* (Curry, 1934; Martin-Löf, 1975; Howard, 1980; Wadler, 2015): the logical connectives and quantifiers are *already present* in type theory as constructions on types. For instance, if A and B are types representing propositions P and Q, respectively, then $A \times B$ represents the conjunction $P \wedge Q$. This is justified because the way we construct an element of $A \times B$—by constructing an element of A and an element of B—corresponds precisely to the way we prove $P \wedge Q$—by proving P and also proving Q. Similarly, the type of functions from A to B (usually denoted $A \to B$) represents the implication $P \to Q$, and so on.

If we interpret logic directly according to this correspondence, we find that just as with the encoding into ZFC, the distinction between construction and proof is destroyed; only this time it is because we must encode proofs as constructions rather than vice versa. Whereas in ZFC we cannot construct objects, only prove that they exist, under Curry–Howard we cannot prove that something exists without constructing it.

The innovation of HoTT/UF is to allow both kinds of existence to coexist smoothly. We follow the overall philosophy of propositions-as-types, but in addition we single out a small but important class of types: those that have at most one point, with no higher equality information.[11] I will call these types *truth values*, since we think of them as representing "false" (if empty) or "true" (if inhabited); they are also often called *propositions* or *mere propositions*. Moreover, we add a rule that for any type A there is a *truncation* $\|A\|$ (also called the *bracket* or *squash*), such that $\|A\|$ is a truth value, and such that given any $a : A$ we have $|a| : \|A\|$. (Since $\|A\|$ is a truth value, $|a|$ doesn't depend on the value of a, only that we have it.)

---

[11] The importance of these types has been particularly advocated by Voevodsky, building on precursors such as Constable et al. (1986) and Awodey and Bauer (2004).

Now we can distinguish between existence proofs and constructions by whether the type of the result is truncated or not. When we construct an element of a type A that is not a truth value, we are defining some specific object; but if we instead construct an element of $\|A\|$, we are "proving" that some element of A exists without specifying it.[12] From this point of view, which is shared by many members of the HoTT/UF community, it is misleading to think of propositions-as-types as "encoding first-order logic in type theory". While this description can serve as a first approximation, it leads one to ask and argue about questions like "should the statement $\exists x : A$ be encoded by the type A or the type $\|A\|$?" We regard this question as invalid, because it implicitly assumes that mathematics has already been encoded into first-order logic, with constructions and pure-existence proofs collapsed into the quantifier $\exists$. We reject this assumption: the proper approach is to encode *mathematics* directly into HoTT/UF, representing a construction of an element of A by the type A itself, and a pure-existence statement by its truncation $\|A\|$.

It is true that due to the ascendancy of ZFC and first-order logic in general, most modern mathematicians "think in first-order logic" and are not used to distinguishing constructions from existence proofs. However, it remains true that some kinds of theorem, such as "A is isomorphic to B", are almost always "proven" by giving a construction, and a careful analysis reveals that such "proofs" must convey more information than mere existence, because frequently one needs to know later on exactly *what* isomorphism was constructed. This is one of the ways in which HoTT/UF represents the actual practice of mathematics more faithfully than other contenders. With a little bit of practice, and careful use of language, we can learn to consciously use this feature when doing mathematics based on HoTT/UF.

By the way, while the distinction between construction and proof is sometimes identified with the opposition between constructive/intuitionistic and classical logic (as is suggested by the shared root "construct"), the relationship between the two is actually limited. On one hand, while it is true that the "natural" logic obtained by Curry–Howard turns out to be intuitionistic, one can add additional axioms that are not "constructive" but can nevertheless be used in "constructions". Indeed, the exceedingly nonconstructive Axiom of Choice asserts exactly that objects which merely exist can nevertheless be assumed to be specified, i.e. "constructed" in a formal sense. In particular, axioms of classical logic can consistently be included in HoTT/UF.

On the other hand, intuitionistic first-order logic includes "pure unspecified existence" just like classical logic does, and constructive/intuitionistic set theory (Beeson, 1985; Aczel and Rathjen, 2000/1) collapses constructions into proofs just like ZFC does. It is true that constructive mathematicians in the tradition of Martin-Löf (1975) do adhere intentionally to the original Curry–Howard interpretation, regarding it as part of their constructivism; but they must also separately refrain from using any

---

[12] The possibility of these two interpretations of existence was actually already noticed by Howard (1980, sect. 12).

nonconstructive principles. That is, a constructive *philosophy* may lead one to prefer "constructions" to proofs, but this is a separate thing from the (intuitionistic) *logic* that it also leads one to prefer. Moreover, Escardó has recently argued that Brouwer himself must have intended some notion of unspecified existence, since his famous theorem that all functions $\mathbb{N}^{\mathbb{N}} \to \mathbb{N}$ are continuous is actually *inconsistent* under unmodified Curry–Howard (Escardó and Xu, 2015).

A last aspect of type theory that deserves mention is its computational character: its rules can also be read as defining a programming language that can actually be executed by a computer. This makes it especially convenient for computer formalization of mathematical proofs, as well as for mathematical verification of computer software. Thus, HoTT/UF is also better-adapted to these purposes than set theory is,[13] and indeed computer formalization has been very significant in the origins and development of HoTT/UF. But this would fill a whole chapter by itself, so reluctantly I will say no more about it here.

3.5 Identifications and Equivalences

So far, I have not really said anything that is unique to HoTT/UF. The description of types, rules, and elements in Section 3.4 applies to any type theory, including Martin-Löf's original one. The approach to logic using truncations is more novel, but it still does not depend on regarding types as ∞-groupoids. However, this kind of logic is particularly appropriate in HoTT/UF, for several reasons.

The first is that, like our considerations in Section 3.2, it drives us inexorably from sets to ∞-groupoids. Namely, if statements are interpreted by types, then in particular for any $x : A$ and $y : A$, the statement "$x = y$" must be *a type*, whose points we refer to as *identifications* of x with y. If A is a set, then this type is a mere truth value, but in general there is no reason for it to be so.

Somewhat magically, it turns out that the "most natural" rule governing the type $x = y$, as first given by Martin-Löf (1975), does *not* imply that it is always a truth value, but *does* imply that it automatically inherits the structure of an ∞-groupoid (Lumsdaine, 2010; van den Berg and Garner, 2011). This rule is related to Leibniz's "indiscernibility of identicals", but its form is rather that of Lawvere (1970), who characterized equality using an adjunction between unary predicates and binary relations. Martin-Löf's version says that if we have a type family $C(x, y, p)$ depending on x, y, and an identification $p : x = y$, and if we want to construct an element of $C(x, y, p)$ for every x, y, and p, then it suffices to construct elements of $C(x, x, \text{refl}_x)$ for every x. (Here refl_x denotes a canonically specified element of $x = x$, called the *reflexivity witness* or the *identity identification*. The standard proofs of transitivity

---

[13] Although finding the best way to extend the computational aspects of type theory to the specific features of HoTT/UF is an active research area.

and symmetry of equality from indiscernibility of identicals become in HoTT/UF constructions of the first level of ∞-groupoid structure.)

In this way, ∞-groupoids become much simpler in HoTT/UF than they are in set theory. We saw in Section 3.2 that a mathematician trying to *define* ∞-groupoids in set theory is led to a rather complicated structure. However, HoTT/UF reveals that *synthetically*, an ∞-groupoid is really quite a simple thing: we might say that we obtain a synthetic theory of ∞-groupoids by (0) starting with type theory, (1) taking seriously the idea that a statement of equality $x = y$ should be a type, (2) writing down the most natural rule governing such a type, and then (3) simply *declining to assert* that all such types are mere truth values.[14]

To be precise, however, this is not quite correct; better would be to say that Martin-Löf's type theory, unlike set theory, is sufficiently general to *permit* its types to be treated as ∞-groupoids, and in HoTT/UF we choose to do so. (This is analogous to how intuitionistic logic, unlike classical logic, is sufficiently general to permit the assumption of topological or computational structure.) Thus, in order to obtain a true synthetic theory of ∞-groupoids, we need to add some rules that are *specific* to them, which in particular will ensure that it is definitely *not* the case that all equality types are truth values.

The principal such rule in use is Voevodsky's *univalence axiom* (Kapulkin and Lumsdaine, 2012). This is formulated with reference to a *universe type* $\mathcal{U}$, whose points are other types. (For consistency, $\mathcal{U}$ cannot be a point of itself; thus one generally assumes an infinite hierarchy of such universes.) Univalence says that for types $A : \mathcal{U}$ and $B : \mathcal{U}$, the type $A = B$ consists of *equivalences* between A and B, the latter being a standard definition imported from higher category theory[15] that generalizes bijections between sets. In particular, if A has any nontrivial automorphisms, then $A = A$ is not a mere truth value.

Univalence is the central topic of Awodey's chapter (see Chapter 4); he concludes that it codifies exactly the principle of structuralism, "isomorphic objects are identical". Indeed, with univalence we no longer need any Fregean abstraction to define "structure"; we can simply consider types themselves (or, more generally, types equipped with extra data) to *be* structures. Fregean abstraction is for forgetting irrelevant facts not preserved by isomorphism, like whether $0 \in 1$, but in HoTT/UF there are no such facts, since isomorphic types are actually already *the same*. Thus, if we wish, we may consider HoTT/UF to be a *synthetic theory of structures*.[16]

---

[14] Many people contributed to this view of Martin-Löf's equality types, but Hofmann and Streicher (1998) and Awodey and Warren (2009) were significant milestones.

[15] Although it requires some cleverness to formulate it correctly in type theory; this was first done by Voevodsky.

[16] This is not in conflict with also calling it a synthetic theory of ∞-groupoids; the two phrases simply emphasize different aspects of HoTT/UF. We could emphasize both aspects at once by calling it a "synthetic theory of ∞-groupoidal structures".

More concretely, univalence ensures that any construction or proof can be transported across an isomorphism (or equivalence): anything we prove about a type is automatically also true about any equivalent type. Here again HoTT/UF captures precisely an aspect of mathematical practice that is often glossed over by set theory. Univalence also implies that two "truth values", as defined in Section 3.4, are equal as soon as they are logically equivalent; thus, they really do carry no more information than a truth value.

A second way that the logic of Section 3.4 is particularly appropriate is that HoTT/UF clarifies the distinction between types and truth values, by placing it on the first rung of an infinite ladder. In fact, for any integer $n \geq -2$ there is a class of types called *n-types*,[17] such that the singleton is the only (-2)-type, the truth values are the (-1)-types, and the sets are the 0-types. Informally, an n-type contains no higher equality information above level n: two elements of a 0-type (i.e. a set) can be equal in at most one way, two *equalities* in a 1-type can be equal in at most one way, and so on. Formally, A is an n-type if for all $x : A$ and $y : A$, the type $x = y$ is an $(n-1)$-type (with the induction bottoming out at $n = -2$).

In addition, for any n we have an *n-truncation* operation: $\|A\|_n$ is an n-type obtained from A by discarding all distinctions between equalities above level n. In particular, $\|A\|_{-1}$ discards all distinctions between *points* of A, remembering only whether A is inhabited; thus, it is the truth-value truncation $\|A\|$ from Section 3.4. The next most important case is the 0-truncation $\|A\|_0$, which makes A into a set by discarding distinctions between equalities between its points, remembering only the truth value of whether they are equal.

At this point we can deal with one of the examples of a groupoid from Section 3.2: sets and cardinalities. In HoTT/UF the *type of sets* is naturally defined as a subtype of the universe $\mathcal{U}$ which contains only the sets (0-types). By univalence, then, for sets A and B, the type $A = B$ is the type of bijections between them. Thus, two sets are *automatically* identical exactly when they are bijective, so it may appear that there is no need to specify the equalities separately from the points in this case.

However, since the type of bijections between sets A and B is itself a set and not (generally) a truth value, the type of sets is a 1-type and not a set. This is an important difference with ZFC, in which the collection of sets (belonging to some universe) *is* itself a set — but it matters little in mathematical practice, which is mostly structural. Indeed, mathematicians familiar with category theory tend to be drawn to this idea: it seems perverse to distinguish between isomorphic sets as ZFC does.[18]

---

[17] This notion is well known in homotopy theory under the name *homotopy n-type* and in higher category theory under the name *n-groupoid*. Its definition in type theory is due to Voevodsky, who calls them "types of h-level $n + 2$".

[18] This should not be confused with distinguishing between *subsets* of some fixed set that may be abstractly isomorphic as sets, such as $\mathbb{N} \subseteq \mathbb{R}$ and $\mathbb{Q} \subseteq \mathbb{R}$, which is common and essential to mathematics. The point is rather that of Benacerraf (1965): there is no reason to distinguish between, say, $\{\emptyset, \{\emptyset\}, \{\{\emptyset\}\}, \ldots\}$ and $\{\emptyset, \{\emptyset\}, \{\emptyset, \{\emptyset\}\}, \ldots\}$ as definitions of "the natural numbers".

On the other hand, mathematicians *are* accustomed to consider the collection of *cardinalities* to form a set (modulo size considerations). Thus, in HoTT/UF it is sensible to define the set of cardinalities to be the 0-truncation of the type of sets. That is, a cardinality is presented by a set, and bijective sets present equal cardinalities, but unlike sets, two cardinalities can be equal in at most one way. One nice consequence is that the subset of *finite* cardinalities is then equal to the natural numbers.

The 0-truncation has many other uses; for instance, it allows us to import the definition of *homotopy groups* from algebraic topology. Given a type X and a point $x:X$, we first define the *loop space* $\Omega(X,x)$ to be the type $x = x$, and the *n-fold loop space* by induction as $\Omega^{n+1}(X,x) = \Omega^n(\Omega(X,x), \mathsf{refl}_x)$. The *n*th *homotopy group* of X based at x is then $\pi_n(X,x) = \|\Omega^n(X,x)\|_0$. If X is an *n*-type, then $\pi_k(X,x)$ is trivial whenever $k > n$; in general, it can be said to measure the nontriviality of the identification structure of X at level n. For instance, if X is a set, then $\pi_k(X,x)$ is trivial for any $k \geq 1$; whereas if $X = \mathcal{U}$ and x is a set A, then $\pi_1(\mathcal{U}, A)$ is the automorphism group of A while $\pi_k(\mathcal{U}, A)$ is trivial for $k > 1$.

3.6 Higher Inductive Types

I mentioned in Section 3.4 that HoTT/UF consists of rules describing operations we can perform on types and their points. In fact, all but a couple of these rules belong to one uniformly specified class, known as *higher inductive types* (HITs), which can be considered a generalization of Bishop's rule for set-construction that takes higher identifications into account.

Higher inductive types include, in particular, *ordinary* inductive types, which have been well known in type theory for a long time (several examples appear already in Martin-Löf (1975)). The simplest sorts of these are *nonrecursive*, in which case the rule says that to define a type X, we specify zero or more ways to construct elements of X. This amounts to stipulating some finite list of functions with codomain X and some specified domain, called the *constructors* of X. For instance, given types A and B, their *disjoint union* $A + B$ is specified by saying that there are two ways to construct elements of $A + B$, namely by injecting an element of A or an element of B; thus, we have two constructors $\mathsf{inl} : A \to A + B$ and $\mathsf{inr} : B \to A + B$.

As recognized by Martin-Löf (1975, sect. 1.1), this is similar to Bishop's rule; the main difference is that we omit the specifying of equalities. How then are we to know when two points of such a type are equal? The answer is that an inductive type should be regarded as *freely generated* by its constructors, in the sense that we do not "put in" anything—whether a point or an identification—that is not *forced* to be there by the constructors. For instance, every point of $A + B$ is either of the form $\mathsf{inl}(a)$ or $\mathsf{inr}(b)$, since the constructors do not force any other points to exist. Moreover, no point of the form $\mathsf{inl}(a)$ is equal to one of the form $\mathsf{inr}(b)$, since the constructors do not force any such identifications to exist. However, if we have $a : A$ and $a' : A$ with $a = a'$, then there *is* an induced identification $\mathsf{inl}(a) = \mathsf{inl}(a')$, since all functions (including inl) must respect equality.

More generally, ordinary inductive types can be *recursive*, meaning that some of the constructors of X can take as input one or more previously constructed elements of X. For example, the natural numbers $\mathbb{N}$ have one nonrecursive constructor $0 : \mathbb{N}$ and one recursive one $s : \mathbb{N} \to \mathbb{N}$. The elements and equalities in such a type are all those that can be obtained by applying the constructors, over and over again if necessary.

Higher inductive types are a generalization of ordinary ones, which were invented by the author and others.[19] The simplest case is a nonrecursive *level-1* HIT, where in addition to specifying ways to construct elements of X, we can specify ways to construct identifications between such elements. Thus, in addition to constructor functions as before (which we now call *point-constructors*), we also have *identification-constructors*.

This is almost the same as Bishop's rule for set-construction, with two differences. First, a HIT need not be a set. Second, the identification-constructors need not form an equivalence relation; e.g. we may specify $x = y$ and $y = z$ but not $x = z$. However, since all types *are* ∞-groupoids, in such a case it will nevertheless be *true* that $x = z$. More precisely, if we have constructors yielding identifications $p : x = y$ and $q : y = z$, then there will be an induced identification $p \cdot q : x = z$, which is forced to exist even though we didn't "put it in by hand".

Suppose now that we *are* in Bishop's situation, i.e. we have a type A and an equivalence relation $\sim$ on it. We can define a HIT X, with one point-constructor $q : A \to X$, and one identification-constructor saying that whenever $a \sim a'$ we have $q(a) = q(a')$. Then X will be close to the quotient of $\sim$, except that it will not generally be a set even if A is. For instance, since $a \sim a$ for any $a : A$, our identification-constructor yields an identification $q(a) = q(a)$; but nothing we have put into X forces this identification to be the same as $\mathsf{refl}_{q(a)}$, and so (by the free generation principle) it is not. Thus, to obtain the usual quotient of $\sim$, we must 0-truncate X; in HoTT/UF we may call this the *set-quotient*. For instance, the set of real numbers could be defined as the set-quotient of the equivalence relation on infinite decimal expansions from Section 3.2. In this way we essentially recover Bishop's set-formation rule.[20]

Higher inductive types can also be recursive: both kinds of constructor can take previously constructed elements of X as inputs. This is very useful—e.g. it yields free algebraic structures, homotopical localizations, and even the n-truncation—but also somewhat technical, so I will say no more about it.

---

[19] Specifically, Lumsdaine, Bauer, and Warren, with further notable contributions by Brunerie and Licata. The basic theory of HITs is still under development by many people; currently, the best general reference is Univalent Foundations Program (2013, ch. 6).

[20] There is one subtle difference: Bishop actually allows us to distinguish between 0.5 and 0.4$\overline{9}$ as long as we speak of an "operation" rather than a "function". In HoTT/UF such an "operation" is just a function defined on decimal expansions, not anything acting on "real numbers".

The reader may naturally wonder *why* we don't ask the identification-constructors to form an equivalence relation. One reason is that for HITs that are not sets, the analogue of an equivalence relation would be an "∞-groupoid" in the exceedingly complicated sense referenced at the beginning of Section 3.3. Forcing ourselves to use such structures would vitiate the already-noted advantages of a *synthetic* theory of ∞-groupoids.

As a concrete example of the usefulness of not requiring equivalence relations *a priori*, if we have two functions $f, g : A \rightrightarrows B$ between sets, we can construct their *set-coequalizer* as the 0-truncation of the HIT with one point-constructor $q : B \to X$ and one identification-constructor saying that for any $a : A$ we have $q(f(a)) = q(g(a))$. In set theory, we would have to first construct the equivalence relation on B freely generated by the relations $f(a) \sim g(a)$ and then take its quotient; HITs automate that process for us. Moreover, if we omit the assumption that A and B are sets and also omit the 0-truncation, we obtain a *homotopy coequalizer*, which would be *much* harder to construct otherwise.

Another reason for considering freely generated ∞-groupoids is that many very interesting ∞-groupoids *are* freely generated, and in most cases a fully explicit description of them *is not known* and is not expected to be knowable. Thus, HITs are the *only* way we can represent them in HoTT/UF.

A simple example of a freely generated ∞-groupoid is the *circle*[21] $\mathbb{S}^1$, which as a HIT has one point-constructor $\mathsf{b} : \mathbb{S}^1$ and one identification-constructor $\ell : \mathsf{b} = \mathsf{b}$. Since nothing forces ℓ to be equal to refl_b, it is not—nor is $\ell \cdot \ell$, or $\ell \cdot \ell \cdot \ell$, and so on. In fact, $\Omega(\mathbb{S}^1, \mathsf{b})$ is isomorphic to the integers $\mathbb{Z}$.[22] Since $\mathbb{Z}$ is a set, this implies that $\pi_1(\mathbb{S}^1, \mathsf{b}) = \mathbb{Z}$ while $\pi_k(\mathbb{S}^1, \mathsf{b}) = 0$ for all $k > 1$, so in this case we do have a fully explicit description. However, there are similar types for which no such characterization is known, particularly when we move onto *level-n* HITs having constructors of "higher identifications". For instance, the *2-sphere* $\mathbb{S}^2$ has one point-constructor $\mathsf{b} : \mathbb{S}^2$ and one level-2 identification-constructor $\mathsf{refl}_\mathsf{b} = \mathsf{refl}_\mathsf{b}$; the *3-sphere* has $\mathsf{b} : \mathbb{S}^3$ with a level-3 $\mathsf{refl}_{\mathsf{refl}_\mathsf{b}} = \mathsf{refl}_{\mathsf{refl}_\mathsf{b}}$; and so on. Analogously to $\mathbb{S}^1$ we have $\pi_n(\mathbb{S}^n) = \mathbb{Z}$,[23] but also for

---

[21] This is a "homotopical" circle, not a "topological" circle such as $\{(x, y) \in \mathbb{R} \times \mathbb{R} \mid x^2 + y^2 = 1\}$. The latter can also be defined in HoTT/UF, of course, but it will be a set, whereas the HIT $\mathbb{S}^1$ is not. The homotopical circle is so-called because it is the *shape* (a.k.a. "fundamental ∞-groupoid") of the topological circle, with continuous paths in the latter becoming identifications in the former; and historically ∞-groupoids were originally studied as shapes of topological spaces. In HoTT/UF the shape ought to be constructible as a HIT, but no one has yet managed to do it coherently at all levels. Unlike classically, not every type in HoTT/UF can be the shape of some space, but we can hope that the HIT $\mathbb{S}^1$ is still the shape of the topological circle.

There is an arguably better approach to such questions called "axiomatic cohesion" (Schreiber and Shulman, 2012; Shulman, 2017), in which the types of HoTT/UF are enhanced to carry intrinsic topological structure in addition to their higher identifications. Unfortunately, space does not permit me to discuss this here, but a brief introduction can be found in Corfield's chapter (Chapter 2).

[22] This is well known in homotopy theory; its first proof in HoTT/UF by Licata and Shulman (2013) was an early milestone in combining HITs with univalence.

[23] Also a standard result in homotopy theory; see Univalent Foundations Program (2013); Licata and Brunerie (2013) for proofs in HoTT/UF.

example, $\pi_3(\mathbb{S}^2) = \mathbb{Z}$, despite the fact that $\mathbb{S}^2$ has no *constructors* of level 3. In general, $\pi_k(\mathbb{S}^n)$ is usually nontrivial when $k \geq n$, but most of its values are not known. Computing them, for classically defined ∞-groupoids, is a major research area which is not expected to ever be "complete".

What does this mean to a philosopher? For one thing, it shows how a simple foundational system can give rise very quickly to deep mathematics. The rules governing HITs are arguably unavoidable, once we have the idea of defining types in such a way, while the spheres $\mathbb{S}^n$ result from quite simple applications of those rules. Moreover, we have seen that even the basic notion of ∞-groupoid arises inescapably from thinking about equality. Thus, there are numerical invariants like $\pi_3(\mathbb{S}^2)$ quite close to the foundations of logic.

3.7 General Covariance

At long last, we return to the third example from Section 3.2: spacetime manifolds. For simplicity, I will consider only *Minkowski* spacetimes, corresponding to special rather than general relativity; similar ideas can be applied to other kinds of gauge invariance/covariance as well.

A modern mathematician defines a Minkowski spacetime to be a 4-dimensional real affine space with a Lorentzian inner product. We can repeat this definition in HoTT/UF, yielding a type Mink whose points are Minkowski spacetimes. Now we can ask what the *identifications* are in Mink. This is a special case of a more general question: what are the identifications in a *type of structured sets*? Recall that univalence ensures that identifications in the type of *all* sets are bijections; this turns out to imply that an identification of structured sets is a bijection which "preserves all the structure", i.e. an *isomorphism* in the appropriate category (see e.g. Univalent Foundations Program, 2013, sect. 9.8). Thus, an identification in Mink is an isometry, as we would hope. In particular, anything we can say in HoTT/UF about Minkowski spacetimes is automatically covariant under isometry.

Note that since isometries form a set, Mink is a 1-type. We could, if we wished, 0-truncate it to obtain a set, as we did with the type of sets in Section 3.5 to obtain the set of cardinalities. However, the hole argument tells us that this would be *wrong*, at least for the purpose of modeling reality: we really do need to remember the nontrivial identifications in Mink.

So far, so good. However, there is another side to the story, which I alluded to briefly in Section 3.2: why did the hole argument seem paradoxical for so long? This can be attributed at least partly to a radically different viewpoint on manifolds, as described by Norton (1993):

our modern difficulty in reading Einstein literally actually stems from a change...in the mathematical tools used....In recent work...we begin with a very refined mathematical entity, an abstract differentiable manifold.... We then judiciously add further geometric objects only as the physical content of the theory warrants.... In the 1910s, mathematical practices in physics were different.... one used number manifolds—$\mathbb{R}^n$ or $\mathbb{C}^n$ for example. Thus Minkowski's 'world'... was literally $\mathbb{R}^4$, that is it was the set of all quadruples of real numbers.

Now anyone seeking to build a spacetime theory with these mathematical tools of the 1910s faces very different problems from the ones we see now. Modern differentiable manifolds have too little structure and we must add to them. Number manifolds have far too much structure... the origin $\langle 0, 0, 0, 0 \rangle$ is quite different from any other point, for example.... The problem was not how to add structure to the manifolds, but how to deny physical significance to existing parts of the number manifolds. How do we rule out the idea that $\langle 0, 0, 0, 0 \rangle$ represents the preferred center of the universe... ?

In brief, *mathematical structuralism* had not yet been invented. Our explanation of the hole argument relied on comfort with the structural idea of an isometry between abstract manifolds. But if one views spacetime as the *specific* manifold $\mathbb{R}^4$, this sort of argument is unavailable; thus, the confusion surrounding the hole argument becomes more understandable.

While structuralism is the modern method of choice to deal with this conundrum, it is not the only possible solution; historically, Klein's *Erlangen* program was used for the same purpose. Here is Norton again:

Felix Klein's *Erlangen* program provided precisely the tool that was needed. One assigns a characteristic group to the theory.... Only those aspects of the number manifold that remain invariant under this group are allowed physical significance.... As one increases the size of the group, one strips more and more physical significance out of the number manifold.

This suggests a different definition of Mink: we could begin with the singleton type $\{\mathbb{R}^4\}$ and *add identification-constructors* making up the desired symmetry group (in this case, the Poincaré group[24]). In other words, we say that there is *one* Minkowski spacetime, namely $\mathbb{R}^4$, and that it can be identified with itself in many ways, such as translations, 3D rotations, and Lorentz boosts. These extra added identifications force everything we say about "Minkowski spacetimes" to be invariant under their action. For example, while in $\mathbb{R}^4$ we can distinguish the point $\langle 0, 0, 0, 0 \rangle$; in a Minkowski spacetime we cannot, because this point is not invariant under translations. However, we can say that a Minkowski spacetime comes with a Lorentzian distance function, since this structure on $\mathbb{R}^4$ *is* preserved by the Poincaré group. This is precisely the point of the *Erlangen* program, which HoTT/UF codifies into the foundations of mathematics by constructing a type that "remembers exactly those aspects of $\mathbb{R}^4$ preserved by the group action".

---

[24] The Poincaré group is usually considered not as a discrete group but as a *Lie* group, with its own manifold structure. This can be incorporated as well using "axiomatic cohesion", mentioned briefly in footnote 21.

Finally, we can show in HoTT/UF that these two definitions of Minkowski spacetime agree. Roughly, this is because two abstract Minkowski spacetimes can always be identified *somehow*, while their automorphisms can be identified with the Poincaré group; thus, the points and the identifications can be matched up consistently. Thus, HoTT/UF could be said to unify the *Erlangen* and structuralist approaches to geometry.

One might argue that these approaches were unified long ago, by the development of category theory. Indeed, as detailed by Marquis (2009), category theory can be seen as a generalization of the *Erlangen* program, where rather than simply having a group act by automorphisms of a single object, we consider isomorphisms, or more generally morphisms, between different objects, and permit as meaningful only those properties that vary appropriately under such transformations (i.e. those that are covariant—or, perhaps, *contravariant*, the dual sort of variation that can be distinguished only once we allow noninvertible morphisms). And category theory is, of course, the language of choice for the modern structuralist.

However, when category theory is built on top of a foundational set theory, one must take the additional step of *defining* the notion of isomorphism as the appropriate "criterion of sameness" and (in principle) *proving* that all properties of interest are invariant under isomorphism. As Marquis says, in the *Erlangen* program:

what is usually taken as a *logical* notion, namely equality of objects, is captured in geometry by motions, or transformations of the given group. (Marquis, 2009, 19; emphasis added)

Moreover, when generalized to higher groupoids and higher categories, this leads to the highly complicated *defined* notion of ∞-groupoid mentioned in Section 3.3. But with univalence and HITs, HoTT/UF places the notion of equality back where it belongs—in logic, or more generally the foundations of mathematics—while maintaining the insights of the *Erlangen* program and category theory.

3.8 Conclusion

There is much more to HoTT/UF than I have been able to mention in this short chapter, but those aspects I have touched on revolve around a single idea, which generalizes Bishop's set-definition principle: whenever we define a collection of objects, we must also ensure that the identifications and higher identifications between them are correctly specified. Sometimes the correct identifications arise "automatically", such as from the univalence axiom; other times we must generate new ones, as with higher inductive types. But in no case must we (or even *can* we) separate those identifications from the objects themselves: with ∞-groupoids as basic foundational objects, every collection carries along with itself the appropriate notion of identification between its objects, higher identification between those, and so on. This can be regarded as the central innovation of HoTT/UF, both for mathematics and for philosophy.

References

Aczel, P., and Rathjen, M. (2000/1). Notes on constructive set theory, *Mathematical Logic* 40, Reports Institut Mittag-Leffler. http://www.ml.kva.se/preprints/meta/AczelMon_Sep_24_09_16_56.rdf.html

Awodey, S. (2012). Type theory and homotopy, in *Epistemology versus Ontology*, Vol. 27 of *Log. Epistemol. Unity Sci.*, 183–201, Springer, Dordrecht. http://dx.doi.org/10.1007/978-94-007-4435-6_9

Awodey, S., and Bauer, A. (2004). Propositions as [types], *J. Logic Comput.* 14(4), 447–71. http://dx.doi.org/10.1093/logcom/14.4.447

Awodey, S., Pelayo, Á., and Warren, M. A. (2013). Voevodsky's univalence axiom in homotopy type theory. *Notices Amer. Math. Soc.* 60(9), 1164–7. http://dx.doi.org/10.1090/noti1043

Awodey, S., and Warren, M. A. (2009). Homotopy theoretic models of identity types. *Math. Proc. Camb. Phil. Soc.* 146(45), 45–55.

Baez, J. (2007). The homotopy hypothesis, http://math.ucr.edu/home/baez/homotopy/Lecture at *Higher Categories and Their Applications*.

Beeson, M. (1985). *Foundations of Constructive Mathematics*. Springer, Dordrecht.

Benacerraf, P. (1965). What numbers could not be. *The Philosophical Review* 74(1), 47–73.

Bishop, E. (1967). *Foundations of Constructive Analysis*. McGraw-Hill series in higher mathematics. McGraw-Hill, New York.

Constable, R. L., Allen, S. F., Bromley, H. M., Cleaveland, W. R., Cremer, J. F., Harper, R. W., Howe, D. J., Knoblock, T. B., Mendler, N. P., Panangaden, P., Sasaki, J. T., and Smith, S. F. (1986). *Implementing Mathematics with the Nuprl Proof Development System*. Prentice-Hall, Upper Saddle River, NJ.

Curry, H. B. (1934). Functionality in combinatory logic. *Proceedings of the National Academy of Science* 20, 584–90.

Escardó, M., and Xu, C. (2015). The inconsistency of a Brouwerian continuity principle with the Curry–Howard interpretation, in T. Altenkirch (ed.), *13th International Conference on Typed Lambda Calculi and Applications, TCLA 2015*, Leibniz International Proceedings in Informatics, vol. 38, pp. 153–64. Dagstuhl Publishing, Saarbrücken.

Ewald, W. (1996). *From Kant to Hilbert: A Source Book in the Foundations of Mathematics*. Clarendon Press, Oxford.

Frege, G. (1884). *Die Grundlagen der Arithmetik: eine logisch-mathematische Untersuchung über den Begriff der Zahl*. W. Koebner, Breslau.

Hilbert, D. (1899). *Grundlagen der Geometrie*. Teubner, Berlin.

Hilbert, D. (1900). Über den Zahlbegriff, *Jahresbericht der deutschen Mathematiker-Vereinigung* 8, 180–4. English translation in Ewald (1996).

Hilbert, D., and Bernays, P. (1934–9). *Grundlagen der Mathematik*. Springer, Dordrecht.

Hofmann, M., and Streicher, T. (1998). The groupoid interpretation of type theory, in *Twenty-Five Years of Constructive Type Theory (Venice, 1995)*, Vol. 36 of Oxford Logic Guides, 83–111. Oxford University Press, New York.

Howard, W. A. (1980). The formulae-as-types notion of construction, in *To H. B. Curry: Essays on Combinatory Logic, Lambda Calculus, and Formalism*, pp. 479–91. Academic Press. Notes originally circulated privately in 1969.

Kapulkin, C., and Lumsdaine, P. L. (2012). The simplicial model of univalent foundations (after Voevodsky). arXiv:1211.2851.

Lawvere, F. W. (1966). The category of categories as a foundation for mathematics, in *Proc. Conf. Categorical Algebra (La Jolla, Calif., 1965)*, pp. 1–20. Springer, New York.

Lawvere, F. W. (1970). Equality in hyperdoctrines and comprehension schema as an adjoint functor, in *Applications of Categorical Algebra (Proc. Sympos. Pure Math., Vol. XVII, New York, 1968)*, pp. 1–14. Amer. Math. Soc., Providence, RI.

Lawvere, F. W. (2005). An elementary theory of the category of sets (long version) with commentary, *Repr. Theory Appl. Categ.* 11, 1–35 (electronic). Reprinted and expanded from *Proc. Nat. Acad. Sci. U.S.A.* 52 (1964), With comments by the author and Colin McLarty.

Lawvere, W. (1994). Cohesive toposes and Cantor's "lauter Einsen". *Philosophia Mathematica* 2(3), 5–15.

Licata, D. R., and Brunerie, G. (2013). $\pi_n(S^n)$ in homotopy type theory, Certified Programs and Proofs: Third International Conference, Proceedings, pp. 1–16. http://dlicata.web.wesleyan.edu/pubs/lb13cpp/lb13cpp.pdf

Licata, D. R., and Shulman, M. (2013). Calculating the fundamental group of the circle in homotopy type theory, in *Proceedings of the 2013 28th Annual ACM/IEEE Symposium on Logic in Computer Science (LICS '13)*, 223–32. IEEE Computer Society, Washington, DC.

Lumsdaine, P. L. (2010). Weak omega-categories from intensional type theory. *Logical Methods in Computer Science* 6(3).

Marquis, J.-P. (2009). *From a Geometrical Point of View: A Study of the History and Philosophy of Category Theory*. Springer, Dordrecht.

Martin-Löf, P. (1975). An intuitionistic theory of types: predicative part, in *Logic Colloquium*. North Holland, Amsterdam.

Norton, J. (1993). General covariance and the foundations of general relativity: eight decades of dispute. *Rep. Prog. Phys.* 56, 791–858.

Pelayo, A., and Warren, M. A. (2014). Homotopy type theory and Voevodsky's univalent foundations. *Bull. Amer. Math. Soc. (N.S.)* 51(4), 597–648.

Rodin, A. (2017). On constructive axiomatic method. arXiv:1408.3519. To appear in *Logique et Analyse*.

Sachs, R. K., and Wu, H. (1977). *General Relativity for Mathematicians*. Springer-Verlag, New York.

Schreiber, U., and Shulman, M. (2012). Quantum gauge field theory in cohesive homotopy type theory, presented at *Quantum Physics and Logic, 10–12 October 2012, Brussels, Belgium (QPL 2012)*. http://ncatlab.org/schreiber/files/QFTinCohesiveHoTT.pdf

Shulman, M. (2017). Brouwer's fixed-point theorem in real-cohesive homotopy type theory. arXiv:1509.07584. To appear in *Mathematical Structures in Computer Science*.

Univalent Foundations Program (2013). *Homotopy Type Theory: Univalent Foundations of Mathematics*, first edn, http://homotopytypetheory.org/book/

van den Berg, B., and Garner, R. (2011). Types are weak ω-groupoids. *Proceedings of the London Mathematical Society* 102(2), 370–94.

Wadler, P. (2015). Propositions as types. *Communications of the ACM* 58(12), 75–84. http://homepages.inf.ed.ac.uk/wadler/papers/propositions-as-types/propositions-as-types.pdf

4

Structuralism, Invariance, and Univalence

Steve Awodey

Recent advances in foundations of mathematics have led to some developments that are significant for the philosophy of mathematics, particularly structuralism. Specifically, the discovery of an interpretation of Martin-Löf's constructive type theory into abstract homotopy theory by Awodey and Warren (2009) suggests a new approach to the foundations of mathematics, one with both intrinsic geometric content and a computational implementation (Univalent Foundations Program, 2013). Leading homotopy theorist Vladimir Voevodsky has proposed an ambitious new comprehensive program of foundations on this basis, including a new axiom with both geometric and logical significance: the *Univalence Axiom* (Awodey et al., 2013). It captures the familiar aspect of informal mathematical practice according to which one can identify isomorphic objects. While it is incompatible with the conventional interpretation of type theoretic foundations, it is a powerful addition to the framework of homotopical type theory.

4.1 The Principle of Structuralism

The following statement may be called the *Principle of Structuralism*:

$$\text{Isomorphic objects are identical.} \qquad \text{(PS)}$$

From one perspective, this captures a principle of reasoning embodied in everyday mathematical practice:

- The fundamental group of the circle $\pi_1(S^1)$ is a free group on one generator, so for the purposes of group theory, these are *the same* group, namely the group $\mathbb{Z}$ of integers.
- The Cauchy real numbers are isomorphic to the Dedekind reals, so as far as analysis is concerned, these are *the same* number field, $\mathbb{R}$.

- The unit interval $[0,1]$ is homeomorphic to the closed interval $[0,2]$, so as a topological space, these can be regarded as *the same* space, say I.

Within a mathematical theory, theorem, or proof, it makes no practical difference which of two "isomorphic copies" are used, and so they can be treated as *the same* mathematical object for all practical, mathematical purposes. This common practice is even sometimes referred to light-heartedly as "abuse of notation", and mathematicians have developed a sort of systematic sloppiness to help them implement this principle, which is quite useful in practice, despite being literally false. It is, namely, incompatible with the conventional approach to foundations of mathematics in set theory, and one of the reasons that some people have advocated turning to category theory instead (McLarty, 1993).

Indeed, from a more literal point of view, it is, of course, just false that isomorphic objects are identical. The fundamental group consists of (equivalence classes of) paths, not numbers. The Cauchy reals are sequences of rationals, i.e. functions $\mathbb{N} \to \mathbb{Q}$, and the Dedekind reals are subsets of rationals, etc.

Mathematical objects are often constructed out of other ones, and thus also have some residual structure resulting from that construction, in addition to whatever structure they may have as objects of interest, i.e. the real number field actually consisting of functions or of subsets. So there is a clear sense in which (PS) is simply false.

What if we try to be generous and read "identity" in a weaker way:

"A is identical to B" means "A and B have all the same properties".

Here we shall want to restrict the "properties" that can occur on the right, however, in order for this to really be a weaker condition. So we now have the reformulation

$$A \cong B \implies \text{for all relevant properties } P, P(A) \implies P(B) \qquad \text{(PS')}$$

The relevant properties will be those pertaining to the subject matter: group-theoretic, topological, etc. However, consider, for example, perfectly reasonable set-theoretic properties like

$$\cap X \in X, \quad \text{or} \quad \emptyset \in X.$$

Such mundane properties of sets (and so of any structures built from sets) do not "respect isomorphism" in the sense stated in (PS'). So the task of precisely determining the "relevant" properties is evidently no simple matter.

Our leading question may now be formulated as follows:

Is there a precise sense in which (PS) can be held true?

Such a reading would legitimize much mathematical practice and support a structuralist point of view. Let us consider the terms of (PS) in turn.

4.2 Isomorphism

What does it mean to say that two things are isomorphic? Some might be tempted to say

$$A \cong B \iff A \text{ and } B \text{ have the same structure.} \tag{4.1}$$

And that would be true.

But it is not the *definition* of isomorphism, because it presumes the notion of structure and an identity criterion for it, and that is putting the cart before the horse. *Structure* is like *color* or *the direction of a line* or *number*: it is an abstract concept. And as Frege wisely taught us, it is determined by an abstraction principle, in this case:

$$\mathsf{str}(A) = \mathsf{str}(B) \iff A \cong B. \tag{DS}$$

The structure of A is the same as the structure of B just in case A and B are isomorphic. So (4.1) is indeed true, but it's the definition of "structure" in terms of isomorphism, not the other way around. (Of course, such an informal "definition" of the notion of structure would require a more explicit formal setting to actually be of any use; we shall take a different course to arrive at the same result.)

So what is the definition of "isomorphism"? Two things A and B are *isomorphic*, written $A \cong B$, if there are structure-preserving maps

$$f : A \to B \quad \text{and} \quad g : B \to A$$

such that

$$g \circ f = 1_A \quad \text{and} \quad f \circ g = 1_B.$$

This standard definition of isomorphism makes reference to "structure-preserving maps", "composition" $\circ$, and "identity" maps 1_A, but these are primitive concepts in category theory, and so need not be further defined. Note that "isomorphism" is always relative to a given category, which determines a kind of structure via a primitive notion of "structure-preserving maps", and not the other way around.

Sometimes structures are built up from other ones, like groups, graphs, and spaces. These are sets equipped with operations, relations, etc. The notion of a structure-preserving map, or *morphism*, is then defined in terms of maps in the underlying category of sets (i.e. functions) that preserve the operations, relations, etc.—i.e. "homomorphisms", in the usual sense. But there are lots of other examples where the notion of morphism is given directly, and that then determines the corresponding notion of "structure"—e.g. the "smooth structure" on a manifold, or the ordering relation in a partially ordered set, regarded as a category.[1]

---

[1] This view of the concept of mathematical structure as determined via category theory—a category determines a species of structure, rather than the other way around—was already presented in Awodey (1996). It results in a notion of structure that differs radically from that presumed in some of the philosophical literature, e.g. Resnik (1997); Shapiro (1997). Here a structure is determined *externally*, as

4.3 Objects

If two objects are isomorphic, they stay that way if we forget some of the structure. For example, isomorphic groups are also isomorphic as sets,

$$G \cong H \;\Rightarrow\; |G| \cong |H|.$$

Therefore, one can *distinguish* different structures by taking them to non-isomorphic underlying sets.

More generally, any functor (not just a forgetful one) preserves isomorphisms, so we can distinguish non-isomorphic structures by taking them to non-isomorphic objects by some functor. For example, consider these two simple partial orders A and B:

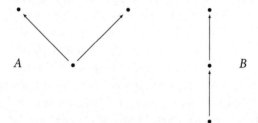

Are they isomorphic? Without even thinking about what order-preserving maps there may be between them, we can simply count the number of order-preserving maps from the simple partial order $I = (\cdot \to \cdot)$, and we see immediately that A has 5 such, and B has 6. Since taking the set $\mathsf{Hom}(I, X)$ of maps from any fixed object (here I) is always a functor of X, it follows immediately that A and B cannot be isomorphic.

A property $P(X)$ like this is called an *invariant*:

$$A \cong B \;\&\; P(A) \;\Rightarrow\; P(B).$$

It is a property that respects isomorphism, i.e. a "structural property". If in our (restricted) Principle of Structuralism (PS') we were to restrict attention to only structural properties P in this sense, we would get the statement

$$A \cong B \;\Rightarrow\; \text{for all structural properties } P,\; P(A) \Rightarrow P(B).$$

That is, "isomorphic objects have all the same structural properties". Now, this statement is trivially true, because of the way we've set up the definitions, but it does indeed capture the way in which (PS) is actually used in practice. Namely,

If $A \cong B$ and $P(X)$ is any structural property such that $P(A)$, then also $P(B)$.

---

it were, by its mappings to and from other objects of the same kind, rather than *internally*, in terms of relations and operations on elements. This solves Benacerraf-style problems directly by rejecting the idea that "mathematical objects" are the elements of structured sets (e.g. particular numbers) in favor of the view that they are structures (e.g. the system of all numbers, together with 0 and the operation of successor). See also, again, McLarty (1993).

But in order for this to actually be *useful* in practice, we obviously need to be able to recognize that $P(X)$ is structural, without going through the trouble of proving in general that it respects isomorphisms! Let us consider one common way of doing just that.

Suppose that we have one or more functors from the category $\mathcal{S}$ of structures of a given kind to the category of sets,

$$F : \mathcal{S} \to \mathsf{Sets}.$$

Then the identity of structures—i.e. isomorphism in $\mathcal{S}$—can be tested via the functors F in terms of the identity of the structure of the underlying "sets", i.e. set isomorphisms. This was the situation in our earlier example of counting the number of homomorphisms from a fixed object: there our F was the functor $\mathsf{Hom}(I, -)$. A typical kind of problem asks whether there are enough such "invariants" F of a certain kind to determine any given structure in $\mathcal{S}$: e.g. is the homotopy type of a space determined by its homotopy groups? is a formula provable if it is true in all models? etc.

Now that we have reduced identity of structures in general to the question of isomorphism in some "base category", like Sets, we just need to know: what are the invariant properties of these base "sets"? We already know that many sentences of conventional set theory express properties that are *not* invariant, like $\emptyset \in X$. One way to obtain invariant properties is by restricting the methods used to specify them to only combinations of ones already known to be structural. As we have already indicated, category-theoretic methods are always invariant, and this is one reason that categorical foundations have a more structural character, but another large and useful class of structural properties are those that can be defined in a foundational system of *type theory*, rather than set theory.[2]

The basic operations of what is called *constructive type theory* (Martin-Lof, 1984), starting with a basic object or "type of individuals" X, are as follows:

$$0,\ 1,\ A + B,\ A \times B,\ A \to B,\ \Sigma_{x:A} B(x),\ \Pi_{x:A} B(x),\ \mathsf{Id}_A(x, y),$$

and these correspond to the logical propositions

$$\bot,\ \top,\ A \vee B,\ A \wedge B,\ A \Rightarrow B,\ \exists x : A.\ B(x),\ \forall x : A.\ B(x),\ x =_A y.$$

The correspondence—called "Curry–Howard" or "Propositions-as-Types"—is given by

$$\text{proof} : \text{Proposition} \approx \text{term} : \text{Type}.$$

That is, the proofs of a proposition are the terms of the corresponding type. We shall return to this idea later.

---

[2] These two different kinds of invariants are actually closely related, as explained by categorical logic.

The system of type theory has the important property that *any* definable property of objects (types) is invariant. Thus, if $P(X)$ is any type that is definable in the system over a basic type X, then the following inference is derivable:

$$\frac{A \cong B \quad P(A)}{P(B)}.$$

The proof is a straightforward induction on the construction of $P(X)$, showing that $A \cong B$ implies that $P(A) \cong P(B)$. Let us record this as the following *Principle of Invariance* for type theory:[3]

> All properties definable in type theory are isomorphism invariant. (PI)

4.4 Identity

What is meant by "the objects A and B are identical"? In set theory it means $\forall x.\ x \in A \Leftrightarrow x \in B$, but we have just decided to opt out of set theory for the sake of invariance. In *impredicative* type theory, one can define identity by Leibniz's Law as

$$A = B := \forall P.\, P(A) \Rightarrow P(B).$$

But this is identity between terms $A, B : X$ of some common type X, and the universal quantifier $\forall P$ is over the type $\mathcal{P}(X)$ of all properties on (or, if you prefer, subsets of) X. So, more explicitly,

$$A =_X B := \forall P : \mathcal{P}(X).\, P(A) \Rightarrow P(B),$$

which is not the relation we are looking for; instead, we want one between *types* X and Y, which are the "mathematical objects" of our foundational system.

In constructive type theory there is a primitive identity relation $\mathsf{Id}_X(a, b)$ on the terms a, b of each type X, and its rules do allow the expected inference,

$$\frac{\mathsf{Id}_X(a, b) \quad P(a)}{P(b)},$$

for any predicate $P(x)$ on X, giving (half of) the effect of the Leibnizian definition. In order to be able to reason about identity of *types*, then, one can add a universe U of all (small) types; this will then have its identity relation, satisfying the desired

$$\frac{\mathsf{Id}_U(A, B) \quad P(A)}{P(B)}$$

for any definable property $P(X)$ of types.

Now that we have found a way to reason about identity of types, we can ask how it is related to isomorphism. We can start by asking whether this extension by a universe

---

[3] The importance of this principle has also been emphasized by Makkai (1998).

is still compatible with the Principle of Invariance. Before we added U, we observed that all definable properties $P(X)$ are invariant, in the sense that

$$\frac{A \cong B \quad P(A)}{P(B)}.$$

If this inference also held for definable properties P involving U, we could set

$$P(X) := \mathsf{Id}_U(A, X),$$

and infer from $\mathsf{Id}_U(A, A)$ that

$$A \cong B \implies \mathsf{Id}_U(A, B),$$

i.e. the Principle of Structuralism.

Thus, we see that in the full system of type theory with a universe, the Principle of Invariance implies the Principle of Structuralism. More specifically, if in the extended system of type theory with a universe it is still the case that all definable properties are isomorphism invariant, then in particular isomorphic objects are identical.

Is that really possible?

4.5 What does "are" mean?

In type theory, every proposition determines a type (namely, the type of proofs of the proposition), and every type can be regarded as a proposition (namely, the proposition that the type has terms). For instance, corresponding to the proposition $A \cong B$ we have the type of isomorphisms between A and B:

$$\text{``}A \cong B\text{''} \approx \mathsf{Iso}(A, B).$$

In order to prove that $A \cong B$, one constructs a term of type $\mathsf{Iso}(A, B)$, which is exactly an isomorphism between A and B.

Similarly, associated to the identity type $\mathsf{Id}_U(A, B)$ there is the proposition $A =_U B$ that A and B are identical (small) types,

$$\text{``}A =_U B\text{''} \approx \mathsf{Id}_U(A, B).$$

The object $\mathsf{Id}_U(A, B)$ can be regarded as the type of "proofs that A is identical B", or the type of "identifications of A and B". (There is a neat geometric interpretation of this type that we unfortunately cannot go into here; see Univalent Foundations Program (2013)). For the remainder of this chapter, we shall follow the usual type-theoretic practice of simply *identifying propositions and types*, and use the more familiar notation $A \cong B$ and $A =_U B$ exclusively.

Now, it is easily shown that there is always a map,

$$(A =_U B) \to (A \cong B),$$

since the relation of isomorphism is reflexive. The Univalence Axiom implies that if A and B are sets, then this map is itself an isomorphism,

$$(A =_U B) \cong (A \cong B).$$

Thus, in particular, there is a map coming back,

$$(A \cong B) \to (A =_U B),$$

which may reasonably be read "isomorphic objects are identical"—that is to say, the Principle of Structuralism. Indeed, this is the inference that we just doubted was even possible.

Let us immediately say that Voevodsky's Univalence Axiom itself actually has a more general form, namely that *identity of objects is equivalent to equivalence*,

$$(A =_U B) \simeq (A \simeq B), \tag{UA}$$

where the notion of "equivalence" is a broad generalization of isomorphism that subsumes homotopy equivalence of spaces, categorical equivalence of (higher-) groupoids, isomorphism of sets and set-based structures like groups and rings, and logical equivalence of propositions (Univalent Foundations Program, 2013).

Since the rules of identity permit substitution of identicals, one consequence of (UA) is the schema that we called the Principle of Invariance,

$$\frac{A \simeq B \quad P(A)}{P(B)},$$

but now this holds for *all* objects A and B and *all* properties $P(X)$. Thus, even in the extended system with a universe and the Univalence Axiom, it still holds that *all definable properties are structural*. Indeed, to maintain the Principle of Invariance we must also add (something like) the Univalence Axiom when we have a universe, else $A =_U B$ would give rise to a non-invariant property. In this sense, (UA) is a very natural assumption, since it specifies the (otherwise under determined) identity relation on the universe in a way that preserves the "invariant" character of the system without a universe.

Rather than viewing UA as "identifying equivalent objects", however, and thus as *collapsing* distinct objects, it is more useful to regard it as *expanding the notion of identity* to that of equivalence. For mathematical purposes, that is, equivalence is then the sharpest notion of identity available; the question whether two equivalent mathematical objects are "really" identical in some stronger, non-logical sense, is thus outside of mathematics.

As fascinating as this result may be from a philosophical perspective, it should be said that it is not the only motivation for the Univalence Axiom, or even the primary one. Univalence is not a philosophical claim but rather a working axiom, with many important mathematical consequences and applications, in a new system of foundations of mathematics (Awodey et al., 2013). For example, it is used heavily in the recently discovered "logical" calculations of some of the homotopy groups of spheres (Univalent Foundations Program, 2013).

In sum, the Univalence Axiom is a new principle of logic that not only makes sense of, but actually implies, the Principle of Structuralism with which we began:

$$\text{Isomorphic objects are identical.} \qquad \text{(PS)}$$

This seems quite radical from a conventional foundational point of view, and it is indeed incompatible with the naive, classical, set theoretic view.[4] It is, however, fully compatible and even quite natural within a type-theoretic foundation, and that is part of the remarkable new insight behind the Univalence Axiom.[5]

4.6 Some Historical Perspective

From a foundational perspective, the Univalence Axiom is certainly radical and unexpected, but it is not entirely without precedent.

The first edition of *Principia Mathematica* used an intensional type theory, but the axiom of reducibility implied that every function has an extensionally equivalent predicative replacement. This also had the effect of spoiling the interpretation of functions as expressions—open sentences—and the substitutional interpretation of quantification, what may be called the "syntactic interpretation", as favored by Russell on the days when he was a constructivist. In the second edition, Russell states a new principle that he attributes to Wittgenstein: "a function can only occur in a proposition through its values", and he says that this justifies an axiom of extensionality, thus doing at least some of the work of the axiom of reducibility. What is going on here is that Wittgenstein has noticed that all functions that actually occur, i.e. that can be explicitly defined, are in fact extensional, and so one can consistently add the axiom of extensionality without destroying the syntactic interpretation.

We have here a somewhat similar move: the Univalence Axiom is in some sense a very general extensionality principle. Indeed, it even implies the usual extensionality laws for propositions and propositional functions:

$$p \leftrightarrow q \;\rightarrow\; p = q$$
$$\forall x (fx \leftrightarrow gx) \;\rightarrow\; f = g.$$

Since nothing in syntax violates UA, we can add it to the system and still maintain the good properties of syntax, like invariance under isomorphism.

---

[4] There is a sophisticated consistency proof of the Univalence Axiom with respect to Zermelo–Fraenkel theory with choice (ZFC), using the theory of simplicial sets from Algebraic Topology (Kapulkin et al., 2012), so care is required in stating the precise sense in which UA is incompatible with set theory (Awodey et al., 2013).

[5] For further information on Homotopy Type Theory and the Univalent Foundations of Mathematics program, see *Homotopy Type Theory* (n.d.); Awodey (2012; forthcoming, b).

Rudolf Carnap was one of the first people to observe that the properties definable in the theory of types are all invariant under isomorphisms (he did so in pursuit of his ill-fated *Gabelbarkeitssatz*). Tarski later proposed this condition as a sort of explication of the concept of a "logical notion". The idea was to generalize Felix Klein's program from geometric to "arbitrary" transformations, in order achieve the most general notion of an "invariant", which would then be a *logical notion*.

The consistency of the Univalence Axiom shows that the entire system of type theory is, in fact, invariant under the even more general notion of *homotopy equivalence*— i.e. "same shape"—which is a much larger class of maps than Tarski's bijections of sets. The Univalence Axiom takes this basic insight,

$$A \simeq B \ \& \ P(A) \ \Rightarrow \ P(B),$$

i.e. "all logical properties are invariant", and turns it into a logical axiom,

$$(A \simeq B) \simeq (A = B).$$

i.e. "logical identity is equivalent to equivalence".[6]

Finally, observe that, as an informal consequence of (UA), together with the very definition of "structure" (DS), we have that two mathematical objects are identical if and only if they have the same structure:

$$\mathsf{str}(A) = \mathsf{str}(B) \ \Leftrightarrow \ A = B.$$

In other words, mathematical objects simply *are* structures. Could there be a stronger formulation of structuralism?

Acknowledgments

Thanks to Peter Aczel for many discussions on the subject of this paper, to Hannes Leitgeb and the Munich Center for Mathematical Philosophy where this work was done and presented, and to *Philosophia Mathematica* where it was originally published.

References

Awodey, S. (1996). Structure in mathematics and logic: A categorical perspective. *Philosophia Mathematica* 3, 209–37.

Awodey, S. (2012). Type theory and homotopy, in P. D. et al. (eds), *Epistemology versus Ontology: Essays on the Philosophy and Foundations of Mathematics in Honour of Per Martin-Löf*, 183–201. Springer, Berlin.

Awodey, S. (forthcoming, *a*). Univalence as a principle of logic. In preparation.

Awodey, S. (forthcoming, *b*). A proposition is the (homotopy) type of its proofs, in E. Reck (ed.), *Essays in Honor of W.W. Tait*. Springer, Berlin.

---

[6] The relationship among univalence, isomorphism invariance, and logical definability is investigated further in Awodey (forthcoming, *a*).

Awodey, S., Pelayo, A., and Warren, M. (2013). Voevodsky's univalence uxiom in homotopy type theory. *Notices of the Amer. Mathem. Soc.* 60(8), 1164–7.

Awodey, S., and Warren, M. (2009). Homotopy-theoretic models of identity types. *Math. Proc. Camb. Phil. Soc.* 146, 45–55.

Homotopy Type Theory (n.d.). http://homotopytypetheory.org

Kapulkin, C., Lumsdaine, P. L., and Voevodsky, V. (2012). Univalence in simplicial sets, *arXiv* 1203.2553.

McLarty, C. (1993). Numbers can be just what they have to. *Noûs* 27(4), 487–98.

Makkai, M. (1998). Towards a categorical foundation of mathematics, in *Logic Colloquium '95'*, Vol. 11 of Lecture Notes in Logic, 53–190. Springer, Berlin.

Martin-Lof, P. (1984). *Intuitionistic Type Theory*, Vol. 17. Bibliopolis, Naples. Notes by Giovanni Sambin.

Resnik, M. (1997). *Mathematics as a Science of Patterns*. Clarendon Press, Oxford.

Shapiro, S. (1997). *Philosophy of Mathematics: Structure and Ontology*. Oxford University Press, Oxford.

Univalent Foundations Program, T. (2013). *Homotopy Type Theory: Univalent Foundations of Mathematics*, Institute for Advanced Study. http://homotopytypetheory.org/book

5
Category Theory and Foundations

Michael Ernst

5.1 Introduction

Category theory was not immediately considered to have foundational importance. In the now famous Eilenberg and Mac Lane (1945), Eilenberg and Mac Lane introduced it as a useful language to help them formulate the notions they needed for their work, those of functors and natural transformations. This is how category theory began, as a 'useful language' for work in parts of mathematics.[1] It was not until much later that a foundational use of category theory was proposed.

The first two categorial foundations ever proposed were in Lawvere (1964) and Lawvere (1966). Lawvere (1964) introduced the Elementary Theory of the Category of Sets (ETCS), which is a theory of sets in the language of category theory. Lawvere (1966) introduced the Category of Categories as a Foundation (CCAF), which is meant to provide a framework for the many categories used in mathematics. These two foundations sparked a debate about whether or not categorial foundations can replace set theory, in the form of Zermelo–Fraenkel set theory with the Axiom of Choice (ZFC), as the foundation for mathematics or at least play the same role.

The debate has focused almost exclusively on two major topics, technical adequacy and autonomy. The technical adequacy of both categorial foundations and ZFC has been called into question. The claim is that ZFC is not capable of supporting parts of the practice of category theory and similarly that categorial foundations are not capable of supporting parts of the practice in a number of fields of mathematics. The question of autonomy, on the other hand, concerns only the acceptability of categorial foundations. The claim is that categorial foundations cannot stand on their own without an appeal to some other theory, such as ZFC, and so cannot provide a foundation for mathematics.

What has been neglected in the debate is one of the primary motivations for advocating categorial foundations in the first place, that they 'would seemingly be much more natural and readily-useable' for work in mathematics (Lawvere, 1966, 1).

---

[1] See Landry and Marquis (2005) for this point developed further.

This is not to say that advocates have not made a positive case for this position, but that it has undergone relatively little scrutiny. This is unfortunate, as what investigation has been done suggests not only that it could reveal a lot about categorial foundations but also about mathematical thinking and mathematical practice more generally.

The goal of this chapter is to provide a condensed summary of the foundational debate up to this point and then to briefly consider this neglected area of interest. To that end, the first two sections introduce the basics of categorial foundations, ETCS, and CCAF, respectively. We privilege ETCS and CCAF because they are the first categorial foundations and the most discussed. In the interest of space this chapter will not address the basics of category theory, readers in need of those details may refer to McLarty (1995) or Awodey (2009). Section 5.3 describes the debate on technical adequacy, in which all the major questions appear to have been settled. Section 5.4 addresses the debate on autonomy, much of which is still ongoing. Finally, Section 5.5 addresses the aforementioned neglected feature of categorial foundations and how investigating it could be fruitful for the philosophy of mathematics.

5.2 Categorial Foundations: ETCS

ETCS was the first categorial foundation to be appear in print, appearing in Lawvere's aptly titled 1964 paper "An Elementary Theory of The Category of Sets".[2] As mentioned in the Introduction, ETCS is a theory of sets. It is an alternative to ZFC as a description of the universe of sets.

It is the perceived need for such an alternative to ZFC that led to Lawvere's creation of ETCS. Lawvere was asked to teach his students foundations prior to teaching them analysis and calculus (Lawvere, 2005, 5). However, he "soon realized that even an entire semester would not be adequate for explaining all the (for a beginner bizarre) membership-theoretic definitions and results, then translating them into operations usable in algebra and analysis" (Lawvere, 2005, 5–6). Lawvere was confident that using ZFC would take too long and be too complicated for his students to grasp in the time available. Lawvere's solution "was to present in a couple of months an explicit axiomatic theory of the mathematical operations and concepts (composition, functionals, etc.) as actually needed in the development of the mathematics" (Lawvere, 2005, 6). This is how ETCS was born; it was intended as a more accessible foundation that provided everything necessary for the analysis and algebra he planned to teach. The fact that this claim and others like it have undergone relatively little scrutiny will be addressed in Section 5.5.

However, this is only part of the stated goals of Lawvere (1964). Lawvere (1964) has two explicit goals:

First, the theory [ETCS] characterizes the category of sets and mappings as an abstract category in the sense that any model for the axioms which satisfies the additional (non-elementary)

---

[2] While the paper first appeared in 1964, the 2005 elaboration is also used as a source for this chapter.

axiom of completeness (in the usual sense of category theory) can be proved to be equivalent to $\mathbb{S}$. Second, the theory provides a foundation for mathematics which is quite different from the usual set theories in the sense that much of number theory, elementary analysis, and algebra can apparently be developed within it even though no relation with the usual properties of $\in$ can be defined. (Lawvere, 2005, 7)

The first goal is to establish a uniqueness result for the category of sets, $\mathbb{S}$, that is characterized by the axioms of ETCS. The idea is that such a result shows that these axioms really have picked out a specific category, namely $\mathbb{S}$, that they describe a determinate concept. This issue of uniqueness has not played a role in the debate between foundational systems and so we will not explore it further here. It plays a much more significant role in debates on categoricity and the determinateness of the continuum hypothesis (e.g., Martin (2001)).

The second explicit goal we have already mentioned. ETCS is supposed to provide a more accessible presentation of the notions necessary to provide a foundation for mathematics. While we will not evaluate its accessibility here (instead postponing such questions to Section 5.5), what we will look at in the remainder of this section is how ETCS is able to provide a theory of sets without a global membership relation.

As a point of clarification we need to be clear on the global membership relation that is used in ZFC. In ZFC the relation $\in$ can apply to any two sets A and B; it makes sense to ask if $A \in B$ or $B \in A$ or neither is the case. This is the sense in which the relation is global. It is from this global membership relation that the entire theory is built. Since ETCS gives up the global membership relation we need to see how it does make sense of membership and how it handles reasoning about sets.

ETCS is intended to describe a category where the objects are sets and the arrows are functions between them. In ETCS the terminal object is the canonical one element set; every set can be mapped to it in exactly one way. We use the terminal object to define elements of an arbitrary set: "x is an *element* of A, denoted $x \in A$, iff $1 \xrightarrow{x} A$" (Lawvere, 2005, 9). The elements of a set A are simply the arrows from 1 to A. Immediately this definition guarantees that 1 has exactly 1 element, namely the identity arrow from 1 to itself. However, it also guarantees that if A and B are distinct sets, then they have no elements in common, which would seem to undercut our ability to take unions or intersections or any other number of operations. In order to get to those operations we need to make sense of the members of a subset, a notion distinct from elements in ETCS.

In ETCS, the subsets of a set are defined as the monomorphisms to that set. If x is a monomorphism from B to A then we say that x is a subset of A. That subsets are monics means we can uniquely identify elements of a subset with elements of its superset. If we consider a monomorphism $x : B \rightarrowtail A$ and look at the elements of B, that is, the arrows from 1 to B, these will uniquely determine elements of A. For example, let $u, v \in B$ if $u \neq v$, then $x \circ u, x \circ v : 1 \rightarrow A$ are not equal because x is monic. In such a situation we say that $x \circ u, x \circ v$ are 'members' of x. This allows us to talk about two subsets of a given set A as having some members in common and some not in common. Thus, using subsets we can make sense of unions and intersections. We can even make sense of when two subsets are subsets of one another, namely when all the members of the first are members of the second.

There are a couple of interesting features of these notions of element and member. First, note that the elements of a set A are members of 1_A, which is a monomorphism from A to A. Thus, if we used the single-type presentation of category theory where there is no object type and instead the identity arrows play the role of the objects, then the notion of an element is simply a special case of the notion of a member. Second, a monomorphism $x : 1 \rightarrowtail A$ can be viewed in two ways, as an element of A (member of 1_A) or as a subset of A. The same arrow plays the role of a singleton subset of A as well as a single element of A.

ETCS supports numerous other categorial constructions beyond those we've considered so far. We can take the product of two sets, construct function sets, take powersets, build characteristic functions, and do many more things. These allow ETCS to play a foundational role:

> The foundational point is: Once you get beyond axiomatic basics, to the level of set theory that mathematicians normally use, ZF and ETCS are not merely inter-translatable. They work just alike. (McLarty, 2005a, 41)

The claim is that in ETCS you can do the same constructions that are carried out in ZFC that are used to provide a foundation for mathematics. Furthermore, while there are slight differences at the beginning, very quickly working in either system to construct the objects used in many fields of mathematics is supposed to look exactly the same. Thus, this is a very real and direct sense in which ETCS is to play the same foundational role as ZFC.

Now, ETCS has certain technical limitations that make it unable to support all of the constructions possible in ZFC. We will address these limitations in Section 5.3. I note them here because of the final comment of Lawvere (1964): "it is the author's feeling that when one wishes to go substantially beyond what can be done in the theory presented here [ETCS], a more satisfactory foundation will involve a theory of the category of categories" (Lawvere, 1964, 1510). Thus, the author of ETCS felt that CCAF would provide an even better foundation for those looking for more than ETCS can provide. However, such a move is not necessary and in Section 5.3 we will consider strengthenings of ETCS that allow it to go beyond the basic theory. Nonetheless, CCAF represents a distinct alternative to ETCS for those considering categorial foundations and so we describe it next.

5.3 Categorial Foundations: CCAF

CCAF was first introduced in Lawvere (1966). It was motivated by contemporaneous mathematical practice:

In the mathematical development of recent decades one sees clearly the rise of the conviction that the relevant properties of mathematical objects are those which can be stated in terms of their abstract structure rather than in terms of the elements which the objects were thought to be made of.   (Lawvere, 1966, 1)

The primary idea is that mathematical objects are described in terms of how they are related to one another, their 'abstract structure', instead of their particular composition. This leads to the following goal addressed in the paper:

The question thus naturally arises whether one can give a foundation for mathematics which expresses wholeheartedly this conviction concerning what mathematics is about, and in particular in which classes and membership in classes do not play any role.   (Lawvere, 1966, 1)

The goal of CCAF is to capture the structural nature of mathematics and to do it without reference to the traditional notion of membership. The traditional notion is thought to be problematic because it involves properties that are considered non-structural. This is very similar to the second goal of Lawvere (1964), where the goal is a foundation where the usual notion of $\in$, i.e. membership, cannot be defined.

While Lawvere (1966) introduced CCAF, some of the stronger axioms in that paper have technical flaws that were pointed out in the review Isbell (1967). For this reason, we use McLarty (1991) as our reference for CCAF. The fundamental treatment of CCAF is the same as in Lawvere (1966) but with the modifications necessary to avoid the problems pointed out by Isbell (1967).

CCAF is intended to describe a category where the objects are categories and the arrows are functors between them. If we have a category A in CCAF, which is just an object in the theory, we need a definition for the objects and arrows of A. Since we want to talk about the objects and arrows of *these* categories, to avoid confusion we refer to the objects and arrows of the category described by CCAF as categories and functors, respectively. As in ETCS, the terminal category 1 is our starting point. It is the canonical one object category and so we define the objects of A as functors from 1 to A.

However, categories have more structure than the possession of objects. We need to make sense of the arrows of categories and the composition of those arrows. To do this, CCAF makes use of two more finite categories progressively more complex than 1, conveniently called 2 and 3. 2 has two objects c and d; i.e. there are two functors $c, d : 1 \to 2$. Furthermore, there are only three functors from 2 to itself, $c o !_2$, $d o !_2$, and 1_2. The arrows of a category A are defined as the functors from 2. We can use d and c to capture the domain and codomain of arrows. Note that for an arrow $a : 2 \to A$ of A, $a \circ d : 1 \to A$ and $a \circ c : 1 \to A$ are objects of A. In fact, $a \circ d$ is the domain of a and $a \circ c$ is the codomain of a.

This guarantees that 2 has three arrows: two identity arrows, $c o !_2$, $d o !_2$; and one non-identity arrow 1_2 (which is also an identity functor). Also, 1 has a single identity arrow $!_2$. Thus, 1 and 2 look like (with the identity arrows suppressed)

$$1_1 \qquad\qquad d \xrightarrow{\;1_2\;} c$$
$$\text{The Category 1} \qquad \text{The Category 2}$$

The only basic notion that remains to be defined is the composition of arrows within a category. The key axiom here is the assertion that the following configuration of categories and functors is a pushout

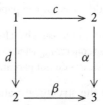

where α and β are new functor constants. Furthermore, we add a constant γ for a functor $\gamma : 2 \to 3$ with $\alpha \circ d = \gamma \circ d$ and $\beta \circ c = \gamma \circ c$. Note that α, β, and γ are all arrows of 3. Furthermore, the codomain of α is equal to the codomain of β, while γ has the same domain as α and the same codomain as β. The idea is that γ is the composition of α and β. 3 looks like

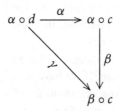

Consider a category A with arrows a_1 and a_2 with $a_1 \circ c = a_2 \circ d$. Then the pushout guarantees a functor $t : 3 \to A$, where $t \circ \alpha = a_1$ and $t \circ \beta = a_2$. Then $t \circ \gamma : 2 \to A$ is an arrow of A with the same domain as a_1 and the same codomain as a_2. This allows us to define the composition in A of the arrows a_1 and a_2 as $t \circ \gamma$. Thus, the category 3 allows us to make sense of the composition of arrows in an arbitrary category A.

This is only the beginning of what can be done in CCAF. We can consider subcategories, product categories, functor categories, and much more. However, even with these, CCAF really only provides "a background theory to be used with axioms on particular categories or functors" (McLarty, 1991, 1259). We can add to CCAF axioms asserting the existence of certain categories. It is with these additional axioms that CCAF provides a foundation for mathematics.

There are a number of options when it comes to these additional axioms. The most common example is an axiom that asserts the existence of a category satisfying the axioms of ETCS (or ETCS plus extensions considered in Section 5.4). Such a category of sets can be used to construct other categories, such as the category of all groups or the category of all graphs. These are built as categories of functors from specific finite categories to the category of sets. Thus, the objects of the resulting functor categories are sets equipped with certain functions and relations, as such objects are usually defined in foundations. While much can be proved in the basic CCAF framework, a lot of its strength depends on these additional axioms. In what follows, we shall consider it equipped with the aforementioned axiom asserting the existence of a category of sets.

5.4 Technical Adequacy

In the debate on foundations, the technical adequacy of ZFC and the technical adequacy of categorial foundations have both been called into question. In a number of places Saunders Mac Lane has pointed to constructions that are desirable or necessary for category theory that cannot be carried out in ZFC (e.g. Mac Lane (1969) or Mac Lane (1971a)). Similarly, A. R. D. Mathias has pointed to the limitations of categorial foundations. He has shown that a number of constructions used in analysis cannot be carried out in ETCS (e.g. Mathias (2001) or Mathias (2000)). In this section we will address both sets of complaints. First, we will look at the shortcomings Mac Lane levels against ZFC and then we will consider the limitations Mathias ascribes to ETCS.

5.4.1 ZFC

The original foundational complaint about ZFC from the categorial perspective is that it has problems forming important categories of interest:

These problems arise in the use of collections such as the category of all sets, of all groups, or of all topological spaces. It is the intent of category theory that the "all" be taken seriously; the usual axiomatizations of set theory do not allow the formation of collections such as the set of all sets, or the set of all groups. (Mac Lane, 1969, 192)

The problem is that in ZFC one cannot form collections large enough to serve as unlimited categories. ZFC is exclusively about sets and on pain of contradiction cannot allow the formation of the set of all sets or even the set of all groups. The natural response to this difficulty is to introduce another notion of collection above that of sets, one which allows for a collection of all sets or a collection of all groups.

The introduction of classes to the theory does exactly that and so appears to successfully deliver the category of all sets or the category of all groups. One does this by forming the class of all sets or the class of all groups and using these for the categories. Thus, a theory like Von Neumann–Bernays–Gödel (NBG) appears to overcome the limitations of ZFC. In fact, the very first paper on category theory, Eilenberg and Mac Lane (1945), takes NBG as one of many possible satisfactory foundations for category theory. However, NBG itself runs into problems. The first is that it "does not allow for the free formation of functor categories" (Mac Lane, 1969, 193). The collection of functors between two large (class-sized) categories is, in general, too big to be a class itself. Furthermore, "for many purposes, ... one would like to have a bigger **cat**, say the category of all large categories or of all categories *überhaupt*" (Mac Lane, 1971a, 234). The problem is that objects of interest appear at the level of classes and we would like to be able to form categories of those as well.[3] A richer theory of these large objects is able to solve at least one of these problems, the problem of providing functor categories.

This theory is provided by strong axioms of infinity, also called large cardinal axioms. For categorial purposes only inaccessible cardinals, the weakest large cardinals, are required. For categorial applications they appear in the form of Grothendieck universes. Reminiscent of Zermelo (1930), the idea is to have an increasing hierarchy of universes of sets, each larger than the last. Starting in a universe U we consider G the set of all groups of U. Now, G is too large to be an element of U but it is an element of a larger universe U'. Thus, in that larger universe one can form the desired functor categories involving G. However, for every universe U one can only form the category of all groups *in* U, or as in Mac Lane (1971a): "Given any universe U', one can always form the category of all categories within U'. This is still not that will-of-the-wisp, the category of all categories *überhaupt*." Larger universes have groups not present in U. So, in such a system one can never form the category of all groups, which means that this extension of ZFC fails in the same way ZFC is considered to fail.[4]

It is worth noting that the assumption of one Grothendieck universe is enough for most applications in category theory (see (Mac Lane, 1971b, I.6)), and is certainly sufficient for the major theorems of category theory (see Mac Lane (1969)). However, the one universe system is not able to do *everything* for exactly the same reasons a system with multiple Grothendieck universes is not able to do everything.[5]

Solomon Feferman has an ongoing project, first in Feferman (1969) and most recently in Feferman (2013), on the prospects of finding a satisfactory foundation that is able to deliver all of the desired features. Feferman has suggested that the desired features are captured by the following three requirements:

- (R1) Form the category of all structures of a given kind, e.g. the category **Grp** of all groups, **Top** of all topological spaces, and **Cat** of all categories.
- (R2) Form the category B^A of all functors from A to B, where A, B are any two categories.
- (R3) Establish the existence of the natural numbers N, and carry out familiar operations on objects $a, b, \ldots$ and collections $A, B, \ldots$, including the formation of $[a, b], (a, b), A \cup B, A \cap B, A - B, A \times B, B^A, \cup A, \cap A, \Pi B_x[x \in A]$, etc.

(Feferman, 2013, 9)

---

[3] This same concern appears also in Mac Lane's later work: "we will have many such occasions to form categories which are not classes. One such category is the category **Cls** of all classes... Another useful category is **Cat**, the category of all large classes. It is not a class" (Mac Lane, 1971b, 23).

[4] There are additional complications with Grothendieck universes that have to do with relating objects between different universes.

[5] Specifically, in a one universe system "we cannot form the category of all sets or of all groups" for the same reasons I have highlighted here (Mac Lane, 1969, 196).

(R1) is the requirement that the foundation can produce the categories of *all* objects of a particular kind (so-called unlimited categories). (R2) and (R3) are primarily concerned with the ability to do mathematical work with both the objects of those categories and the categories themselves.

Feferman's most advanced attempt to meet all three requirements is based on New Foundations with Urelements (NFU). The basic system is able to satisfy both (R1) and (R2). However, a number of familiar operations cannot be carried out within the system. For that reason, he extends the system to include further special operators, one for pairing and one for universal choice. The problem is that even those operators are insufficient to fully satisfy (R3) as the enriched system is still not capable of forming arbitrary cartesian products.[6] Thus, for this project "[t]he big question that remains is whether any workable system that meets (R1)-(R3) in full can be shown to be consistent relative to some accepted system of set theory" (Feferman, 2013, 13). Feferman (2013) considers it possible that some as yet unconceived foundation will be able to satisfy all three requirements.

That big question is taken up in Ernst (2015). The primary result of that paper is that no foundation will be capable of satisfying all three requirements. This is done by using the category of all graphs and deriving a contradiction directly from the requirements, instead of working in a particular foundational system. This provides conclusive evidence that we will never be able to capture that will-of-the-wisp, the category of *all* categories.[7] Thus, it looks like some hierarchy of sizes is unavoidable.

However, this is not really the problem it might first appear to be. It turns out that some notion of small/large or set/class is an important part of category theory. In fact, "unlike in most fields of mathematics outside of set theory, questions of *size* play an essential role in category theory" (Shulman, 2008, 1). For example, the Adjoint Functor Theorem, one of the central results of pure category theory, requires such a size distinction. Thus, the fact that we end up with a such a hierarchy when extending ZFC to handle cases important to category theory is a welcome effect.

We learn at least two important things from Mac Lane's complaints about ZFC as a foundation. First, the result of Ernst (2015) shows us that a distinction between large and small objects is unavoidable, even if for some reason we wanted to avoid it (as Shulman (2008) suggests categorists should not). Mac Lane's category of all categories *überhaupt* is unattainable, no foundation can provide it without giving up something essential. Second, while there is a sense that Mac Lane is right and that the most basic formulation of ZFC does not provide a satisfactory foundation, the natural extension of ZFC through the addition of universes does provide such a foundation. Furthermore, universes (in the form of inaccessible cardinals) are a standard addition

---

[6] McLarty (1992) illustrates why any system based on New Foundations is going to run into problems of exactly this kind.

[7] While this may appear to provide a problem for CCAF it does not because CCAF does not intend to capture *all* categories. For one thing, the category described by CCAF is not an object in itself.

to ZFC today and are on the weak end of such principles used and investigated in set theory.[8]

5.4.2 Categorial foundations

Mathias' complaints are focused on ETCS, specifically as it is presented in Mac Lane (1986).[9] His primary complaint is that ETCS does not allow for important constructions of interest. As a terminological note, Mathias discusses a number of systems that are equivalent to ETCS, specifically systems he calls MAC (for Mac Lane set theory) and ZBQC (Zermelo set theory with bounded quantifiers and choice).[10] The fact that only bounded quantifiers are available in the scheme of separation in these systems restricts what can be achieved.

Mathias locates the shortcomings of ETCS in two areas. First, in a particular part of mathematics:

I suggest that an area ill supported by Mac Lane's system ZBQC is that of iterative constructions. We know from the work of Cantor onwards that there are processes which need more than ω steps to terminate; of which examples may be found even within traditional areas of mathematics. (Mathias, 1992, 114)

One of the primary problems with ETCS (in the form of ZBQC) is that there are iterative constructions that cannot be carried out. For example, using it one "cannot prove that each initial segment of the function $n \mapsto \omega + n$ is a set" (Mathias, 2001, 115). The system is unable to produce finite initial segments of the $\beth$ sequence (the sequence of powersets beginning with ω). In many cases these problems arise because of the failure of the induction scheme for natural numbers in ZBQC.

Furthermore, Mathias emphasizes that these problems are not restricted to the field of mathematical logic. The latter example concerning the $\beth$ sequence is "an illustration drawn from algebra of the inadequacy of MAC" (Mathias, 2001, 115). It is this focus on how strong logical principles show up in all different fields of mathematics that raises Mathias' second technical concern. ETCS does not support iterated constructions that can be carried out in ZFC, and it also does not support such constructions that go beyond ZFC.

Mathias "exhibit[s] four 'natural assertions', A, B, C, D, about certain sets of real numbers, each of these assertions is provably equivalent (in ZFC) to an assertion that certain ordinals possess large-cardinal properties in certain inner models" (Mathias, 2000, 515). The idea is that a variety of statements within analysis are intimately connected to large cardinals and so large cardinals are not a bizarre uniquely set-theoretic concept but in fact pervade mathematics. This means that the failure to

---

[8] For more information are such cardinals, their justification, and how they are used in set theory see Maddy (1988a, b).

[9] This section is focused almost exclusively on ETCS because that is where the focus of the debate has been. We will consider some brief comments about CCAF at the end of the section.

[10] These systems are spelled out in detail in Mathias (2001).

provide for large cardinals is a failure to support many different parts of mathematics. Mathias' is not an isolated result: "Harvey Friedman and his collaborators on the necessary use of abstract set theory show how set-theoretic questions concerning large notions of infinity may arise even in apparently innocent mathematical statements concerning only finite structures" (Mathias, 2001, 227). Thus, large cardinals represent another aspect of the shortcomings of ETCS. These aspects can be addressed separately, limitation with respect to iterative constructions that can occur in ZFC and limitation with respect to large cardinals.

While ETCS as traditionally stated cannot handle these cases, there are extensions of ETCS that can handle both. We will consider each in turn, beginning with the limitations concerning iterative constructions in ZFC. In both cases we find the needed extensions in McLarty (2005a).[11]

There is a replacement axiom that can be added to ETCS to create the aptly named ETCS+R, which is mutually interpretable with ZFC. ETCS+R can do everything that can be done by ZFC. Categorial replacement takes as input a set A and a relation $R(a, S)$ of arrows to sets (expressible in ETCS) that matches each $x \in A$ with a set S_x that is unique up to isomorphism. If these conditions are met then there is a set S and an arrow $f : S \to A$ such that for any $x \in A$ and S_x such that $R(x, S_x)$ there is an arrow from S_x to S that makes the following a pullback:

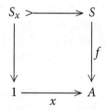

This makes S "a disjoint union of the S_x and F gives the structure of a set of sets $\{S_x | x \in A\}$ as each S_x is the preimage of its x" (McLarty, 2005a, 48). Since the addition of categorial replacement makes ETCS+R as strong as ZFC, ETCS+R can support the aforementioned iterative constructions.[12] What remains is making sense of large cardinals in ETCS (or more appropriately now in ETCS+R).

Large cardinals are even easier to add to the ETCS framework. The key observation is that "Large-cardinal axioms are usually given, by ZF set theorists, in isomorphism-invariant form. They are homophonically ETCS axioms in the first place" (McLarty, 2005a, 50–1). Large cardinal axioms do not make explicit use of the global membership relation, the local membership relation of ETCS being sufficient. As a result, they can be understood directly as statements made in the language of ETCS; no translation or

---

[11] In Mac Lane (2000), a brief response that immediately follow Mathias (2000), Mac Lane himself is unmoved by these considerations as he does not think they are representative of normal mathematics.

[12] Osius (1974) provides a different axiom to play the role of categorial replacement, but the resulting system is equivalent. The proof of equivalence with ZFC given in McLarty (2005a) draws on the proof from Osius (1974).

change is needed. They can be added to ETCS in exactly the same way that they are added to ZFC.

The primary difference is that while large cardinal axioms are a standard addition to ZFC, the investigation of such axioms and their consequences being one of the central investigations in mathematical logic, they are not standard components in the ETCS framework. This is because many advocates of that perspective consider them unnecessary for normal mathematics, contra Mathias and Friedman (e.g. Mac Lane (2000)). Hence, while ETCS can be extended to accommodate the technical shortcomings noted by Mathias such extension is resisted by some of its advocates.

The case of CCAF is significantly less discussed than that of ETCS. Part of the reason is that the difficulties in Lawvere (1966) revealed by Isbell (1967) meant that a fully formulated version of CCAF arrived much later on the scene than ETCS. In addition, as noted in the section on CCAF, one of the primary axioms that adds strength to CCAF is the assertion that there is a category satisfying ETCS (or ETCS+R with the addition of large cardinals). Thus, with respect to technical strength CCAF is very much tied to ETCS.

5.4.3 Results on adequacy

We have shown that there are some similarities for both ZFC and categorial foundations in terms of technical adequacy but also important differences. One of the purported shortcomings of ZFC, that it could not provide unlimited categories, turns out not to be a shortcoming at all. That requirement is impossible to satisfy. Thus, at least some of what was asked of ZFC turns out to be an empty criticism; ZFC cannot do the impossible.

Both ZFC and ETCS must be extended by large cardinals in order to meet the demands placed upon them by their detractors. However, there is a difference in the strength of the large cardinals demanded and the perspectives on their addition. To overcome the limitations posed by Mac Lane, ZFC need only be supplemented by inaccessible cardinals, which are a weak form of large cardinal and a standard addition to ZFC as a normal part of mathematical practice. To meet the limitations posed by Mathias, ETCS needs supplementation by categorial replacement and large cardinal axioms, including ones stronger than inaccessible cardinals.[13] These additions are less standard and are resisted by some advocates of ETCS as a foundational system. Ultimately, through the extensions we have considered both systems are capable of providing adequate technical foundations to cover mathematical practice. The primary difference is only in the resistance by some advocates of categorial foundations toward making the necessary additions.

---

[13] Of course to really be satisfactory to *everyone* and not just Mac Lane, ZFC also needs these stronger large cardinals.

5.5 Autonomy

The original objections to categorial foundations on the grounds of autonomy come from Feferman (1977).[14] It argues that "the notions of operation and collection are prior to all structural notions", which includes the notion of categories (Feferman, 1977, 149). The idea is that categorial foundations cannot stand on their own because they are theories about structural notions and so must depend on prior theories of operation and collection. Linnebo and Pettigrew (2011) consider the objections raised by Feferman (1977) and also introduces a new objection concerning the justification of axioms. To be autonomous from ZFC (or extensions thereof) a foundation "must be able to justify its assertions without appealing to orthodox set theory [ZFC], or to any aspect of the justification of orthodox set theory that belongs primarily to that theory" (Linnebo and Pettigrew, 2011, 244). The question is whether or not the axioms of a foundation can be *justified* without appealing to other foundations or their justifications. We will consider both of these objections and responses in turn. However, first, we address a confusion in the literature about what are the candidate categorial foundations.

5.5.1 *Assertiveness*

The primary confusion is thinking that the category axioms are intended as a foundation for mathematics. It is correctly noted that these axioms "*assert nothing. They merely tell us what it is for something to be a structure of a certain kind.*" (Hellman, 2003, 134). The category axioms allow us to recognize categories when they arise but do not tell us whether there are any categories. This leads to the conclusion that category theory cannot provide a foundation because "*it lacks an external theory of relations, and it lacks substantive axioms of mathematical existence*" (Hellman, 2003, 138). The category axioms do not appear to provide either of these things, meaning they are not foundation material.

The mistake lies in thinking that the category axioms are intended as a foundation at all. While the category axioms themselves may not make assertions or answer questions of existence "each specific categorical foundation offers various quite strong existence axioms" (McLarty, 2005a, 42). As we have seen, both ETCS and CCAF assert the existence of many different objects. In ETCS there is the singleton set 1 and also infinite sets and products of sets and many other constructions. In Section 5.4.2, we saw how ETCS (and in a similar fashion CCAF) could be extended to include even stronger existence assumptions, in the way ZFC is traditionally supplemented. Thus, it becomes very important in the debate on foundations to be clear about what is actually be proposed as a foundation.

---

[14] It should be noted that Feferman (1977) does not take ZFC to be an acceptable foundation either, only closer to what is necessary.

5.5.2 Prior theory

Feferman (1977) argues that operation and collection must be prior to categorial foundations because they are required to make sense of them. First, "*when explaining the general notion of structure and of particular kinds of structures... we implicitly presume as understood* the ideas of *operation* and *collection*" (Feferman, 1977, 150). Anytime a structure is under consideration, operation and collection are involved. Also, as we build new structures on top of others "at each step we must make use of the unstructured notions of operation and collection to explain the structural notions to be studied" (Feferman, 1977, 150). The fundamental idea is that these unstructured notions of operation and collection are necessary precursors to any account of structures. Thus, Feferman (1977) concludes that "a theory whose objects are supposed to be highly structured and which does not explicitly reveal assumptions about operations and collections cannot claim to constitute a foundation" (Feferman, 1977, 150). Operation and collection as fundamental underlying assumptions must be made sense of for a system to be a foundation. ZFC does not suffer from these objections because it is not a theory whose objects are supposed to be highly structured but is itself a theory of collections and so is intended to explicitly reveal the underlying assumptions.

Feferman (1977) goes on to clarify that operation and collection are logically and psychologically prior. Logical priority concerns "the order of definition of concepts, in the cases where certain of these *must* be defined before others" (Feferman, 1977, 152). On the other hand, "psychological priority' has to do with the natural order of understanding" (Feferman, 1977, 153). Thus, logical priority has to do with the underlying logical structure, whereas psychological priority has to do with our understanding of the concepts involved. In the terminology of Linnebo and Pettigrew (2011), logical priority concerns issues of logical autonomy while psychological priority concerns issues of conceptual autonomy.

There have been a range of responses to these objections. Here we will only consider those that maintain a foundational role for categorial concepts. Landry (2013) claims that there are two different notions of foundation, genetic and axiomatic, and "even if the genetic method requires the notions of operation and collection, the axiomatic method does not". (Landry, 2013, 24).[15] Landry (2013) claims that category theory is intended to provide an axiomatic foundation and not a genetic one, whereas ZFC provides a genetic foundation. Furthermore, since the considerations of Feferman (1977) are only relevant to genetic foundations the objections do not apply to categorial foundations.[16]

Marquis (2013) agrees with Feferman (1977) that "priority must lie with notions of operation and collection", but goes on to argue that "categories play an indispensable

---

[15] Landry (2013) is a continuation of a larger project that includes Landry (2006) and Landry (2011).
[16] Marquis (1995) canvases a number of different ways in which a theory might be considered a foundation.

role in such a foundational framework" (Marquis, 2013, 51). Thus, even though the arguments of Feferman (1977) are accepted, the claim is that any account of operation or collection will necessarily include categorial concepts. Furthermore, Marquis (2013) argues that this requires us to view foundations in a new way. We will not go into the details of these latter two responses here because they both involve a change in the view of foundations. While such a change may be in order, we are here considering how categorial foundations are supposed to play the same foundational role as has been played by ZFC. Namely, foundations like ZFC are supposed "to provide a generous, unified arena to which all local questions of coherence and proof can be referred" (Maddy, 2011, 34). The traditional foundational role of ZFC is to tell us what mathematical notions are coherent and can be legitimately considered to exist.

The final major response is to agree with Feferman (1977) that foundations must begin with a theory of operation and collection and then argue that categorial foundations do exactly that. For example, McLarty (2005b) says, "Obviously I agree with Feferman that foundations of mathematics should lie in a general theory of operations and collections, only I say the currently best general theory of those calls them *arrows* and *objects*" (McLarty, 2005b, 49). Thus, according to responses of this kind it is not that Feferman (1977) is wrong about the importance of operation or collection, but that it does not appreciate aspects of the theories provided by categorial foundations. We consider how specific versions of this response are made in the particular cases of CCAF and ETCS.[17] We begin with the consideration of ETCS and its logical autonomy.

Linnebo and Pettigrew (2011) observe that while ETCS "is itself a theory of sets and functions, it does not depend on a *prior* theory of these entities. Rather, it provides such a theory" (Linnebo and Pettigrew, 2011, 233). The key here is that ETCS is not a theory whose objects are highly structured. ETCS is a theory about sets, which are unstructured objects. The objections of Feferman (1977) are specifically leveled against theories of structured objects and ETCS is not such a theory.[18] Thus, ETCS provides a theory of collections, i.e. sets, and operations, i.e. functions, exactly as required and so is logically autonomous.

To establish conceptual autonomy for ETCS, it needs to be possible to explain and understand ETCS without reference to ZFC or some other theory of operations and collections. We noted in Section 5.2 that ETCS was originally developed by Lawvere to teach to students instead of ZFC. For this reason, Lawvere's explanations of ETCS, especially in Lawvere (2003) are specifically designed not to make use of other theories. Similar explanations can be found in Mac Lane (1986), Linnebo and Pettigrew (2011), McLarty (1995), and elsewhere. The point is that "these introductory glosses are autonomous with respect to any notions that belong peculiarly to orthodox set theory"

---

[17] McLarty (2005b) appears to be talking about category theory more generally, but we will look at specific arguments given by others for ETCS and CCAF.

[18] Feferman (2013) argues that the objection of Feferman (1977) still applies, but it appears to do this by arguing that a theory of sets and functions (ETCS) requires the notion of a category.

(Linnebo and Pettigrew, 2011, 243). There are many explications of ETCS that do not seem to depend on ZFC or other theories at all.

In the case of CCAF, the logical and psychological priority objections are intertwined. Linnebo and Pettigrew (2011) claim that in order to make sense of the meaning of CCAF we must make sense of what categories are. That is, for conceptual autonomy we need a way to understand CCAF that does not appeal to other theories. The claim is that categories are made sense of either in a viciously circular fashion or in terms of some prior notion of collection. Thus, CCAF either fails to have real content because of the circularity or is not logically autonomous since it would be defined in terms of a theory of collections. In the latter case, it is that theory of collections that would provide the real foundation.

Logan (2014) argues that this has simply neglected another way of looking at CCAF. It argues that CCAF can be understood as a general theory of operations, in which case "it seems to do Feferman one better by relying only on a previous understanding of the general concept of an operation" (Logan, 2014, 5–6).[19] Thus, if Logan (2014) is correct, we can make sense of CCAF on its own, avoiding psychological priority problems, and the content of its assertions are not dependent on some other notion of collection, avoiding problems of logical priority.

Thus, with respect to logical and psychological priority the debate currently turns on whether or not these presentations achieve what is claimed. The case of ETCS is more developed because there are many thorough presentations of ETCS, while the case of CCAF is less developed but still appears to have at least one possibly satisfactory presentation.

5.5.3 *Justification*

In order to enjoy justificatory autonomy, the axioms of a categorial foundation must be justified on grounds of their own. We here consider ETCS exclusively as that is where the bulk of the debate is focused. For Linnebo and Pettigrew (2011) the canonical example of an acceptable and autonomous justification for a mathematical theory is the use of the iterative conception as a justification of ZFC (e.g. Boolos (1971)). However, such a view is overly narrow and misses much of how axioms are actually justified (e.g. Maddy (2011)). That said, we can still evaluate how ETCS and its extensions fare in this particular part of the question of justification. We can see whether there is some counterpart to the iterative conception that can play the same role for them. The goal is not to provide some prior theory that justifies ETCS, but something like an intuitive picture of sets that can help justify it. To do this, we will first consider some specific questions of justification before considering ETCS and extensions as a whole.

---

[19] This primacy for operations is interesting because "operators are prior to collections in the foundations of Feferman's theory" (McLarty, 2005*b*, 49).

Hellman (2003) questions the justification of categorial replacement and large cardinals in ETCS. The objection to categorial replacement is that the advocate of ETCS+R "fails to provide . . . any independent category-theoretic reason for 'believing' the axiom" (Hellman, 2003, 144). It appears that categorial replacement is motivated in the same way replacement in ZFC has been motivated. The response to this objection is to look at the origins of replacement. The claim is that for both categorial replacement and replacement in ZFC "[w]e all have Cantor's original motive which predates the membership-based ZF version by decades" (McLarty, 2005a, 49). ETCS is supposed to capture Cantor's view of sets (as ETCS is a theory of sets) and so it has Cantor's motivation for replacement (and so becoming ETCS+R). Furthermore, since Cantor's motivation for replacement predates the ZFC view of sets it cannot be dependent upon it.[20] This argument is a special case of the more general argument made in Lawvere (1994).

Lawvere (1994) is devoted to describing Cantor's picture of set, which it argues is distinct from what is captured in ZFC. Lawvere (1994) claims that Cantor had a notion of abstract sets that "may be conceived of as a bag of dots which are devoid of properties apart from mutual distinctness" (Lawvere, 1994, 5). The expression in Cantor is *lauter Einsen* which "could mean roughly 'nothing but many units' " (Lawvere, 1994, 5). This is meant to be a notion of set distinct from the one advocated and described by Zermelo, one which Zermelo explicitly rejected (Lawvere, 1994, 6). It is supposed to be what is captured by ETCS. Thus, it would appear ETCS has a picture analogous to the iterative conception to help justify its axioms.

The primary objection Linnebo and Pettigrew (2011) raise to the *lauter Einsen* view is that it attributes some properties to sets that mathematicians do not attribute and that it fails to attribute properties to sets that mathematics do attribute.[21] Both of these become issues of what actually occurs in practice.

After rejecting the *lauter Einsen* view, Linnebo and Pettigrew (2011) again turn to questions of practice. They consider the possibility for a naturalistic justification of ETCS. However, the paper questions both whether the evidence for such a justification is present and whether or not that kind of a justification is acceptable. Thus, the focus of the debate turns on what actually occurs in mathematical practice and the connection between practice and justification.

As for the substance of mathematical practice, the objections of Linnebo and Pettigrew (2011) are based on the content of mathematical textbooks. Linnebo and Pettigrew (2011) claim that "many textbooks that introduce elementary areas of mathematics, such as algebra, analysis, and number theory, include an introductory section

---

[20] Additionally, the iterative conception is typically viewed as not providing a justification for replacement and so here defenders of ETCS+R could potentially claim an advantage.

[21] We do not consider here the discussion in Linnebo and Pettigrew (2011) that considers how the pure existence axioms of ETCS are justified. That discussion appears out of place, as earlier the paper explicitly decides to bracket the justification of existence and claims that the iterative conception cannot provide it (Linnebo and Pettigrew, 2011, 245).

surveying the elements of set theory, and this set theory is explicitly orthodox set theory" (Linnebo and Pettigrew, 2011, 249). Set theory of large parts of mathematical practice apparently does not agree with the set theory of ETCS. McLarty (2012) disagrees and provides examples of many textbooks that do not "make any serious distinction between categorial and iterative set theory" (McLarty, 2012, 112). It looks as if all of the textbooks considered can be viewed as using ETCS without difficulty. This is a factual disagreement about the content of mathematical textbooks. I am not in a position to settle this question; I can only point to the examples provided by McLarty (2012) and investigate any further examples as they are provided.

As for the connection between mathematical practice and justification, Linnebo and Pettigrew (2011) claim that for ETCS "its justificatory autonomy turns on whether or not mathematical theories can be justified by appeal to mathematical practice" (Linnebo and Pettigrew, 2011, 227). McLarty (2012) points out that this would seem impossible to deny as it would require philosophers "to claim to know a better mathematics than mathematicians practice" (McLarty, 2012, 113). This is a topic much more complicated than simply the content of mathematical textbooks. Furthermore, it is too large for consideration in a book chapter, let alone a section of such a chapter. Readers interested in this issue should consider Maddy (1997), Leng (2002), or Carter (2008).

5.5.4 Results on autonomy

Thus, the debate on the autonomy of categorial foundations comes mostly to this: ETCS can be considered autonomous to the extent to which it is a theory of operations and collections and the extent to which the *lauter Einsen* view provides a picture of sets independent of the iterative conception. There are further questions about the content and connection to practice that bear upon these questions. The case for CCAF is similar, in that it also has been defended as its own independent theory of operations. However, that viewpoint is less developed than that for ETCS.

Between the questions of technical adequacy and autonomy we have surveyed the vast bulk of the debate concerning categorial foundations as an alternative to set-theoretic foundations. I would like now to turn briefly to some of the quite interesting issues raised by categorial foundations that have been mostly overlooked.

5.6 Accessibility and Mathematical Thought

One of the major motivations for categorial foundations is that they "would seemingly be much more natural and readily-useable than the classical one when developing such subjects as algebraic topology, functional analysis, model theory of general algebraic systems, etc." (Lawvere, 1966, 1). There are fields in which the language and methods of category theory are extremely useful and so such fields are better served by a categorial foundation.

This is a view promoted by nearly every paper we have considered here advocating for categorial foundations. This view of the effectiveness of categorial methods also occurs in many papers not devoted to categorial foundations, such as Awodey (1996).[22] Categorial methods provide a particularly effective way of thinking for much of mathematics. While many positive cases have been made for this position, it has undergone relatively little scrutiny within the debate.

While there "are portions of mathematical reasoning for which the language of categories gives a very smooth presentation...there are portions for which it is clumsy" and similarly for the language of set theory (Mathias, 2001, 227). For example, "[c]ategories and functors are everywhere in topology and in parts of algebra, but they do not as yet relate well to most of analysis" (Mac Lane, 1986, 407). While category theory is particularly effective in many areas of mathematics, there are other areas in which it is not.[23] The interesting question is what we can learn from these cases where categorial methods falter. One tantalizing suggestion by Mathias (2001) is that in the particular case it considers this reveals that the theories in question "capture two distinct modes of thought" (Mathias, 2001, 228). While category theory provides an effective way of thinking for many fields in mathematics, those areas where it falters suggest that a different kind of thinking is in play.

It would seem that looking at both the successes and failures of categorial methods then can tell us something about the nature of mathematical thought in different fields of mathematics. This possibility, especially for the case of failures, has been almost entirely neglected and yet may reveal much about mathematics.

5.7 Conclusion

We have explored the two major aspects of the foundational debate, technical adequacy and autonomy. Both ZFC and categorial foundations have technically adequate extensions, the only difference being the perspective taken on those extensions. Categorial foundations, especially ETCS, have well-developed independent characterizations that at least suggest autonomy, though that debate is still ongoing. Finally, there is the quite interesting and under-explored suggestion that investigating both where categorial methods succeed and where they fail can help to reveal some of the nature of mathematical thought.

---

[22] Awodey (1996) does not advocate categorial foundations because Awodey advocates a rejection of the entire foundational viewpoint.

[23] For example, Brown et al. (2008) uncover a systematic mistake in the construction of the category of undirected graphs which suggests graph theory may be one of these poorly served areas.

References

Awodey, S. (1996). Structure in mathematics and logic: A categorical perspective. *Philosophia Mathematica* 4, 209–37.

Awodey, S. (2009). *Category Theory*. Oxford University Press, New York.

Boolos, G. (1971). The iterative conception of set. *Journal of Philosophy* 68, 215–31.

Brown, R., Morris, I., Shrimpton, J., and Wensley, C. D. (2008). Graphs of morphisms of graphs. *Electronic Journal of Combinatorics* 15(1), Article 1, 28.

Carter, J. (2008). Structuralism as a philosophy of mathematical practice. *Synthese* 163(2), 119–31.

Eilenberg, S., and Mac Lane, S. (1945). General theory of natural equivalences. *Transactions of the American Mathematical Society* 58, 231–94.

Ernst, M. (2015). The prospects of unlimited category theory: doing what remains to be done. *Review of Symbolic Logic* 1–22.

Feferman, S. (1969). Set-theoretical foundations of category theory, in *Reports of the Midwest Category Seminar. III*, 201–47. Springer, Berlin.

Feferman, S. (1977). Categorical foundations and foundations of category theory, in *Logic, Foundations of Mathematics and Computability Theory (Proc. Fifth Internat. Congr. Logic, Methodology and Philos. of Sci., Univ. Western Ontario, London, Ont., 1975), Part I*, 149–69. Reidel, Dordrecht.

Feferman, S. (2013). Foundations of unlimited category theory: what remains to be done. *Review of Symbolic Logic* 6(1), 6–15.

Hellman, G. (2003). Does category theory provide a framework for mathematical structuralism? *Philosophia Mathematica* 11(2), 129–57.

Isbell, J. (1967). Review of Lawvere 1966. *Mathematical Reviews* 34, 1354–5.

Landry, E. (2006). Category theory as a framework for an in re interpretation of mathematical structuralism, in J. Benthem, G. Heinzmann, M. Rebuschi, and H. Visser (eds), *The Age of Alternative Logics*, 163–79. Springer, Dordecht.

Landry, E. (2011). How to be a structuralist all the way down. *Synthese* 179(3), 435–54.

Landry, E. (2013). The genetic versus the axiomatic method: responding to feferman 1977. *Review of Symbolic Logic* 6(1), 24–51.

Landry, E., and Marquis, J. (2005). Categories in context: historical, foundational, and philosophical. *Philosophia Mathematica* 13(1), 1.

Lawvere, F. (1964). An elementary theory of the category of sets. *Proceedings of the National Academy of Sciences of the U.S.A.* 52(6), 1506.

Lawvere, F. (1966). The category of categories as a foundation for mathematics. *Proceedings of the La Jolla Conference on Categorical Algebra*.

Lawvere, F. (1994). Cohesive toposes and cantor's "lauter einsen". *Philosophia Mathematica* 2(1), 5–15.

Lawvere, F. (2005). An elementary theory of the category of sets (long version) with commentary. *Reprints in Theory and Applications of Categories* 11, 1–35.

Lawvere, F. W., and Rosebrugh, R. (2003). *Sets for Mathematics*. Cambridge University Press, Cambridge, UK.

Leng, M. (2002). Phenomenology and mathematical practice. *Philosophia Mathematica* 10(1), 3–25.

Linnebo, Ø., and Pettigrew, R. (2011). Category theory as an autonomous foundation. *Philosophia Mathematica* 19(4), 227–54.

Logan, S. (2014). Category theory is a contentful theory. *Philosophia Mathematica* p. nku030.

Mac Lane, S. (1969). One universe as a foundation for category theory, in *Reports of the Midwest Category Seminar. III*, 192–200. Springer, Berlin.

Mac Lane, S. (1971a). Categorical algebra and set-theoretic foundations, in *Axiomatic Set Theory (Proc. Sympos. Pure Math., Vol. XIII, Part I, Univ. California, Los Angeles, Calif., 1967)*, 231–40. American Mathematical Society, Providence, RI.

Mac Lane, S. (1971b). *Categories for the Working Mathematician*. Springer-Verlag, New York.

Mac Lane, S. (1986). *Mathematics, Form and Function*. Springer-Verlag, New York.

Mac Lane, S. (2000). Contrary statements about mathematics. *Bulletin of the London Mathematical Society* 32(5), 527.

McLarty, C. (1991). Axiomatizing a category of categories. *Journal of Symbolic Logic* 56, 1243–60.

McLarty, C. (1992). Failure of cartesian closedness in nf *Journal of Symbolic Logic* 57(2), 555–6.

McLarty, C. (1995). *Elementary Categories, Elementary Toposes*. Clarendon Press, Oxford.

McLarty, C. (2005a). Exploring categorical structuralism. *Philosophia Mathematica* 12(1), 37–53.

McLarty, C. (2005b). Learning from questions on categorical foundations. *Philosophia Mathematica* 13(1), 44–60.

McLarty, C. (2012). Categorical foundations and mathematical practice. *Philosophia Mathematica* 20(1), 111–13.

Maddy, P. (1988a). Believing the axioms. i. *Journal of Symbolic Logic* 53(02), 481–511.

Maddy, P. (1988b). Believing the axioms. ii. *Journal of Symbolic Logic* 53(03), 736–64.

Maddy, P. (1997). *Naturalism in Mathematics*. Oxford University Press, Oxford.

Maddy, P. (2011). *Defending the Axioms*. Oxford University Press, Oxford.

Marquis, J. (1995). Category theory and the foundations of mathematics: philosophical excavations. *Synthese* 103(3), 421–47.

Marquis, J. (2013). Categorical foundations of mathematics: or how to provide foundations for *abstract* mathematics. *Review of Symbolic Logic* 6(1), 51–75.

Martin, D. (2001). Multiple universes of sets and indeterminate truth values. *Topoi* 20(1), 5–16.

Mathias, A. (1992). What is mac lane missing?, in *Set Theory of the Continuum (Berkeley, CA, 1989)*, 113–18. Springer, New York.

Mathias, A. (2000). Strong statements of analysis. *Bulletin of the London Mathematical Society* 32(5), 513–26.

Mathias, A. (2001). The strength of Mac Lane set theory. *Annals of Pure and Applied Logic* 110(1–3), 107–234.

Osius, G. (1974). Categorical set theory: a characterization of the category of sets. *Journal of Pure and Applied Algebra* 4(1), 79–119.

Shulman, M. (2008). Set theory for category theory. arXiv preprint arXiv:0810.1279.

Zermelo, E. (1930). On boundary numbers and domains of sets, in W. Ewald (ed.), *In from Kant to Hilbert: A Source Book in the Foundations of Mathematics*. Oxford University Press, New York.

6

Canonical Maps

Jean-Pierre Marquis

> Every science is of itself a system; and it is not enough that in it we build in accordance with principles and thus proceed technically; rather, in it, as a freestanding building, we must also work architectonically, and treat it not like an addition and as a part of another building, but as a whole by itself, although afterwards we can construct a transition from this building to the other or vice versa.
>
> (Kant, *Critique of the Power of Judgment*, AA 05:381, quoted by Gava, 2014, 385)

6.1 Introduction

The foundational status of category theory has been challenged as soon as it has been proposed as such.[1] The philosophical literature on the subject is roughly split into two camps: those who argue against category theory by exhibiting some of its shortcomings and those who argue that it does not fall prey to these shortcomings.[2] Detractors argue that it supposedly falls short of some basic desiderata that any foundational framework ought to satisfy: logical, epistemological, ontological or psychological. To put it bluntly, it is sometimes claimed that category theory fails to rest on simple notions, or its objects are too complicated, or it presupposes more basic concepts, or it can only be understood after some prior notion has been assimilated, and, of course, these disjunctions are not exclusive. In this chapter, I want to reverse this perspective completely. I want to present what I take to be one of the main virtues of category theory as a foundational framework. More precisely, I claim that only a categorical framework, or a framework that encodes the same features as those

---

[1] The first publication presenting the category of categories as a foundational framework is Lawvere (1966). As far as I know, the first printed reaction against such an enterprise is Kreisel's appendix in Mac Lane (1969).

[2] Here are some of the standard references on the topic: Feferman (1977), Bell (1981), Hellman (2003), Awodey (2004), McLarty (2004, 2005), Landry and Marquis (2005), Shapiro (2005), Pedroso (2009), Linnebo and Pettigrew (2011), McLarty (2011), Shapiro (2011), Landry (2013), Marquis (2013*a*), and Makkai (2014).

I am going to present here, can capture an essential and fundamental aspect of contemporary mathematics, namely the existence of canonical maps and their pivotal role in the practice and development of mathematics, or should I say their pivotal role in mathematics, period. Moreover, not only can this be seen as a virtue of category theory, but it should also be thought as constituting a drawback of purely extensional set theories, e.g. ZF(C). I should emphasize immediately that I am *not* arguing against set theory in general, but only a specific formalization of it. In fact, sets still play a crucial role in a categorical foundational framework.[3] To be entirely clear: I will not be arguing in favor of a *particular* categorical framework in this chapter. The point I will be trying to make is simply that any foundational framework written in the language of category theory *necessarily* exhibits essential conceptual features of mathematical practice.

From a philosophical and a practical point of view, set theory has the advantage of offering an ontological unification: all mathematical objects can be defined or ought to be defined as sets. Thus, number systems, functions, relations, geometrical spaces, topological spaces, Banach spaces, groups, rings, fields, categories, etc., can all be defined as sets. This is very well known. The epistemological aspects of mathematics are supposed to be taken care of by the logical machinery. This also provides a form of unification.[4] I want to underline that the analysis of mathematics via first-order logic and set theory, which used to be called metamathematics, offers clear and important payoffs. As is also very well known, category theory itself challenges the ontological and methodological unifications provided by set theory and first-order logic. It is not so much the notion of large categories which is the problem, but rather the inescapable usage of functor categories and functors between them that raises the issue. There *are* various technical solutions to the problem and we won't discuss them here.[5] The philosophical limitations of the purely extensional set-theoretical framework are too familiar for us to mention them here.[6]

One of the advantages of category theory is that it puts morphisms on a par with objects, both ontologically and epistemologically. As such, it does not seem to differ that much from one formulation of set theory. After all, Von Neumann has given an

---

[3] How this can be is explained in Marquis (2013a) and Makkai (2014). See also Univalent Foundations Program (2013).

[4] Category theory also provides a form of unification, in fact many forms of unification. The most obvious and in some sense superficial is the fact that almost any kind of mathematical structure together with its morphisms form a category. Then, there are deeper forms of unification, forms that are revealed after serious mathematical work has been done. To mention but one example, the notion of Grothendieck topos provides a deep unification between the continuous and the discrete, a unification which leads to the development of arithmetic geometry. Our main claim in this chapter is based on a different form of unity. The various forms of unification inherent to category theory and their philosophical advantages will be explored elsewhere.

[5] There is a vast literature on the subject of large categories and set-theoretical foundations for the latter. See, for instance, Feferman (1969, 2013) and Blass (1984).

[6] See, for instance, Manders (1987), McLarty (1993), Makkai (1999), and Marquis (2013b).

axiomatic set theory based on a notion of function.[7] What does a categorical framework add to the foundational picture, apart from organizing mathematics differently?

Let me start with a loose and informal sketch. Suppose you have laid out in front of view a network of objects with multiple arrows between them. At first sight, the whole thing looks like a messy graph. The fact that stands out from the development of mathematics from the past fifty years or so is that if this graph represents a (potentially large) portion of mathematics, then not all arrows in the graph have the same role nor the same status. When we use a road map or any geographical map, there are conventions underlying the representation that allow us to see immediately which roads are highways or which portions of the map are major rivers, mountains, etc. The latter representations work because the language used for the construction of these maps contains a code that captures these elements. The point I want to make here is extremely simple: category theory, and not just its language, provides us with the proper code to represent the map of mathematical concepts. To use another metaphor: when the graph is illuminated with the proper lighting, some mappings stand out as having singular properties. It is as if one lights the given network with a blacklight which makes some of the mappings become apparent through fluorescence. The interesting thing is that this can now be reflected in the foundational domain and thus acquire a philosophical interpretation. I claim that the illuminated mappings occupy a privileged position: they provide, to use yet another metaphor, the basic routes along which concepts can be moved around. They constitute the highway system of mathematical concepts. This is already significant, but that is not all. Perhaps even more important is the fact that these basic, even elementary, roads open up the way for other conceptually important roads that are in some sense built upon them.[8] Thus, not only do they provide the roads, they also provide the frame upon which the other concepts are constructed or erected. These morphisms are usually called "canonical maps". Once the latter have been identified or recognized for what they are, morphisms that are not canonical but that are important acquire a new meaning too. To use an analogy here, one could say that in the same way that symmetries are fundamental in the sciences, broken symmetries are just as important as long as one has understood the importance of symmetry in the first place. From a set-theoretical point of view, canonical maps are merely maps like all the others. They can be defined as sets and they are not highlighted in any special manner. From a categorical point of view, the situation is quite different. The language and the concepts of category theory provide the blacklight. These maps show up as having a special character. They are singled out as being significant. Some mathematicians have decided to use them as much as possible, once they realized the role they played in the architecture of mathematics.

---

[7] A referee pointed out to me that the notion of function used by Von Neumann is not the usual notion of function as it is used in algebra, analysis, or topology. See Von Neumann (1967).

[8] Allow me to abuse the metaphor: in the same way that cities are built along rivers, mathematical theories are built along these roads.

Thus, the very practice of mathematics is modified by the conscious recognition of these maps and their status. And their status is not innocuous. Fundamental mathematical theorems rest on the existence and the properties of these morphisms. It then seems reasonable to have a way to reflect this fundamental character in a foundational framework and that is precisely what a categorical framework does.

Here is another metaphor: among all the morphisms, some are given right from the start. Mathematics is, in some sense, born with them. They come with the objects, or rather, the objects come with them. For, in some sense, they are inseparable from the objects themselves and it is hard to say which is which, that is, whether we are dealing with the objects or we are dealing with the morphisms. In fact, there is no conundrum: we are in fact dealing with both. However, there are *other* morphisms between objects. The latter are defined and used for specific purposes, whereas canonical maps are constitutive of the framework itself. Thus, canonical morphisms provide the blueprint of mathematics. It is perhaps not totally inappropriate to develop an analogy with genetics. It is unquestionable that genes play a key role in the development of an organism. But as is well known, the environment plays a key role too. It is a different role and the important point is that the same genetic pool can yield organisms with significantly different phenotypes in different environments. There is subtle form of dualism at work here and we propose that the same kind of subtle dualism is revealed by the presence and the role of canonical morphisms. One of the main points to be made is that canonical morphisms are built in a given framework. They are not an option. And once they are seen as such, mathematics develops around them in a completely organic manner. One of the main functions of category *theory* is precisely to reveal these maps and their properties. In turn, the nature and the role of these maps allow mathematicians to explain various mathematical facts.

6.2 Canonical Maps: The Folklore

In the mathematical vernacular, a canonical map f is usually said to be a map defined without any arbitrary decision. What can this possibly mean? It suggests that the definition of the map is somehow read off directly from what is given in a specific context. This is of course very rough, imprecise and approximate. We should be careful and beware of the folklore. The ability to read off directly from a context is far from being a clear and transparent notion. There are, in all cases, too much presupposed knowledge and abilities involved in the act of *reading off directly* from something.

The following operational characterization is also part of the folklore: a map f is canonical whenever any two mathematicians having to define f give the same definition. This is another attempt at capturing the idea that it is made without any arbitrary choice.[9]

---

[9] I should emphasize immediately that this has nothing to do with the axiom of choice at this point. I should also emphasize that I am focusing on canonical *maps*, but that I am not talking about morphisms of

Another, slightly different, informal characterization says that a map f is canonical if the definition of f depends exclusively on structural elements of the context in which the definition is made. As to how one determines what is a structural component or that only structural components are used in a definition, this is left unspecified. I suppose that it is assumed that professional mathematicians are taken to be able to identify the structural components of a situation.

Categories were introduced for the sole purpose of clarifying the informal notion of being 'natural' for a map. In that respect, it was entirely successful. Every mathematician now knows what it means for a map to be a natural transformation. What was an informal and vague term became a purely technical term that captured in a satisfactory manner an important and general mathematical notion. This usage, in turn, helps us see that the usage of "natural" does not coincide with the usage of "canonical":

Occasionally, we use the equal sign to mean a "canonical" isomorphism, perhaps not, strictly speaking, an equality. The word "canonical" is often used for the concept for which the word "natural" was used before category theory gave the word a precise meaning. That is, "canonical" certainly means natural when the latter has meaning, but it means more: that which might be termed "God-given." We shall make no attempt to define that concept precisely. (Thanks to Dennis Sullivan for a theological discussion in 1969.) (Bredon, 1997, vii)

Bredon makes two considerably different claims in the preceding quote. First, canonical maps should be natural, when the latter has meaning. Although I am not quite sure I understand this first claim, I believe it can be given a very plausible reading. I take it to mean that whenever it is meaningful for a canonical map to be a natural transformation, then it is one. More precisely, this says that if a given canonical map can be put in the right commutative diagram, then it should turn out to be a natural transformation in that context. If this interpretation is correct, then it is indeed a very reasonable claim. However, since we do not have a definition of what it is to be canonical for a map, I am not quite sure what to make of this claim. I submit that it should be taken as a desiderata for any characterization of the notion of being canonical for a map. Second, canonical maps might be termed "God-given". Of course, there is certainly a form of humor at work here. There is a reference to the tables given by God and it is well known that the term "canon" was also used to designate mathematical tables.[10] For the term "canonical" literally means "required by canon law" or coming from a sacred text. We would therefore be in front

---

symplectic manifolds which are also called "canonical transformations". In the latter case, the terminology refers to a completely different concept, since a symplectomorphism is a transformation of canonical *coordinates*. Finally, I am not considering other usages of the word "canonical" in mathematics and logic, for instance when one talks about canonical models. I will say more about the historical usage of the terms "canon" and "canonical" involved in Sections 6.3 and 6.4.

[10] For instance, Napier's book of logarithmic tables was called *Mirifici Logarithmorum Canonis Descriptio* in Latin. I thank Daniel Otero for this remark. Another example is provided by Jacobi's book entitled *Canon Arithmeticus*, which is a book of tables of numbers n and i such that, for each prime $p \leq 1000$ and

of a piece of mathematics that comes from "The Book". Of course, there is a wink towards Kronecker's infamous statements about the natural numbers and Erdös' ideas that some proofs are taken directly from God's book of mathematical proofs (see, for instance, Aigner and Ziegler, 2010). There is nonetheless room for interpretation. Here are a few obvious options: canonical maps are inherent to mathematics; they are completely independent of human choices and representations; we discover them. These are all possible interpretations. Another important connotation to the term is that it serves as a standard or a reference in a given field, but one that is not arbitrary. This usage is common in aesthetics where one finds the expression "canon of beauty".[11] I will come back to this very important dimension at the end of the chapter. I first must establish that some maps, which mathematicians dub "canonical", have a very special property that separates them from other maps. Furthermore, we must try to see what it is about these maps that one is inclined to qualify them as being God-given.

One of the claims I want to make here is that some of these canonical maps arise spontaneously when mathematics is developed in a categorical framework. They are inevitable. They are omnipresent. And they are significant. In a very nice paper on the notion of equality in mathematics, the mathematician Barry Mazur makes a similar claim:[12]

> One of the templates of modern mathematics, category theory, offers its own formulation of *equivalence* as opposed to *equality*, the spirit of category theory allows us to be content to determine a mathematical object, as one says in the language of that theory, *up to canonical isomorphism*. The categorical viewpoint is, however, more than merely "content" with the inevitability that any particular mathematical object tends to come to us along with the contingent scaffolding of the specific way in which it is presented to us, but has this inevitability built in to its vocabulary, and in an elegant way, makes profound use of this.
>
> (Mazur, 2008, 2–3)

As the quote from Brendon's book indicates, even in the context of category theory, canonical maps are not necessarily the same as natural transformations. Some natural transformations are not canonical. In fact, one often sees the expression "a canonical natural isomorphism" in the literature, emphasizing the fact that the natural isomorphism *is* canonical. Not all canonical maps are natural transformations. Some functors are canonical. Furthermore, there may be canonical maps that appear elsewhere or escape the categorical framework. This latter possibility does not affect my claim. The claim that all canonical maps, informally understood, are captured by

---

$1 \leq n \leq p - 1$ and $1 \leq i \leq p - 1$, $n = \rho^i$, for some primitive root ρ modulo p. The first such tables were published by Gauss in his *Disquisitiones*.

[11] There is some interesting circularity going on here, since even the Greek sculptor Polykeitos linked his canon to mathematical proportions and symmetry. There may be more links between being canonical in this sense and the nature of beauty in mathematics, but we leave this topic for another paper.

[12] Mazur's paper goes in the same direction as ours. I unfortunately read his paper long after my paper was finished, but I could not refrain from adding a few elements from his analysis in the final version.

the categorical language is a thesis akin to the Church–Turing Thesis and I do not want to make it here.

The explicit presence of canonical maps is certainly one of the reasons why one would want to develop mathematics in a categorical framework and, in fact, one of the main reasons why one would want to develop a foundational framework in categorical terms. The purpose of this paper is to explore the nature and status of these canonical maps and, from there, establish their significance in mathematics in general. Once the latter is granted, one can use these facts to argue that a foundational framework ought to reflect the presence and the role of these canonical maps. We will try to provide some of the arguments at the end of this paper. Thus, our plan here is simple: we will first sketch how, from the historical point of view, the expression "canonical map" arose naturally in a categorical framework and was used extensively by certain mathematicians (but not all!), then give some non-trivial examples of these maps and their roles in certain proofs, move to their presence in a foundational framework, and end with some speculation on the links between canonical maps and mathematical cognition.

6.3 Examples

Before going any further, I will give four simple examples of maps considered to be canonical by mathematicians. Other examples will be given in a later section. I simply want to provide four rudimentary illustrations immediately to set up the stage.

What seems to be the very first map to have been called "canonical" is the usual quotient map. This construction appears everywhere in mathematics, from topology to algebra. In the set-theoretical framework, one starts with a set X and an equivalence relation $R \subset X \times X$ on X. An equivalence set is defined as usual to be $[x] =_{df} \{y : R(x, y)\}$ and we can now consider the set of equivalence sets $X/R =_{df} \{[x] : x \in X\}$. There is then a special map, called the quotient map, $\pi : X \to X/R$ defined by $\pi(x) = [x]$. This map is always called canonical. One could say that it is the canonical canonical map. In a way, if you follow every step of the construction, the language itself tells you how to define it. There are other maps from a set to a quotient of that set by an equivalence relation, but π is, in some sense, given by the construction itself. There is no choice being made. The notation itself takes care of it. Whenever there are more data involved, e.g. when X is a topological space or a group, the quotient map has the "right" properties; i.e. it is continuous, a homomorphism, etc. A very well-known special case of this construction is the canonical map $\mathbb{N} \rightarrowtail \mathbb{Z}$ defined by $n \to [(n, 0)]$ with the usual equivalence relation.

Although the preceding example is given in set-theoretical terms, it turns out to be but one more example of what we get when we move to the categorical context. As I said, category theory shows immediately what the foregoing map has in common with other cases of canonical maps. I assume the reader knows the basic constructions of category theory, e.g. products, pullbacks, equalizers, terminal object, and limits in general, their dual as well as the notions of functors, natural transformations, and natural isomorphisms.

Let C be a category with pullbacks and binary products. Consider the following pullback diagram:

Then, it is easy to show that there is a canonical *monomorphism* $h : P \rightarrowtail X \times Y$, where $X \times Y$ is the product of X and Y. Category theory tells us why. The product $X \times Y$ comes with two maps and a universal property. This is how the notion of product is *defined* in category theory. The two maps inherent to a product are the so-called *projections* $p_X : X \times Y \to X$ and $p_Y : X \times Y \to Y$. They are also called the *canonical projections* in the literature. Thus, when described in category-theoretical terms, the notion of a product automatically comes with canonical maps.

The universal property is expressed as follows: given an object T *together with* two maps $f : T \to X$ and $g : T \to Y$, then there is *a unique* map $u : T \to X \times Y$ such that $p_X \circ u = f$ and $p_Y \circ u = g$. In turn, these data can be represented by the following commutative diagram:

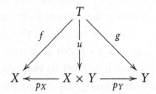

With these information at hand, the reader can now see why and how the foregoing map h automatically shows up in the commutative square above. The fact that h is a monomorphism also follows directly from the data. The map h and the fact that it is a monomorphism are in a precise sense given. There is absolutely no choice made. If this is a pullback square and if the category C has binary products, then h exists and it is uniquely defined by the data. Of course, as usual, there are many other monomorphisms in general between these objects, but the map h depends entirely on specific features of the situation.[13]

---

[13] An interesting variation on this situation is this: suppose there is a monomorphism $Z \rightarrowtail T$ and that, as earlier, P is the pullback of $X \xrightarrow{e} Z \xleftarrow{m} Y$ and that $X \times_T Y$ is the obvious other pullback; then it is easy to show that there is a canonical isomorphism $P \to X \times_T Y$.

The quotient presented above can be given a similar presentation. It becomes a special case of what is called a *coequalizer* in category theory (see Borceux, 1994, 49–50).

The third example is slightly more involved and corresponds to familiar mathematical situations. Let C by a category with finite products, finite coproducts, and a null object 0, which amounts to the existence of a null morphism $0 : X \to Y$ between any two objects of C, defined by the composition of morphisms $X \to 0 \to Y$. Note that the null morphism is canonical, but that is not the example we want to give. For, in this case there is a canonical map

$$X_1 \sqcup \cdots \sqcup X_n \to X_1 \times \cdots \times X_n.$$

Again, this map comes for free. It is there as soon as the category C has the right categorical properties. It can then be used to prove other properties, construct other morphisms, perhaps specific to a particular category, e.g. the category of abelian groups.

Let me finish with a slightly different example, namely the example of a canonical functor. Let (X, x_0) be a pointed topological space, namely a topological space X with a specified point x_0 and let (S^1, p) denote the unit circle with the usual topology with the point p being the north pole. Then, there is a functor $\pi_1 :$ **Top** $\to$ **Grp** from the category of topological spaces and continuous functions to the category of groups with group homomorphisms defined by $(X, x_0) \mapsto Hom((S^1, p), (X, x_0))$, which is the fundamental group of the pointed space. It follows directly from general facts that it *is* a functor. The fact that it is a group also follows from underlying general facts.

These maps appear for purely general reasons in the context of category theory. It is tempting to say that they appear for structural reasons. But since we do not have a general theory of structures and it is hard to pinpoint exactly what is meant by "structural", I will not use this terminology for the moment. Although it is certainly not obvious from this very short discussion, the reader must take our word for it when we claim that these maps are literally everywhere in mathematics. As soon as someone needs to make various constructions in a mathematical context, these maps show up when the proper theoretical tools are used adequately. They are built in the constructions in the categorical language.

6.4 Canonical Maps: Brief Historical Remarks

Canonical maps permeate category theory. However, the terminology does not originate with the field and it took some time before it became accepted and commonly used.[14] In the categorical context, one observes that the term was not used by the

---

[14] As we have already alluded to in preceding footnotes, the term "canonical" has been used in different contexts in mathematics, for instance to circumscribe certain matrix forms in the nineteenth century or curves in geometry also at the same time. As far as I can tell, the term "canonical matrix" was introduced

American school of category theorist in the 1950s and the early 1960s and that it clearly was under the influence of the French school, more specifically Grothendieck's school who used category theory to develop new foundations of algebraic geometry that the terminology acquired a clear status. Here is a sketch of this development. A full story will have to be told elsewhere.

Although Eilenberg and Mac Lane's goal was to give a precise general theory of natural equivalences—or as we now say "natural isomorphisms"—and natural transformations, along the way they stumbled upon examples of canonical maps, but did not give them that name, nor did they underline their presence. For some natural isomorphisms are certainly canonical maps. The very first example given to motivate their whole paper about general equivalences is a canonical map: it is the canonical natural isomorphism between a finite-dimensional vector space V and double its dual V^{**}. Interestingly enough, Eilenberg and Mac Lane do emphasize what could be taken as being one difference between a natural transformation and a canonical map. The first part expresses what we have presented as being now part of the folklore:

For the iterated conjugate space $T(T(L))$, on the other hand, it is well known that one can exhibit an isomorphism between L and $T(T(L))$ *without* using any special basis in L.
(Eilenberg and Mac Lane, 1945, 232)

This is a well-known fact, but extremely important conceptually. There *is* an isomorphism between a finite-dimensional vector space V and its dual space V^*, but in order to define it, one has to fix a basis in V. In fact, there are many of them. But when one moves to the double dual V^{**}, the isomorphism is defined independently of the chosen basis. This is a remarkable fact: one does not have to *choose* a particular basis. The map is given by the construction of V^{**} (their $T(T(L))$) and its links to V. The foregoing passage then shifts to the specific property of a natural transformation:

This exhibition of the isomorphism $L \cong T(T(L))$ is "natural" in that it is given *simultaneously* for *all* finite-dimensional vector spaces L.   (Eilenberg and Mac Lane, 1945, 232)

And there lies one difference between the two concepts. What the technical notion of natural transformation captures is the fact that the transformation is given *uniformly* for *all* the objects of the category and it is a map between functors. Naturality is a global phenomenon, so to speak, whereas canonicity need not be general in that sense. Furthermore, a natural transformation is not necessarily canonical.[15] This is, however, a nice example of a canonical map that satisfied the desiderata mentioned

---

by Sylvester in the mid-nineteen century, following apparently a suggestion made by Hermite. As I have indicated, the terminology was used earlier in various contexts. It would be interesting to explore its history and usage by different mathematicians and philosophers, for instance in Descartes and Leibniz.

[15] There is at least one paper whose title is revealing: "Non-canonical Isomorphisms". One has to be careful, for in this case, the author, the category theorist Steve Lack, equates being canonical in a categorical framework with being a universal morphism. See Lack (2012).

in Section 6.2: it is a canonical map that can be put into a commutative diagram expressing the property of being a natural transformation and it is one.

Various constructions presented in their 1942 paper and in their 1945 paper include what they call "induced maps" and the latter are canonical.

The terminology simply does not show up in the American school in the 1950s and early 1960s. The word appears in Mac Lane's paper on duality for groups published in 1950 to designate the map from a group to its quotient by a normal subgroup and the decomposition of a group homomorphism into an epimorphism and a monomorphism. There is nothing in Kan's paper on adjoint functors in 1958 and the accompanying paper. There is no occurrence of the word in Eckmann and Hilton's paper in 1962 and no occurrence of the term in Freyd's book *Abelian Categories* published in 1964. In Mac Lane's paper on categorical algebra, published in 1965, one finds twenty-one occurrences of the word "canonical". They are not all used to denote maps and sometimes it designates the same map, but it is nonetheless a sign that the vocabulary was getting established in the English-speaking community. It is somewhat surprising to observe that the terminology is not used in Freyd's *Aspects of Topoi* in 1972.

Things change radically when one moves to the French school of category theory, in particular Grothendieck's school of algebraic geometry.

The origin of the terminology clearly goes back to Bourbaki. It was probably intended whimsically at first, typical of Bourbaki's humor. As far as I can tell, it appears in Weil's version of Bourbaki's book on set theory, where one finds the notion of a "canonical correspondence" between two sets: given two sets X and Y, there is a canonical map $f : X \times Y \longrightarrow Y \times X$. In his thesis published in 1951, Serre uses the term twenty-four times, not only to denote the usual quotient map, but many isomorphisms as well.

Given that Grothendieck was a member of Bourbaki in the 1950s and given the influence of Serre on Grothendieck, it is no surprise to find Grothendieck use the term extensively in his work. Indeed, the terminology of canonical maps is omnipresent in Grothendieck's writings. It is already present in his first writings in functional analysis. It is, however, less predominant than in his subsequent writings developed in a categorical context. Those who have read the *Éléments de géométrie algébrique* will have seen the word "canonical" sometimes more than ten times in a single page![16] The omnipresence of the terminology in Grothendieck's work and thereafter is certainly a significant fact. Grothendieck systematically searches for and uses canonical maps in his work. But it is not so much the terminology that is interesting. It is what it reveals about a certain class of morphisms and their role in the mathematical landscape.

---

[16] It would be interesting to explore the number of times the expression appears in the works of various mathematicians, especially those who use category theory. For instance, it appears often in Jacob Lurie's work, no less than 749 times in his book *Higher Algebra* and 139 times in his book *Higher Topos Theory*.

6.5 Canonical Maps: Taking Stock

Our claim is that category theory allows us to identify certain morphisms as being canonical. As we saw in Section 6.3, when one introduces certain constructions in category theory, the constructed objects *always* come equipped with canonical maps. The general notions involved are those of limits and colimits of a diagram in a category. I will give the general definition of a limit of a diagram J in a category C. This will allows us to circumscribe more precisely what we want to include in our characterization of canonical maps.

Let J be a small category and C be a category. A *diagram* of shape J in a category C is a functor $\delta : J \to C$.[17] Informally, a diagram of shape J can be pictured as being composed of objects $\delta(X)$ of C together with morphisms $\delta(X) \to \delta(Y)$ between these objects. A *cone over the diagram* is an object c of C together with morphisms $p_X : c \to \delta(X)$ for each X in J such that for each morphism $i : X \to Y$ in J, $\delta(i) \circ p_X = p_Y$.

A *limit*, denoted by $\varprojlim J$ or simply by $\lim J$, for the diagram J is a *universal cone*. Thus, it is a cone over the given diagram. As such, and this point is fundamental for us, it comes with morphisms $\pi_X : \varprojlim J \to \delta(X)$. Second, a limit satisfies the following universal property: given any cone c over the diagram, there is a unique morphism $u : c \to \varprojlim J$ such that $\pi_X \circ u = p_X$ for all objects X in J.

We should note that a limit for a diagram is unique up to a unique isomorphism. This means that, given a diagram of shape J in C, if both L_1 and L_2 are limits for this diagram, then there is a unique isomorphism between them. This fact follows immediately from the definition of a limit.

The notion of a *colimit* is the dual of the notion of a limit. Thus, everything we will say about limits and their canonical morphisms applies to colimits and their canonical morphisms.

We are now in a position to circumscribe more precisely what we want to include in the notion of canonical morphisms or maps.

1. Morphisms that are part of the data of a limit are canonical morphisms; for instance, the projection morphisms that are part of the notion of a product;
2. The unique morphism from a cone to a limit determined by a universal property is a canonical morphism: and
3. In particular, the unique isomorphism that arise between two candidates for a limit is a canonical morphism.

Of course, this is not a definition of the notion of canonical morphism. I will not even attempt to give such a definition. I do not need to. My point is simply this: no matter how one defines the notion of canonical morphism, the proposed definition will have to include these cases as generic instances.

---

[17] I am being a bit loose here. The informed reader knows how to fix the definition. To the uninformed reader, the differences are uninformative.

One important remark about my presentation via the notion of limit. There are well-known alternatives. Grothendieck used the notion of representable functors to capture these cases systematically.[18] We could have used the notion of adjoint functors. The fact that these different approaches are all equivalent is another important result of category theory.[19]

We are now ready to turn to more philosophical issues.

6.6 Canonical Maps, Category Theory and the Architectonic of Mathematics

By an **architectonic** I understand the art of systems. Since systematic unity is that which makes ordinary cognition into science, i.e., makes a system out of a mere aggregate of it, architectonic is the doctrine of that which is scientific in our cognition in general, and therefore necessarily belongs to the doctrine of method.

(Kant, *Critique of Pure Reason*, A 832/B 860, translation Guyer and Wood).

By using the term "architectonic", I want to emphasize the fact that mathematical concepts as developed in a categorical framework form a system whose nature is different from the systems we are accustomed to.[20] And I claim that this new architectonic has philosophical traction. I won't be able to develop these claims as well as they should in such a short chapter. Needless to say, many concepts and claims must be clarified and a lot of background material must be assimilated.[21] The challenge is, at the very least, threefold: (1) to explicate precisely the *kind* of system I have in mind; (2) to say how that kind differs from other kinds of systems; (3) what makes that particular kind of system philosophically relevant. Given that the type of architectonic I am presenting depends upon the presence of canonical morphisms, I will call it "the Canonical Architectonic" from now on. I will briefly develop these three challenges in turn.

6.6.1 *The Canonical Architectonic: what kind of underlying principles?*

Let me start with what most people have in mind when they think about category theory and the system of mathematical concepts. It is well known that mathematical concepts, like the notions of monoids, groups, rings, fields, topological spaces, metric spaces, vector spaces, and partial orders, all form categories. Thus, one has the category

---

[18] This explains in part why the Yoneda lemma and the Yoneda embedding are central in his work.

[19] David Ellerman has recently suggested a slightly more general notion, which he calls an heteromorphic theory of adjoint functors, and that he has applied to the life sciences. See Ellerman (2007a, b, 2015). Be that as it may, the central concept is clearly the notion of a universal morphism.

[20] The term "architectonic" has a rich and complex history. Although usually associated with Kant, as I do here, it was already in usage before him and can be traced back to Aristotle. I could and perhaps should have added a section on the very notion of an "architectonic", since there are some very interesting, surprising, and rich claims associated with the term that resonate with what I am trying to do here. Suffice it to mention that the term was used by Baumgarten as a synonym for an ontology, a *scientia architectonica*. For more on the history of the term, see, for instance, Manchester (2003, 2008).

[21] For some of the background material, I refer the reader to Krömer (2007) and Marquis (2006, 2009).

of groups and group homomorphisms, the category of topological spaces and continuous maps, the category sets and functions, etc. This is the usual understanding of what mathematicians have in mind when they think about the organizational power of category theory. Mathematics is thus presented as a system of *concrete* categories with functors between them (and natural transformations between the latter).[22] For most mathematicians and logicians, that is all there is to it. It is not that this picture is false nor that it is useless to mathematicians. It does some work, but it is philosophically completely uninteresting and it is mathematically and from a foundational point of view radically incomplete. The unifying principle underlying this view of category theory in mathematics and its foundations is simply the fact that these notions and their morphisms satisfy the axioms of a category and, by doing so, it allows for a certain organization of mathematics. Furthermore, this organization is seen to be at a very high, algebraic level, useful for the mathematician doing algebraic geometry, algebraic topology, or what have you, but completely useless for foundational purposes and philosophical analyses.[23] The standpoint in this case is that category theory and its concepts are *applied* to various situations in fruitful ways. These successes are seen, so to speak, after the fact and it is as if one would not have good theoretical reasons to develop mathematics in categories in the first place. This situation is general in philosophy of mathematics and was already put forward by Ken Manders more than twenty-five years ago:

> suppose we could attribute "fruitfulness-engendering" structural properties to ways of treating a subject matter, . . . , instead of judging the outcome of attempts to treat the subject after the fact. Then there could be reasons for success and failure: x succeeds, because x has α and having α is inherently fruitful. Then there would be a normative fact of the matter, with theoretical rather than just practical standing, whether a treatment was correct, preferable, or appropriate, based on intrinsic merits, and regardless of attempts to carry it out.   (Manders, 1989, 554)

The problem with the foregoing view of category theory as applied mathematics is that it misses crucial elements of the picture, in particular the principles underlying the new unity. In accordance with Kant's idea, the use of the term "architectonic" is justified inasmuch as the system is developed according to certain principles and, in some sense, in a coherent or cohesive fashion.[24] The latter requirement, interpreted appropriately, is central to our claim.

In the foregoing picture, categories are introduced *after* some given structures are provided by other means, usually by an axiomatic presentation in first-order logic interpreted in the universe of sets. In the picture I am proposing, the underlying

---

[22] A *concrete* category is usually thought of as being a category of structured sets of some sort. A more technical definition was provided by Freyd (1970, 1973) where a concrete category is a category that has a full and faithful functor into the category of sets. Freyd even gives necessary and sufficient conditions for a category to be concrete.

[23] And we are back to the opening paragraph of this paper . . .

[24] Kant would say *a priori*. This is precisely what we want to claim here, but the term *a priori* is philosophically so charged that I hesitate to use it explicitly.

principles *precede* particular structures and specific categories.[25] Again, I am not arguing in favor of one of singular, complete, technical development of the category of categories, but rather I am saying that the presence of canonical morphisms ought to be included in any such picture, since they already provide the underlying new principle.[26]

By an architectonic, I informally understand a global plan of a domain, a plan whose organization is given by underlying principles such that (1) the unity of the whole is given by these underlying principles; and (2) the whole is logically prior to its parts. An architectonic in this sense has to be opposed to an agglomeration, a whole whose parts are not linked to one another by underlying principles.[27] Thus, canonical morphisms provide the architectonic of mathematics.[28]

> Human reason is by nature architectonic, i.e., it considers all cognitions as belonging to a possible system, and hence it permits only such principles as at least do not render an intended cognition incapable of standing together with others in some system or other.
>
> (Kant, *Critique of Pure Reason*, A474, translation Guyer and Wood)

Canonical morphisms are not *merely* morphisms that connect various "objects" together. In the three types that we have identified in this chapter, the nature of the connections is revealing. The first group of morphisms, those that are constitutive of the notions of limits or colimits, already gives us an underlying network between concepts of a certain sort. Thus, by stipulating that a certain category has limits or colimits of a certain kind, one immediately specifies an underling system of concepts that determines a global network.

It is, however, the second kind of canonical morphisms that is even more important. The universal property satisfied by limits and colimits is attached to a universal and canonical morphism. This says that limits and colimits occupy a privileged position in the given system of concepts. By stipulating that one is working in a category with these limits or colimits, one automatically knows that there are routes in the network that have a special status. By travelling through these routes, one obtains important and often crucial information.

---

[25] It is in this sense that the underlying principles are *a priori*. They do not depend on any previous formal system. They might be explained by informal concepts and pictures, as I am trying to do here.

[26] I will nonetheless say some things about such technical developments in the conclusion of this chapter.

[27] I believe that this very rough and informal definition is consistent with the traditional use of the term, particularly in the eighteenth century.

[28] Bourbaki explicitly wrote about the *architecture* of mathematics and talked about three mother structures: algebraic structures, order structures and topological structures. See Bourbaki (1950). Bourbaki's paper is fascinating and it constitutes a clear and concise presentation of what mathematicians understood and, to a large extent, still understand by the axiomatic method and how it unifies mathematics. It should be clear that I am going in a different direction altogether. I am not claiming that there are "mother structures". I am claiming that there is a global architectonic of mathematical concepts, a global conceptuel unity revealed by canonical morphisms. The term has also been used by Balzer and Moulines in the title of their book about the structuralist program in philosophy of science. See Balzer et al. (1987). Although there might be links to their program, I am not making a structuralist claim in this paper.

Often in mathematics, one has to prove that two independently given constructions are in fact isomorphic or equivalent in an appropriate sense. When the constructions are done in a categorical framework, they often come with canonical maps and the isomorphism or the equivalence that one is searching for can sometimes fall automatically from these canonical maps.[29] In other cases, the canonical maps provide the starting steps in the construction and the required isomorphism or equivalence amounts to putting together the given canonical maps together with specific maps inherent to the situation at hand. This is just not handy and useful for a mathematician. One uses all the given resources at hand, all the underlying capacities of the situation.

The last type of canonical morphisms simply reinforces the second point. The unique isomorphism between two limits for the same diagram of a certain shape can be thought of as an identity. This last fact ought to be interpreted as a special case of Leibniz's principle of identity: if $P(X)$ and $X \simeq Y$, then $P(Y)$, where X and Y are objects in a given category $\mathcal{C}$ and $P(-)$ is a property of the theory.[30]

The canonical architectonic arises from the intrinsic properties of the system. Clearly, it is an *abstract* system and I claim that it is precisely this abstract character that makes it philosophically interesting *and* problematic.

6.6.2 The Canonical Architectonic: how does it differ from other systems?

Mathematics constitutes an organic whole of concepts, methods and theories crossing each other and interdependent in such a way that any attempt to isolate one branch, even for the sake of providing a foundation for the whole body, would be in vain; nor would it make sense to break this organic solidarity by looking for an external foundation (be it natural science or logic or some psychological reality).   (Cavaillès quoted in Drossos (2006, 104))

Cavaillès underlines the *organic* whole of concepts, methods, and theories and their interdependence. One could argue that Kant had a similar view of scientific knowledge when he decided to characterize it as being architectonic. As we have already indicated, it is precisely this specific aspect of the canonical architectonic that I want to put forward. Some would like to talk about its holistic nature. I prefer to avoid this terminology altogether. It is systematic and the system forms a whole in the same way that an organism forms a whole.

As I have already indicated in the opening section of this chapter, set theory and first-order logic, taken together, already provide a unity to mathematical knowledge. What are the underlying principles of this unity? There is no need to elaborate this

---

[29] Some sort of epistemological ideal can be drawn here: a canonical proof would be a proof in which all the steps or the important steps are given by the existence of canonical morphisms. The interest in such a proof is that, in some sense, the proof would follow directly and automatically from the data involved and without any choice. One could argue that Grothendieck was always striving for such proofs and that all the main notions he developed were more or less introduced with this goal in mind.

[30] We are in this particular case dealing with a *unique* isomorphism. It should be noted that the principle ought to apply to cases where there are many isomorphisms or equivalences between given objects. This is a topic in itself that will certainly be addressed in this book by other authors.

picture extensively. The axioms of set theory, say Zermelo-Fraenkel theory with choice (ZFC), when interpreted in the cumulative hierarchy, are the fundamental principles and are supposed to be self-evident, at least with respect to the cumulative hierarchy.[31] However, the system is in this case purely deductive. Logic provides the links between various mathematical propositions. The epistemological standpoint is the usual foundationalist position. The gain is a fine-grained look at justification of propositions. On the practical side, axioms are given for various structures. They are interpreted in the universe of sets. But the latter does not reflect the various relationships between the concepts, e.g. the numerous links between topological spaces, groups, more generally modules.

Conceptual unification, understandability, clarity, even length of proofs, fall outside the narrow justificational concern of foundational epistemology. That concern, we should now see with the benefit of a hundred years' hindsight, fails to capture the intellectual enterprise of mathematics. Any dispassionate look at mathematical sciences should teach that the "mathematical way of knowing" seeks as much to render things understandable as it does to establish theorems or avoid error. The process of establishing deductive relations is subsidiary to the larger goal of rendering understandable. (Manders, 1989, 562)

To put it bluntly: the universe of sets is conceptually flat, but combinatorially rich. I propose we look at the dual picture.

In the picture proposed here, mathematics is developed within a universe of categories. The usual mathematical concepts, e.g. groups and topological spaces, can be defined in that universe.[32] Canonical morphisms are built in that universe and can be used to develop various theories and connections between them. This is what is usually and unfortunately called "pure category theory". In the latter, one finds theorems about additive categories, abelian categories, triangulated categories, exact categories, regular categories, toposes, monoidal categories, braided categories, Quillen model categories, Waldhausen categories, etc., to mention but the most well known.

Results in these "pure" categories are seldom by themselves equivalent to what mathematicians are directly interested in, although, in some cases, they are.[33] However, one can then bring in what was seen as the "applied" dimension of categories, e.g. more specific structures or data, required by the problem or theory at hand. Thus, the applied dimension finds its full expression when the built-in morphisms have been identified and used.

We end up with a picture of mathematics that is not quite standard: it is a new form of dualism. There is the canonical part of mathematics which seems entirely written in the concepts themselves and various contraptions that we invent and manipulate for more specific needs and purposes. From the point of view of mathematicians, it is a methodological dualism and it has been developed and propounded by category

---

[31] The standard presentation of this position can be found in Shoenfield (1977). Historically, the situation is considerably more delicate and intricate. See, for instance, Ferreirós (2007) and Moore (2013).

[32] See, for instance, Pedicchio (2003).

[33] For instance, many duality theorems find their natural expressions at this level.

theorists from the 1960s onwards. The strategy is to used the canonical morphism to obtain as much information as possible at the most abstract level and, then, add the specific, particular, or concrete ingredients of the situation one is interested in to prove the wanted results.

Even though this is a form of dualism, it is not a form of platonism in the standard sense of that expression.

6.6.3 The Canonical Architectonic: why is it philosophically relevant?

The philosophical content of the presence of canonical morphisms is multifaceted and deep.

First, to work at the level of canonical morphisms is to work at an abstract level. One uses only the properties of canonical morphisms to obtain various results. For instance, when a mathematician obtains a result in a topos using purely topos theoretical properties, it is certainly not yet a result in, say, algebraic geometry. One could wonder what is the gain to work this way. For one thing, it is similar to the gain associated with the axiomatic method and that led Bourbaki to talk about the architecture of mathematics. Whereas Bourbaki thought that there were three fundamental mathematical structures, I see a global organization of mathematical concepts deriving from the presence of canonical morphisms. The analogy with an organism seems to be apt and adequate. The unity of an organism depends upon the presence of various organs that are organized into a system, that is, relationships between these organs. Categories defined via the existence of canonical morphisms roughly correspond to these organs and results obtained from properties of these canonical morphisms correspond to the relationships. An architectonic provides us with a plan of mathematical knowledge and an understanding of its fundamental organization.

Second, results proved in categories defined via canonical morphisms are valid in all applications where these morphisms exist. This provides a vast unification of mathematical concepts and results. Associated with this new form of unification, one finds also a simplification of various results. This might appear paradoxical, since category theory is usually seen as being difficult and complicated. To wit:

Any relation in the language of derived categories and functors gives rise to assertions formulated in the more traditional language of cohomology groups, filtrations, spectral sequences.... Of course, these can frequently be proved without explicitly mentioning derived categories so that we may wonder why we should make the effort of using this more abstract language. The answer is that the simplicity of the phenomena, hidden by the notation in the old language, is clearly apparent in the new one. The example of the Kenneth relations... serves to illustrate this point. (Keller, 1996, 672)

Or, again:

The first part of the paper, on which everything else depends, may perhaps look a little frightening because of the abstract language that it uses throughout. This is unfortunate, but there is no way out. It is not the purpose of the abstract language to strive for great generality.

The purpose is rather to simplify proofs, and indeed to make some proofs understandable at all. The reader is invited to run the following test: take theorem 2.2.1 (this is about the worst case), translate the complete proof into not using the abstract language, and then try to communicate it to somebody else. (Waldhausen, 1985, 318)

Note the explicit claim made by Waldhausen that the usage of the categorical language is not to strive for great generality. One too often reads that this is the reason why some mathematicians use category theory.[34] These two quotes put to the fore two very different reasons: simplicity and understandability. In both cases, canonical morphisms play a key role.

Third, one of the indications of the fundamental nature of canonical morphisms is the fact that basic logical rules are captured in a categorical framework by canonical morphisms. It is a striking fact that basic logical rules arise from elementary canonical morphisms via adjointness. Thus, one merely needs some rudimentary canonical morphisms and the notion of adjoint situation for the logical rules to organize themselves systematically into deductive systems. The pervasiveness of adjoint situations and the unificatory power of the concept came after the fact. It was an empirical discovery. It is, literally, an awesome discovery. It is hard to imagine the unifying power of the concept of adjoint situation, thus of canonical morphisms. The diversity of the situations covered by adjointness is simply baffling. Once it is understood and once one realizes what it yields, it is hard not to conclude that the notion of adjoint situation has a particular status, some sort of *a priori* character to it. It even became a prescription in the field in the 1960s and 1970s: look for adjoints. In this sense, canonical morphisms provide direction for reason, for the development of mathematics. This prescription remains entirely relevant today.

Let us come back to Bredon's quote. It is now common in textbooks on algebraic geometry to replace *canonical* isomorphisms by a sign of equality. This might seem innocuous, but I believe that it indicates an important cognitive aspect of the situation.

Canonical morphisms are interesting only when they arise systematically from a theoretical framework. Otherwise, they would not deserve to be called architectonic. Once they are seen as such, the cognitive gain is substantial. The introduction of the arrow notation $X \to Y$ acquires all its power once the presence of canonical morphisms has been recognized and is used explicitly. This fact must be underlined too. It was already emphasized by Eilenberg and Steenrod in their book on the foundations of algebraic topology.

Certain diagrams occur repeatedly in whole or as parts of others. Once the abstract properties of such a diagram have been established, they apply each time it recurs.

---

[34] This was mentioned again and again in the homages to Grothendieck after his death in November 2014. I think that these quotes reflect more adequately Grothendieck's motivations underlying his use of category theory.

The diagrams incorporate a large amount of information. Their use provides extensive savings in space and in mental effort. In the case of many theorems, the setting up of the correct diagram is the major part of the proof. (Eilenberg, 1952, xi)

Not only the new notation, the explicit mark denoting an arrow, and more specifically a canonical map, provides a substantial cognitive gain, but, as is often the case with a good notational system, it allows for the creation of new objects and new frameworks. This is particularly clear when one looks at monoidal categories and their applications.[35]

6.7 Conclusion

It is not possible to end without pointing towards higher-dimensional category theory. As I am writing this, there is still no adopted definition of higher-dimensional categories. There is, however, a reasonably good understanding of $(\infty, 1)$-categories via various models, for instance quasi-categories, although a full-fledged theory of $(\infty, 1)$-categories, which should form a $(\infty, 2)$-category, is still to be given.[36] There is no need to go in the technical details of what these objects are. A fundamental point can nonetheless be made already. The *language* and the *core* concepts of category theory can be used *as they are* for quasi-categories. The computations are far from trivial, but one can show that the main concepts and theorems of category theory still hold for quasi-categories. Thus, and this is the moral I want to make, the notion of canonical morphism is directly lifted to this new context.

It is clear that, no matter what the universe of categories will look like, if it is taken as a foundational framework, then canonical morphisms will play a central role. And that is all we need for our claims to hold.

Acknowledgments

The author gratefully acknowledges the financial support of the SSHRC of Canada while this work was done.

References

Aigner, M., and Ziegler, G. M. (2010). *Proofs from The Book*, fourth edn. Springer-Verlag, Berlin. http://dx.doi.org/10.1007/978-3-642-00856-6

Awodey, S. (2004). An answer to G. Hellman's question: "Does category theory provide a framework for mathematical structuralism?" [*Philosophia Mathematica (3) 11 (2003)*,

---

[35] For a nice example, I invite the reader to look at Street (2007).

[36] See, for instance, Joyal (2002), Lurie (2009), and Cisinski (2015). A full theory of higher-dimensional categories will be a theory of ∞-categories. The particular feature of $(\infty, 1)$-categories is that morphisms of dimension strictly greater than 1 are invertible.

no. 2, 129–57; mr1980559], *Philosophia Mathematica (3)* 12(1), 54–64. http://dx.doi.org/10.1093/philmat/12.1.54

Balzer, W., Moulines, C. U., and Sneed, J. D. (1987). *An Architectonic for Science: The Structuralist Program*. Reidel, Dordrecht.

Bell, J. L. (1981). Category theory and the foundations of mathematics. *British Journal of Philosophy of Science* 32(4), 349–58. http://dx.doi.org/10.1093/bjps/32.4.349

Blass, A. (1984). The interaction between category theory and set theory, in *Contemporary Mathematics*, 5–29. American Mathematical Society, Providence, RI.

Borceux, F. (1994). *Handbook of Categorical Algebra. 1*, Vol. 50 of *Encyclopedia of Mathematics and Its Applications*. Cambridge University Press, Cambridge, UK. Basic category theory.

Bourbaki, N. (1950). The architecture of mathematics. *American Mathematical Monthly* 57, 221–32.

Bredon, G. E. (1997). *Sheaf Theory*, Vol. 170 of Graduate Texts in Mathematics, second edn. Springer-Verlag, New York. http://dx.doi.org/10.1007/978-1-4612-0647-7

Cisinski, D.-C. (2015). Catégories supérieures et théorie des topos. *Séminaire Bourbaki* (1097), 1–58.

Drossos, C. A. (2006). Sets, categories and structuralism, in G. Sica (ed.), *What is Category Theory?* Polimetrica, Milan.

Eilenberg, S., and Mac Lane, S. (1945). General theory of natural equivalences. *Transactions of the American Mathematical Society* 58(2), 231.

Eilenberg, S., and Steenrod, N. (1952). *Foundations of Algebraic Topology*. Princeton University Press, Princeton, NJ.

Ellerman, D. (2007a). Adjoint functors and heteromorphisms. arXiv:0704.2207

Ellerman, D. (2007b). Adjoints and emergence: applications of a new theory of adjoint functors. *Axiomathes* 17(1), 19–39.

Ellerman, D. (2015). On adjoint and brain functors. *Axiomathes* 1–21.

Feferman, S. (1969). Set-theoretical foundations of category theory, in *Reports of the Midwest Category Seminar. III,*' no. 106 in Lecture Notes in Mathematics, 201–47. Springer-Verlag, Berlin.

Feferman, S. (1977). Categorical foundations and foundations of category theory, in *Logic, Foundations of Mathematics and Computability Theory (Proc. Fifth Internat. Congr. Logic, Methodology and Philos. of Sci., Univ. Western Ontario, London, Ont., 1975), Part I*, Vol. 9 of Univ. Western Ontario Ser. Philos. Sci., 149–69. Reidel, Dordrecht.

Feferman, S. (2013). Foundations of unlimited category theory: what remains to be done. *Review of Symbolic Logic* 6(1), 6–15. http://dx.doi.org/10.1017/S1755020312000111

Ferreiros, J. (2007). *Labyrinth of Thought: A History of Set Theory and Its Role in Modern Mathematics*. Birkhäuser Verlag, Basel.

Freyd, P. (1970). Homotopy is not concrete, in F. Peterson (ed.), *The Steenrod Algebra and Its Applications: A ...*, Vol. 168 of Springer Lecture Notes in Mathematics, 25–34. Springer-Verlag, New York.

Freyd, P. (1973). Concreteness. *Journal of Pure and Applied Algebra* 3(2), 171–91.

Gava, G. (2014). Kant's definition of science in the architectonic of pure reason and the essential ends of reason. *Kant Studien* 105(3), 372–93.

Hellman, G. (2003). Does category theory provide a framework for mathematical structuralism? *Philosophia Mathematica (3)* 11(2), 129–57. http://dx.doi.org/10.1093/philmat/11.2.129

Joyal, A. (2002). Quasi-categories and kan complexes. *Journal of Pure and Applied Algebra* 175(1–3), 207–22.

Keller, B. (1996). Derived categories and their uses, in M. e. a. Hazewinkel (ed.), *Handbook of Algebra*, Vol. 1, 671–701. Elsevier, London.

Krömer, R. (2007). *Tool and Object*, Vol. 32 of Science Networks. Historical Studies. Birkhäuser Verlag, Basel. A history and philosophy of category theory.

Lack, S. (2012). Non-canonical isomorphisms. *Journal of Pure and Applied Algebra* 216(3), 593–7.

Landry, E. (2013). The genetic versus the axiomatic method: responding to Feferman 1977. *Review of Symbolic Logic* 6(1), 24–50. http://dx.doi.org/10.1017/S1755020312000135

Landry, E. and Marquis, J.-P. (2005). Categories in context: historical, foundational, and philosophical, *Philosophia Mathematica (3)* 13(1), 1–43. http://dx.doi.org/10.1093/philmat/nki005

Lawvere, F. W. (1966). The category of categories as a foundation for mathematics, in *Proc. Conf. Categorical Algebra (La Jolla, Calif., 1965)*, 1–20. Springer, New York.

Linnebo, Ø., and Pettigrew, R. (2011). Category theory as an autonomous foundation. *Philosophia Mathematica (3)* 19(3), 227–54. http://dx.doi.org/10.1093/philmat/nkr024

Lurie, J. (2009). *Higher Topos Theory*. Princeton University Press, Princeton, NJ.

Mac Lane, S. (ed.) (1969). *Reports of the Midwest Category Seminar. III*, Lecture Notes in Mathematics, No. 106. Springer-Verlag, New York.

Makkai, M. (1999). On structuralism in mathematics, in *Language, Logic, and Concepts*, 43–66. Bradford Book, MIT Press, Cambridge, MA.

Makkai, M. (2014). The theory of abstract sets based on first-order logic with dependent types. http://www.math.mcgill.ca/makkai/Various/MateFest2013.pdf

Manchester, P. (2003). Kant's conception of architectonic in its historical context. *Journal of the History of Philosophy* 41(2), 187–207.

Manchester, P. (2008). Kant's conception of architectonic in its philosophical context. *Kant Studien* 99(2), 1–19.

Manders, K. L. (1987). Logic and conceptual relationships in mathematics, in *Logic Colloquium '85 (Orsay, 1985)*, Vol. 122 of Stud. Logic Found. Math., 193–211. North-Holland, Amsterdam. http://dx.doi.org/10.1016/S0049-237X(09)70554-3

Manders, K. L. (1989). Domain extension and the philosophy of mathematics, *Journal of Philosophy* 86(10), 553–62.

Marquis, J.-P. (2006). What is category theory?, in G. Sica (ed.), *What is Category Theory*, Vol. 3, 221–55, Polimetrica, New York.

Marquis, J.-P. (2009). *From a Geometrical Point of View: A Study of the History and Philosophy of Category Theory*, Vol. 14 of Logic, Epistemology, and the Unity of Science. Springer Science & Business Media, New York.

Marquis, J.-P. (2013*a*). Categorical foundations of mathematics or how to provide foundations for abstract mathematics. *Review of Symbolic Logic* 6(1), 51–75.

Marquis, J.-P. (2013*b*). Mathematical forms and forms of mathematics: leaving the shores of extensional mathematics. *Synthese* 190(12), 2141–64. http://dx.doi.org/10.1007/s11229-011-9962-0

Mazur, B. (2008). When is one thing equal to some other thing?, in *Proof and Other Dilemmas*, 221–41. MAA Spectrum, Math. Assoc. America, Washington, DC.

McLarty, C. (1993). Numbers can be just what they have to, *Noûs* 27(4), 487–98. http://dx.doi.org/10.2307/2215789

McLarty, C. (2004). Exploring categorical structuralism. *Philosophia Mathematica (3)* 12(1), 37–53. http://dx.doi.org/10.1093/philmat/12.1.37

McLarty, C. (2005). Learning from questions on categorical foundations. *Philosophia Mathematica (3)* 13(1), 44–60. http://dx.doi.org/10.1093/philmat/nki006

McLarty, C. (2011). Recent debate over categorical foundations, in *Foundational Theories of Classical and Constructive Mathematics*, Vol. 76 of *West. Ont. Ser. Philos. Sci.*, 145–54. Springer, Dordrecht. http://dx.doi.org/10.1007/978-94-007-0431-2_7

Moore, G. H. (2013). *Zermelo's Axiom of Choice: Its Origins, Development, and Influence*. Dover Books on Mathematics, London.

Pedicchio, M. C., and Tholen, W. (ed.) (2003). *Categorical Foundations: Special Topics in Order, Topology, Algebra, and Sheaf Theory*, Vol. 97 of *Encyclopedia of Mathematics and its Applications*. Cambridge University Press, Cambridge, UK.

Pedroso, M. (2009). On three arguments against categorical structuralism. *Synthese* 170(1), 21–31. http://dx.doi.org/10.1007/s11229-008-9346-2

Shapiro, S. (2005). Categories, structures, and the Frege-Hilbert controversy: the status of meta-mathematics. *Philosophia Mathematica (3)* 13(1), 61–77. http://dx.doi.org/10.1093/philmat/nki007

Shapiro, S. (2011). Foundations: structures, sets, and categories, in *Foundational Theories of Classical and Constructive Mathematics*, Vol. 76 of *West. Ont. Ser. Philos. Sci.*, 97–110. Springer, Dordrecht. http://dx.doi.org/10.1007/978-94-007-0431-2_4

Shoenfield, J. (1977). Axioms of set theory, in J. e. Barwise (ed.), *Handbook of Mathematical Logic*, 321–44. North-Holland, Amsterdam.

Street, R. (2007). *Quantum Groups: A Path to Current Algebra*, Vol. 19 of Autralian Mathematical Society Lecture Series. Cambridge University Press, Cambridge, UK.

Univalent Foundations Program, T. (2013). *Homotopy Type Theory: Univalent Foundations of Mathematics*. Institute for Advanced Studies, New York. http://homotopytypetheory.org/book

von Neumann, J. (1967). An axiomatization of set theory, in J. van Heijenoort (ed.), *From Frege to Gödel: A Source Book in Mathematical Logic, 1879–1931*, 393–413. Harvard University Press, Cambridge, MA.

Waldhausen, F. (1985). Algebraic k-theory of spaces, in *Algebraic and Geometric Topology*, Vol. 1126 of *Lecture Notes in Mathematics*, 318–419. Springer-Verlag, Berlin.

7

Categorical Logic and Model Theory

John L. Bell

7.1 Introduction

Category theory was invented by Eilenberg and Mac Lane in 1945, and by the 1950s its concepts and methods had become standard in algebraic topology and algebraic geometry. But few topologists and geometers had much interest in logic. And although the use of algebraic techniques was long-established in logic, starting with Boole and continuing right up to Tarski and his school, logicians of the day were, by and large, unacquainted with category theory. The idea of recasting the entire apparatus of logic—semantics as well as syntax—in category-theoretic terms, creating thereby a full-fledged "categorical logic", rests with the visionary category-theorist F. William Lawvere. Lawvere's pioneering work, which came to be known as *functorial semantics*, was first presented in his 1963 Columbia thesis and summarized that same year in the *Proceedings of the National Academy of Sciences of the U.S.A.* (Lawvere, 1963a; see also Lawvere, 1965). Lawvere identified an *algebraic theory* as a category of a particular kind, an interpretation of one theory in another as a functor between them, and a model of such a theory as a (product-preserving) functor on it to the category of sets. He also began to extend the categorical description of algebraic (or equational) theories to full first order, or *elementary* theories, building on his observation that existential and universal quantification can be seen as left and right adjoints, respectively, of substitution (Lawvere, 1966, 1967). In the seminal paper Lawvere (1969) the idea of grounding mathematics itself on the concept of adjoint functors is explored.

The description of quantifiers led to the introduction of the notion—one which was to prove of great importance in later developments—of an object of *truth values* so as to enable relations and partially defined operations to be described to the higher-order case. The idea of a truth-value object was further developed in Lawvere (1969) and Lawvere (1970a); the latter paper contains in particular the observation that the presence of such an object in a category enables the comprehension principle to be reduced to an elementary statement about adjoint functors.

Lawvere was convinced that category theory, resting on the bedrock concepts of map and map composition, could serve as a foundation for mathematics reflecting its essence in a direct and faithful fashion unmatchable by its "official" foundation, set theory. A first step taken toward this goal was Lawvere's publication, in 1964, of an "elementary" description of the category of sets (Lawvere, 1964) followed in 1966 by a similar description of the category of categories. In the former paper is to be found the first published appearance of the categorical characterization of the natural number system, later to become known as the *Peano–Lawvere* axiom. In another direction, reflecting his abiding interest in physics, Lawvere hoped to develop a foundation for continuum mechanics in category-theoretic terms. In lectures given in several places during 1967 he suggested that such a foundation should be sought "on the basis of a direct axiomatization of the essence of differential topology using results and methods of algebraic geometry". And to achieve this, in turn, would require "[an] axiomatic study of categories of smooth sets, similar to the topos of Grothendieck"; in other words, an elementary axiomatic description of *Grothendieck toposes*—categories of sheaves on a site, that is, a category equipped with a Grothendieck topology.

In 1969–70 Lawvere, working in collaboration with Myles Tierney, finally wove these strands together (see Lawvere, 1970a; and Tierney, 1972, 1973). Lawvere had previously observed that every Grothendieck topos had an object of truth values, and that Grothendieck topologies are closely connected with self-maps on that object. Accordingly, Lawvere and Tierney began to investigate in earnest the consequences of taking as an axiom the existence of an object of truth values: the result was a concept of amazing fertility, that of *elementary topos*—a cartesian closed category equipped with an object of truth values.[1] In addition to providing a natural generalization of elementary (i.e., first-order) theories, the concept of elementary topos, more importantly, furnished the appropriate elementary notion of a category of sheaves; in Lawvere's powerful metaphor, a topos could be seen as a category of *variable sets* (varying over a topological space, or a general category). The usual category of sets (itself, of course, a topos) could then be thought of as being composed of "constant" sets.

The fact that a topos possesses an object of truth values made it apparent that logic could be "done" in a topos; from the fact that this object is not normally a Boolean but a Heyting algebra it followed that a topos's "internal" logic is, in general, *intuitionistic*. (A *Boolean* topos, e.g., the topos of sets, is one whose object of truth values is a Boolean algebra; in that event its internal logic is classical.) It was William Mitchell (1972) who first put forward in print the idea that logical "reasoning" in a topos should be performed in an explicit language resembling that of set theory, the topos's *internal language*. (The internal language was also invented, independently, by André Joyal, Jean Bénabou, and others.) This language and its logic soon became identified as a

---

[1] Lawvere and Tierney's definition of elementary topos originally required the presence of finite limits and colimits—these were later shown to be eliminable. See, e.g., Bell (1988).

form of *intuitionistic type theory*, or *higher-order intuitionistic logic*, and underwent rapid development at many hands. Just as each topos determines a type theory, so each type theory generates a topos.

A major stimulus behind Lawvere and Tierney's development of elementary topos theory was the desire to provide a categorical formulation of Cohen's (1966) proof of the independence of the continuum hypothesis from the axioms of set theory. The connection between elementary topos theory and the set-theoretic independence proofs, in which Cohen's notion of *forcing* plays a pivotal role, led Joyal to observe in the early 1970s that the various notions of forcing (which in addition included those of Robinson and Kripke) could all be conceived as instances of a general truth-conditional scheme for sentences in a (sheaf) topos. This concept, which first explicitly appeared in print in Osius (1975), came to be known as *Kripke–Joyal semantics, sheaf semantics,* or *topos semantics*. Topos semantics showed, in particular, that the rules for Cohen forcing coincide with the truth conditions for sentences in the topos of presheaves over the partially ordered set of forcing conditions, thus explaining the somewhat puzzling fact that, while Cohen's independence proofs employ only classical models of set theory, his (original) notion of forcing obeys *intuitionistic* rules.

R. Diaconescu (1975) established the important fact, conjectured by Lawvere, that, in a topos, the axiom of choice implies that the topos is Boolean. This led to the recognition that, in intuitionistic set theory IZF, the axiom of choice implies the law of excluded middle.

Of major significance in the development of topos theory was the emergence of the concept of *classifying topos* for a first-order theory, that is, a topos obtained by freely adjoining a model of the theory to the topos of constant sets. The roots of this idea lie in the work of the Grothendieck school and in Lawvere's functorial semantics, but it was Joyal and Reyes (see Reyes, 1974) who, in 1972, identified a general type of first-order theory, later called a *coherent* or *geometric* theory, which could be shown always to possess a classifying topos. This work was later extended by Reyes and Makkai to infinitary geometric theories: they showed that any topos is the classifying topos of such a theory (Makkai and Reyes, 1977). In 1973 Lawvere (see Lawvere, 1975) pointed out that, in virtue of Joyal and Reyes's work, a previous theorem of Deligne (see Artin et al., 1964, VI 9.0) on coherent toposes—the classifying toposes of (finitary) geometric theories—was equivalent to the Gödel–Henkin completeness theorem for geometric theories. He also observed that, analogously, the theorem of Barr (1974) on the existence of "enough" Boolean toposes was equivalent to a "Boolean-valued" completeness theorem for infinitary geometric theories.

A further strand within the evolution of categorical logic derives from the *algebraic analysis of deductive systems*. In the 1960s, J. Lambek and M. E. Szabo promulgated the view that a deductive system is just a graph with additional structure, or equivalently, a category with missing equations (see, e.g. Lambek, 1968, 1969; Szabo, 1978; and Lambek and Scott, 1986). In his work Lambek also systematically exploited the inverse view that, by assimilating its objects to statements or formulas, and its arrows to proofs or deductions, a category can be presented "equationally" as a certain kind of deductive system with an equivalence relation imposed on proofs.

As touched on earlier in connection with toposes, categories can also be viewed as *type theories*. From this perspective, the objects of a category are regarded as *types* (or sorts) and the arrows as *mappings* between the corresponding types. In the 1970s, Lambek established that, viewed in this way, cartesian closed categories correspond to the typed λ-calculus (see Lambek, 1972, 1974, 1980; and Lambek and Scott, 1986). Seely (1984) proved that locally cartesian closed categories correspond to Martin-Löf, or predicative, type theories (see Martin-Löf, 1984; Beeson, 1985; and the papers collected in Sambin and Smith, 1998).

7.2 Categories and Deductive Systems

The relationship between category theory and logic is most simply illustrated in the case of the (intuitionistic) propositional calculus. Accordingly let S be a set of sentences in a propositional language L: S gives rise to a category C_S specified as follows. The objects of C_S are the sentences of L, while the arrows of C_S are of two types. First, for sentences p, q of L we count as an arrow $\pi : p \to q$ any pair $\langle P, p \rangle$ where P is a proof of q from $S \cup \{p\}$ in the intuitionistic propositional calculus. And second, for each sentence p the pair $\langle \emptyset, p \rangle$ is to count as an arrow $p \to p$. Given two arrows $\pi = \langle P, p \rangle : p \to q$ and $\sigma : \langle Q, q \rangle : q \to r$ the composite $\sigma \circ \pi : p \to r$ is defined to be the pair $\langle P \bigstar Q, p \rangle$, where $P \bigstar Q$ is the proof obtained by concatenating P and Q. The identity arrow on p is taken to be $\langle \emptyset, p \rangle$.

The category C_S is called the *syntactic category* determined by S. Syntactic categories have a number of special properties. To begin with, they are cartesian closed, with terminal object given by the identically true proposition ⊤, the product of two sentences p, q given by the conjunction $p \wedge q$ (with canonical proofs of conjuncts from conjunctions as projection arrows), and the implication sentence $p \Rightarrow q$ playing the role of the exponential object p^q. (This last follows from the fact that, for any r, there is a natural bijection between proofs of q from $S \cup \{r \wedge p\}$ and proofs of $p \Rightarrow q$ from $S \cup \{r\}$.) Moreover, C_S has an initial object in the form of the identically false proposition ⊥ and coproducts given by disjunctions in the same way as products correspond to conjunctions. We may sum all this up in by saying that syntactic categories are *bicartesian closed*.

So a certain kind of deductive system—intuitionistic propositional calculi—can be regarded as a category of a special sort. Conversely, it is possible to furnish the concept of deductive system with a definition of sufficient generality to enable every category to be regarded as a special sort of deductive system. This is done by introducing the concept of a *graph*. A (*directed*) *graph* consists of two classes: a class A whose members are called *arrows*, and a class O whose members are called *objects* (or *vertices*), together with two mappings from A to O called *source* and *target*. For $f \in A$ we write $f : A \to B$ to indicate that $source(f) = A$ and $target(f) = B$. Now a *deductive system* is a graph in which with each object A there is associated an arrow $1_A: A \to A$, called the *identity arrow* on A, and with each pair of arrows $f: A \to B$ and $g: B \to C$ there is associated an arrow $g \circ f: A \to C$ called the *composite* of f and g. It is natural for a logician to think of the objects of a deductive system as *statements* and of the arrows as *deductions* or *proofs*. In this spirit the arrow composition operation

$$\frac{f: A \to B \quad g: B \to C}{g \circ f: A \to C}$$

may be thought of as a *rule of inference*.

A *category* may then be defined as a deductive system in which the following equations hold among arrows:

$$f \circ 1_A = f = 1_B \circ f \qquad (h \circ g) \circ f = h \circ (g \circ f) \qquad (*)$$

for all $f: A \to B, g: B \to C, h: C \to D$. Thus each category is a certain kind of deductive system. Conversely, by identifying proofs in such a way as to make equations (*) hold, any deductive system engenders a unique category.

7.3 Functorial Semantics

As has been observed, Lawvere took the first step in providing the theory of models with a categorical formulation by introducing *algebraic theories*, the categorical counterparts of equational theories. Here the key insight was to view the logical operation of substitution in equational theories as composition of arrows in a certain sort of category. Lawvere showed how models of such theories can be naturally identified as functors of a certain kind, so launching the development of what has come to be known as *functorial semantics*.

An algebraic theory T is a category whose objects are the natural numbers and which for each m is equipped with an m-tuple of arrows, called *projections*,

$$\pi_i: m \to 1 \quad i = 1, \ldots, m,$$

making m into the m-fold power of 1: $m = 1^m$. (Here 1 is not a terminal object in T.)

In an algebraic theory the arrows $m \to 1$ play the role of m-ary operations. Consider, for example, the algebraic theory **Rng** of *rings*. To obtain this, one starts with the usual (language of) the first-order theory Rng of rings and introduces, for each pair of natural numbers (m, n), the set $P(m, n)$ of n-tuples of polynomials in the variables $x_1, \ldots, x_m$. The members of $P(m, n)$ are then taken to be the arrows $m \to n$ in the category **Rng**. Composition of arrows in **Rng** is defined as substitution of polynomials in one aother. The projection arrow $\pi_i: m \to 1$ is just the monomial x_i considered as a polynomial in the variables $x_1, \ldots, x_m$. Each polynomial in m variables, as an arrow $m \to 1$, may be regarded as an m-ary operation in **Rng**.

In a similar way every equational theory—groups, lattices, Boolean algebras—may be assigned an associated algebraic theory.

Now suppose given a category E with finite products. A *model* of an algebraic theory T in E, or a *T-algebra* in E, is defined to be a finite product preserving functor $A: T \to E$. The full subcategory of the functor category E^T whose objects are all T-algebras is called the *category of* T-*models* or T-*algebras* in E, and is denoted by **Alg**(T, E).

When T is the algebraic theory associated with an equational theory S, it is not hard to see that the category of T-models in **Set**, the category of sets, is equivalent to the category of algebras axiomatized by S.

Lawvere later extended functorial semantics to *first-order* logic. Here the essential insight was that existential and universal quantification can be seen as left and right adjoints, respectively, of substitution.

To see how this comes about, consider two sets A and B and a map $f: A \to B$. The power sets *PA* and *PB* of *A* and *B* are partially ordered sets under inclusion, and so can be considered as categories. We have the map ("preimage") $f^{-1}: PB \to PA$ given by

$$f^{-1}(Y) = \{x : f(x) \in Y\},$$

which, being inclusion-preserving, may be regarded as a *functor* between the categories *PB* and *PA*. Now define the functors $\exists_f, \forall_f: PA \to PB$ by

$$\exists_f(X) = \{y: \exists x(x \in X \wedge f(x) = y)\} \quad \forall_f(X) = \{y: \forall x(f(x) = y \Rightarrow x \in X)\}.$$

These functors $\exists_f$ ("image") and $\forall_f$ ("coimage"), which correspond to the existential and universal quantifiers, are easily checked to be, respectively, left and right adjoint to f^{-1}; that is, $\exists_f(X) \subseteq Y \Leftrightarrow X \subseteq f^{-1}(Y)$ and $f^{-1}(Y) \subseteq X \Leftrightarrow Y \subseteq \forall_f(X)$. Now think of the members of *PA* and *PB* as corresponding to *attributes* of the members of *A* and *B* (under which the attribute corresponding to a subset is just that of belonging to it), so that inclusion corresponds to entailment. Then, for any attribute *Y* on *B*, the definition of $f^{-1}(Y)$ amounts to saying that, for any $x \in A$, x has the attribute $f^{-1}(Y)$ just when $f(x)$ has the attribute *Y*. That is to say, the attribute $f^{-1}(Y)$ is obtained from *Y* by "substitution" along f. This is the sense in which quantification is adjoint to substitution.

Lawvere's concept of *elementary existential doctrine* (Lawvere, 1969) presents this analysis of the existential quantifier in a categorical setting. Accordingly, an elementary existential doctrine is given by the following data: a category T with finite products—here the objects of T are to be thought of as *types* and the arrows of T as *terms*—and for each object *A* of T a category **Att**(*A*) called the *category of attributes* of *A*. For each arrow $f: A \to B$ we are also given a functor **Att**$(f):$ **Att**$(B) \to$ **Att**(A), to be thought of as substitution along f, which is stipulated to possess a left adjoint $\exists_f$—existential quantification along f.

The category **Set** provides an example of an elementary existential doctrine: here for each set *A*, the category of attributes **Att**(*A*) is just *PA* and for $f: A \to B$,

$A(f)$ is f^{-1}. This elementary existential doctrine is *Boolean* in the sense that each category of attributes is a Boolean algebra and each substitution along maps a Boolean homomorphism.

Functorial semantics for elementary existential doctrines is most simply illustrated in the Boolean case. Thus, a (set-valued) *model* of a Boolean elementary existential doctrine (T, Att) is defined to be a product preserving functor $M: T \to$ **Set** together with, for each object A of T, a Boolean homomorphism $\text{Att}(A) \to P(MA)$ satisfying certain natural compatibility conditions.

This concept of model can be related to the usual notion of model for a first-order theory T in the following way. First, one introduces the so-called "Lindenbaum" doctrine of T: this is the elementary existential doctrine (T, A) where T is the algebraic theory whose arrows are just projections among the various powers of 1 and in which $\text{Att}(n)$ is the Boolean algebra of equivalence classes modulo provable equivalence from T of formulas having free variables among $x_1, \ldots, x_n$. For $f: m \to n$, the action of $\text{Att}(f)$ corresponds to syntactic substitution, and in fact $\exists_f$ can be defined in terms of the syntactic $\exists$. It is not difficult to see that each model of T in the usual sense gives rise to a model of the corresponding elementary existential doctrine (T, A).

7.4 Local Set Theories and Toposes

We have described the categorical counterparts to equational and first-order logic. It is natural now to ask: what sort of category corresponds to *higher-order* logic? As already remarked the answer to this question—an *elementary topos*—was provided in 1969 by Lawvere and Tierney.

For our present purposes a(n) (elementary) *topos* will be defined as a category possessing a terminal object, products, a truth-value object, and power objects. Here a *truth-value object* is an object Ω together with an arrow $t: 1 \to \Omega$ such that associated with each monic $m: A \rightarrowtail B$ there is a unique arrow $\chi(m): B \to \Omega$—the characteristic arrow of A—such that the diagram

$$\begin{array}{ccc} A & \longrightarrow & 1 \\ m \downarrow & & \downarrow t \\ B & \underset{\chi(m)}{\longrightarrow} & \Omega \end{array}$$

is a pullback, and conversely any diagram of the form

$$B \xrightarrow{u} \Omega \xleftarrow{t} 1$$

has a pullback. (In **Set**, Ω is the set $2 = \{0, 1\}$, t is the map $1 = \{0\} \to 2$ that sends 0 to 1, and $\chi(m)$ is the characteristic function of the subset $\{m(x): x \in A\}$ of B.)

Arrows $B \to \Omega$ are in natural bijective correspondence with *subobjects* of an object B. Here a subobject of B is defined as follows. Call two monic arrows $m: A \rightarrowtail B$

and $m: C \rightarrowtail B$ equivalent, and write $m \sim n$, if there is an isomorphism $i: C \rightarrowtail B$ such that $m \circ i = n$. Clearly $\sim$ is an equivalence relation. A *subobject* of B is defined to be an equivalence class under $\sim$ of monics into B.

In a topos, just as in set theory, one can give natural definitions of *inclusion* ($\subseteq$), *union* ($\cup$), and *intersection* ($\cap$) of subobjects.

A *power object* of an object X is an object PX together with an arrow ("evaluation") $e_X: X \times PX \to \Omega$ such that, for any $f: X \times PX \to \Omega$, there is a unique arrow $\bar{f}: YPX$ such that the diagram

$$\begin{array}{ccc} & X \times Y & \\ {}_{1_X \times \bar{f}} \downarrow & \searrow^{f} & \\ X \times PX & \xrightarrow[e_X]{} & \Omega \end{array}$$

commutes. (In **Set**, PX is the power set of X and e_X the characteristic function of the membership relation between X and PX.)

The system of higher-order logic associated with a topos is most conveniently formulated as a generalization of classical set theory within intuitionistic logic: *intuitionistic type theory*. The system to be described here, *local set theory*, is a modification, due to Zangwill (1977), of that of Joyal and Boileau, later published as their (1981). A full account of local set theory may be found in Bell (1988).

The category of sets is a prime example of a topos, and the fact that it is a topos is a consequence of the axioms of classical set theory. Similarly, in a local set theory the construction of a corresponding "category of sets" can also be carried out and shown to be a topos. In fact *any* topos is obtainable (up to equivalence of categories) as the category of sets within some local set theory. Toposes are also, in a natural sense, the *models* or *interpretations* of local set theories. Introducing the concept of *validity* of an assertion of a local set theory under an interpretation, such interpretations are *sound* in the sense that any theorem of a local set theory is valid under every interpretation validating its axioms and *complete* in the sense that, conversely, any assertion of a local set theory valid under every interpretation validating its axioms is itself a theorem. The basic axioms and rules of local set theories are formulated in such a way as to yield as theorems precisely those of higher-order intuitionistic logic. These basic theorems accordingly coincide with those statements valid under *every* interpretation.

In a local set theory the set concept, as a primitive, is replaced by that of *type*. A type in this sense may be thought of as a *natural kind* or *species* from which sets are extracted as subspecies. The resulting set theory is *local* in the sense that, for example, the inclusion relation will obtain only among sets which have the same type, i.e., are subspecies of the same species.

A local set theory, then, is a type-theoretic system built on the same primitive symbols $=, \in, \{:\}$ as classical set theory, in which the set-theoretic operations of forming products and powers of types can be performed, and which in addition contains a "truth value" type acting as the range of values of "propositional functions" on types.

A local set theory is determined by specifying a collection of *axioms* formulated within a *local language* defined as follows.

A *local language* L has the following basic symbols:

- **1** (*unity type*), Ω (*truth value type*)
- S, T, U, ... (*ground types*: possibly none of these)
- f, g, h, ... (*function symbols*: possibly none of these)
- $x_A, y_A, z_A, \ldots$ (variables of each type A, where a *type* is as defined below).

The *types* of L are defined recursively as follows:

- **1**, Ω are types
- any ground type is a type
- $A_1 \times \ldots \times A_n$ is a type whenever $A_1 \ldots A_n$ are, where if $n = 1$, $A_1 \times \ldots \times A_n$ is A_1, while if $n = 0$, $A_1 \times \ldots \times A_n$ is **1** (*product types*)
- PA is a type whenever A is (*power types*).

Each function symbol **f** is assigned a *signature* of the form $A \to B$, where A, B are types; this is indicated by writing $f: A \to B$.

Terms of L and their associated *types* are defined recursively as follows. We write $\tau: A$ to indicate that the term τ has type A.

Term: type	Proviso
$\star: \mathbf{1}$	
$x_A: A$	
$f(\tau): B$	$f: A \to B \quad \tau: A$
$\langle \tau_1, \ldots, \tau_n \rangle: A_1 \times \ldots \times A_n$, where $\langle \tau_1, \ldots, \tau_n \rangle$ is τ_1 if $n = 1$ and $\star$ if $n = 0$	$\tau_1: A_1, \ldots, \tau_n: A_n$
$(\tau)_i: A_i$ where $(\tau)_i$ is τ if $n = 1$	$\tau: A_1 \times \ldots \times A_n, 1 \leq i \leq n$
$\{x_A: \alpha\}: PA$	$\alpha: \Omega$
$\sigma = \tau: \Omega$	σ, τ of same type
$\sigma \in \tau$	$\sigma: A, \tau: PA$ for some type A.

Terms of type Ω are called *formulas*, *propositions*, or *truth values*. Notational conventions we shall adopt include the following:

$\omega, \omega', \omega'', \omega'''$	variables of type Ω
α, β, γ	formulas
$x, y, z \ldots$	$x_A, y_A, z_A \ldots$
$\tau(x/\sigma)$ or $\tau(\sigma)$	result of substituting σ at each free occurrence of x in τ: an occurrence of an occurrence of x is *free* if it does not appear within $\{x: \alpha\}$

A term is *closed* if it contains no free variables; a closed term of type Ω is called a *sentence*.

The *basic* axioms for L are s follows:

Unity : $x_1 = \star$

Equality $x = y, \alpha(z/x) : \alpha(z/y)$ (x, y free for z in α)

Products : $(\langle x_1, \ldots, x_n \rangle)_i = x_i$

: $x = \langle (x)_1, \ldots, (x)_n \rangle$

Comprehension : $x \in \{x : \alpha\} \leftrightarrow \alpha$

The *rules of inference* for L are the following:

Thinning $\dfrac{\Gamma : \alpha}{\beta, \Gamma : \alpha}$

Restricted cut $\dfrac{\Gamma : \alpha \quad \alpha, \Gamma : \beta}{\Gamma : \beta}$ (any free variable of α free in Γ or β)

Substitution $\dfrac{\Gamma : \alpha}{\Gamma(x/\tau) : \alpha(x/\tau)}$ (τ free for x in Γ and α)

Extensionality $\dfrac{\Gamma : x \in \sigma \leftrightarrow x \in \tau}{\Gamma : \sigma = \tau}$ (x not free for x in Γ and α)

Equivalence $\dfrac{\alpha, \Gamma : \beta \quad \beta, \Gamma : \alpha}{\Gamma : \alpha \leftrightarrow \beta}$

These axioms and rules of inference yield a system of *natural deduction* in L. If S is any collection of sequents in L, we say that the sequent $\Gamma : \alpha$ is *deducible from S*, and write $\Gamma \vdash_S \alpha$ provided there is a deduction of $\Gamma : \alpha$ using the basic axioms, the sequents in S, and the rules of inference. We shall also write $\Gamma \vdash_S \alpha$ for $\Gamma \vdash_\emptyset \alpha$ and $\vdash_S \alpha$ for $\emptyset \vdash_S \alpha$.

A *local set theory* in L is a collection S of sequents closed under deducibility from S. Any collection of sequents S *generates* the local set theory S^* comprising all the sequents deducible from S. The local set theory in L generated by $\emptyset$ is called *pure* local set theory in L.

Logical operations in L can be introduced as follows:

Logical operation	Definition
$\top$ (true)	$\star = \star$
$\alpha \wedge \beta$	$\langle \alpha, \beta \rangle = \langle \top, \top \rangle$
$\alpha \to \beta$	$(\alpha \wedge \beta) \leftrightarrow \alpha$
$\forall x\, \alpha$	$\{x : \alpha\} = \{x : \top\}$
$\bot$ (false)	$\forall \omega.\, \omega$
$\neg \alpha$	$\alpha \to \bot$
$\alpha \vee \beta$	$\forall \omega[(\alpha \to \omega \wedge \beta \to \omega) \to \omega]$
$\exists x\, \alpha$	$\forall \omega[\forall x(\alpha \to \omega) \to \omega]$

We also write $x \neq y$ for $\neg(x = y)$, $x \notin y$ for $\neg(x \in y)$, and $\exists!x\,\alpha$ for $\exists x[\alpha \wedge \forall y \alpha(x/y) \to x = y]$.

It can be shown that the logical operations on formulas just defined satisfy the axioms and rules of free intuitionistic logic. (For this reason local set theories are also known as *intuitionistic type theories*.)

We can now introduce the concept of *set* in a local language. A *set-like* term is a term of power type; a *closed* set-like term is called an (L-) *set*. We shall use upper case italic letters $X, Y, Z, \ldots$ for sets, as well as standard abbreviations such as $\forall x \in X.\alpha$ for $\forall x(x \in X \to \alpha)$. The set-theoretic *operations* and *relations* are defined as follows. Note that in the definitions of $\subseteq, \cap,$ and $\cup$, X and Y must be of the same type:

Operation	Definition
$\{x \in X : \alpha\}$	$\{x : x \in X \wedge \alpha\}$
$X \subseteq Y$	$\forall x \in X.\, x \in Y$
$X \cap Y$	$\{x : x \in X \wedge x \in Y\}$
$X \cup Y$	$\{x : x \in X \vee x \in Y\}$
$x \notin X$	$\neg(x \in X)$
U_A or A	$\{x_A : \top\}$
$\emptyset_A$ or $\emptyset$	$\{x_A : \bot\}$
$E - X$	$\{x : x \in E \wedge x \notin X\}$
PX	$\{u : u \subseteq X\}$
$\bigcap U\ (U : PPA)$	$\{x : \forall u \in U.\, x \in U\}$
$\bigcup U\ (U : PPA)$	$\{x : \exists u \in U.\, x \in U\}$
$\bigcap_{i \in I} X_i$	$\{x : \forall i \in I.\, x \in X\}$
$\bigcup_{i \in I} X_i$	$\{x : \exists i \in I.\, x \in X\}$
$\{\tau_1, \ldots, \tau_n\}$	$\{x : x = \tau_1 \vee \ldots \vee x = \tau_n\}$
$\{\tau : \alpha\}$	$\{z : \exists x_1 \ldots \exists x_n(z = \tau \wedge \alpha)\}$
$X \times Y$	$\{\langle x, y \rangle : x \in X \wedge y \in Y\}$
$X + Y$	$\{\langle \{x\}, \emptyset \rangle : x \in X\} \cup \{\langle \emptyset, \{y\} \rangle : y \in Y\}$
$Fun(X, Y)$	$\{u : u \subseteq X \times Y \wedge \forall x \in X \exists! y \in Y.\, \langle x, y \rangle \in u\}$

The standard facts concerning the set-theoretic operations and relations now follow as straightforward consequences of their definitions.

Given a term τ such that

$$\langle x_1, \ldots, x_n \rangle \in X \vdash_S \tau \in Y,$$

we write $(\langle x_1, \ldots, x_n \rangle \mapsto \tau)$ or simply $x \mapsto \tau$ for

$$\{\langle \langle x_1, \ldots, x_n \rangle, \tau \rangle : \langle x_1, \ldots, x_n \rangle \in X\}.$$

Clearly we have

$$\vdash_S ((\langle x_1, \ldots, x_n \rangle \mapsto \tau)) \in Fun(X, Y),$$

and so we may think of $(\langle x_1, \ldots, x_n \rangle \mapsto \tau)$ as the function from X to Y determined by τ.

We now show that each local set theory determines a topos. Let S be a local set theory in a local language L. Define the relation $\approx_S$ on the collection of all L-sets by

$$X \approx_S \text{ iff } \vdash_S X = Y.$$

This is an equivalence relation. An *S-set* is an equivalence class $[X]_S$—which we normally identify with X—of L-sets under the relation $\approx_S$. An *S-map* $f \colon X \to Y$ is a triple (f, X, Y)—normally identified with f—of S-sets such that $\vdash_S f \in Y^X$. X and Y are, respectively, the *domain* dom(f) and the *codomain* cod(f) of f. It is now readily shown that the collection of all S-sets and maps forms a category $\mathbf{C}(S)$, the *category of S-sets*, in which the composite of two maps $f \colon X \to Y$ and $g \colon Y \to Z$ is given by

$$g \circ f = \{\langle x, z \rangle \colon \exists y (\langle x, y \rangle \in f \land \langle y, z \rangle \in.$$

In fact, $\mathbf{C}(S)$ is a *topos*, the *topos of sets determined by* S. It has terminal object U_1; the product of two objects (S-sets) X, Y is the S-set $X \times Y$, with projections given by

$$\pi_1 = (\langle x, y \rangle \mapsto x) \colon X \times Y \to X, \quad \pi_1 = (\langle x, y \rangle \mapsto y) \colon X \times Y \to Y;$$

its truth-value object is (Ω, t), where $t \colon 1 \to \Omega$ is the S-map $\{\langle \bigstar, \top \rangle\}$; and the power object of an object X is (PX, e_X), where $e_X \colon X \times PX \to \Omega$ is the S-map $\langle x, z \rangle \mapsto x \in z$. All this is proved in much the same way as for classical set theory.

We next show how to interpret a local language in a topos.

Let L be a local language and E a topos. An *interpretation* I of L in E is an assignment:

- to each type **A**, of an E-object $\mathbf{A}_I$ such that:

  $(\mathbf{A}_1 \times \ldots \times \mathbf{A}_n)_I = (\mathbf{A}_1)_I \times \ldots \times (\mathbf{A}_n)_I$,
  $(P\mathbf{A}_I) = P\mathbf{A}_I$,
  $\mathbf{1}_I = 1$, the terminal object of E,
  $\mathbf{\Omega}_I = \Omega$ the truth value object of E.

- to each function symbol **f**: **A** → **B**, an E-arrow $f_I \colon \mathbf{A}_I \to \mathbf{B}_I$.

We usually write A_E or just A for $\mathbf{A}_I$.

We extend I to terms of L as follows. If $\tau \colon \mathbf{B}$, write x for $(x_1, \ldots, x_n)$, any sequence of variables containing all variables of τ (and call such sequences *adequate* for τ). Define the E-arrow

$$[\![\tau]\!]_x \colon A_1 \times \ldots \times A_n \longrightarrow B$$

recursively as follows:

$$[\![\bigstar]\!]_x = A_1 \times \ldots \times A_n \longrightarrow 1$$
$$[\![x_i]\!]_x = \pi_i \colon A_1 \times \ldots \times A_n \longrightarrow A_i$$

$$[\![f(\tau)]\!]_x = f_I \circ [\![\tau]\!]_x$$
$$[\![\langle \tau_1, \ldots, \tau_n \rangle]\!]_x = \langle [\![\tau_1]\!]_x, \ldots, [\![\tau_n]\!]_x \rangle$$
$$[\![(\tau)_i]\!]_x = \pi_i \circ [\![\tau]\!]_x$$
$$[\![\{y : \alpha\}]\!] = [\![\alpha(y/u)]\!]_{ux} \circ can,$$

where in this last clause u differs from $x_1, \ldots, x_n$ and is free for y in α, y is of type **C** (so that B is of type **PC**), can is the canonical isomorphism $C \times (A_1 \times \ldots \times A_n) \cong C \times A_1 \times \ldots \times A_n$, and $\hat{f}$ is as defined for power objects. To understand why, consider the diagrams

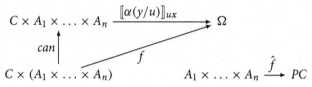

In set theory, $\hat{f}(a_1, \ldots, a_n) = \{y \in C : \alpha(y, a_1, \ldots, a_n)\}$, so we take $[\![\{y : \alpha\}]\!]_x$ to be $\hat{f}$.

Finally,

$[\![\sigma = \tau]\!]_x = eqc \circ [\![\langle \sigma, \tau \rangle]\!]_x$ (with σ, τ: **C** and where eqc is the characteristic arrow of the diagonal arrow $C \to C \times C$)

$[\![\sigma \in \tau]\!]_x = e_c \circ [\![\langle \sigma, \tau \rangle]\!]_x$ (with σ: **C**, τ: **PC** and where e_c is as defined for power objects).

If τ: B is closed, then x may be taken to be the empty sequence $\emptyset$. In this case we write $[\![\tau]\!]$ for $[\![\tau]\!]_\emptyset$; this is an arrow $1 \to B$.

Writing T for the characteristic arrow of an identity arrow $1_A : A \to A$, we note that

$$[\![T]\!]_x = [\![\star = \star]\!]_x = eq \circ \langle [\![\star]\!]_x, [\![\star]\!]_x \rangle = T.$$

For any finite set $\Gamma = \{\alpha_1, \ldots, \alpha_m\}$ of formulas write

$$[\![\Gamma]\!]_{I,x} \text{ for } [\![\alpha_1]\!]_{I,x} \cap \ldots \cap [\![\alpha_m]\!]_{I,x}, \text{ if } m \geq 1 \text{ or } T \text{ if } m = 0.$$

Given a formula α let $x = (x_1, \ldots, x_n)$ list all the free variables in $\Gamma \cup \{\alpha\}$; write

$$\Gamma \vDash_I \alpha \text{ for } [\![\Gamma]\!]_{I,x} \subseteq [\![\alpha]\!]_{I,x}.$$

$\Gamma \vDash_I \alpha$ is read: "Γ is *valid* under the interpretation I in E." If S is a local set theory, we say that I is a *model* of S if every member of S is valid under I. Note that

$$\vDash_I \beta \text{ iff } [\![\beta]\!]_x = T.$$

In particular, if I is an interpretation in a *degenerate* topos, i.e., a topos possessing just one object up to isomorphism, then $\vDash_I \alpha$ for all α. Now write

$$\Gamma \vDash \alpha \text{ for } \Gamma \vDash_I \alpha \text{ for every } I$$
$$\Gamma \vDash_S \alpha \text{ for } \Gamma \vDash_I \alpha \text{ for every model } I \text{ of } S.$$

It can be shown (laboriously) that the basic axioms and rules of inference of any local set theory are valid under every interpretation. This yields the following theorem.

Soundness Theorem.
$$\Gamma \vdash \alpha \Rightarrow \Gamma \vDash \alpha \qquad \Gamma \vdash_S \alpha \Rightarrow \Gamma \vDash_S \alpha.$$

A local set theory S is said to be *consistent* if it is not the case that $\vdash_S \top$. The soundness theorem yields the following:

Corollary. *Any pure local set theory is consistent.*

Proof. Set up an interpretation I of L in the topos Finset of finite sets as follows: $1_I = 1$, $\Omega_I = \{0,1\} = 2$, for any ground type A, A_I is any nonempty finite set. Extend I to arbitrary types in the obvious way. Finally, $f_I : A_I \to B_I$ is to be any map from A_I to B_I.

If $\vdash \bot$, then $\vdash \alpha$, so $\vDash_I \alpha$ for any formula α. In particular $\vDash_I u = v$, where u, v are variables of type P1. Hence $[\![u]\!]_{I,uv} = [\![v]\!]_{I,uv}$, that is, the two projections P1 × P1 → P1 would have to be identical, a contradiction.

Given a local set theory S in a language L, define the *canonical interpretation* $C(S)$ of L in $C(S)$ by
$$A_{C(S)} = U_A \qquad f_{C(S)} = (x \mapsto f(x)): U_A \to U_B \qquad \text{for } f: A \to B.$$

A straightforward induction establishes
$$[\![\tau]\!]_{C(S),x} = (x \mapsto \tau).$$

This yields
$$\Gamma \vDash_{C(S)} \alpha \Leftrightarrow \Gamma \vdash_S \alpha. \tag{*}$$

For
$$\begin{aligned}
\vDash_{C(S)} \alpha &\Leftrightarrow [\![\alpha]\!]_{C(S)} x = T \\
&\Leftrightarrow (x \mapsto \alpha) = (x \mapsto T) \\
&\Leftrightarrow \vdash_S \alpha = T \\
&\Leftrightarrow \vdash_S \alpha.
\end{aligned}$$

Since $\Gamma \vdash_S \alpha \Leftrightarrow \vdash_S \gamma \to \alpha$, where γ is the conjunction of all the formulas in Γ, the special case yields the general one.

Equivalence (*) may be read as asserting that $C(S)$ is a *canonical model* of S. This fact yields the following:

Completeness Theorem.
$$\Gamma \vDash \alpha \Rightarrow \Gamma \vdash \alpha \qquad \Gamma \vDash_S \alpha \Rightarrow \Gamma \vdash_S \alpha$$

Proof. We know that C(S) is a model of S. Therefore using (*),

$$\Gamma \vDash_S \alpha \Rightarrow \Gamma_{C(S)}\alpha \Rightarrow \Gamma \vdash_S \alpha.$$

The soundness and completeness theorems proved in this section are the counterparts, for interpretations of local set theories in toposes, of the soundness and completeness theorems for first-order theories in the usual set-based structures.

7.5 Model Theory of First-Order Languages in Categories

A model of a first-order theory in the usual (Tarski) sense may be regarded as an interpretation of the language of the theory within the category **Set** of sets. This is readily extended to the idea of an interpretation of a first-order language, and so also of a model of a first-order theory, within an arbitrary topos.

A (many-sorted) *first-order language* L is equipped with the following:

- sorts A, B, C, ...
- function symbols f, g, h, ..., each of which is assigned a pair $(\vec{A}, B)$ called its *signature*, where $\vec{A} = (A_1, \ldots, A_n)$ is a finite (possibly empty) sequence of sorts and B is a sort. We write f: $(\vec{A}, B)$.
- relation symbols R, S, ... each of which is assigned a signature $\vec{A}$: we write R: $\vec{A}$
- logical symbols $\wedge, \vee, \Rightarrow, \top, \bot$
- quantifiers $\forall, \exists$
- equality symbol $=$
- variables $x_A, y_A, \ldots$ for each sort A.

Terms of L and their corresponding *sorts* are defined inductively as follows:

- each variable x_A is a term of sort A.
- if f: $(\vec{A}, B)$ and $\vec{\tau} = (\tau_1, \ldots, \tau_n)$ is a sequence of terms of sorts $(A_1, \ldots, A_n)$, then $f(\vec{\tau})$ is a term of sort B.

Finally, the *formulas* of L are given inductively by the following clauses:

- $\top, \bot$ are formulas
- if σ, τ are terms of the same sort, then $\sigma = \tau$ is a formula.
- if R: $\vec{A}$ is a relation symbol and $\vec{\tau}$ is a sequence of terms of sorts $(A_1, \ldots, A_n)$, then $R(\vec{\tau})$ is a formula
- if φ, ψ are formulas, then so are $\varphi \wedge \psi, \varphi \vee \psi, \varphi \Rightarrow \psi$
- if φ is a formula and x is a variable, then $\forall x \varphi$ and $\exists x \varphi$ are formulas.

As usual we write $\neg \varphi$ and $\varphi \Leftrightarrow \psi$ for $\varphi \Rightarrow \top$ and $(\varphi \Rightarrow \psi) \wedge (\psi \Rightarrow \varphi)$, respectively.

We assume that L is equipped with the deductive machinery of free intuitionistic predicate logic, in which the inference rule *modus ponens* takes the restricted form

$$\frac{\varphi, \varphi \Rightarrow \psi}{\psi},$$

provided every variable free in ψ is free in φ.

A set of formulas of L is called a *theory* in L. If T is a theory in L and φ a formula of L, we write $T \vdash \varphi$ for "φ is deducible from T in free intuitionistic logic". We write $\vdash \varphi$ for $\emptyset \vdash \varphi$.

Now let C be a category with finite products. An *interpretation* I of L in C assigns the following:

- to each sort A, an E-object A_I
- to each function symbol f: $(\vec{A}, B)$ an E-arrow

$$f_I \colon \vec{A}_I = (A_1)_I \times \ldots \times (A_n)_I \to B_I$$

- to each relation symbol R: $\vec{A}$ a subobject $[\![R]\!]_I$ of $\vec{A}_I$

A *morphism of interpretations* $m\colon I \to J$ is an assignment to each sort A of an E-arrow $m_A \colon A_I \to A_J$ in such a way that:

- for each f: $(\vec{A}, B)$ the diagram

$$\begin{array}{ccc} \vec{A}_I & \xrightarrow{m_{A_1} \times \ldots \times m_{A_n}} & \vec{A}_J \\ f_I \downarrow & & \downarrow f_J \\ B_I & \xrightarrow{m_B} & B_J \end{array}$$

commutes;
- for each relation symbol R: $\vec{A}$ there is a commutative diagram of the form

$$\begin{array}{ccc} [\![R]\!]_I & \longrightarrow & \vec{A}_I \\ \downarrow & & \downarrow m_{A_1} \times \ldots \times m_{A_n} \\ [\![R]\!]_J & \longrightarrow & \vec{A}_J \end{array}$$

This gives rise to the *category of interpretations* L(E) of L in E.

Suppose given an interpretation I of L in a category C with finite products. Then by interpreting substitution of terms for variables as composition of arrows, each term of sort A can be interpreted as an E-arrow with codomain A_I.

It is natural to extend the interpretation to formulas so that each formula φ with free variables $x_1, \ldots, x_n$ of sorts $A_1, \ldots, A_n$ is interpreted as a subobject $[\![\varphi]\!]_I$ of $\vec{A}_I$.

Here $\top, \bot$ are to be interpreted as the maximal and minimal subobjects of 1; $=$ is interpreted as the diagonal subobject of $A_I \times A_I$; conjunction and disjunction of formulas as intersection and union of subobjects; and existential and universal quantification as image and coimage arising from projection arrows. For this to be possible C must carry the appropriate additional structure—a category carrying this additional structure is called a *Heyting category*. To be precise, a Heyting category is a category that has finite limits, unions of subobjects which are preserved by inverse image functors, image and coimage functors for arbitrary arrows, and in which epis are preserved under pullback.

The formula φ is *true* under I—written $\vDash_I \varphi$—if $[\![\varphi]\!]_I$ is the maximal subobject of $\vec{A}_I$. I is a *model* in C of a theory T (or simply a T-model) if each formula in T is true under I. If $\vDash_I \varphi$ for any T-model I in a Heyting category, we write $T \vDash \varphi$. We then have the following:

Soundness Theorem. $T \vdash \varphi$ implies that $T \vDash \varphi$.

A familiar concept of traditional model theory is that of the *Lindenbaum-Tarski algebra* of an (intuitionistic) first-order theory T. This is a Heyting algebra $H(T)$ obtained from the set of formulas of L: it comes equipped with a natural assignment to each formula φ of L of an element $[\![\varphi]\!]$ of $H(T)$—its "truth value"—in such a way that $[\![\varphi]\!] = 1$, the top element of $H(T)$.

In categorical model theory this idea is extended to that of a *generic model* of a first-order theory. For a given first-order theory T, a category $\mathbf{Syn}(T)$—the *syntactic category* of T—is constructed and within it is identified a T-model $M(T)$—the *generic model* of T—with the property that the formulas true in $M(T)$ are precisely those deducible from T.

Just as $H(T)$ is a Heyting algebra, so $\mathbf{Syn}(T)$ is a Heyting category, with just enough structure to allow for the interpretation of arbitrary first-order formulas. $\mathbf{Syn}(T)$ is defined as follows.

Write $\vec{x}, \vec{y}$, etc. for finite lists of variables of L, and $\varphi(\vec{x})$, etc. to indicate that the free variables of φ are among $\vec{x}$.

The objects of $\mathbf{Syn}(T)$ are *formal class terms*, i.e., symbols of the form $\{\vec{x} \mid \varphi(\vec{x})\}$ where φ is a formula of L whose free variables occur in the list $\vec{x}$, or $\{\varphi\}$, where φ has no free variables. An *arrow* between $\{\vec{x} \mid \varphi(\vec{x})\}$ and $\{\vec{y} \mid \psi(\vec{y})\}$ is defined to be an equivalence class, with respect to T-provable equivalence, of formulas $\theta(\vec{x}, \vec{y})$ which (T-provably) define functions with domain $\{\vec{x} \mid \varphi(\vec{x})\}$ and codomain $\{\vec{y} \mid \psi(\vec{y})\}$. We denote the arrow defined by $\theta(\vec{x}, \vec{y})$ by $[\vec{x} \mapsto \vec{y} \mid \theta(\vec{x}, \vec{y})]$.

The resulting category $\mathbf{Syn}(T)$ can then be shown to be a Heyting category.

The generic T-model $M(T)$ in $\mathbf{Syn}(T)$ is the interpretation of L defined as follows: to each sort A it assigns the object $\{x_A : \top\}$, to a function symbol f: $(\vec{A}, B)$ it assigns the arrow $[\vec{x} \mapsto y \mid (f(\vec{x}) = y)]: \{\vec{x} \mid \varphi(\vec{x})\} \to \{y : \top\}$, and to a relation symbol $R: \vec{A}$ it assigns the subobject of $\{x_A : \top\}$ whose domain is $\{\vec{x} \mid R(\vec{x})\}$.

From the existence of generic models the Completeness Theorem follows immediately, namely: $T \vDash \varphi$ implies that $T \vdash \varphi$.

We mention here for future reference that a certain *covering* system or *Grothendieck topology* K_T on $\mathbf{Syn}(T)$ can be introduced. First, a finite family of arrows

$$[\vec{x}_i \mapsto \vec{y} \mid \theta_i(\vec{x}_i, \vec{y})]: \{\vec{x}_i \mid \varphi(\vec{x}_i)\} \to \{\vec{y} \mid \psi(\vec{y})\} \quad (i = 1, \ldots, n)$$

is said to be *T-provably epic* on $\{\vec{y} \mid \psi(\vec{y})\}$ if

$$T \vdash \forall \vec{y}[\psi(\vec{y}) \Rightarrow \bigvee_{i=1}^{n} \exists \vec{x}_i \theta_i(\vec{x}_i, \vec{y}_i)].$$

For each object $\{\vec{y} \mid \psi(\vec{y})\} = A$ we define $K_T(A)$ to consist of all cosieves U on A (i.e., families of arrows to A closed under composition on the right) such that U contains a provably epic family. This yields a covering system on **Syn**(T). The site (**Syn**(T), K_T) is called the *syntactic site* associated with T.

7.6 Models in Toposes: Geometric Theories and Classifying Toposes

We have seen that the generic model of a first-order theory T "lives" in the Heyting category **Syn**(T). It is natural to ask whether such generic models can be found in *toposes*. While this is actually the case for arbitrary first-order theories, a much stronger notion of generic model can be given for the so-called *geometric* theories.

A (finitary) *geometric* formula of L is one which does not contain $\Rightarrow$ or $\forall$ (and hence not $\neg$ either). A *geometric implication* is a sentence of the form $\forall \vec{x}(\varphi \Rightarrow \psi)$ where φ and ψ are geometric formulas, and a theory consisting solely of geometric implications is called a *geometric theory*. As we shall see, geometric theories have particularly nice model-theoretic properties in a categorical setting. Most of the usual algebraic theories are in fact geometric.

If T is a geometric theory, the *geometric category* **Geom**(T) associated with T is constructed in the same way as the syntactic category, except that throughout attention is confined to *geometric* formulas. We again write K_T for the corresponding covering system, defined as at the end of the previous section. (**Geom**(T), K_T) is called the *geometric site* associated with T.

If A is an object of a category D, a family $\{B_i \xrightarrow{f_i} A : i \in I\}$ is *jointly epic* if, for any arrows $g, h: A \to C$, $g \circ f_i = h \circ f_i$ for all $i \in I$ implies that $g = h$.

If (C, K) is a site, a functor $F: \mathbf{C} \to \mathbf{D}$ *preserves* (K−) *covers* if, for any C-object A and any cosieve $S \in K(A)$, the family $\{Ff: f \in S\}$ is jointly epic in D. A functor $F: \mathbf{C} \to \mathbf{E}$ to a topos E is *geometric* if it is left-exact (i.e., preserves products and equalizers) and preserves K-covers.

It can now be shown that, for any geometric theory T and any topos E, *there is a natural bijective correspondence between models of T in E and geometric functors* **Geom**(T) $\to$ E. In view of this, a geometric functor **Geom**(T) $\to$ E is itself called a (functorial) *model* of T in E. Given a model F of T in E, a geometric implication φ is *true* in F if it is true under the corresponding interpretation in E: clearly, if $T \vdash \varphi$, then φ is true in F.

Given a geometric theory T, let **Set**[T] be the topos of K_T − *sheaves* on **Geom**(T), and write Y for the Yoneda embedding **Geom**(T) $\to$ **Set**$^{\mathbf{Geom}(T)^{op}}$ (the Yoneda embedding of a category C into the presheaf category **Set**$^{\mathbf{C}^{op}}$ sends each object A of C to the contravariant hom-functor determined by A). If L is the sheafification functor $\mathrm{Set}^{\mathrm{Geom}(T)^{op}} \to \mathrm{Set}[T]$, it can be shown that the composite functor $U(T)$

$$\mathrm{Geom}(T) \xrightarrow{Y} \mathrm{Set}^{\mathrm{Geom}(T)^{op}} \xrightarrow{L} \mathrm{Set}[T]$$

is geometric and is in fact a model of T.

The model $U(T)$ of T in $\mathrm{Set}[T]$ has the following remarkable *universal property*. Calling a topos of sheaves on a site a *Grothendieck topos* (in particular $\mathrm{Set}[T]$ is a Grothendieck topos), then for any Grothendieck topos E and any T-model F in E, there is a unique geometric morphism (the natural arrows in the category of toposes) $\gamma \colon \mathrm{Set}[T] \to \mathrm{E}$ such that the diagram

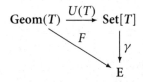

commutes. $U(T)$ is called the *universal model* of T and $\mathrm{Set}[T]$ the *classifying topos* of T.

This explains the notation $\mathrm{Set}[T]$: for it may be thought of as the topos obtained by adjoining a universal model of T to Set.

Since geometric morphisms are known to preserve the truth of geometric formulas (which incidentally explains the use of the term "geometric formula"), it follows from the universal property of $U(T)$ that, if we call a model in a Grothendieck topos (Grothendieck) *topos-based*, then *the geometric implications true in $U(T)$ are precisely those which are true in every topos-based T-model*. It can also be shown that $U(T)$ is a generic model of T in the sense that a geometric implication φ is true in $U(T)$ if and only if $\Sigma \vdash \varphi$.

From this we deduce the following:

Completeness Theorem for Geometric Theories Given a geometric theory T and a geometric implication φ, $T \vdash \varphi$ if and only if φ is true in every topos-based T-model.

Somewhat surprisingly, this can be strengthened, by a result due to Barr (1974), to assume a "classical" form in which "topos-based" is replaced by "*Boolean*-topos-based". (A topos is *Boolean* if the classical law of excluded middle holds in its internal logic.) It follows that *a geometric implication classically deducible from a geometric theory is also intuitionistically deducible from it*.

For an arbitrary (intuitionistic) first-order theory T, it is not possible in general for a classifying topos in the above sense to exist, but by modifying the construction of $\mathrm{Set}[T]$, a Grothendieck topos $\mathrm{E}(T)$ containing a generic model of T can be obtained. Remarkably, if the language of T contains just countably many symbols, then $\mathrm{E}(T)$ can be taken to be the topos of sheaves over the Cantor discontinuum. This is a categorical version of the old theorem that the theorems of the first-order intuitionistic calculus are precisely those which are true under all interpretations in closed subspaces of the Cantor discontinuum.

In conclusion here are some examples of geometric theories and their classifying toposes.

- *The theory* Ob *of objects*. Here the appropriate language has a single sort and no relation or function symbols, and the theory itself no axioms. A model of Ob in a topos E is then just an object of E. The classifying topos **Set**[Ob] is the functor category **Set**^{finord}, where **finord** is the category of finite ordinals and all maps between them.
- *The theory* D *of discrete objects*. Here the appropriate language has a single sort and a binary predicate $\neq$ and D has the axioms

$$\forall x \forall y (\top \Rightarrow x = y \vee x \neq y) \quad \forall x (x \neq x \Rightarrow \bot).$$

A model of D in a topos is a discrete object in E, i.e., an object A for which the diagonal arrow $\Delta: A \to A \times A$ has a complement. The classifying topos **Set**[D] is the functor category **Set**$^{\text{finord}^*}$, where **finord*** is the category of finite ordinals and injective maps between them.
- *The theory* Rng *of rings*. Here the appropriate language has one sort A, two function symbols $+, \cdot$ with signatures $((A, A), A)$ representing addition and multiplication on A, and one function symbol 0 with signature $((\,), A)$. The theory Rng has as axioms geometric implications asserting that $(A, +, \cdot, 0)$ is a ring. A model of Rng in a topos E is a ring object in E. The classifying topos **Set**[Rng] is the functor category **Set**^{FPRng}, where **FPRng** is the category of *finitely presented rings*, whose objects are quotients of rings of polynomials over the integers.
- *The theory* Re *of a Dedekind real number*. Here the language has sorts N and Q corresponding to the natural numbers and rationals and two relation symbols U, V of signature (Q). The axioms of Re are geometric implications asserting that a pair (U, V) is a Dedekind cut in Q. A model of Re in a (Grothendieck) topos E is a Dedekind real in E. The classifying topos **Set**[Re] of Re is **Sh**($\mathbb{R}$), the category of sheaves over the real numbers with the usual topology. This is the topos obtained by adjoining a generic real number to **Set**.

References

Artin, M., Grothendieck, A., and Verdier, J.-L. (1964). Séminaire de géometrie algébrique, iv: Théorie des topos, 269–70, in *Lecture Notes in Mathematics*. Springer, Berlin.

Barr, M. (1974). Toposes without points. *Journal of Pure and Applied Algebra* 5(3), 265–80.

Beeson, M. J. (1985). Foundations of constructive mathematics, 6, in *Ergebnisse der Mathematik und ihrer Grenzgebiete*, third series. Springer-Verlag, Berlin.

Bell, J. L. (1988). *Toposes and Local Set Theories: An Introduction*. Clarendon Press, Oxford. Reprinted by Dover, 2008.

Boileau, A. (1975). Types vs. Topos. PhD thesis, Univesité de Montreal.

Boileau, A., and Joyal, A. (1981). La logique des topos. *Journal of Symbolic Logic* 46(1), 6–16.

Butz, C., and Johnstone, P. (1998). Classifying toposes for first-order theories. *Annals of Pure and Applied Logic* 91(1), 33–58.

Cohen, P. (1966). *Set Theory and the Continuum Hypothesis (Lecture Notes Given at Harvard University, Spring 1965)*. W. A. Benjamin, New York.

Coste, M. (1974). Logique d'ordre supérieur dans les topos élémentaires. *Séminaire dirigé par Jean Bénabou*. Université Paris-Nord, Paris.

Diaconescu, R. (1975). Axiom of choice and complementation. *Proceedings of the American Mathematical Society* 51, 176–8.

Eilenberg, S., and MacLane, S. (1945). General theory of natural equivalences. *Transactions of the American Mathematical Society* 58(2), 231–94.

Fourman, M. P. (1977). The logic of topoi. *Studies in Logic and the Foundations of Mathematics* 90, 1053–90.

Freyd, P. J. (1966). The theory of functors and models, in J. W. Addison, L. Henkin, and A. Tarski (eds), *The Theory of Models: Proceedings of the 1963 International Symposium at Berkeley*, 107–20. North-Holland, Amsterdam.

Johnstone, P. T. (1977). *Topos Theory*. Academic Press, San Diego, CA.

Johnstone, P. T. (2002). *Sketches of an Elephant: A Topos Theory Compendium*, Vols. 1 and 2. Oxford University Press, Oxford.

Kock, A., and Reyes, G. E. (1977). Doctrines in categorical logic, in *Handbook of Mathematical Logic*, 284–313. North Holland, Amsterdam.

Lambek, J. (1968). Deductive systems and categories i. *Mathematical Systems Theory* 2(4), 287–313.

Lambek, J. (1969). Deductive systems and categories ii, in *Category Theory, Homology Theory, and their Applications, Lecture Notes in Mathematics*, 76–122. Springer, Berlin.

Lambek, J. (1972). Deductive systems and categories iii: Cartesian closed categories, intuitionist propositional calculus and combinatory logic, in *Toposes, Algebraic Geometry and Logic, Lecture Notes in Mathematics*, 57–82. Springer, Berlin.

Lambek, J. (1974). Functional completeness of cartesian categories. *Annals of Mathematical Logic* 6(3), 259–92.

Lambek, J. (1995). Some aspects of categorical logic. *Studies in Logic and the Foundations of Mathematics* 134, 69–89.

Lambek, J., and Scott, P. J. (1986). *Introduction to Higher-Order Categorical Logic*. Cambridge University Press, Cambridge, UK.

Lawvere, F. W. (1963a). Functorial semantics of algebraic theories. *Proceedings of the National Academy of the Sciences* 50, 869–72.

Lawvere, F. W. (1963b). Functorial semantics of algebraic theories. Dissertation, Columbia University.

Lawvere, F. W. (1964). An elementary theory of the category of sets. *Proceedings of the National Academy of the Sciences* 52, 1506–11.

Lawvere, F. W. (1965). Algebraic theories, algebraic categories, and algebraic functors, in *The Theory of Models: Proceedings of the 1963 International Symposium at Berkeley*, 413–18. North Holland, Amsterdam.

Lawvere, F. W. (1966). Functorial semantics of elementary theories. *Journal of Symbolic Logic* 31(2), 294–5.

Lawvere, F. W. (1967). Theories as categories and the completeness theorem. *Journal of Symbolic Logic* 32(4), 562.

Lawvere, F. W. (1968). Some algebraic problems in the context of functorial semantics of algebraic structures, in S. Mac Lane (ed.), *Reports of the Midwest Category Seminar, II*, Vol. 61 of *Lecture Notes in Mathematics*. Springer, Berlin.

Lawvere, F. W. (1969). Adjointness in foundations. *Dialectica* 23, 281–96.

Lawvere, F. W. (1970a). Equality in hyperdoctrines and the comprehension schema as an adjoint functor, in A. Heller (ed.), *Proceedings of the New York Symposium on Applications of Categorical Algebra*, 1–14. American Mathematical Society, Providence, RI.

Lawvere, F. W. (1970b). Quantifiers and sheaves, in *Actes du congres international des mathematiciens, Nice* 1, 329–34.

Lawvere, F. W. (1972). Introduction, in *Toposes, Algebraic Geometry and Logic*, 1–12. Springer, Berlin.

Lawvere, F. W. (1975). Continuously variable sets: Algebraic geometry = geometric logic, in R. H. E. and S. J. C. (eds), *Logic Colloquium 73*, 135–56. North Holland, Amsterdam.

Lawvere, F. W. (1976). Variable quantities and variable structures in topoi, in A. Heller and M. Tierney (eds), *Algebra, Topology, and Category Theory: A Collection of Papers in Honor of Samuel Eilenberg*, 101–31. Academic Press, San Diego, CA.

Lawvere, F. W., Maurer, C. and Wraith, G. C. (eds) (1975). *Model Theory and Topoi*, Vol. 445 of *Lecture Notes in Mathematics*. Springer, Berlin.

Linton, F. E. J. (1969a). Applied functorial semantics ii, in *Seminar on Triples and Categorical Homology Theory (ETH, Zurich, 1966/67)*, 53–74. ETH, Zurich.

Linton, F. E. J. (1969b). An outline of functorial semantics, in *Seminar on Triples and Categorical Homology Theory (ETH, Zurich, 1966/67)*, 7–52. ETH, Zurich.

Linton, F. E. J. (1970). Applied functorial semantics i. *Annali di Matematica Pura ed Applicata* 86(4), 1–13.

Mac Lane, S., and Moerdijk, I. (1992). *Sheaves in Geometry and Logic: A First Introduction to Topos Theory*. Springer-Verlag, Berlin.

Makkai, M. (1981). The topos of types, in *In Logic Year 1979–80*, Vol. 859 of *Lecture Notes in Mathematics*, 157–201. Springer, Berlin.

Makkai, M., and Reyes, G. E. (1976). Model-theoretic methods in the theory of topoi and related categories. I. *Bull. Acad. Polon. Sci. Sér. Math. Astronom. Phys.* 24(6), 379–84.

Makkai, M., and Reyes, G. E. (1977). *First-Order Categorical Logic*, Vol. 611 of *Lecture Notes in Mathematics*. Springer, Berlin.

Martin-Löf, P. (1984). *Inuitionistic Type Theory: Notes by Giovanni Sambin of a Series of Lecture Given in Padova, June 1980*. Bibliopolis.

Mitchell, W. (1972). Bocategories and the theory of sets. *Journal of Pure and Applied Algebra* 2, 261–74.

Moerdijk, I. (1995). *Classifying Spaces and Classifying Topoi*, Vol. 1616 of *Lecture Notes in Mathematics*. Springer, Berlin.

Osius, G. (1975). A note on kripke-joyal semantics for the internal language of topoi, in F. W. Lawvere, C. Maurer and G. C. Wraith (eds), *Model Theory and Topoi*. Springer, Berlin.

Reyes, G. E. (1974). From sheaves to logic. *Studies in Logic and the Foundations of Mathematics*. MAA Studies in Math 9. Springer, Berlin.

Reyes, G. E. (1977). Sheaves and concepts: a model theoretic interpretation of grothendieck topoi. *Cahiers de topologie et géométrie différentielle* 18(2), 105–37.

Sambin, G., and Smith, J. M. (1998). *Twenty Five Years of Constructive Type Theory*, Vol. 36. Oxford University Press, Oxford.

Ščedrov, A. (1984). *Forcing and Classifying Topoi*, Vol. 48 of *Memoirs of the American Mathematical Society*. American Mathematical Society, Providence, RI.

Seely, R. A. (1984). Locally closed categories and type theory, in *Mathematical Proceedings of the Cambridge Philosophical Society*, Vol. 95, pp. 33–48. Cambridge University Press, Cambridge, UK.

Seely, R. A. (1987). Categorical semantics for higher order polymorphic lambda calculus. *Journal of Symbolic Logic* 52, 969–89.

Szabo, M. E. (1978). *Algebra of Proofs*, vol. 88 of *Studies in Logic and the Foundations of Mathematics*. North Holland, Amsterdam.

Tierney, M. (1972). Sheaf theory and the continuum hypothesis, in *Toposes, Algebraic Geometry and Logic*, Vol. 274 of *Lecture Notes in Mathematics*, 13–42. Springer, Berlin.

Tierney, M. (1973). Axiomatic sheaf theory: some constructions and applications, in *Categories and Commutative Algebra (C. I. M. E, III Ciclo Varenna, 1971)*, 249–326. Edizioni Cremonese, Florence.

Tierney, M. (1976). Forcing topologies and classfying topoi, in *Algebra, Topology, and Category Theory (A Collection of Papers in Honor of Samuel Eilenberg)*, 211–19. Academic Press, San Diego, CA.

Zangwill, J. (1977). Local set theory and topoi. M.Sc. thesis, Bristol University.

8

Unfolding FOLDS: A Foundational Framework for Abstract Mathematical Concepts

Jean-Pierre Marquis

8.1 Introduction

One of the greatest intellectual accomplishments of the twentieth century is the creation and the development of a *science* of the foundations of mathematics. The roots of this development go back to the nineteenth century.[1] The development of first-order logic together with the development of set theory lead to the constitution of a scientific foundational framework.[2] On the one hand, a precise formal system with definite properties in which axioms for various mathematical systems can be written was developed. On the other hand, a precise and mathematically defined universe of entities, e.g. a universe of pure sets, in which these axiomatic presentations can be interpreted and for which various results and theorems can be proved was constructed. The latter was conceived as the ontological component of the discipline, linked to the formal system by a rigorous semantic. These precise formal developments constituted a science in the following sense. First, both these components are taken as capturing significant properties of *already given* components of mathematical knowledge, e.g. a systematic but informal language and a system of informal mathematical concepts. Second, the precise and rigorous framework and its accompanying theoretical means

---

[1] One could possibly go back to Leibniz to look for preliminary, informal ideas on the subject, but I believe that a reasonable demarcation point is the constitution of precise technical tools and in this regard, it is hard to deny that the twentieth century provided the first real, precise, and technical developments. However, the *scientific* character of the foundational enterprise must be underlined and that feature came about only in the course of the twentieth century. For a foundational framework to be considered scientific, it must satisfy certain properties. I believe that this is a crucial component of Makkai's view on the subject.

[2] The history of the field is, as almost all histories are, quite convoluted. In particular, it is important to understand that the status that first-order logic acquired is directly linked to its role in the development of axiomatic set theory. See, for instance, Moore (1980, 1987, 1988); Ferreirós (1996, 2001, 2007); Mancosu et al. (2009); Schiemer and Reck (2013).

allow for the rigorous and exact proofs of many important and significant results, e.g. completeness and compactness of first-order logic, Gödel's incompleteness results, the consistency and the independence of the continuum hypothesis with the usual axioms of set theory, to mention but the most obvious. One should also keep in mind the creation and development of model theory and proof theory which can be seen as specific articulations and emphasis of various components of this picture.

The success of this intellectual enterprise should also be seen as giving us a set of norms for *any scientific* foundational framework.[3] Thus, *any scientific* foundational framework should be based on the following components:

1. A precise and explicit mathematical syntax, also known as a formal system or sometimes a language, $\mathcal{L}$, given by recursive rules, together with a deductive structure, also known as a logic; the latter can be given independently or it may be inherent to the formal system right from the start;[4]
2. A mathematical construction or definition of a universe $\mathbb{U}$ of mathematical objects; and
3. A systematic interpretation of a theory written in the language in the constructed universe such that some of the propositions of the theory, usually called "axioms", become "obviously" true under that interpretation.

As we have said, in the present state of affairs, the formal system is first order logic with the usual axioms or rules and the theory is given by the axioms of Zermelo-Fraenkel set theory. The specific universe intended is provided by the cumulative hierarchy, thus the universe of pure or regular sets.[5]

It should be clear from our opening sentence that we believe that the standard framework has much to commend. Our goal in this chapter is to present a different foundational framework which captures another, complementary aspect of mathematical concepts that the logician Michael Makkai has been developing. It is not that the standard framework must be rejected. It does what it does very well. But it suffers from a conceptual limitation: it does not model an aspect of mathematical knowledge that emerged concurrently with the development of the foundational framework itself.

---

[3] I emphasize the scientific character of the enterprise once again, for some might argue that the foundations must ultimately rest upon simple informal ideas or conceptions and that, for that reason, *formal* set theory or *formal* category theory cannot provide "real" foundations. I take it that this is or is sufficiently close to Mayberry's view as expressed in Mayberry (2000). By underlining the scientific character of the discipline, I want to emphasize the fact that the methods used in the science of foundations are the same as in the other sciences. Thus, there is, as I have said, an informal or intended collection of facts that the foundational system is supposed to capture, reflect, and illuminate. In that respect and as in the natural sciences, there may be a chasm between the scientific image resulting from the theoretical work and the pretheoretical ideas used and known.

[4] An instance of the latter case is now given by Homotopy Type Theory. The standard presentation of first-order logic separates the purely linguistic component from the logical system.

[5] Of course, there numerous, incompatible models of the universe of set theory and it is part of the scientific foundational enterprise to build and investigate the properties of these universes to adjudicate their values and merits.

As is often the case in the history of ideas, it is only when an alternative theoretical framework is sufficiently developed that one can clearly see a "flaw", "limitation", or an "anomaly" in the accepted theory.[6]

The "anomaly" of the current set-theoretical picture has to do with the fact that it fails to capture the *abstract* character of contemporary mathematical concepts. With the advent of the abstract method and the resulting development of conceptual mathematics, a large portion of contemporary mathematics has become resolutely abstract. Of course, what the latter expression means is debatable.[7] Be that as it may, the anomaly comes from the codification of the abstract part of contemporary mathematics into the universe of pure sets. The abstract nature of the concepts involved is simply lost by the codification into pure sets. To see this, we need to delve into the nature of abstract mathematical concepts and pure sets. Once this is done, we will sketch the basic components of the alternative foundational framework that is emerging and present some of its philosophical implications.

8.2 Some generalities about abstract mathematical concepts

8.2.1 Contrast and compare: pure sets

Let us start with a recapitulation of the universe of pure sets, for this will allow us to exhibit the main conceptual differences between the latter and the picture we are introducing. Informally, a pure set is a collection of pure sets. This apparently circular characterization can be explained analytically or synthetically. Analytically, it means that when one decomposes a pure set into its elements, one again finds pure sets and this process repeats itself uniquely until at the end of each branch of the decomposition one finds the empty set. Synthetically, it means that one starts with the empty set, which is a pure set, and constructs sets from there by various well-known (possibly infinitary) operations.[8]

---

[6] Thus, instead of seeing set theory and category theory as rivals, one could draw an analogy with the case of Newtonian physics and relativity theory or classical physics and quantum physics.

[7] One thing *is* sure. It certainly does not have the usual ontological meaning. Mathematicians certainly do *not* mean, when they talk about abstract algebra, that that kind of mathematics has no spatio-temporal coordinates or that it is causally inert, etc. They do not have in mind the traditional philosophical distinction between the abstract and the concrete. I, for one, am convinced that they have an epistemological distinction in mind. See, for instance, Marquis (2014). The underlying view of what it is to be abstract is here rooted in the way it is understood in the development of mathematics, particularly in the twentieth century. It is certainly close to some views defended in the philosophical literature, for instance Nodelman and Zalta (2014). However, we refrain from using the term "structuralism" in this context and prefer to concentrate on the abstract character of the concepts. The term "structuralism" as it is used by most philosophers is appropriate to but a small fragment of the universe we have in mind here.

[8] Needless to say, this is very informal and one must be careful. In the analytic description, carelessness will lead to non-well-founded sets, whereas in the synthetic description, one must make sure to have infinitary operations at hand; otherwise, the universe will contain only hereditarily finite sets. The latter are certainly important, but they clearly do not provide the proper picture of the usual universe of sets.

There is no need to present and discuss the universe of pure sets in great detail, for our point here is very simple.[9] I claim that pure sets do not encode abstract mathematical concepts properly.[10] The introduction of pure sets in the foundational landscape provides a clean and tidy house: every object has a definite structure, has a definite nature, and can be identified uniquely by its elements. If pure sets had physical properties, they would be prototypically concrete entities. To clarify this latter statement, consider an arbitrary pure set X and an element $x \in X$. The element x is itself a pure set and it is therefore itself entirely determined by its own elements. In that sense, x has an independent identity from X. In fact, x has its own unique identity in the universe of pure sets. To illustrate this, consider a specific pure set; e.g. let X be the pure set $\{\{\emptyset\}, \{\emptyset, \{\emptyset\}\}\}$ and let x be $\{\emptyset\}$. Both x and X are pure sets and $x \in X$. Although it cannot be said that pure sets are concrete in the standard ontological sense of that expression—that is, they do not possess spatio-temporal coordinates nor are they causally efficacious, at least in the way molecules are—they certainly are concrete in an epistemic/semantic sense. We have a complete picture of the components and the composition of a pure set. Furthermore, all this information is contained in the notation itself. When we write down a specific pure set, e.g. $\{\{\emptyset\}, \{\emptyset, \{\emptyset\}\}\}$, we can say exactly what are its components and how these are put together to yield that particular, unique pure set. Last, but not least, this also allows us to see why the axiom of extensionality holds for pure sets. The latter only confirms the fact that pure sets—in fact, any system of sets satisfying the Zermelo-Fraenel (ZF) axioms—should be thought of as being particulars or individuals.[11] Indeed, the axiom of extensionality constitutes the criterion of identity for ZF-sets: two sets X and Y are the same if and only if they have the same elements. The beauty of pure sets is that the criterion of identity is homogeneous or uniform: it works all the way down on the elements of pure sets; that is, to determine when two given elements of X and Y are the same is still an internal matter. Let us now contrast this with the way mathematicians talk about abstract mathematical concepts.

---

[9] Category theorists have been critical of the underlying conception of sets provided by ZF from the 1960s onwards. These criticisms are, in fact, pointing in the direction we are engaging in, without emphasizing the abstract nature of the notion of sets defended. In fact, many opponents to the views articulated then thought that category theorists were simply trying to eliminate sets altogether, which lead to a profound misunderstanding between the two groups. For some of the critical arguments and alternative theory presented, see, for instance, McLarty (1993, 2004); Leinster (2014).

[10] In a sense, one could interpret Benacerraf's well-known argument as saying just that. In fact, as Makkai has already mentioned, FOLDS provides an elegant and direct solution to that problem. See, for instance, Makkai (1998, 1999).

[11] I prefer this terminology to the term "concrete". In fact, the term "abstract" should perhaps be replaced by the term "universals" as the latter was used in philosophy. The problem is not so much with the term "abstract", but more with the term "concrete". It is more difficult to talk about concrete mathematical entities. The term "abstract" in mathematics is linked to the abstract method or to a method of abstraction and, as such, should be opposed to a method of representation. For more on the relevance of these distinctions in the present context, see Ellerman (1988); Marquis (2000).

8.2.2 Compare and contrast: being abstract

"Let G be an abstract group."

This is a common way of talking in contemporary mathematics, say in group theory or in representation theory. If every mathematical object is a set, then the abstract group G ought to be a set or have an underlying set. Can this underlying set be a *pure* set? Suppose it is. Then, it should be a particular pure set X_G with specific properties, e.g. having specific elements, e.g. $\{\{\emptyset\}, \{\emptyset, \{\emptyset\}\}\} \in X_G$. But the specification that the group G is an *abstract* group is given *precisely* to avoid this situation; that is, the elements and the properties of the underlying set as a specific, concrete set are completely irrelevant. What mathematicians want to say in this context is that the abstract group has an underlying *abstract* set and the properties of the latter that ought to be considered are those that a set possesses as an abstract set and nothing else.

The situation is even more strange when one considers *operations* on abstract groups.[12] Thus, let G and H be abstract groups. Now, suppose a mathematician was to present the following construction: consider the underlying sets X_G and Y_H of G and H and take their intersection $X_G \cap Y_H$. And now suppose that the mathematician suggests to work with the latter set-theoretical construction to get some results about abstract groups. Now, if the underlying sets X_G and Y_H are pure sets, then they have a well-defined and unique intersection which is a pure set. Clearly, the latter has nothing to do with the group structure of G and H. The construction presented would certainly be judged awkward and irrelevant.

When a mathematician says "let G be an abstract group", she probably has two features of G in mind: (1) The precise nature of the elements of G is left unspecified, and (2) the only properties that she is interested in are those attributable to G as a group. More specifically, the first feature means that the elements of G can be presented in various ways and be of various types, the nature of which can vary considerably from one embodiment to the other and the properties of these elements are irrelevant. The second feature means that our mathematician knows what it is to be a group-theoretical property. A mathematician certainly learns to identify the relevant group-theoretical properties and distinguish them from, say, set-theoretical properties. In the vernacular language, a mathematician can identify a group-theoretical property when she sees one. Is there a way to formally express what the latter means? Structure-preserving maps, in this case group homomorphisms, preserve some group-theoretical properties. As is well known, this is not enough. Some important algebraic properties are not preserved by homomorphisms.[13] The proper answer is this: the relevant properties are those that are preserved under the right criterion of identity for the entities considered. In the case of groups, this means that the relevant properties are those invariant under group isomorphisms.[14]

---

[12] This example is given by Makkai at various places.

[13] For instance, an ideal of a ring is not necessarily preserved by an arbitrary ring homomorphism.

[14] This invariance property has also been associated with abstraction in modern mathematics in the literature. The first person to explicitly make the connection is, to my knowledge, Hermann Weyl in Weyl (1949). It was discussed and developed by Stephen Pollard in two articles. See Pollard (1987, 1988).

Our mathematician certainly thinks that the abstract group G has an underlying *abstract* set.[15] An abstract set is basically a set whose elements have no structure. It is therefore made up of faceless, or in the words of Fréchet, abstract individuals. The *only* information we have about such elements is that it is possible to differentiate them somehow; that is, each set X comes equipped with an identity relation that allows us to tell, for two elements x and y of X, what is the truth value of the proposition "$x =_X y$". Since the elements x and y are abstract, they have no independent existence outside of X. They are given by X.

If an abstract set has abstract elements, to what extent do these elements determine the identity of that set? There is no reason to believe that they should. The identity of an abstract set should not depend on the identity of its elements. One key observation leads us towards the solution. Abstract elements together with an identity relation allow us to take the notion of *function* between abstract sets as primitive. Thus, abstract sets are taken to form a system. After all, if these abstract sets have to capture the fact that certain abstract mathematical concepts have underlying abstract sets and if these abstract mathematical concepts are naturally connected to one another or exhibit some form of conceptual dependence, then it seems reasonable to expect that some of these dependences are already noticeable at the level of sets. However, once the notion of function is available and given that there is a criterion of identity for functions, then there is a natural criterion of identity that emerges naturally, namely the notion of bijection between abstract sets. In fact, it makes perfect sense to say that two abstract sets are identical if there is a bijection between them. For this is indeed a criterion of identity that applies to a set as a unit, independently of the nature of its elements, as it should be.

The question naturally arises at this stage as to whether one can construct a theory of abstract sets as a special case of a theory of abstract mathematical concepts. Indeed, such a theory can be developed and has been developed. I refer the reader to Makkai (2013) for details.[16]

The foregoing discussion contrasting pure sets with abstract sets allows us to postulate two basic principles about abstract mathematical concepts in general. First, we posit that one of the specific properties of abstract concepts is that they are known via their instances, that is something that is seen or conceived as being an instance of

---

[15] The French mathematician Maurice Fréchet, who was one of the pionners of the abstract method in mathematics, described abstract sets thus:

> In modern times it has been recognized that it is possible to elaborate full mathematical theories dealing with elements of which the nature is not specified, that is with abstract elements. A collection of these elements will be called an *abstract set*. (...)
> It is necessary to keep in mind that these notions are *not of a metaphysical nature*; that when we speak of an abstract element we mean that the nature of this element is indifferent, but *we do not mean at all that this element is unreal*. Our theory will apply to all elements; in particular, applications of it may be made to the natural sciences. (Fréchet, 1951, 147)

This is a specific quote by a mathematician that specifies that the property of being abstract is epistemological rather than ontological.

[16] The question as to which mathematical concepts ought to have underlying abstract sets can also receive a precise mathematical answer.

that concept. It is important to understand that the last claim means that the instance is given as an instance, not as something that has an independent identity. It certainly can, but in the context where the abstract concept is given, the instance is dependent on that concept and is not and cannot be considered independently of the concept. Thus, abstract concepts come *together with* and are *inseparable from* instances. They are *not*, however, *identified* with these instances. We think about the abstract concepts with the instances and never solely with the concepts. I believe that this is a fundamental cognitive aspect of abstract mathematical thinking.

Second, abstract mathematical concepts are not given individually, independently of one another. There is a natural order, a natural organization of concepts. In fact, some concepts depend upon others, previously given, concepts. The easiest way to illustrate the idea is via the idea of dependent variable in the natural sciences. When various concepts are linked to one another by a functional relation, then certain concepts depend upon others. In these cases, a function f represents a certain dependency between concepts and the function is tied to the conceptual context. It might be mathematically simple and, in a direct sense, detachable from that context as a mathematical operation, but it appears in the conceptual framework and plays a role as such in that framework. The surprising fact is that something similar occurs in mathematics and in such a way that it ought to be captured by a foundational framework.

Although this is not quite a form of conceptual holism, it is certainly at odds with a long standing tradition in analytical philosophy, namely logical atomism.[17] It is not a form of holism simply because there is an organization which allows one to separate some parts from the others and consider certain components independently of others. However, the latter cannot be done arbitrarily nor can it be done completely, as if one would decompose a living cell down to its atoms and thus believe to completely understand what life is. These informal, general, and imprecise remarks will hopefully become clearer once we will have introduced the framework more formally.

8.2.3 Informal remarks about the syntax and the logic for abstract concepts

Both of these basic epistemological tenets are reflected in the grammar of the theory itself. The first tenet is encoded by adopting the following convention: we write "$x : X$" to declare that x is an instance of the concept X. Since this is a declaration, the expression "$x : X$" is *not* a proposition. It cannot be true or false. We can immediately infer that the string of symbols "$\neg(x : X)$" will not be well-formed in the syntax of the system. The instance x comes with the concept X.[18] The instance x cannot be

---

[17] It is certainly not a coincidence that the doctrine of logical atomism made its appearance soon after the acceptance of the doctrine of atomism by physicists. One wonders how atomism in general influenced the thinking about the foundations of mathematics and the establishment of set theory as such a foundation.

[18] Some would say that it is constructed from X. For the time being, we want to stay away from that terminology. The main point is simply that, to know X, one must consider x and x naturally comes with X by a certain, unspecified process. The latter process can be clarified in the semantics. For instance, in a

considered by itself. It is only seen as an instance of X, thus bringing to the front some of its features and pushing in the background other features. In contrast with the case of set theory, we do not have x on the one hand, and the concept X on the other and verify that indeed x has the right properties associated with X. Thus, abstract mathematics starts with concepts together with instances of the latter. This is an ontological shift with respect to the standard set-theoretical picture. In the latter, a set is built up from its elements, whereas in the picture I am presenting, the instances are always presented as such, that is as representations of a given abstract concept.[19]

The second tenet is expressed by the fact that we are introducing a language with dependent sorts.[20] These dependencies put strict constraints on the grammar of the language, as we will see.

Let us now come back to the development of abstract mathematical concepts. While a science of the foundations of mathematics was put on firm grounds, mathematics itself was undergoing profound changes. The abstract method played a key role in these changes. With its help, mathematicians started to talk and theorize about monoids, groups, rings, fields, vector spaces, topological spaces, differential manifolds, Banach spaces, Hilbert spaces, partially ordered sets, lattices, categories, homology and cohomology theories, abelian categories, triangulated categories, derived categories, monoidal categories, etc. Each one of these abstract concepts comes with its own criterion of identity, different from the criterion of identity for sets. In fact, the criterion of identity for these concepts is extracted from the concepts. It is not given *a priori*. For example, the criterion of identity for groups is the notion of group isomorphism, the criterion of identity for topological spaces is given by the notion of homeomorphism, and the criterion of identity for categories is the notion of equivalence of categories. We take it that this facet of abstract mathematical concepts ought to be reflected directly in the foundational framework. In particular, the framework should not have a universal identity relation, usually denoted by the equality sign "$=$". The criterion of identity relevant for the entities at hand must be determined by the abstract concepts themselves. The criterion of identity should be *derived* from the concept themselves.

Another key feature of abstract mathematical concepts is that although the concept determines the notion of identity for its instances, there can be many different identities between two given instances. Thus, there can be many different identities,

---

category C, an instance $x : X$ can be given by a morphism $x : 1 \to X$, where 1 denotes the terminal object of the category, when it has one. It can also be what is called a *generalized element* $x : U \to X$ from an arbitrary object U of the category.

[19] Thus, in some sense, we are trying to reintroduce a certain aspect of the comprehension principle. Of course, the principle itself is not introduced, but the constraints put on the grammar are such that there are no independently given atoms from which the universe can be built. To repeat what was already said in the previous section: concepts and their instances are woven together right from the start.

[20] This is not new, nor is the previous grammatical convention with sorted variables. These features have been used by logicians and computer scientists for more than forty years now. In particular, the specific language with dependent sorts or types, as they are also known, was introduced by Martin-Löf in 1970. There are new formal elements introduced by Makkai, as we will see in the next sections.

that is isomorphisms, between two given groups and even between one and the same group. This might sound strange and surprising, but it is as it should be for instances of abstract concepts. For two instances of an abstract concept are identical (or should we say "equivalent") whenever they have the same properties determined by the abstract concept they are instances of. Since these properties are preserved by any isomorphism between them, each and any one of these isomorphism constitute a way of being identical as instances of the given abstract concept. In fact, knowing ways of being identical, even self-identical, reveals a lot of information about the concept itself.

Furthermore, an instance of an abstract concept will have the properties determined by the concept, but it will also have other properties inherent to the mode of presentation used. Thus, in the context of pure sets, when groups are presented, they not only have group theoretical properties as they should, but they also have irrelevant set-theoretical properties. This may bring in a certain confusion. An adequate language for an abstract concept should allow us to write only properties that are relevant for the concept. The notion of relevance is here determined by the criterion of identity. Only the properties invariant under the given criterion of identity should be expressible in the language. More precisely, given a language $\mathcal{L}$, a signature $\mathcal{S}$ in $\mathcal{L}$, and a derived criterion of identity $x \simeq_\mathcal{S} y$, if $P(x)$ and $x \simeq_\mathcal{S} y$, then we should have $P(y)$. It should be possible to *prove* this invariance principle in the foundational framework for abstract mathematical concepts.

These remarks indicate that the language developed will have different properties from the standard syntax of the language for set theory.

8.2.4 *Informal remarks about the universe of abstract mathematical concepts*

The universe of abstract mathematical concepts will also differ considerably from the universe of pure sets, e.g. the cumulative hierarchy. Let us emphasize immediately one point they have in common: in both cases, we deal with hierarchies. However, this common feature is in fact very superficial, for the hierarchies are deeply different, both from an ontological point of view and from an epistemological point of view. In the case of abstract mathematical concepts, the hierarchy is based on the introduction of levels of abstraction, that is systems of different kinds, irreducible to one another.[21] In the case of sets, the universe is composed of a unique kind. The hierarchy in the cumulative hierarchy is determined by the *rank* of a set, that is the least ordinal number greater than the rank of any member of the set.[22]

---

[21] Contemporary mathematicians commonly talk about levels of abstraction. See, for instance, Marquis (2016), for a preliminary exploration of the idea. Introducing such a hierarchy seems to be a key idea in contemporary artificial intelligence. See, for instance, LeCun et al. (2015).

[22] Assuming the axiom of foundation, of course.

The basic structure of the universe can be given informally as follows.[23] The first level is made up of abstract sets.[24] As we have said, each abstract set X comes with an identity relation $=_X$ for its elements. The identity criterion for sets is given by the notion of bijection between sets. The totality of sets is not a set; it is a *category*. Most of the abstract mathematical concepts introduced in the last quarter of the nineteenth century and the first half of the twentieth century can be described and studied at this level. Thus, the category of monoids, groups, rings, fields, topological spaces, vector spaces, etc., form categories in this sense.[25]

Let us be a little bit more precise: abstract sets are connected by functions. The latter compose and satisfy the expected equalities and each abstract set has an identity function whose composition satisfies obvious equalities. Thus, the system of abstract sets is a category. Identities between sets are given by the notion of isomorphism or bijection between sets. Recall how one defines the notion of isomorphism between sets: an isomorphism between sets X and Y is a function $f : X \to Y$ such that there is a function $g : Y \to X$ satisfying $f \circ g = 1_Y$ and $g \circ f = 1_X$. Note the presence of the identity symbol between functions. The identity symbol is in fact coherent, since we have assumed that every abstract set X possesses an internal identity relation $=_X$ and it is these relations that are at work here. In fact, we should write $f \circ g =_Y 1_Y$ and $g \circ f =_X 1_X$ to be exact. This notion of isomorphism within a category works perfectly well for the usual set-based notions: isomorphism for groups, homeomorphism for topological spaces, diffeomorphism for manifolds, etc. are the criteria of identity for these concepts. So far so good.

We now have categories. Since there is no identity relation between the objects of a category—isomorphisms as defined in a given category play that role—an abstract category is not an abstract set. Hence a category is a new kind of object or system. Whereas sets are connected by functions, categories are connected by *functors*. When categories are defined as sets, a functor $F : \mathbf{C} \to \mathbf{D}$ is given by two functions, a function $F_O : Ob(\mathbf{C}) \to Ob(\mathbf{D})$ that sends objects of $\mathbf{C}$ to objects of $\mathbf{D}$ and a function $F_M : Mor(\mathbf{C}) \to Mor(\mathbf{D})$ sending morphisms of $\mathbf{C}$ to morphisms of $\mathbf{D}$ such that $F_M(1_X) = 1_{F_O(X)}$ for all objects X of $\mathbf{C}$ and $F_M(f \circ g) = F_M(f) \circ F_M(g)$.[26] This seems innocuous: it simply says that a functor preserves the structure of composition of morphisms of a category. Thus, a functor is a structure preserving morphism between categories. However, the definition is inconsistent with the facts—I am tempted to

---

[23] A more technical description will be given in Section 8.4.

[24] There is some fluctuation still. One could start with a different first level. But we won't get into these options here.

[25] Ironically, these categories, the category of monoids, groups, rings, fields, posets, lattices, topological spaces, manifolds, vector spaces, etc., that is categories of structured sets, are called "concrete categories" in the literature! The latter term has a precise technical meaning: a category **C** is said to be *concrete* whenever there is a faithful functor to the category of sets. It can be shown that there are non-concrete categories in this sense. See, for instance, Freyd (1973, 2004).

[26] For a covariant functor.

say "the reality"—of category *theory*.[27] The inconsistency can be seen from different angles. Here is a simple one. Within a category and from the perspective of category theory, any object isomorphic to a given one does the same categorical work. In other words, in a category, mathematics is done up to isomorphism. Now, when a functor F is defined as in the foregoing definition, it assigns to a given object X of **C** a unique object $F_O(X)$ of **D**. But, and this is the inconsistency introduced by the usual set-theoretical definition, any object isomorphic to $F_O(X)$ would do and should do. A functor cannot identify, in the usual sense of that word, two objects X_1 and X_2 of **C**, i.e. one cannot write $F_O(X_1) = F_O(X_2)$, since there are no identities between objects in a category.

Consider now the effect this fact has on the identity of categories themselves. In their original paper on categories, Eilenberg and Mac Lane treated the question of the identity of categories in a standard algebraic fashion for the period: they stipulated that two categories are identical whenever there is an isomorphism between them. An isomorphism between categories is defined in the expected way: a functor $F: \mathbf{C} \to \mathbf{D}$ is an isomorphism if there is a functor $G : \mathbf{D} \to \mathbf{C}$ such that $G \circ F = 1_\mathbf{C}$ and $F \circ G = 1_\mathbf{D}$. These equalities hold also for the *objects* of the categories involved, e.g. $G(F(X)) = 1_\mathbf{C}(X)$ for all objects X of **C**. Since there are no equalities between objects, the notion of isomorphism of categories is inadequate. Ironically, Eilenberg and Mac Lane's motivation for the introduction of categories was to provide the proper mathematical setting to express the notion of natural transformation, that is the appropriate notion of morphism between functors, and the latter is the key to the concept of identity for categories.

Indeed, to obtain the right notion of identity for categories, one must replace the identities between the compositions and the identity functors by natural transformations that are isomorphisms; i.e. the foregoing equations are replaced by the existence of two natural isomorphisms $\eta: G \circ F \simeq 1_\mathbf{C}$ and $\mu: F \circ G \simeq 1_\mathbf{D}$, satisfying certain obvious conditions. Thus, when one starts with an object X, coming back in the category via the composition of the functors G and F yields an object *isomorphic* to X. Whenever these functors and natural isomorphisms exist, the categories involved are said to be *equivalent*. This is the correct notion of identity for categories.

Thus, a system of categories is a new kind of system. It consists of categories, functors, *and* natural transformations. In other words, such a system is composed of

---

[27] I should emphasize that this fact was not and could not be obvious to Eilenberg and Mac Lane. The reason for this is simple: one could argue that although Eilenberg and Mac Lane introduced the concepts of category, functor, and natural transformation, they did not develop category *theory* in their original paper. For some of the core concepts of the theory were introduced 10 years later by Kan and Grothendieck. (For more on this, see, for instance, Krömer, 2007; Marquis, 2009.) The inconsistency as it is presented here was observed already in the 1960s and was made explicitly by Jean Bénabou and by G. Maxwell Kelly and probably many others. In Bénabou's mind, it gave rise to the development of an idea introduced by Lawvere, namely what are called distributors and it can be shown that a locally representable distributor is an anafunctor. Kelly introduced what essentially became the notion of anafunctor in Makkai's writings, but did not develop the theory. See Kelly (1964); Bénabou (1967). For the sake of completeness, I should mention that these notions are linked to the notions of profunctors and pseudofunctors in the literature. These remarks do not do justice to the history of the subject, but we will leave it at that.

objects, morphisms between objects, called 1-morphisms, and morphisms between morphisms, called 2-morphisms. Informally, it seems to be simple enough, but the complexity of the notion follows from the various ways these morphisms compose. There is an operation of composition for 1-morphisms and another one for 2-morphisms and these two operations necessarily interact and these interactions have to be coherent with one another. It is tempting to say that a system of categories must be a set and must be a category. From the point of view we are trying to develop, both claims are false. That is, from a specific theoretical point of view, a system of categories does *not* have an underlying abstract set. Furthermore, since a category only has (1-)morphisms and since a system of categories has 2-morphisms that play a key role in the structure of the system, a system of categories cannot be merely a category. Thus, a system of categories is an entity of a new kind, usually called a bicategory or a weak 2-category.[28] It takes more than one page to give the formal definition of a weak 2-category. The difficulty is not conceptual, but combinatorial, so to speak. We have now introduced a new kind of entity: a weak 2-category. One can easily convince oneself that there are many weak 2-categories and that there are morphisms between them. Thus, given the latter, we have that a system of weak 2-categories is made up of objects, 1-morphisms, 2-morphisms, and, these are now new, 3-morphisms between 2-morphisms. All these morphisms compose and these compositions must respect certain "laws". Once again, they form a system and one has to find a proper criterion of identity for weak 2-categories. At this stage, the reader will not be surprised by the claim that a system of weak 2-categories cannot be a set, a category, or a weak 2-category. It is a weak 3-category.

There is a clear pattern emerging. Informally, one expects this organization to go on and to consider weak n-categories which would constitute a totality of weak $(n-1)$-categories. The totality of all these would then be an ω-category and at this stage, it is possible that the latter would itself be an ω-category.[29]

One of the reasons motivating the formulation of FOLDS is precisely to provide an appropriate formal language to describe this hierarchy directly and properly. As we will briefly indicate, one of its central notions can be used to give the fundamental components of the universe we are trying to grasp.

The informal picture of the universe of abstract mathematical concepts is clear enough. The precise, technical mathematical picture is also becoming clear. We now

---

[28] There is also a notion of strict 2-category, but we will ignore the subtle differences between these notions.

[29] I should underline that there is no a priori necessity involved here. It could be that at some level, these notions stabilize into a unique notion. For instance, it is possible to show that every weak 2-category is biequivalent to a 2-category. This latter result can be interpreted as saying that the notion of weak 2-category is reducible to the notion of 2-category. However, this result does not hold for the notions of weak 3-categories and 3-categories. I will not give the definitions of all these mathematical concepts here. In fact, there are various different definitions in the literature, but their study is evolving rapidly. See Leinster (2002, 2004), for instance. I will sketch the formal approach for one of these concepts in Section 8.4.

have the barebones of a formal system, a universe of mathematical objects in which the language can be interpreted systematically. Let us now put some flesh on these bones.

8.3 The Formal System: FOLDS

We now come to the first component of the extended science of foundations: the formal system. We will not present the system of FOLDS in all its details. We will sketch some of the main features of the language.[30]

8.3.1 FOLDS-signatures

FOLDS is an extension of first-order logic (FOL). Thus, it has the usual quantifiers and propositional logical connectives, to which we add two propositional constants $\top$, $\bot$. The language comes with dependent sorts (or types) and variables are sorted. Thus, the quantifiers are always bounded since the variables are sorted.

All textbooks in model theory start by defining what is called a *language*. The latter is not arbitrary: a language $\mathcal{L}$ is designed to describe what is called a mathematical *structure*. Sometimes, an author specifies a *signature* for a given type of structure, but most of the time the notions of language and signature are identified.

Recall that a first-order language $\mathcal{L}$ or a $\mathcal{L}$-signature is given by the following:

1. a set of function symbols $\mathcal{F}$ and positive integers n_f for each $f \in \mathcal{F}$;
2. a set of relation symbols $\mathcal{R}$ and positive integers n_R for each $R \in \mathcal{R}$; and
3. a set of constant symbols $\mathcal{C}$;

where n_f and n_R are the usual functions giving the arities of the function symbols and relation symbols in both sets. Any of these sets can be empty, and it is enough, although in many cases unnatural, to work with a set of relation symbols only. This latter remark is important since this is how one can show that FOLDS is an extension of FOL.

FOLDS departs from FOL by introducing a different notion of $\mathcal{L}$-signature. This is as it should be given that the "structures" we are interested in are of different kinds than the usual structures based on abstract sets. As we have just remarked, this new notion of signature still covers the usual notion of signature. To understand the notion of

---

[30] FOLDS is a type theory and like all type theories, it tries to avoid the paradoxes not by stipulating certain axioms, but by constraining the linguistic resources right from the start. One of the underlying ideas is that the linguistic constraints should reflect a certain ontology. For a short history of type theory, see Kamareddine et al. (2002, 2012). When Makkai introduced FOLDS twenty years ago, Bart Jacobs' book on categorical logic had not been published. Dependent type theories were well known to computer scientists and logicians working in theoretical computer science. Jacobs gives a presentation of first-order dependent type theory in Chapter 10 of his book. His presentation and development of the framework is completely different from Makkai's development. Makkai's motivation is closer to Martin-Löf's motivation, although the latter was trying to provide a foundational framework for *constructive* mathematics, whereas I hope it is clear by now, Makkai is trying to articulate a foundational framework for *abstract* mathematics. See Jacobs (1999) for Jacobs presentation. For Martin-Löf's work, see Martin-Löf (1975, 1984).

FOLDS signature, observe that an n-ary relation symbol sorted as $R \subset X_1 \times \ldots \times X_n$ can be replaced by a new sort R together with operations $R \xrightarrow{p_i} X_i$. The latter symbol with the (sorted) operations is a one-way graph, that is a graph whose arrows are all going in the same direction. Thus, in the language of FOLDS, an n-ary relational symbol becomes a one-way graph with two levels and n operations. A FOLDS-signature is a generalization of that situation; that is, there can be (usually finitely) many levels and the morphisms between the levels compose. Instead of considering (one-way) graphs, it turns out to be simpler to use the notion of (one-way) category directly. Let us fix the terminology for the remaining sections of the paper: a *proper morphism* is a morphism different from an identity morphism. Before we give the definition of a FOLDS-signature, we need the following formal definition.

Definition 8.1 *A category $\mathcal{L}$ is said to be* one-way *if it satisfies the following conditions:*

1. *$\mathcal{L}$ is small;*
2. *$\mathcal{L}$ has the finite fan-out property: for any object K of $\mathcal{L}$, there are only finitely many morphisms with domain K; and*
3. *$\mathcal{L}$ is reversed well-founded: there is no infinite ascending chain $\langle K_n \xrightarrow{f_n} K_{n+1}\rangle_{n \in \mathbb{N}}$ of composable proper morphisms ($f_n \neq 1_{K_n}$).*

It follows from the definition that in any one-way category, the only morphism from an object K to itself is the identity morphism 1_K. The last condition of the definition implies that there are no cycles in a one-way category; that is there are no cycles $K_0 \to K_1 \to \ldots \to K_n$ of proper morphisms with $K_0 = K_n$.

Definition 8.2 *An $\mathcal{L}$-signature for FOLDS is a one-way category.*

The standard example of a FOLDS signature is the $\mathcal{L}_{Cat}$-signature for the concept of abstract category. Here is a representation of the one-way category with all the non-composite morphisms displayed and the identity morphisms omitted:

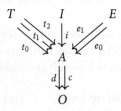

Since the signature is a category, the morphisms compose and there are identities between some of them. In fact, we must have the identities

$$d \circ t_0 = d \circ t_2 \qquad d \circ i = c \circ i \qquad d \circ e_0 = d \circ e_1$$
$$c \circ t_0 = d \circ t_1 \qquad \qquad \qquad c \circ e_0 = c \circ e_1$$
$$c \circ t_1 = c \circ t_2$$

There is a natural informal interpretation of this signature. The symbol 'O' stands for the sort of objects, 'A' for the sort of arrows, 'I' for the sort of identity arrows, 'E' for the equalities between arrows, and 'T' for the commutative triangles of a category. Given this informal interpretation, the identities between the composites become obvious. The morphisms between the sorts denote the dependencies involved. Reading from bottom to top, we understand that the morphisms depend upon the objects; more precisely, a morphism must be given with a domain and a codomain. Thus, in order to write down that the symbol 'f' is an arrow, we must first give '$x : O$', '$y : O$' and then write '$f : A(x,y)$'. The notation indicates that the symbol 'f' is of type 'A', for arrow, but the latter depends on the type 'O' of objects with two parameters, 'x' denoting the domain of 'f' and 'y' denoting its codomain. In fact the number of morphisms, including the composites, from a sort to another sort gives us the arity of the former sort and its dependence structure. Thus, the dependency structure yields constraints on the grammar of the language.

The definition of an $\mathcal{L}$-signature for FOLDS is deceptively simple, but the example illustrates some of the complexities that arise. There are clearly levels in a FOLDS-signature. The objects in the bottom level, which we call L_0, are not the domain of any proper morphism. When we go up, that is for levels $i > 0$, the sorts are made up of objects x for which all proper morphisms have a codomain in lower levels, e.g. in L_j, for $j < i$, and for which there is at least one proper morphism with codomain in the next lower level L_{i-1}. Thus, all proper morphisms go from a level to a lower level. We will call the objects of a $\mathcal{L}$-signature its *kinds* and, clearly, each kind K has a level. For any morphism p, K_p will denote the codomain of p. For any kind K, the (finite) set of all proper morphisms $p : K \to K_p$ will be denoted by $K|\mathcal{L}$.

It is not obvious to see how, given a FOLDS signature $\mathcal{L}$, one writes formulas and propositions in $\mathcal{L}$. When we are given a FOL-signature, we understand how terms are constructed, how atomic formulas are constructed, and how arbitrary formulas are constructed. As we have already said, the introduction of dependent sorts modifies how the grammar of the language works. Although the ensuing discussion is more technical, it is important to understand the impact this new notion of signature has on the structure of the syntax.

Let us now fix an arbitrary FOLDS-signature $\mathcal{L}$. The *sorts* and *variables* are defined recursively as follows. Let $n \in \mathbb{N}$ be a natural number and suppose that sorts of kinds of level less than n have been defined, as well as variables of such sorts.

Definition 8.3 *Let K be a kind of level n. A sort of kind K is a formal set*[31] $\langle 1, K, \langle x_p \rangle_{p \in K|\mathcal{L}} \rangle$, *which we will denote by $K(\langle x_p \rangle_{p \in K|\mathcal{L}})$, such that:*

1. *For every $p \in K|\mathcal{L}$, x_p is a variable of sort $K_p(\langle x_{p,q} \rangle_{q \in K_p|\mathcal{L}})$;*
2. *For every $q \in K_p|\mathcal{L}$, $x_{p,q} = x_{qp}$.*

---

[31] A formal set is a syntactic entity. Sets are here used to codify syntactic entities.

Roughly, a sort is obtained by filling the pth place of a kind K, for any p in the arity of $K|\mathcal{L}$ of K, by an appropriate variable x_p.

Definition 8.4 *Given a sort $X = K(\langle x_p \rangle_{p \in K|\mathcal{L}})$, a variable x of sort X is a formal set $\langle 2, X, \alpha \rangle$, where α is a formal set called the* parameter of x. *We write $x : X$ to indicate that x is a variable of sort X.*

We see how, even at the formal, syntactical level, a variable "comes" from a sort. A variable is not merely a symbol having a specific syntactical status and function. A variable in FOLDS always carries its origin upfront.

Definition 8.5 *For a sort $X = K(\langle x_p \rangle_{p \in K|\mathcal{L}})$, the* variables of X, $\mathrm{Var}(X)$ *are defined as $\{x_p : p \in K|\mathcal{L}\}$. If $x : X$, we write $\mathrm{Dep}(x)$ for $\mathrm{Var}(X)$.*

Looking at the signature $\mathcal{L}_{Cat}$ given in the standard example following Definition 8.2 and the foregoing definitions, the reader can easily convince herself that $O|\mathcal{L} = \varnothing$, $A|\mathcal{L} = \{d, c\}$, and, for instance, $T|\mathcal{L} = \{d \circ t_0, d \circ t_1, c \circ t_2, t_0, t_1, t_2\}$. Thus, $O(*)$, where $*$ denotes the empty sequence of variables, is a sort. So is $A(x_d, x_c)$, whenever $x_d : O, x_c : O$. Similarly, given $x : O, y : O, z : O$ and $f : A(x, y), g : A(y, z), h : A(x, z)$, then $T(x, y, z, f, g, h)$ is a sort of kind T.

The reader will have noticed that sorts and variables of a given sorts are, in fact, formal sets determined by the given signature. The strategy is extended to cover the notions of a free variable and of a formula for a FOLDS signature.

Definition 8.6 *The* free variables of ϕ, $\mathrm{Var}(\phi)$, *is defined as follows:*

1. *Base case, already defined: for a kind K, $\mathrm{Var}(K(\langle x_p \rangle_{p \in K|\mathcal{L}})) = \{x_p : p \in K|\mathcal{L}\}$;*
2. *$\mathrm{Var}(\top) =_{def} \mathrm{Var}(\bot) =_{def} \varnothing$;*
3. *If $\mathrm{Var}(\phi)$ and $\mathrm{Var}(\psi)$ are defined, then:*
   (a) *$\mathrm{Var}(\neg \phi) =_{def} \mathrm{Var}(\phi)$, where $\neg \phi$ is the abbreviation of the formal set $\langle 3, \neg, \phi \rangle$;*
   (b) *$\mathrm{Var}(\phi \to \psi) =_{def} \mathrm{Var}(\phi) \cup \mathrm{Var}(\psi)$, where $\phi \to \psi$ is the abbreviation of the formal set $\langle 3, \to, \phi, \psi \rangle$;*
   (c) *$\mathrm{Var}(\phi \wedge \psi) =_{def} \mathrm{Var}(\phi \vee \psi) =_{def} \mathrm{Var}(\phi) \cup \mathrm{Var}(\psi)$, where $\phi \wedge \psi$ is the abbreviation of the formal set $\langle 3, \wedge, \phi, \psi \rangle$ and $\phi \vee \psi$ is the abbreviation of $\langle 3, \vee, \phi, \psi \rangle$;*[32]
4. *If $\mathrm{Var}(\phi)$ is defined, $x : X$, and there is no $y : Y \in \mathrm{Var}(\phi)$ such that $x \in \mathrm{Dep}(y)$, then*

$$\mathrm{Var}(\forall x : X.\phi) =_{def} \mathrm{Var}(\exists x : X.\phi) =_{def} (\mathrm{Var}(\phi) \cup \mathrm{Var}(\psi)) - \{x\},$$

where $\forall x : X.\phi$ and $\exists x : X.\phi$ stand for $\langle 3, \forall, \{x\}, \phi \rangle$ and $\langle 3, \exists, \{x\}, \phi \rangle$.

---

[32] Of course, the definition can be adapted for infinitary conjunctions and disjunctions in the obvious way.

The formulas of the language $\mathcal{L}_{\omega,\omega}$ is the least class of formal sets containing $\top, \bot$ and such that Var(ϕ) is defined according to Definition 8.6, for every ϕ in the language.[33] Not surprisingly, a *sentence* in $\mathcal{L}$ is a formula ϕ with no free variables, that is, such that Var(ϕ) = $\varnothing$.

It is extremely important to give examples of formulas and sentences in FOLDS, for the grammar obtained from the foregoing definition is not immediate. Here is a formula in $\mathcal{L}_{Cat}$

$$\exists \tau : T(x, y, z, f, g, h).\top$$

with free variables $\{x, y, z, f, g, h\}$. Here is a sentence in $\mathcal{L}_{Cat}$

$$\forall x : O.\forall y : O.\forall z : O.\forall f : A(x, y).\forall g : A(y, z).\exists h : A(x, z).\exists \tau : T(x, y, z, f, g, h).\top.$$

This sentence says that, whenever two arrows are composable, then there is a composite arrow.

What is striking about formulas and sentences in FOLDS is the role played by the dependencies themselves. It constrains the grammar considerably. The following expression, for instance, is not a formula of the language $\mathcal{L}_{Cat}$:

$$\forall x : O.\exists \tau : T(x, y, z, f, g, h).\top,$$

for to assert the existence of a triangle τ, one would have to quantify on the variables for arrows before quantifying on objects. In a sense, a FOLDS-signature provides an underlying ontology built in the language itself and such that certain expressions are excluded from the language.[34] FOLDS incorporates the idea in the syntax that the universe of mathematics is not ontologically homogeneous, uniform, and isotropic. It reflects in the syntax itself the fact that mathematics is built up of entities of different kinds and that the mathematical universe is heterogenous. Of course, one quickly introduces abbreviations so that the grammar of formulas and sentences resemble more the usual grammar.

At this stage, in textbooks on model theory, authors introduce the notion of an interpretation for a language in the usual way: a domain of interpretation is fixed, constant symbols become elements of a set, functional symbols become functions and relational symbols become relations. The system used to interpret a language is called a *structure*. The same can be done for FOLDS, with the required adjustments.

Definition 8.7 *Given a FOLDS-signature $\mathcal{L}$, an $\mathcal{L}$-structure is a functor*

$$M : \mathcal{L} \to \mathbf{Set}.$$

---

[33] Once again, it is easy to give the definition for infinitary first-order languages.
[34] I am using the word "ontology" in the same way as it is used in information science. As we will see, a FOLDS-signature is intimately related to a criterion of identity, thus specifying when two objects should be seen as being the same. The signature therefore provides a clear and strong ontological constraint.

This is once more deceptively simple, although one of the reasons is precisely because a FOLDS-signature is a category. Of course, this is a semantics valued in the category of sets. It is possible to replace the latter with a category C, say a category with finite limits. Note that the totality of $\mathcal{L}$-structures, denoted by $Str_{Set}(\mathcal{L})$, form a category. Its objects are the **Set**-valued $\mathcal{L}$-structures and the morphisms, the natural transformations between them.

We will not give the complete, general description of FOLDS semantics. It would be unnecessarily technical. We do have to add a few additional ingredients nonetheless. In particular, we need to describe more explicitly how *valuations* are defined in FOLDS, for some of the elements involved will be referred to later.

A *context* (of variables) is a finite set $\mathcal{Y}$ of variables such that, for all $y \in \mathcal{Y}$, $Dep(y) \subset \mathcal{Y}$. Note that for any formula ϕ, $Var(\phi)$ is a context. Recall that from Definition 8.3, for a variable $y \in \mathcal{Y}$, the sort of y is written $y : K_y(\langle x_{p,q}\rangle_{p\in K_y | \mathcal{L}})$. The next definition connects, so to speak, the relations of dependence correctly, thus giving the legitimate valuations of a context of variables $\mathcal{Y}$ in M.

Definition 8.8 *Let M be a $\mathcal{L}$-structure and K be a kind of $\mathcal{L}$.*

1.

$$M[\mathcal{Y}] = \left\{ \langle a_y \rangle_{y \in \mathcal{Y}} \in \prod_{y \in \mathcal{Y}} M(K_y) : (Mp)(a_y) = a_{x_{y,p}} \text{ for all } y \in \mathcal{Y}, p \in K_y | \mathcal{L} \right\}.$$

2. *The set $M[K]$ of valuations of K in M is defined similarly:*

$$M[K] = \left\{ \langle a_p \rangle_{p \in K | \mathcal{L}} \in \prod_{p \in K | \mathcal{L}} M(K_p) : (Mq)(a_p) = a_{qp} \text{ for all } q \in K_p | \mathcal{L} \right\}.$$

The elements of the set $M[K]$ are called the *contexts* for K in M.

It is now possible to define the notion of an interpretation of a formula ϕ in a structure M in the context $\mathcal{Y}$. We will not give the complete definition here, for it is done as usual by recursion on the complexity of the formula ϕ. Here are some of the simplest cases. Let ϕ be a formula in a given $\mathcal{L}$-signature, M an $\mathcal{L}$-structure, and $\mathcal{Y}$ a context such that $Var(\phi) \subset \mathcal{Y}$. The *interpretation* of ϕ in M in the context $\mathcal{Y}$, $M[\mathcal{Y} : \phi]$ is defined thus:

$$M[\mathcal{Y} : \top] =_{df} M[\mathcal{Y}];$$
$$M[\mathcal{Y} : \bot] =_{df} \emptyset;$$
$$\langle a_y \rangle_{y \in \mathcal{Y}} \in M[\mathcal{Y} : \psi \wedge \theta] =_{df} \langle a_y \rangle_{y \in \mathcal{Y}} \in M[\mathcal{Y} : \psi] \text{ and } \langle a_y \rangle_{y \in \mathcal{Y}} \in M[\mathcal{Y} : \theta].$$

And the other connectives are defined in a similar manner.

We refer the reader to Makkai's papers for the complete description of the semantics for FOLDS.

We must describe and explain the notion of equivalence in FOLDS, since it is, after the notion of signature, the original element of the theory and arguably the most important.

8.3.2 Equivalences in FOLDS

FOLDS was designed by Makkai for the following purpose: to define languages such that all statements in such a language are invariant under the equivalence appropriate for the kind of structures described by the given language. This is feasible only if it is possible to capture basic facts about the different kinds of equivalences involved in the universe of abstract mathematical concepts. We have here a crucial and significant departure from the traditional logical analysis in which the identity relation is presented as an *a priori* and universal or global notion. In FOLDS, an adequate notion of identity is *derived* from an $\mathcal{L}$-signature. Thus, the identity relation comes *a posteriori* or, in a slightly different vocabulary, it is a local relation. This order of presentation reflects the historical order of introduction of abstract mathematical concepts. Indeed, if abstract mathematical concepts are *abstracted* from various mathematical contexts, then the identity criterion should also be abstracted from the basic properties that define these new abstract concepts and this is indeed how it happened historically.[35] What is surprising is that it is possible to find a way to extract a criterion of identity from a given signature. Here is roughly how it works.

Let $\mathcal{L}$ be a given FOLDS-signature and let M and P be $\mathcal{L}$-structures and $h : P \to M$ be a natural transformation between $\mathcal{L}$-structures. For each kind $K \in \mathcal{L}$ and each $x \in P[K]$, there is an induced map

$$h_{K,x} : PK(x) \to MK(hx),$$

where hx is an abbreviation for $\langle h_{K_p} x_p \rangle_{p \in K | \mathcal{L}}$. The natural transformation is said to be *fiberwise surjective* whenever $h_{K,x}$ is surjective for all kinds $K \in \mathcal{L}$ and all $x \in P[K]$.

Definition 8.9 *Let $\mathcal{L}$ be a given FOLDS-signature and let M and N be $\mathcal{L}$-structures. M and N are said to be $\mathcal{L}$-equivalent, written $M \sim_\mathcal{L} N$, if there is an $\mathcal{L}$-structure P together with natural transformations $m : P \to M$ and $n : P \to N$ such that m and n are fiberwise surjective.*

This is certainly a surprising way of introducing an equivalence. It is not even clear that it *is* an equivalence relation. Let us try to unpack it. First, in any category C, given objects X and Y of C, a *span* on X and Y is an object S of C together with morphisms

---

[35] This is interesting in itself and, historically, the situation is more complex than one might expect from how we learn these notions. Contemporary mathematicians are usually convinced that the criterion of identity for various abstract concepts came simultaneously with the concepts themselves. For instance, category theorists are often surprised to learn that the notion of categorical equivalence *was not* in Eilenberg and Mac Lane's original paper in category theory published in 1945. Eilenberg and Mac Lane defined the notion of *isomorphism* of categories. The correct criterion of identity for categories was introduced by Grothendieck in his paper on homological algebra published in 1957.

Spans can be thought of as being generalizations of relations between X and Y. Indeed, a relation $R \subset X \times Y$ of X and Y is a span, with the projections $\pi_X : R \to X$ and $\pi_Y : R \to Y$. Thus, the $\mathcal{L}$-structure P together with the natural transformations m and n is a span on M and N. It is not a new notion. When the category $\mathbf{C}$ has a minimum of structure, for instance pullbacks, then it is easy to show that spans compose in the obvious way. Given the latter fact, it is easy to verify that the foregoing relation is indeed an equivalence relation.

It is a remarkable fact that this definition, together with the definition of FOLDS-signature, yields the appropriate notion of identity for the concepts captured by the signature. Thus, for a signature corresponding to a classical first-order signature, the definition yields the usual notion of isomorphism of structure.[36] In the case of the signature for categories given earlier, one gets the notion of equivalence of categories. It is also possible to use the simplex category Δ to define a FOLDS-signature $\mathcal{L}$ such that for Kan complexes X, Y, X and Y are homotopy equivalent if and only if they are equivalent as $\mathcal{L}$-structures.[37] In other words, it is possible to give a FOLDS-signature such that the notion of equivalence for that language is equivalent to the notion of homotopy equivalence.

Moreover, it can then be *proved* that the derived notion of equivalence satisfies the following invariance principle:[38] given a FOLDS-signature $\mathcal{L}$, M and N $\mathcal{L}$-structures, for any FOLDS-formula ϕ of $\mathcal{L}$,[39]

$$M \models \phi \wedge M \simeq_{\mathcal{L}} N \implies N \models \phi.$$

Again, I want to emphasize that the notion of equivalence is not given *a priori*. The theorem must be proved for each particular case. It basically says that, in FOLDS, it is not possible to make irrelevant claims about the entities talked about in a given language.

---

[36] This is not trivial and requires some care in the statement of the theorem and its proof.

[37] The simplex category Δ can be described thus: its objects are the ordinals $[n] = \{0, 1, ..., n\}$, for all $n \in \mathbb{N}$, and the morphisms $[m] \to [n]$ are the order-preserving functions. To construct the FOLDS-signature, define the category $\Delta^{\uparrow}$ with the same objects as the simplex category but restrict the morphisms to the injective functions of Δ. The FOLDS-signature is then $(\Delta^{\uparrow})^{op}$, the opposite of the previous category. It can be shown that given this notion of FOLDS-signature, an $\mathcal{L}$-structure is, in that case, a simplicial set and, in particular, a Kan complex.

[38] We give a loose and imprecise formulation of the theorem. The precise formulation would not yield more insight at this stage.

[39] The proof is by induction on the complexity of ϕ.

8.4 The Universe: Higher-Dimensional Categories

We can now move to the universe $\mathcal{U}$ of mathematical objects in which the language can be interpreted systematically. It is the universe of higher-dimensional categories. The informal target is clearly identified. Historically, bicategories or, equivalently, weak 2-categories came on the mathematical scene rather early in the development of category theory (Bénabou, 1967), although they appeared as a conceptual curiosity at first. It became progressively clear, mostly in the 1980s, that bicategories (weak 2-categories) and tricategories (weak 3-categories) had to be used in various contexts. Thus, the necessity of having a clear picture of the universe of higher-dimensional categories (HDC) imposed itself on the community.[40] Thus, in the 1990s, category theorists started to articulate a formal, theoretical analysis of the universe and, interestingly enough, various and different formalizations appeared quickly on the scene. It brought to the fore the mathematical question of proving that these various mathematical frameworks were equivalent, a problem still evolving today. From a philosophical point of view, the question as to how the mathematical community will determine which formal framework is the most appropriate is interesting. It is a philosophical case study of the development of mathematical knowledge.[41]

I will give some of the key ideas underlying the definitions of the universe of HDC. The next step is to give one precise, rigorous, technical definition of the universe, the universe $\mathbb{U}$ of Section 8.1, simply to illustrate how it is done and what it looks like.

The basic ingredients of category theory are simple: there are categories, functors, and natural transformations between functors. These are the building blocks of the theory. Thus, at first sight, one might conclude that the universe of categories is and cannot be anything else than a category, that is, the category of categories. However, as we have already said in Section 8.2.4, it turns out that new kinds of entities are required and show up naturally in certain contexts. When categories are put up together, something else emerges and when these entities are put up together, then, again, something else emerges, and so on and so forth.

At the heart of the description of higher-dimensional categories one finds certain geometric shapes which are introduced to capture how various morphisms compose. Thus, in the formal definitions, one finds a collection of basic abstract shapes that determine the underlying structure of the universe. The universe in then given by

---

[40] I should add that many people still doubt the necessity of introducing these levels of abstraction. It is interesting to note that Lawvere introduced a universe of category as a foundations of mathematics in the early 1960s. See Lawvere (1966). However, this universe is simply a category: it is the category of categories. The picture now emerging is considerably more intricate and radically different from the category of categories. As we have said and as Makkai has emphasized in many occasions, the bicategory of categories is *not* a category with additional structure. It does not even have an underlying category. It *is* a genuinely new kind of entity.

[41] A parallel with the development of set theory and its formalization might be interesting also, as well as a parallel with Maddy's work on the choice of axioms in set theory. See Maddy (1988a,b). Quasicategories are now fashionable, but they are not the only concepts used. For the definition of quasicategories, see Joyal (2002). For its role in the development of higher-dimensional algebra, see Lurie (2009, 2016).

specifying certain properties of these shapes. It should be noted that Makkai has used FOLDS and the concept of FOLDS-signature to present these shapes and the universe. See Makkai (2004). We will merely sketch one way of thinking about the constructions involved.

Consider the following common situations in basic category theory. First, a functor F from a category C to a category D is represented by an arrow:

$$F : C \to D.$$

The situation that more or less gave rise to category theory is given by two parallel functors $F, G : C \to D$ with a natural transformation $\eta : F \to G$. This situation is sometimes depicted as follows:

$$C \Downarrow D$$

Abstracting from the specific elements, we have a geometric form composed of points, directed lines, and the double line which can be thought of as representing a surface that is stretched from the upper arrow to the lower one. The representation becomes:

$$\bullet \Downarrow \bullet$$

Geometrically, this can be thought of as a disk. But the image is dynamic—stretching a surface from one directed line to the other directed line—and this aspect is crucial and must be kept in mind when reading it and similar cases. Since functors compose and natural transformations compose, this basic geometric form can be pasted with other basic forms, yielding new complex forms. For instance, the composition of two functors with natural transformations between them yields the form

$$\bullet \Downarrow \bullet \Downarrow \bullet$$

Of course, the composition of the natural transformations together with the functors is not entirely clear from the diagram itself. Similarly, the composition of two natural transformations between three functors, the so-called vertical composition, yields another form:

It is now easy to see that these basic forms can be pasted together in various ways and the question as to how to determine which ones can be transformed into one another immediately arises.[42] And as should be clear from informal geometric knowledge, these forms are 2-dimensional. The basic 3-dimensional case can be depicted thus:

Thus, the informal description of the universe of higher-dimensional categories is understood, in the same way that the informal universe of sets is understood or was understood by Cantor and others before the introduction of a formal axiomatic framework with a rigorous and precise description of the cumulative hierarchy. As we have said, but should perhaps be repeated at this point, the move to FOLDS and the universe of higher-dimensional categories corresponds to the move to first-order logic and the universe of sets. In the latter case, the universe of sets was crystallized in the definition of the cumulative hierarchy. There are now various definitions that can be used to define a universe of higher-dimensional categories, the universe $\mathbb{U}$.

To describe the universe of higher-dimensional categories, one therefore needs a language in which the basic geometric forms, also called the k-cells, are given. Once these have been described, the theory must specify how these compose and which ones are equivalent. In the end, that is after many pages of formal definitions and clarifications, the definition of the universe itself holds in a short paragraph. To wit, here is Makkai's definition.

Definition 8.10 *A multitopic ω-category is a multitopic set S such that for every multitope σ and every σ-shaped pasting diagram α in S, there is at least one cell a parallel to α such that, for $\theta = c\sigma$, the Mlt$\langle\theta\rangle$-structures $S\langle a\rangle$ and $S\langle\alpha\rangle$ are Mlt$\langle\theta\rangle$-equivalent by an equivalence span that extends the identity on Mlt.*

Without the appropriate definitions and results, the definition is incomprehensible. Many notions involved must be explained and defined. For instance, the multitopes are here, the basic forms used. (See Hermida et al. (2000) for the definition of multitopes.) Makkai gives a FOLDS-signature for the multitopic ω-category. Two elements must be mentioned. First, the required properties for a multitopic set to be a multitopic ω-category are all of the form that certain composites, defined by universal properties, are to exist. Second, the notion of FOLDS-equivalence plays a key role in the definition.

All these technicalities are neither surprising nor a problem. After all, the definition of the cumulative hierarchy underlying the set-theoretical foundational framework is also technical and requires the certain sophisticated set-theoretical notions, e.g. transfinite ordinals, be understood.

---

[42] Of course, I did not write "equal" in the last sentence. This is the key element of the situation. Composing various transformations does not yield, in the general case, equalities between the morphisms. At best, there is a equivalence between them.

8.5 Conclusion

Once the community has a good grasp of the various parts constituting the new foundational space, logicians will be able to extend and develop the science of the foundations of mathematics. This is clearly Makkai's goal. Thus, the point here is not to defend an ideological position regarding the nature of mathematical knowledge, for the whole project starts from the observation that abstract mathematical concepts are now part and parcel of contemporary mathematics. Nor is it to develop a formal system that will lead to automated proof checking or automated theorem proving. The goal is to extend in a specific direction the science of the foundations of mathematics, thus to obtain results about certain structures, results with an intrinsic conceptual value and that can be relevant for mathematics and for philosophy.

Why should philosophers care about FOLDS or, more generally, about a science of the foundations of mathematics?[43] Why should they care about a foundational framework for *abstract* mathematics? Most philosophers nowadays probably think that the time when they had to know and understand technical issues related to the foundations of mathematics is now behind them. Issues related to the foundations of mathematics were philosophically relevant, so the argument goes, in as much as they were related to the fact that mathematical knowledge could no longer be *rationally justified*. Once a reasonable solution to the paradoxes of set theory had been found, the technical developments leading to the creation of model theory and proof theory seemed to be of interest only to a handful of technically sensitive philosophers of logic and mathematics.

I will not try to argue for the importance of a science of the foundations of mathematics for philosophy in general. This is a topic for a whole book. I will here concentrate on a few aspects that are inherent to FOLDS.

The first element worth underlining is the fact that FOLDS is aimed at *abstract* mathematical concepts. This in itself is central to epistemology and ontology in general. Indeed, a better understanding of what is, how we know, and how we understand abstract mathematical concepts is bound to open the way to a better understanding of abstractness in general. The latter notion is omnipresent, from aesthetic to ethics. Specific results about FOLDS could be read as exhibiting singular aspects of abstract mathematical concepts.

Second, philosophers should be interested in a language that can only express invariant properties of abstract objects. This is in itself a remarkable original feature. There is no need to underline the importance of what is called Benacerraf's problem in philosophy of mathematics and how FOLDS leads to a direct and simple solution to this problem, as Makkai has already seen himself.

---

[43] We take it for granted here that we do not have to convince logicians about the importance of FOLDS. We may be wrong about this and, if we are, it is an interesting and intriguing fact.

Third, FOLDS and the universe of HDC force us to think differently about the notion of structure and various kinds of structuralism in mathematics and the sciences in general. For, in most cases, the debate surrounding structuralism is based on a set-theoretical notion of structure.[44]

Finally, specific technical results of FOLDS will also have a philosophical impact. Certain model theoretical properties, e.g. existential closure or model completeness, have a direct philosophical interpretation which sheds an important light on the nature of mathematical knowledge and mathematical understanding.[45] Similar results for theories written in FOLDS will allow a better analysis of certain aspects of the development of contemporary mathematics and its peculiar features, e.g. the use of abstract concepts.

In the end, FOLDS is a starting point for anyone who wants to develop a foundation for abstract mathematical concepts and, perhaps, for any abstract concepts.

Acknowledgments

The technical apparatus and the formal definitions given in this paper have been presented by Michael Makkai over the past twenty years or so in various forms and in various places. The definitions are taken directly from various papers, talks, and presentations given by Makkai. I have given specific references in the text. I want to thank Michael Makkai for his help. I also want to thank two anonymous referees for their comments, criticisms, and suggestions. They allowed me to clarify important points and avoid some mistakes. All remaining mistakes, errors, and omissions are mine.

References

Bénabou, J. (1967). Introduction to bicategories, in *Reports of the Midwest Category Seminar*, 1–77. Springer, Berlin.

Ellerman, D. P. (1988). Category theory and concrete universals. *Erkenntnis* 28(3), 409–29.

Ferreirós, J. (1996). Traditional logic and the early history of sets, 1854–1908. *Archive for History of Exact Sciences* 50(1), 5–71. http://dx.doi.org/10.1007/BF00375789

Ferreirós, J. (2001). The road to modern logic—an interpretation. *Bulletin of Symbolic Logic* 7(4), 441–84. http://dx.doi.org/10.2307/2687794

Ferreirós, J. (2007). *Labyrinth of Thought*, second edn. Birkhäuser Verlag, Basel. A history of set theory and its role in modern mathematics.

Fréchet, M. (1951). Abstract sets, abstract spaces and general analysis. *Mathematics Magazine* 24, 147–55.

---

[44] I should hasten to add: on a concrete notion of set. Bourbaki, in his notion of structure, included the isomorphisms and thus, one could argue, on an abstract notion of set as it should. Indeed, one could use Makkai's notion of abstract set to reconstruct Bourbaki's notion of structure. But again, as I have already indicated, this is but one small fragment of the universe we are discussing here.

[45] See Manders (1989).

Freyd, P. (2004). Homotopy is not concrete. *Reprints in Theory and Applications of Categories* (6), 1–10. Reprinted from *The Steenrod Algebra and its Applications, Lecture Notes in Mathematics*, 168, 25–34. Springer, Berlin, 1970 [MR0276961].

Freyd, P. J. (1973). Concreteness, *Journal of Pure and Applied Algebra* 3, 171–91.

Hermida, C., Makkai, M., and Power, J. (2000). On weak higher dimensional categories. I., in *Category Theory and Its Applications* (Montreal, QC, 1997). *Journal of Pure and Applied Algebra* 154(1–3), 221–46. http://dx.doi.org/10.1016/S0022-4049(99)00179-6

Jacobs, B. (1999). *Categorical Logic and Type Theory*, Vol. 141 of *Studies in Logic and the Foundations of Mathematics*. North-Holland, Amsterdam.

Joyal, A. (2002). Quasi-categories and Kan complexes. *Journal of Pure and Applied Algebra* 175(1), 207–22.

Kamareddine, F., Laan, T., and Nederpelt, R. (2002). Types in logic and mathematics before 1940. *Bulletin of Symbolic Logic* 8(2), 185–245. http://dx.doi.org/10.2307/2693964

Kamareddine, F., Laan, T., and Nederpelt, R. (2012). A history of types, in D. M. Gabbay, A. Kanamori, and J. Woods (eds), *Logic: A History of its Central Concepts* Vol. 11 of *Handbook of History of Logic*, 451–511, North-Holland, Amsterdam.

Kelly, G. M. (1964). Complete functors in homology. I. Chain maps and endomorphisms. *Proceedings of the Cambridge Philosophical Society* 60, 721–35.

Krömer, R. (2007). *Tool and Object*, Vol. 32 of *Science Networks. Historical Studies*. Birkhäuser Verlag, Basel. A history and philosophy of category theory.

Lawvere, F. W. (1966). The category of categories as a foundation for mathematics, in *Proceedings of the Conference on Categorical Algebra (La Jolla, Calif. 1965)*, 1–20. Springer-Verlag, New York.

LeCun, Y., Bengio, Y. and Hinton, G. (2015). Deep learning. *Nature* 521(7553), 436–44.

Leinster, T. (2002). A survey of definitions of n-category. *Theory and Applications of Categories* 10, 1–70 (electronic).

Leinster, T. (2004). *Higher Operads, Higher Categories*, Vol. 298 of *London Mathematical Society Lecture Note Series*. Cambridge University Press, Cambridge, UK. http://dx.doi.org/10.1017/CBO9780511525896

Leinster, T. (2014). Rethinking set theory. *American Mathematical* 121(5), 403–15. http://dx.doi.org/10.4169/amer.math.monthly.121.05.403

Lurie, J. (2009). *Higher Topos Theory*, number 170. Princeton University Press, Princeton, NJ.

Lurie, J. (2016). Higher algebra. http://www.math.harvard.edu/~lurie/papers/HA.pdf

Maddy, P. (1988a). Believing the axioms. I. *Journal of Symbolic Logic* 53(02), 481–511.

Maddy, P. (1988b). Believing the axioms. II. *Journal of Symbolic Logic* 53(03), 736–64.

Makkai, M. (1998). Towards a categorical foundation of mathematics, in *Logic Colloquium '95 (Haifa)*, Vol. 11 of *Lecture Notes in Logic*, 153–90, Springer, Berlin. http://dx.doi.org/10.1007/978-3-662-22108-2_11

Makkai, M. (1999). On structuralism in mathematics, in *Language, Logic, and Concepts*, Bradford Book, 43–66. MIT Press, Cambridge, MA.

Makkai, M. (2004). The multitopic ω-category of all multitopic ω-categories. http://www.math.mcgill.ca/makkai/mltomcat04/mltomcat04.pdf

Makkai, M. (2013). The theory of abstract sets based on first-order logic with dependent types. http://www.math.mcgill.ca/makkai/Various/MateFest2013.pdf

Mancosu, P., Zach, R., and Badesa, C. (2009). The development of mathematical logic from Russell to Tarski, 1900–1935, in *The Development of Modern Logic*. Oxford University Press, Oxford. http://dx.doi.org/10.1093/acprof:oso/9780195137316.003.0029

Manders, K. (1989). Domain extension and the philosophy of mathematics. *Journal of Philosophy* 86(10), 553–62. http://www.jstor.org/stable/pdf/2026666.pdf?seq=1#page_scan_tab_contents

Marquis, J.-P. (2000). Three kinds of universals in mathematics? *Logical Consequence: Rival Approaches and New Studies in Exact Philosophy: Logic, Mathematics and Science* 2, 191–212.

Marquis, J.-P. (2009). *From a Geometrical Point of View: A Study of the History and Philosophy of Category Theory*, Vol. 14 of *Logic, Epistemology, and the Unity of Science*. Springer Science & Business Media, Berlin.

Marquis, J.-P. (2014). Mathematical abstraction, conceptual variation and identity, in P. S.-H. G. H. W. H. P. E. Bour (ed.), *Logic, Methodology and Philosophy of Science—Proceedings of the 14th International Congress (Nancy)*, 299–322. College Publications, London.

Marquis, J.-P. (2016). Stairway to heaven: levels of abstraction in mathematics, *The Mathemtical Intelligencer* 38(3), 41–51.

Martin-Löf, P. (1975). An intuitionistic theory of types: predicative part, in *Logic Colloquium '73 (Bristol, 1973), Studies in Logic and the Foundations of Mathematics*, Vol. 80, 73–118. North-Holland, Amsterdam.

Martin-Löf, P. (1984). *Intuitionistic Type Theory*, Vol. 1 of *Studies in Proof Theory*. Lecture Notes. Bibliopolis, Naples. Notes by Giovanni Sambin.

Mayberry, J. (2000). *The Foundations of Mathematics in the Theory of Sets*. Cambridge University Press, Cambridge, UK.

McLarty, C. (1993). Numbers can be just what they have to. *Noûs* 27(4), 487–98. http://dx.doi.org/10.2307/2215789

McLarty, C. (2004). Exploring categorical structuralism. *Philosophia Mathematica (3)* 12(1), 37–53. http://dx.doi.org/10.1093/philmat/12.1.37

Moore, G. H. (1980). Beyond first-order logic: the historical interplay between mathematical logic and axiomatic set theory, in *History and Philosophy of Logic*, Vol. 1, 95–137. Abacus, Tunbridge Wells.

Moore, G. H. (1987). A house divided against itself: the emergence of first-order logic as the basis for mathematics, in *Studies in the History of Mathematics*, Vol. 26 of *MAA Studies in Mathematics*, 98–136. Mathematics Association. America, Washington, DC.

Moore, G. H. (1988). The emergence of first-order logic, in *History and Philosophy of Modern Mathematics (Minneapolis, MN, 1985), Minnesota Studies in the Philosophy of Science*, XI. University of Minnesota Press, Minneapolis, MN.

Nodelman, U., and Zalta, E. N. (2014). Foundations for mathematical structuralism. *Mind* 1–42.

Pollard, S. (1987). What is abstraction? *Noûs* 21(2), 233–40. http://dx.doi.org/10.2307/2214916

Pollard, S. (1988). Weyl on sets and abstraction, *Philosophical Studies* 53(1), 131–40. http://dx.doi.org/10.1007/BF00355680

Schiemer, G., and Reck, E. H. (2013). Logic in the 1930s: type theory and model theory. *Bulletin of Symbolic Logic* 19(4), 433–72. http://projecteuclid.org/euclid.bsl/1388953941

Weyl, H. (1949). *Philosophy of Mathematics and Natural Science. Revised and Augmented English Edition Based on a Translation by Olaf Helmer*. Princeton University Press, Princeton, NJ.

9

Categories and Modalities

Kohei Kishida

9.1 Syntax, Semantics, and Duality

Category theory provides various guiding principles for modal logic and its semantics. One is the so-called syntax–semantics duality. Stone duality and its extensions have been a prominent theme in algebraic semantics of modal logic (see Venema, 2007). McKinsey, Tarski, and Jónsson, among others, extended Stone duality to Boolean algebras with operators, and in effect provided a semantics for modal logic that heralded Kripke semantics (see Goldblatt, 2006). Categorical methods show how the following two kinds of semantics are related to each other. (Here we use the terminology for propositional logic, but the same idea extends to predicate logic.)

- Algebraic semantics: They concern the entailment relation $\vdash$ between propositions. A family of propositions forms an *algebra*.
- What may be called "spatial" semantics, e.g. Kripke semantics: They concern the satisfaction relation $\vDash$ between a proposition, on the one hand, and a model or a world, on the other. A family of models or worlds forms a *space*, e.g. a Kripke frame.

The relationship to be shown is not just some technical connection between conceptually disparate modes of semantics; indeed, category theory brings to light a conceptual unity that algebras lie behind spatial semantics and spaces behind algebraic semantics. In this section, we review this idea of duality, taking as an example the case of classical propositional logic (leaving the modal part to Section 9.2).[1]

9.1.1 Syntax as algebras

Perhaps the most elementary of spatial semantics is the semantics via truth tables, which concerns the satisfaction relation between a sentence and a row of a table. Even at this elementary level, algebras are already behind the semantics. The conventional story for truth tables goes like this. Given a set V of propositional variables, write

---

[1] See Johnstone (1982) for a thorough exposition of Stone duality.

$Sn_\tau(V)$ for the set of sentences obtained from V using the set τ of connectives, which contains $\wedge, \vee, \Rightarrow, \neg$ and possibly 0-ary $\top$ and $\bot$. Then, given any row w of a truth table for V, i.e. any function $w : V \to 2$ assigning truth values $w(p) = 1, 0 \in 2$ to $p \in V$, it extends values $\bar{w}(\varphi) \in 2$ to all the sentences $\varphi \in Sn_\tau(V)$ by inductive clauses such as

$$\bar{w}(\varphi \wedge \psi) = 1 \iff \bar{w}(\varphi) = 1 \text{ and } \bar{w}(\psi) = 1. \tag{9.1}$$

Now note that (9.1) is indeed a statement of algebraic nature, which becomes obvious once we use $\wedge : 2 \times 2 \to 2$ and rewrite (9.1) as

$$\bar{w}(\varphi \wedge \psi) = \bar{w}(\varphi) \wedge \bar{w}(\psi),$$

which states that $\bar{w}$ preserves $\wedge$, thereby interpreting $\wedge$ of $Sn_\tau(V)$ with $\wedge$ of 2. Similarly, the other inductive clauses amount to $\bar{w}$ preserving the other connectives. In short, $\bar{w}$ is a homomorphism. The upshot is *not* that an alternative semantics is available using algebras; it is that interpretations simply *are* homomorphisms of algebras (or categories more generally)—one of the core insights of categorical logic.

To be more precise, let us define a "τ-algebra" as any set S equipped with an operation $\circledast : S^{\text{ar}(\circledast)} \to S$ for each connective $\circledast \in \tau$ of arity $\text{ar}(\circledast)$, so that $Sn_\tau(V)$ and 2 are both τ-algebras.[2] So the inductive clauses altogether amount to

(9.2) $\bar{w} : Sn_\tau(V) \to 2$ is a homomorphism of τ-algebras.

Then the fact that the induction extends a function w uniquely to a homomorphism $\bar{w}$, i.e.

$$\text{Hom}_{\text{Sets}}(V, 2) \underset{\text{restrict}}{\overset{\text{extend}}{\underset{\cong}{\rightleftarrows}}} \text{Hom}_{\tau\text{-Alg}}(Sn_\tau(V), 2) \tag{9.3}$$

for the category τ-**Alg** of τ-algebras, is just an instance of the fact that $Sn_\tau(V)$ is the free τ-algebra on V; i.e. that the functor $Sn_\tau : \textbf{Sets} \to \tau\text{-}\textbf{Alg}$ is left adjoint to the forgetful functor:

$$\text{Hom}_{\text{Sets}}(V, A) \cong \text{Hom}_{\tau\text{-Alg}}(Sn_\tau(V), A) \qquad \textbf{Sets} \underset{U}{\overset{Sn_\tau}{\rightleftarrows}} \tau\text{-}\textbf{Alg}. \tag{9.4}$$

The kind of algebra normally used in the algebraic semantics, e.g. *Boolean algebras* for classical logic and *Heyting algebras* for intuitionistic logic (see Johnstone, 1982), is a τ-algebra that moreover has an order expressing a *binary* entailment relation. For example, Boolean and Heyting algebras A have a partial order (i.e. reflexive, transitive and antisymmetric relation) $\leqslant$ in addition to the τ-algebra structure $\wedge, \neg, \ldots$. The τ-algebra $A = Sn_\tau(V)$ of sentences we saw above may appear to lack such an order, but we can take a natural one, e.g. the entailment relation $\vdash$ satisfying the laws of

---

[2] The τ-algebras are the algebras of the polynomial functor $P_\tau : \textbf{Sets} \to \textbf{Sets} :: S \mapsto \sum_{\circledast \in \tau} S^{\text{ar}(\circledast)}$.

classical logic but no "extra-logical" ones (e.g. that $p \vdash q$ for a particular pair $p, q \in V$). This makes $F(A) = (A, \vdash)$ a Boolean *pre*algebra—"pre" means "minus antisymmetry", as in "preorder"—because it satisfies the Boolean laws; and, since it satisfies no other laws, it is indeed the weakest, *free* Boolean prealgebra on A. That is,

$$\mathrm{Hom}_{\tau\text{-Alg}}(A, B) = \mathrm{Hom}_{\mathbf{BpA}}(F(A), B) \quad (9.5)$$

for every Boolean prealgebra B, where we write **BpA** for the category of Boolean prealgebras.

Boolean algebras are the Lindenbaum–Tarski algebras $\mathrm{Sk}(B)$ of Boolean prealgebras B; instead of sentences $\varphi \in B$, we take their entailment-equivalence classes $[\varphi] \in \mathrm{Sk}(B)$. Categorically put, $\mathrm{Sk}(B)$ is the skeleton of B; although B and $\mathrm{Sk}(B)$ may not be isomorphic, they are equivalent (as categories). Therefore, although algebraic semantics may often appear to involve abstract entities as elements of algebras, it can equivalently be done using sets of sentences. (So we will henceforth use the word "proposition" loosely, to refer either to a sentence or to an algebra element representing sentences—e.g. an equivalence class of sentences, or a subset of worlds.)

Indeed, $\mathrm{Sk} : \mathbf{BpA} \to \mathbf{BA}$ is left adjoint to the inclusion of the category **BA** of Boolean algebras into **BpA**. Thus, we have a series of three adjoint pairs (U are forgetful and i is the inclusion):

$$\mathbf{Sets} \xrightleftharpoons[U_1]{\mathrm{Sn}_\tau} \tau\text{-}\mathbf{Alg} \xrightleftharpoons[U_2]{F} \mathbf{BpA} \xrightleftharpoons[i]{\mathrm{Sk}} \mathbf{BA}$$

This encapsulates the point that even the truth-table semantics has the structures of algebraic semantics behind it.

9.1.2 Semantics as homomorphisms

The clause (9.2) expresses the satisfaction relation between a single row w of a truth table and sentences in $A = \mathrm{Sn}_\tau(V)$. Extending it to the entire set $X = \mathrm{Hom}_{\mathbf{Sets}}(V, \mathbf{2})$ of rows, we have a relation $\vDash \subseteq X \times A$; we may write $\vDash : X \times A \to \mathbf{2}$ by identifying subsets and characteristic functions. In a more general setting, $w \in X$ may be "possible worlds" rather than truth-table rows; A may also be a prealgebra equipped with $\vdash$ and not just a τ-algebra.

Now note that the relations $R \subseteq S_1 \times S_2$, i.e.

$$R : S_1 \times S_2 \to \mathbf{2},$$

correspond naturally, by "currying", to the functions

$$\bar{R} : S_1 \to (S_2 \to \mathbf{2}),$$

i.e. $\bar{R} : S_1 \to \mathcal{P}S_2$; we say R and $\bar{R}$ "transpose" each other.[3] Hence the relation $\vDash$: $X \times A \to 2$ has two transposes:

- $\llbracket - \rrbracket : A \to \mathcal{P}X$ assigns to each proposition $\varphi \in A$ its "truth set", $\llbracket \varphi \rrbracket = \{ w \in X \mid w \vDash \varphi \}$.
- $\mathbb{T} : X \to \mathcal{P}A$ assigns to each world $w \in X$ its theory, $\mathbb{T}(w) = \{ \varphi \in A \mid w \vDash \varphi \}$,

$$w \vDash \varphi \iff w \in \llbracket \varphi \rrbracket \iff \varphi \in \mathbb{T}(w).$$

We say $\llbracket - \rrbracket$ and $\mathbb{T}$ respectively give the "denotational" and "axiomatic" presentations of the semantics.[4]

Crucially, the three maps $\vDash$, $\llbracket - \rrbracket$, and $\mathbb{T}$ carry the same information. Each expresses the semantic rules of logic with constraints on it, but the constraints on the three maps are equivalent to one another. For instance, modifying (9.1) and (9.2), we have

(9.6) $\left\{ \begin{array}{l} w \vDash \varphi \wedge \psi \\ w \in \llbracket \varphi \wedge \psi \rrbracket \\ \varphi \wedge \psi \in \mathbb{T}(w) \end{array} \right\}$ iff $\left\{ \begin{array}{l} w \vDash \varphi \text{ and } w \vDash \psi \\ w \in \llbracket \varphi \rrbracket, \llbracket \psi \rrbracket \\ \varphi, \psi \in \mathbb{T}(w) \end{array} \right\}$; so $\left\{ \begin{array}{l} \llbracket \varphi \wedge \psi \rrbracket = \llbracket \varphi \rrbracket \cap \llbracket \psi \rrbracket \\ \mathbb{T}(w)(\varphi \wedge \psi) = \mathbb{T}(w)(\varphi) \wedge \mathbb{T}(w)(\psi) \end{array} \right\}$.

(We use both types $\mathbb{T}(w) \subseteq A$ and $\mathbb{T}(w) : A \to 2$.) This states that we interpret $\wedge$ with $\cap$; to express this we may write $\llbracket \wedge \rrbracket = \cap$, so that $\llbracket \varphi \wedge \psi \rrbracket = \llbracket \varphi \rrbracket \llbracket \wedge \rrbracket \llbracket \psi \rrbracket$. Similarly, $\llbracket \neg \rrbracket = X \setminus -$ with

(9.7) $\left\{ \begin{array}{l} w \vDash \neg \varphi \\ w \in \llbracket \neg \varphi \rrbracket \\ \neg \varphi \in \mathbb{T}(w) \end{array} \right\}$ iff $\left\{ \begin{array}{l} w \nvDash \varphi \\ w \notin \llbracket \varphi \rrbracket \\ \varphi \notin \mathbb{T}(w) \end{array} \right\}$; so $\left\{ \begin{array}{l} \llbracket \neg \varphi \rrbracket = X \setminus \llbracket \varphi \rrbracket \\ \mathbb{T}(w)(\neg \varphi) = \neg \mathbb{T}(w)(\varphi) \end{array} \right\}$.

Furthermore, to model any (extra-logical) laws of A, we assume a version of soundness:

(9.8) If $\varphi \vdash \psi$ for $\varphi, \psi \in A$, then $\left\{ \begin{array}{l} w \vDash \varphi \\ w \in \llbracket \varphi \rrbracket \\ \varphi \in \mathbb{T}(w) \end{array} \right\}$ implies $\left\{ \begin{array}{l} w \vDash \psi \\ w \in \llbracket \psi \rrbracket \\ \psi \in \mathbb{T}(w) \end{array} \right\}$; so $\left\{ \begin{array}{l} \llbracket \varphi \rrbracket \subseteq \llbracket \psi \rrbracket \\ \mathbb{T}(w)(\varphi) \leqslant \mathbb{T}(w)(\psi) \end{array} \right\}$.

In sum, extending (9.2), the denotational and axiomatic presentations of semantics respectively have

(9.9) $\llbracket - \rrbracket : A \to \mathcal{P}X$ is an order-preserving homomorphism of τ-algebras (i.e. interpreting $\wedge, \neg, \ldots, \vdash$ of A with $\cap, X \setminus -, \ldots, \subseteq$ on $\mathcal{P}X$).

(9.10) $\mathbb{T} : X \to \mathcal{P}A$ maps $w \in X$ to an order-preserving homomorphism $\mathbb{T}(w) : A \to 2$ of τ-algebras.

---

[3] The idea that Stone duality arises from transposes of matrices is laid out in Pratt (1995).
[4] The names derive from denotational and axiomatic semantics in computer science. The insight that both presentations are essential components of syntax–semantics duality is due to Abramsky (1991).

The theme thus recurs that semantic interpretations are algebraic homomorphisms.[5]

9.1.3 Syntax–semantics duality

Since the subsets $S \subseteq X$ are the functions $S : X \to 2$, we have $\mathrm{Hom}_{\mathbf{Sets}}(X, 2) = \mathcal{P}X$, the Boolean algebra of all subsets. On the other hand, a subset u of a Boolean (pre)algebra A is called an *ultrafilter* of A if it satisfies (9.6)–(9.8) in place $\mathbb{T}(w)$, i.e. if $u : A \to 2$ is a Boolean homomorphism; when $(A, \vdash)$ is a prealgebra of sentences, its ultrafilters are precisely the maximal $\vdash$-consistent sets. So let us write

$$\mathrm{Ult}\, A = \mathrm{Hom}_{\mathbf{BpA}}(A, 2) \subseteq \mathcal{P}A$$

for the set of ultrafilters of A. These $\mathrm{Hom}(-, 2)$ extend to contravariant functors that work on arrows by "precomposition": e.g. for $f : X \to Y$, $\mathrm{Hom}_{\mathbf{Sets}}(f, 2)$ sends a subset $S : Y \to 2$ to another $S \circ f : X \to 2$; so it amounts to the inverse-image map

$$\mathrm{Hom}_{\mathbf{Sets}}(f, 2) = \mathcal{P}f = f^{-1} : \mathcal{P}Y \to \mathcal{P}X :: S \mapsto f^{-1}[S] = \{w \in X \mid f(w) \in S\}.$$

Now, for a Boolean prealgebra A, (9.9)–(9.10) respectively mean

$$[\![-]\!] \in \mathrm{Hom}_{\mathbf{BpA}}(A, \mathrm{Hom}_{\mathbf{Sets}}(X, 2)),$$
$$\mathbb{T} \in \mathrm{Hom}_{\mathbf{Sets}}(X, \mathrm{Hom}_{\mathbf{BpA}}(A, 2)).$$

The correspondence between these transposes means that the two functors are dual adjoint to each other.

$$\mathbf{Sets} \xrightleftharpoons[\mathrm{Ult} = \mathrm{Hom}_{\mathbf{BpA}}(-, 2)]{\mathcal{P} = \mathrm{Hom}_{\mathbf{Sets}}(-, 2)} \mathbf{BpA}^{\mathrm{op}}. \tag{9.11}$$

One may call this a "syntax–semantics dual adjunction". While it fails to give duality in the strict sense of $\mathbf{Sets} \simeq \mathbf{BpA}^{\mathrm{op}}$, it displays syntax–semantics duality in a broad sense.

---

[5] A remark may be needed concerning the recursive definition of $\vDash$, since one may think we should
  (i) Use any relation $\vDash_V : X \times V \to 2$ (or function $[\![-]\!]_V : V \to \mathcal{P}X$) as primitive and define $\vDash : X \times \mathrm{Sn}_\tau(V) \to 2$ (or $[\![-]\!] : \mathrm{Sn}_\tau(V) \to \mathcal{P}X$) by recursion.
Yet (9.4)—$\mathrm{Sn}_\tau(V)$ being free on V—implies that

$$\mathrm{Hom}_{\mathbf{Sets}}(V, \mathcal{P}X) \underset{\mathrm{restrict}}{\overset{\mathrm{extend}}{\xrightleftharpoons{\cong}}} \mathrm{Hom}_{\tau\text{-}\mathbf{Alg}}(\mathrm{Sn}_\tau(V), \mathcal{P}X)$$
$$[\![-]\!]_V \qquad\qquad\qquad\qquad\qquad [\![-]\!]$$

just the way (9.2) does (9.3). Therefore (i) is equivalent to the following.
  (ii) Regard (9.9) as a constraint and take any $[\![-]\!] : \mathrm{Sn}_\tau(V) \to \mathcal{P}X$ satisfying it as primitive.
Yet (ii) can be applied more generally to any prealgebra and not just the free $\mathrm{Sn}_\tau(V)$.

To illustrate syntax–semantics duality in general terms, generalize (9.11) and assume we have

$$\text{Hom}_A(A, FX) \cong \text{Hom}_X(X, GA) \qquad X \underset{G}{\overset{F}{\rightleftarrows}} A^{\text{op}}.$$
$$[\![-]\!] \qquad \mathbb{T}$$

When we use a space X of worlds to semantically interpret an algebra A of propositions, an arrow $[\![-]\!]$ of **A** presents the semantics by taking the algebra FX associated with X, whereas an arrow $\mathbb{T}$ of **X** presents the same by taking the space GA associated with A. The dual adjunction between F and G means that these presentations in **A** and **X** are equivalent. Now, the duality between the space X and the algebra A, if it holds, is an additional feature, that they are associated with each other in the sense above, i.e. $A \cong FX$ and $X \cong GA$. Then there are isomorphisms $[\![-]\!] : A \to FX$ and $\mathbb{T} : X \to GA$ that identify $\varphi \in A$ with $[\![\varphi]\!] \in FX \subseteq \mathcal{P}X$ and $w \in X$ with $\mathbb{T}(w) \in GA \subseteq \mathcal{P}A$. In other words, the duality consists in a sort of stability between the following positions:[6]

(9.12) Given the space X of worlds, its subsets in FX can be identified with the propositions.

(9.13) Given the algebra A of propositions, its subsets in GA can be identified with the worlds.

For example, for $G = \text{Ult}$, the subsets in GA are the maximal $\vdash$-consistent sets; they are what we expect the theories of worlds to be. Now the stability between (9.12) and (9.13) goes like this: One applies (9.12) to a space X of worlds, obtaining an algebra $FX \cong A$ of propositions, and then applies (9.13) to it, but the duality gives back the original space $GFX \cong GA \cong X$; similarly for $FGA \cong FX \cong A$.

Several dualities (in the strict sense) are obtained by modifying the categories and functors in (9.11). One is the Stone duality between **BA** and the category of Stone spaces. Another is the *Lindenbaum–Tarski duality*, between **Sets** and the category **CABA** of "CABAs", complete atomic Boolean algebras:

$$\text{Sets} \underset{\text{Atm} = \text{Hom}_{\text{CABA}}(-, 2)}{\overset{\mathcal{P} = \text{Hom}_{\text{Sets}}(-, 2)}{\rightleftarrows}} \text{CABA}^{\text{op}} \qquad (9.14)$$

Instead of all ultrafilters, this duality uses "principal" ones, or equivalently "atoms" of A—i.e. instead of all homomorphisms in $\text{Hom}_{\text{BA}}(A, 2)$, ones in $\text{Hom}_{\text{CABA}}(A, 2)$ preserving arbitrary meets and joins. From the "interpretations as homomorphisms" perspective, this means that (9.14) is the duality between syntax and semantics of classical logic with infinitary conjunction and disjunction.

---

[6] For instance, see Montague (1969) for propositions as sets of worlds and Adams (1974) for worlds as maximal consistent sets of propositions.

9.2 A Categorical Look at Kripke Semantics

In this section, we discuss the most prevalent semantics that has been given to modal logic, i.e. Kripke semantics—or possible-world semantics with "accessibility" relations—for propositional, classical modal logic. Nevertheless, the categorical insights we laid out in Section 9.1 apply equally to Kripke semantics. Extending the tradition since McKinsey, Tarski, and Jónsson, we show how closely Kripke semantics is tied to algebras of modal logic. This perspective, as we demonstrate, provides structural explication for many conceptions of what Kripke modalities are, as well as many facts in the model theory of Kripke semantics.

9.2.1 Basics of Kripke semantics

Modal logic is obtained by adding connectives called modal operators to any given logic; the latter, the "non-modal base", may be propositional, first-order, or higher-order, may be typed or untyped, and may be classical or otherwise. The paradigms of modal operators are a pair of unary $\Box$ and $\Diamond$ for "necessarily" and "possibly"; but they can be read variously for various applications, and moreover a single language of modal logic can have more than one pair of modal operators. There can also be modal operators $\circledast$ of $\mathrm{ar}(\circledast) > 1$; for instance, since the origin of modern modal logic (Lewis, 1918), various binary operators have been studied for implication, entailment, and conditionals. In this section, we discuss propositional modal logic with a classical base and with unary modal operators.

The grammar we consider adds a pair of unary modal operators $\Box$ and $\Diamond$, or a family of pairs $[i], \langle i \rangle$ ($i \in I$), to the set τ of connectives. The set of sentences is then a free τ-algebra $\mathrm{Sn}_\tau(V)$ for the new, modal τ, i.e. equipped with $\Box, \Diamond : \mathrm{Sn}_\tau(V) \to \mathrm{Sn}_\tau(V)$. So we define a *Boolean (pre)algebra with operators* to be a Boolean (pre)algebra A equipped with $\Box, \Diamond : A \to A$. Then possible-world semantics for classical-based modal logic is obtained by equipping $\mathcal{P}X$ with $\Box, \Diamond : \mathcal{P}X \to \mathcal{P}X$ and modifying (9.9) into

(9.15) $[\![-]\!] : A \to \mathcal{P}X$ is a homomorphism of Boolean prealgebras with operators; in particular, it preserves $\Box$ and $\Diamond$.

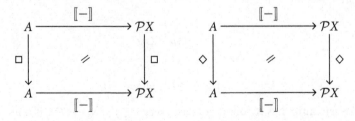

The definition of Kripke semantics boils down to how to define $\Box, \Diamond : \mathcal{P}X \to \mathcal{P}X$.

A *Kripke frame* is a set X equipped with a binary relation $R \subseteq X \times X$. We often read Rwv as "a possible world v is accessible from another w". It interprets a Boolean prealgebra A with operators with a relation $\vDash : X \times A \to 2$ satisfying (9.6)–(9.8), etc., and moreover

(9.16) $\left\{\begin{array}{c} w \vDash \Box\varphi \\ w \in [\![\Box\varphi]\!] \end{array}\right\}$ iff $\left\{\begin{array}{c} v \vDash \varphi \\ v \in [\![\varphi]\!] \end{array}\right\}$ for all $v \in X$ such that Rwv, i.e. iff $\bar{R}(w) \subseteq [\![\varphi]\!]$.

(9.17) $\left\{\begin{array}{c} w \vDash \Diamond\varphi \\ w \in [\![\Diamond\varphi]\!] \end{array}\right\}$ iff $\left\{\begin{array}{c} v \vDash \varphi \\ v \in [\![\varphi]\!] \end{array}\right\}$ for some $v \in X$ such that Rwv, i.e. iff $[\![\varphi]\!] \cap \bar{R}(w) \neq \emptyset$.

These can be rewritten as $[\![-]\!]$ preserving $\Box$ and $\Diamond$, as in (9.15), once we define $\Box_R, \Diamond_R : \mathcal{P}X \to \mathcal{P}X$ in a manner parallel to (9.16)–(9.17): for each $S \subseteq X$,

(9.18) $w \in \Box_R(S)$ iff $v \in S$ for all $v \in X$ such that Rwv.

(9.19) $w \in \Diamond_R(S)$ iff $v \in S$ for some $v \in X$ such that Rwv.

So we have (9.15), with $[\![\Box]\!] = \Box_R$ and $[\![\Diamond]\!] = \Diamond_R$. Thus, Kripke semantics is given by homomorphisms $[\![-]\!]$ to (complete atomic) Boolean algebras $(\mathcal{P}X, \Box_R, \Diamond_R)$ with operators $\Box_R, \Diamond_R$ defined by (9.18), (9.19). Since there are other semantics for $\Box$ and $\Diamond$, let us call $\Box_R$ and $\Diamond_R$ as in (9.16)–(9.19) the "Kripke necessity" and "possibility" of the relation R. A *Kripke model* is defined as a Kripke frame paired with a valuation, i.e. any function $[\![-]\!]_V : V \to \mathcal{P}X$; nevertheless, Kripke models can simply be identified with homomorphisms $[\![-]\!]$.[7] Accordingly, the notion of validity can be defined so that a sentence $\varphi \in A$ is valid in a model $[\![-]\!] : A \to \mathcal{P}X$ if $[\![\varphi]\!] = X$; and, as usual, φ is valid in the frame X if $[\![\varphi]\!] = X$ for all $[\![-]\!] : A \to \mathcal{P}X$.

The (finitary) logic of Kripke semantics is the celebrated **K**. It contains $\Diamond = \neg\Box\neg$, usually as a definition. Its other laws, the "distribution" axiom and the "necessitation" rule

$$\Box(\varphi \Rightarrow \psi) \vdash \Box\varphi \Rightarrow \Box\psi \tag{K}$$

$$\frac{\top \vdash \psi}{\top \vdash \Box\psi}, \tag{N}$$

have a more algebraic formulation stating that $\Box$ is monotone (i.e. preserves order) and preserves finite meets:

$$\frac{\varphi \vdash \psi}{\Box\varphi \vdash \Box\psi} \tag{M}$$

$$\top \vdash \Box\top \tag{N$_{\text{ax}}$}$$

$$\Box\varphi \wedge \Box\psi \vdash \Box(\varphi \wedge \psi). \tag{C}$$

Kripke semantics has **K** sound because every $\Box_R : \mathcal{P}X \to \mathcal{P}X$ is a monotone map preserving all intersections. Indeed, this algebraic property is connected precisely to binary relations via duality.

---

[7] See footnote 5 for why.

9.2.2 Kripke modalities

For a binary relation $R \subseteq X \times Y$, let us write $R : X \twoheadrightarrow Y$ to emphasize its direction. We write $R^\dagger : Y \twoheadrightarrow X$ for the opposite of R, i.e. $R^\dagger vw$ iff Rwv. Now, given any relation $R : X \twoheadrightarrow Y$, define monotone maps $\forall_R, \exists_R : \mathcal{P}X \to \mathcal{P}Y$ by

$$\forall_R(S) = \{v \in Y \mid w \in S \text{ for all } w \in X \text{ such that } Rwv\},$$
$$\exists_R(S) = \{v \in Y \mid w \in S \text{ for some } w \in X \text{ such that } Rwv\}.$$

This is a common generalization of the following:

(9.20) Quantifiers. Let D be a domain of individuals, and a relation $p : D^{n+1} \twoheadrightarrow D^n$ be the function $p : (a_1, \ldots, a_n, b) \mapsto (a_1, \ldots, a_n)$. Then $\forall_p$ and $\exists_p$ interpret quantifiers. For example, given an $(n+1)$-ary property $\varphi \subseteq D^{n+1}$, $\forall_p(\varphi)$ is the n-ary property such that

$$(a_1, \ldots, a_n) \text{ satisfies } \forall_p(\varphi) \iff (a_1, \ldots, a_n, b) \text{ satisfies } \varphi \text{ for all } b \in D.$$

(We will explain this more precisely in Section 9.5.2.)

(9.21) Inverse-image maps. A relation $f : X \twoheadrightarrow Y$ is a function iff $\forall_{f^\dagger} = \exists_{f^\dagger}$, in which case $\forall_{f^\dagger} = \exists_{f^\dagger} = f^{-1}$. It is worth adding that $\exists_f$ is then the direct-image map

$$\exists_f : \mathcal{P}X \to \mathcal{P}Y :: S \mapsto f[S] = \{f(w) \in Y \mid w \in S\}.$$

(9.22) Kripke modalities. For any relation $R : X \twoheadrightarrow X$ on a set X, (9.18)–(9.19) mean that $\Box_R = \forall_{R^\dagger}$ and $\Diamond_R = \exists_{R^\dagger}$.

This supports the conception of Kripke modalities that they are a generalization of quantifiers and inverse-image maps. The *universal modality*, "for all (or some) worlds", is essentially a case of (9.20); other Kripke modality generalizes it by appending "that are accessible".[8] The aspect of (9.21) is prominent in *dynamic logic* (see Harel et al., 2000), in which a relation $R_\alpha wv$ stands for "performing the action α at the state w may bring the system to the state v", so that $\Box_{R_\alpha}\varphi$ (or $\Diamond_{R_\alpha}\varphi$, respectively) means "$\varphi$ will" (or "may") "hold after α".

A fundamental insight of categorical logic is to connect (9.20) and (9.21) using the fact that $\exists_f \dashv f^{-1} \dashv \forall_f$ for any function f. This adjunction is, in fact, a consequence of a fact concerning relations in general, that $\exists_R \dashv \forall_{R^\dagger}$ for any R (which implies that $\exists_f \dashv \forall_{f^\dagger} = f^{-1} = \exists_{f^\dagger} \dashv \forall_f$ for a function f). Thus $\exists_R$ and $\forall_R$ are left and right adjoints and preserve all unions and all intersections, respectively.

Hence, in particular, every Kripke necessity $\Box_R = \forall_{R^\dagger}$ (or possibility $\Diamond_R = \exists_{R^\dagger}$) is a right (or left) adjoint, preserving all conjunctions (or disjunctions), whether finite or infinite. Thus, every Kripke modality is an "adjoint modality". This idea manifests itself in *temporal logic* (see Venema, 2001), which has two opposite directions of modalities,

---

[8] The idea that $\Box$ and $\Diamond$ are a relational generalization of $\forall_f$ and $\exists_f$ is categorically laid out in Hermida (2011), for a more general setting using categories of relations and spans.

viz. past and future. Yet every Kripke modality comes implicitly with such a pair of directions.[9] To see this, take any $R : X \nrightarrow X$; it corresponds—exactly, as we will see in Section 9.2.3—to an adjoint pair $\exists_R \dashv \forall_{R^\dagger} = \Box_R$. Note that $\exists_R = \Diamond_{R^\dagger}$, i.e. it is the Kripke possibility of the opposite $R^\dagger$ of R; it is sometimes written $\blacklozenge_R$. Similarly, R also corresponds to the pair $\Diamond_R = \exists_{R^\dagger} \dashv \forall_R$, so we write $\blacksquare_R = \forall_R$ for the Kripke necessity of $R^\dagger$. In sum, a single relation R corresponds to two adjoint pairs $\blacklozenge_R \dashv \Box_R$ and $\Diamond_R \dashv \blacksquare_R$.[10]

A salient example is past and future modalities in temporal logic: Write Rwv for "a point of time v lies in the (possible) future of another w". Then (9.16) applied to $\Box$ and $\blacksquare$ renders $\Box_R$ the modality "will always be" and $\blacksquare_R$ the opposite, "has always been"; similarly, $\Diamond_R$ and $\blacklozenge_R$ signify the future and past possibilities. They satisfy the axioms

$$\varphi \vdash \Box\blacklozenge\varphi \qquad\qquad \blacklozenge\Box\varphi \vdash \varphi \qquad\qquad (B_{\blacklozenge\Box})$$

$$\varphi \vdash \blacksquare\Diamond\varphi \qquad\qquad \Diamond\blacksquare\varphi \vdash \varphi \qquad\qquad (B_{\Diamond\blacksquare})$$

(diagonal ones entail each other by $\Diamond = \neg\Box\neg$), which are usually understood to express the fact that the future $\Box$, $\Diamond$ and past $\blacksquare$, $\blacklozenge$ are opposite to each other. Another example is a modal logic with the so-called "Brouwersche axiom",[11]

$$\varphi \vdash \Box\Diamond\varphi \qquad\qquad \Diamond\Box\varphi \vdash \varphi, \qquad\qquad (B)$$

which is obtained from $(B_{\blacklozenge\Box})$ and $(B_{\Diamond\blacksquare})$ by adding $\Box = \blacksquare$. This equality corresponds, via (9.16), to R being symmetric, i.e. being opposite to itself. Thus, $(B_{\blacklozenge\Box})$ and $(B_{\Diamond\blacksquare})$ mean the opposite directions of $\Box$ and $\blacksquare$, but they are indeed the unit and counit laws of the adjunctions $\blacklozenge \dashv \Box$ and $\Diamond \dashv \blacksquare$.

Now, in most other modal logics, the vocabulary only has $\Box$ and $\Diamond$ but not $\blacksquare$ or $\blacklozenge$. Yet even then Kripke modalities are adjoint modalities nonetheless. This is semantically because for every relation R its Kripke $\Box_R$ and $\Diamond_R$ preserve all intersections and all meets, respectively. Axiomatically, it is reflected in the fact that the (modal) axioms and rules of K are immediate consequences of the adjunctions $\blacklozenge \dashv \Box$ and $\Diamond \dashv \blacksquare$. Indeed, K is precisely the $\Box$-and-$\Diamond$-only fragment of the (finitary) logic obtained by adding $\blacklozenge \dashv \Box$ and $\Diamond \dashv \blacksquare$ (as well as $\Diamond = \neg\Box\neg$ and $\blacklozenge = \neg\blacksquare\neg$) to classical logic.

9.2.3 Higher duality of relations and modalities

It takes the language of category theory to make precise the fact that a binary relation R corresponds to adjoint pairs $\blacklozenge_R \dashv \Box_R$ and $\Diamond_R \dashv \blacksquare_R$. So let us write **Rel** for the category of sets and binary relations. The composition $R_2 \circ R_1 : X \nrightarrow Z$ of two relations $R_1 : X \nrightarrow Y$ and $R_2 : Y \nrightarrow Z$ is given by

$$(R_2 \circ R_1)wu \iff \text{there is } v \in Y \text{ such that } R_1wv \text{ and } R_2vu.$$

---

[9] The idea was made explicit regarding temporal modalities in von Karger (1998), and also given an epistemic interpretation in Sadrzadeh and Dyckhoff (2009).

[10] See e.g. Goré and Tiu (2007) and Conradie and Palmigiano (2012) for the notation $\blacksquare$, $\blacklozenge$ as well as for the use of the adjunctions in the proof theory and correspondence theory of modal logic. An adjoint pair between posets is also called a *Galois connection*; see ch. 7 of Davey and Priestley (2002).

[11] See Hughes and Cresswell (1996), 70f.

Sending $R : X \nrightarrow Y$ to $R^\dagger : Y \nrightarrow X$ gives a contravariant functor $\dagger : \mathbf{Rel}^{op} \to \mathbf{Rel}$ such that $X^\dagger = X$ and $\dagger \circ \dagger = 1_{\mathbf{Rel}}$, making $\mathbf{Rel}$ a self-dual, "dagger category".

On the other hand, write $\mathbf{CABA}_\wedge$ and $\mathbf{CABA}_\vee$ for the categories of CABAs—i.e. Boolean algebras of the form $\mathcal{P}X$ for a set X, as in (9.14)—with all-$\wedge$- (or $\cap$-) preserving maps and with all-$\vee$- ($\cup$-) preserving maps, respectively. Then, as seen earlier, sending $R : X \nrightarrow Y$ to $\forall_R, \exists_R : \mathcal{P}X \to \mathcal{P}Y$ gives functors $\forall_- : \mathbf{Rel} \to \mathbf{CABA}_\wedge$ and $\exists_- : \mathbf{Rel} \to \mathbf{CABA}_\vee$. In fact, every map $h : \mathcal{P}X \to \mathcal{P}Y$ in $\mathbf{CABA}_\wedge$ (or $\mathbf{CABA}_\vee$, respectively) has the form $\forall_R$ (or $\exists_R$) for a unique relation $R : X \nrightarrow Y$, so $\forall_-$ and $\exists_-$ are equivalences of categories.[12] Now, $\forall_- \circ \dagger : \mathbf{Rel}^{op} \to \mathbf{CABA}_\wedge$, a duality followed by an equivalence, is a duality, and sends $R : X \nrightarrow Y$ to $\forall_{R^\dagger} : \mathcal{P}Y \to \mathcal{P}X$, which, as in (9.22), equals $\square_R : \mathcal{P}X \to \mathcal{P}X$ when $X = Y$. Similarly, $\exists_- \circ \dagger : \mathbf{Rel}^{op} \to \mathbf{CABA}_\vee$ is a duality sending $R : X \nrightarrow X$ to $\diamond_R$. This is how a binary relation and an all-$\wedge$- (or all-$\vee$-) preserving map are dual to each other.

It is worth observing that, since $\forall_-$ and $\exists_-$ generalize inverse-image maps as in (9.21), the dualities $\forall_- \circ \dagger$ and $\exists_- \circ \dagger$ are two extensions of the Lindenbaum-Tarski duality (9.14), whereas $\exists_- : \mathbf{Rel} \to \mathbf{CABA}_\vee$ extends the covariant powerset functor $\mathcal{P}_{\mathrm{cov}}$.

The story goes deeper, or "higher", since in $\mathbf{Rel}$ each $\mathrm{Hom}_{\mathbf{Rel}}(X, Y) = \mathcal{P}(X \times Y)$ is a Boolean algebra, the order $\subseteq$ comparing two arrows $R_1, R_2 : X \nrightarrow Y$. Indeed, $\mathbf{Rel}$ is a "2-category":[13] the order $\subseteq$ is composable not just "vertically" (i.e. transitive) but also "horizontally":

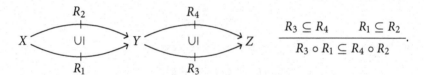

The functor $\dagger : \mathbf{Rel}^{op} \to \mathbf{Rel}$ is then a "1-cell duality", flipping the direction of relations but preserving $\subseteq$; in fact, $R_1 \subseteq R_2$ iff $R_1^\dagger \subseteq R_2^\dagger$. Similarly, $\mathbf{CABA}_\vee$ is also a 2-category, with an order $\leq$ defined on each $\mathrm{Hom}_{\mathbf{CABA}_\vee}(\mathcal{P}X, \mathcal{P}Y)$ by

$$h_1 \leq h_2 \iff h_1(S) \subseteq h_2(S) \text{ for all } S \subseteq X.$$

Moreover, the equivalence $\exists_- : \mathbf{Rel} \to \mathbf{CABA}_\vee$ induces order isomorphisms, because $R_1 \subseteq R_2$ iff $\exists_{R_1} \leq \exists_{R_2}$. On the other hand, $\mathbf{CABA}_\wedge$ is a 2-category with the same definition of order, but now $\forall_-$ gives order anti-isomorphisms: $R_1 \subseteq R_2$ iff $\forall_{R_2} \leq \forall_{R_1}$. Composing $\exists_-$ and $\forall_-$ with $\dagger$, we obtain a 1-cell duality $\exists_- \circ \dagger : \mathbf{Rel}^{op} \to \mathbf{CABA}_\vee$ and a "biduality" $\forall_- \circ \dagger : \mathbf{Rel}^{coop} \to \mathbf{CABA}_\wedge$, which flips the directions of both relations and $\subseteq$.

These higher dualities yield many correspondence results in a straightforward fashion, because, on the one hand, many conditions on a relation $R : X \nrightarrow X$ can be

---

[12] $\mathbf{Rel}$, $\mathbf{CABA}_\wedge$, and $\mathbf{CABA}_\vee$ are different expressions of the "Kleisli category" of $\mathcal{P}_{\mathrm{cov}}$ as a monad in $\mathbf{Sets}$. See Jacobs (2015).

[13] $\mathbf{Rel}$ is the archetype of category called an "allegory"; see Freyd and Scedrov (1990); and Johnstone (2002). It is also a "bicategory of relations" (Carboni and Walters, 1987).

expressed in **Rel**, and, on the other, the validity of many axioms can be expressed by the order between operators on CABAs—i.e. when $[\![\circledast_1]\!], [\![\circledast_2]\!] : (\mathcal{P}X)^n \to \mathcal{P}X$ interpret connectives $\circledast_1$ and $\circledast_2$, they satisfy $[\![\circledast_1]\!] \leqslant [\![\circledast_2]\!]$ iff the axiom $\circledast_1(\varphi_1, \ldots, \varphi_n) \vdash \circledast_2(\varphi_1, \ldots, \varphi_n)$ is valid in the Kripke frame X. Here are examples: A relation $R : X \twoheadrightarrow X$ on X is

(9.23) Reflexive iff $1_X \subseteq R$, iff $\square_R = \forall_{R^\dagger} \leqslant \forall_{1_X^\dagger} = 1_{\mathcal{P}X}$, or, as an axiom,

$$\square\varphi \vdash \varphi. \tag{T}$$

(9.24) Transitive iff $R \circ R \subseteq R$, iff $\square_R \leqslant \square_R \circ \square_R$, i.e.

$$\square\varphi \vdash \square\square\varphi. \tag{S4}$$

(9.25) Symmetric iff $R \subseteq R^\dagger$, iff $\blacklozenge_R = \exists_R \leqslant \exists_{R^\dagger} = \lozenge_R$, iff $1_{\mathcal{P}X} \leqslant \square_R \circ \lozenge_R$, i.e. (B), $\varphi \vdash \square\lozenge\varphi$.

Indeed, these are all instances of the following.

Theorem 9.1 (Lemmon and Scott, 1977) *Write R^n, $\square^n$, and $\lozenge^n$ for the n-fold compositions of $R : X \twoheadrightarrow X$ and $\square, \lozenge : \mathcal{P}X \to \mathcal{P}X$ (the 0-fold ones are the identities). Then, for any $R : X \twoheadrightarrow X$, the condition*

$$R^n \circ (R^\dagger)^m \subseteq (R^\dagger)^\ell \circ R^k$$

corresponds to the axiom

$$\lozenge^m \square^k \varphi \vdash \square^n \lozenge^\ell \varphi.$$

Proof. We have the following series of equivalences:

$$R^n \circ (R^\dagger)^m \subseteq (R^\dagger)^\ell \circ R^k$$

$$\overline{\exists_R{}^n \circ \exists_{R^\dagger}{}^m \leqslant \exists_{R^\dagger}{}^\ell \circ \exists_R{}^k}$$

$$\overline{\blacklozenge_R{}^n \circ \lozenge_R{}^m \leqslant \lozenge_R{}^\ell \circ \blacklozenge_R{}^k}$$

$$\overline{\lozenge_R{}^m \leqslant \square_R{}^n \circ \lozenge_R{}^\ell \circ \blacklozenge_R{}^k}$$

$$\overline{\lozenge_R{}^m \circ \square_R{}^k \leqslant \square_R{}^n \circ \lozenge_R{}^\ell} \qquad \square$$

This proof demonstrates the straightforwardness in which the higher dualities $\exists_- \circ \dagger : \mathbf{Rel}^{op} \to \mathbf{CABA}_\vee$ and $\forall_- \circ \dagger : \mathbf{Rel}^{coop} \to \mathbf{CABA}_\wedge$ connect the semantics of relations R to the logical behavior of $\square$ and $\lozenge$.

9.2.4 Categories of Kripke frames

The dualities just observed extend further to syntax–semantics dualities between categories of Kripke frames and categories of CABAs with certain operators.

Given Kripke frames (X, R_X) and (Y, R_Y), there are two senses in which we can say a function $f : X \to Y$ preserves their structure. One, the weaker, is f being monotone, i.e. $R_X wv$ implying $R_Y f(w) f(v)$; this is equivalent to $f \circ R_X \subseteq R_Y \circ f$ in **Rel**, as in (9.26). The stronger is $f \circ R_X = R_Y \circ f$, as in (9.27); this characterizes the concept of *bounded morphism*.[14]

$$\begin{array}{ccc} X & \xrightarrow{f} & Y \\ R_X \downarrow & \subseteq & \downarrow R_Y \\ X & \xrightarrow{f} & Y \end{array} \qquad (9.26)$$

$$\begin{array}{ccc} X & \xrightarrow{f} & Y \\ R_X \downarrow & = & \downarrow R_Y \\ X & \xrightarrow{f} & Y \end{array} \qquad (9.27)$$

So let us write $\mathbf{Krip}_{\text{monot}}$ and $\mathbf{Krip}_{\text{bdd}}$ for the categories of Kripke frames with monotone maps and with bounded morphisms, respectively.

By (9.21), applying the dualities $\forall_- \circ \dagger$ and $\exists_- \circ \dagger$ to (9.26) and to (9.27) results in

$$\begin{array}{ccc} \mathcal{P}X \xleftarrow{f^{-1}} \mathcal{P}Y \\ \Box R_X \uparrow \quad \supseteq \quad \uparrow \Box R_Y \\ \mathcal{P}X \xleftarrow{f^{-1}} \mathcal{P}Y \end{array} \qquad \begin{array}{ccc} \mathcal{P}X \xleftarrow{f^{-1}} \mathcal{P}Y \\ \Diamond R_X \uparrow \quad \subseteq \quad \uparrow \Diamond R_Y \\ \mathcal{P}X \xleftarrow{f^{-1}} \mathcal{P}Y \end{array} \qquad (9.28)$$

$$\begin{array}{ccc} \mathcal{P}X \xleftarrow{f^{-1}} \mathcal{P}Y \\ \Box R_X \uparrow \quad = \quad \uparrow \Box R_Y \\ \mathcal{P}X \xleftarrow{f^{-1}} \mathcal{P}Y \end{array} \qquad \begin{array}{ccc} \mathcal{P}X \xleftarrow{f^{-1}} \mathcal{P}Y \\ \Diamond R_X \uparrow \quad = \quad \uparrow \Diamond R_Y \\ \mathcal{P}X \xleftarrow{f^{-1}} \mathcal{P}Y \end{array} \qquad (9.29)$$

Thus (9.27) is equivalent to each of (9.29), i.e. $f^{-1} : \mathcal{P}Y \to \mathcal{P}X$ being a homomorphism preserving $\Box$ or $\Diamond$. This means that the syntax–semantics duality of (9.14) between **Sets** and **CABA** extends to that between $\mathbf{Krip}_{\text{bdd}}$ and a category of CABAs with operators. While this can be formulated variously, define a CABAO as a CABA A

---

[14] See sects 2.1 and 5.4 of Blackburn et al. (2001) for the definition of bounded morphisms and a proof of the statement.

with $\Box, \Diamond : A \to A$ such that $\Box$ is from $\mathbf{CABA}_\wedge$ and $\Diamond = \neg\Box\neg$; i.e. it has the form $(\mathcal{P}X, \Box_R, \Diamond_R)$.[15] Then, for the category **CABAO** of CABAOs and (complete) homomorphisms, (9.14) extends to the duality

$$\mathbf{Krip}_{\mathrm{bdd}} \underset{\mathrm{Atm}}{\overset{\mathcal{P}}{\underset{\sim}{\rightleftarrows}}} \mathbf{CABAO}^{\mathrm{op}}$$

between (X, R) and $(\mathcal{P}X, \Box_R, \Diamond_R)$.[16] This is a syntax–semantics duality: restricting (9.15) to **CABAO** gives

$$[\![-]\!] \in \mathrm{Hom}_{\mathbf{CABAO}}((A, \Box, \Diamond), \mathcal{P}(X, R)),$$

whereas the adjunct of this under the duality is

$$\mathbb{T} \in \mathrm{Hom}_{\mathbf{Krip}_{\mathrm{bdd}}}((X, R), \mathrm{Atm}(A, \Box, \Diamond)).$$

The duality between (9.27) and (9.29) immediately explains many facts in the model theory of Kripke semantics. An example is that bounded morphisms "preserve satisfaction". Take $A = \mathrm{Sn}_\tau(V)$, the free prealgebra of sentences composed of propositional variables $p \in V$, and let $f : X \to Y$ be a bounded morphism. Then a usual statement of the fact, in terms of $\vDash_Z : Z \times A \to \mathbf{2}$ ($Z = X, Y$), is the following:

(i) Suppose $w \vDash_X p$ iff $f(w) \vDash_Y p$ for all $p \in V$. Then $w \vDash_X \varphi$ iff $f(w) \vDash_Y \varphi$ for all $\varphi \in A$.

This means, in terms of $[\![-]\!]_Z : A \to \mathcal{P}Z$ and the restrictions $[\![-]\!]_{Z,V} : V \to \mathcal{P}Z$,

(ii) $[\![-]\!]_{X,V} = f^{-1}[\![-]\!]_{Y,V}$ entails $[\![-]\!]_X = f^{-1}[\![-]\!]_Y$.

($[\![-]\!]_X = f^{-1}[\![-]\!]_Y$ implies that if $[\![\varphi]\!]_Y = Y$ then $[\![\varphi]\!]_X = f^{-1}[\![Y]\!]_Y = X$; moreover, if f is surjective then $[\![\varphi]\!]_X = X$ iff $[\![\varphi]\!]_Y = Y$ since the dual f^{-1} is injective.) Statement (ii) is obvious once we note how $[\![-]\!]_Z$ interprets compound sentences. For example,

---

[15] One can also define a CABAO with just $\Box$ from $\mathbf{CABA}_\wedge$ or just $\Diamond$ from $\mathbf{CABA}_\vee$; the resulting categories are all isomorphic. The duality $\mathbf{Krip}_{\mathrm{bdd}} \simeq \mathbf{CABAO}$ was first observed in Thomason (1975) (for CABAOs with just $\Diamond$).

[16] Another description of $\mathbf{Krip}_{\mathrm{bdd}}$ and Kripke semantics is worth mentioning. A Kripke frame X with a relation $R : X \twoheadrightarrow X$, or its transpose $\bar{R} : X \to \mathcal{P}X$, is exactly a coalgebra of the covariant $\mathcal{P}_{\mathrm{cov}} : \mathbf{Sets} \to \mathbf{Sets}$. Indeed, X with a family of relations $(\bar{R}_i)_{i \in I} : I \to (X \to \mathcal{P}X)$ is a coalgebra $\xi : X \to (I \to \mathcal{P}X)$ of the endofunctor $S \mapsto (\mathcal{P}S)^I$. Even a Kripke model is a coalgebra, of $S \mapsto (\mathcal{P}S)^I \times \mathcal{P}V$, as a Kripke frame $\xi : X \to (\mathcal{P}X)^I$ paired with a valuation $[\![-]\!]_V : V \to \mathcal{P}X$ or transpositively $\alpha : X \to \mathcal{P}V$. Furthermore, the arrows of the categories of these coalgebras are exactly the bounded morphisms (or those respecting valuation), rewriting (9.27). Thus coalgebras give a proper generalization of Kripke semantics that subsumes many more semantics—e.g. neighborhood semantics, which we will review in Section 9.3.1. This has led to the research program called *coalgebraic logic* (Moss, 1999), which treats a wide range of modalities in a unifying fashion that facilitates the application of categorical methods.

CATEGORIES AND MODALITIES 177

$$[\![\neg p \vee \Box q]\!] = [\![\neg]\!][\![p]\!][\![\vee]\!][\![\Box]\!][\![q]\!] = [\![\vee]\!] \circ ([\![\neg]\!] \times [\![\Box]\!])([\![p]\!], [\![q]\!]);$$

$$
\begin{array}{ccccc}
\mathcal{P}Z \times \mathcal{P}Z & \xrightarrow{[\![\neg]\!] \times [\![\Box]\!]} & \mathcal{P}Z \times \mathcal{P}Z & \xrightarrow{[\![\vee]\!]} & \mathcal{P}Z \\
([\![p]\!], [\![q]\!]) & \longmapsto & ([\![\neg p]\!], [\![\Box q]\!]) & \longmapsto & [\![\neg p \vee \Box q]\!].
\end{array}
$$

Then, given $[\![r]\!]_X = f^{-1}[\![r]\!]_Y$ for $r = p, q$, (9.29) shows that $[\![\neg p \vee \Box q]\!]_X = f^{-1}[\![\neg p \vee \Box q]\!]_Y$ by a diagram chase:

$$
\begin{array}{ccccc}
([\![p]\!]_Y, [\![q]\!]_Y) & \xrightarrow{[\![\neg]\!]_Y \times [\![\Box]\!]_Y} & ([\![\neg p]\!]_Y, [\![\Box q]\!]_Y) & \xrightarrow{[\![\vee]\!]_Y} & [\![\neg p \vee \Box q]\!]_Y \\
{\scriptstyle f^{-1} \times f^{-1}} \downarrow & \mathrel{/\!/} & {\scriptstyle f^{-1} \times f^{-1}} \downarrow & \mathrel{/\!/} & \downarrow {\scriptstyle f^{-1}} \\
([\![p]\!]_X, [\![q]\!]_X) & \xrightarrow[{[\![\neg]\!]_X \times [\![\Box]\!]_X}]{} & ([\![\neg p]\!]_X, [\![\Box q]\!]_X) & \xrightarrow[{[\![\vee]\!]_X}]{} & [\![\neg p \vee \Box q]\!]_X
\end{array} \quad (9.30)
$$

The syntax–semantics duality $\mathbf{Krip}_{bdd} \simeq \mathbf{CABAO}^{op}$ thus illuminates the essential connection between certain structures in Kripke semantics and modal logic.[17]

On the other hand, the other category of Kripke frames, $\mathbf{Krip}_{monot}$, also plays a crucial role in Kripke semantics. Long story short, it is a "topological category",[18] so that many fundamental constructions in the model theory of Kripke semantics arise canonically in $\mathbf{Krip}_{monot}$. For instance, the "subframe" of a Kripke frame (X, R_X) on a subset $S \subseteq X$—with $R_S w v$ iff $R_X w v$—is characterized equivalently by either of the following:

- R_S is the largest relation on S that makes the inclusion map $i : S \hookrightarrow X :: w \mapsto w$ monotone.
- i is an equalizer in $\mathbf{Krip}_{monot}$.

9.3 Topology and Modality

One important use of modal logic is to compare different logics: using modal vocabulary, we translate one logic into another and elucidate the former in terms of the latter. The archetype of this is a translation of intuitionistic logic into classical-based modal logic S4. Indeed, the elucidation provided by this translation also illustrates two essential perspectives on the conceptual makeup of the semantics of the two logics.

---

[17] Another fundamental concept in Kripke semantics to which the same idea applies is *bisimulation*, which captures a sense of "behavioral equivalence" between Kripke models—see Venema (2007)—and a diagram doubling (9.30) demonstrates why.

[18] See Adámek et al. (1990) for the concept of topological category, as well as the fact that $\mathbf{Krip}_{monot}$ (which is written **Rel** in Adámek et al., 1990) is one, and related concepts.

9.3.1 Neighborhood semantics

Possible-world semantics for classical modal logic, in its full generality, has (9.15) with any pair of functions $\Box, \Diamond : \mathcal{P}X \to \mathcal{P}X$ (satisfying $\Diamond = \neg\Box\neg$). Kripke semantics is a special case, entailing the rather strong properties of $\Box, \Diamond$ observed in Section 9.2.2. Note that, whereas accessibility relations $R : X \twoheadrightarrow X$ in Kripke semantics are spatial structure on the space X, operators $\Box$ and $\Diamond$ in (9.15) are algebraic structure on the algebra $\mathcal{P}X$. So a natural question is, instead of accessibility relations, what spatial structure on the space X achieves the full generality of possible-world semantics.

The solution is provided by *neighborhood semantics* (Montague, 1968; Scott, 1970; Segerberg, 1971). The idea turns out rather simple once we have the concept of transpose. As we saw in Section 9.1.3, the transposition between $\vDash : X \times A \to 2$, $[\![-]\!] : A \to \mathcal{P}X$, and $\mathbb{T} : X \to \mathcal{P}A$ (see Section 9.1.2) is the basis of the duality between spaces and algebras. A similar transposition,

$$\mathcal{N} : X \times A \to 2 \qquad \Box : A \to \mathcal{P}X \qquad \bar{\mathcal{N}} : X \to \mathcal{P}A$$
$$(w, S) \mapsto \mathcal{N}wS \qquad S \mapsto \{w \in X \mid \mathcal{N}wS\} \qquad w \mapsto \{S \in A \mid \mathcal{N}wS\}$$
$$\mathcal{N}wS \iff w \in \Box S \iff S \in \bar{\mathcal{N}}(w),$$

where we write A for $\mathcal{P}X$, shows that the relation $\mathcal{N} \subseteq X \times \mathcal{P}X$ or its transpose $\bar{\mathcal{N}} : X \to \mathcal{P}\mathcal{P}X$ is the spatial structure corresponding to a given function $\Box : \mathcal{P}X \to \mathcal{P}X$ on the algebra $\mathcal{P}X$. So we refer to a set X equipped with a function $\bar{\mathcal{N}} : X \to \mathcal{P}\mathcal{P}X$ as a *neighborhood frame*,[19] and it interprets a Boolean algebra A with operators with $\vDash \subseteq X \times A$ or $[\![-]\!] : A \to \mathcal{P}X$ using (9.6)–(9.8), etc., and moreover:

(9.31) $\quad w \in [\![\Box\varphi]\!]$ iff $[\![\varphi]\!] \in \bar{\mathcal{N}}(w)$; i.e., $[\![\Box\varphi]\!] = \Box[\![\varphi]\!]$.

The sets $S \in \bar{\mathcal{N}}(w)$ are called the *neighborhoods* of w; they are the proposition "necessarily true" at the world w. We call $\bar{\mathcal{N}}$ a *neighborhood map* on X. (Henceforth we write $\mathcal{N}$ for $\bar{\mathcal{N}}$.)

The logic of this semantics is obtained by $\Diamond = \neg\Box\neg$ and the rule

$$\frac{\varphi \dashv\vdash \psi}{\Box\varphi \dashv\vdash \Box\psi} \tag{E}$$

(where $\dashv\vdash$ signifies the entailment-equivalence), which means that $[\![\varphi]\!] = [\![\psi]\!]$ implies that $\Box[\![\varphi]\!] = \Box[\![\psi]\!]$. On top of this, some conditions on $\mathcal{N} : X \to \mathcal{P}\mathcal{P}X$ correspond to axioms on $\Box : \mathcal{P}X \to \mathcal{P}X$. For example,

(9.32) $\quad$ If $S_1 \subseteq S_2 \subseteq X$ then $\left\{\begin{array}{c} S_1 \in \mathcal{N}(w) \\ w \in \Box S_1 \end{array}\right\}$ implies that $\left\{\begin{array}{c} S_2 \in \mathcal{N}(w) \\ w \in \Box S_2 \end{array}\right\}$, i.e. $\Box S_1 \subseteq \Box S_2$,

corresponds to $\Box$ being monotone, and hence the rule (M).[20]

---

[19] In other words, a neighborhood frame is a coalgebra of the monad $\mathcal{P}\mathcal{P}$: **Sets** $\to$ **Sets**; see Hansen et al. (2009, 2014).

[20] For more, see Chellas (1980) (in which neighborhood models are called "minimal models").

Structure-preserving maps of neighborhood frames can be defined in two ways (extending the two kinds of maps of Kripke frames—see Section 9.2.4). A function $f : X \to Y$ from a neighborhood frame $(X, \mathcal{N}_X)$ to another $(Y, \mathcal{N}_Y)$ is

(9.33) *Continuous* if $\left\{ \begin{array}{c} S \in \mathcal{N}_Y(f(w)) \\ f(w) \in \Box_Y S \end{array} \right\}$ implies that $\left\{ \begin{array}{c} f^{-1}[S] \in \mathcal{N}_X(w) \\ w \in \Box_X f^{-1}[S] \end{array} \right\}$, i.e. if

$$\begin{array}{ccc} X \xrightarrow{f} Y & \qquad & \mathcal{P}X \xleftarrow{f^{-1}} \mathcal{P}Y \\ \mathcal{N}_X \downarrow \ \ \leqslant\ \ \downarrow \mathcal{N}_Y & \qquad & \Box_X \uparrow \ \ \leqslant\ \ \uparrow \Box_Y \\ \mathcal{P}\mathcal{P}X \xrightarrow[\mathcal{P}\mathcal{P}f]{} \mathcal{P}\mathcal{P}Y & \qquad & \mathcal{P}X \xleftarrow[f^{-1}]{} \mathcal{P}Y \end{array}$$

($g \leqslant h$ for $g, h : X \to \mathcal{P}\mathcal{P}Y$ means that $g(w) \subseteq h(w)$ for all $w \in X$).

(9.34) *Open* if we have $\left\{ \begin{array}{c} S \in \mathcal{N}_Y(f(w)) \\ f(w) \in \Box_Y S \end{array} \right\}$ iff $\left\{ \begin{array}{c} f^{-1}[S] \in \mathcal{N}_X(w) \\ w \in \Box_X f^{-1}[S] \end{array} \right\}$, i.e. if

$$\begin{array}{ccc} X \xrightarrow{f} Y & \qquad & \mathcal{P}X \xleftarrow{f^{-1}} \mathcal{P}Y \\ \mathcal{N}_X \downarrow \ \ =\ \ \downarrow \mathcal{N}_Y & \qquad & \Box_X \uparrow \ \ =\ \ \uparrow \Box_Y \\ \mathcal{P}\mathcal{P}X \xrightarrow[\mathcal{P}\mathcal{P}f]{} \mathcal{P}\mathcal{P}Y & \qquad & \mathcal{P}X \xleftarrow[f^{-1}]{} \mathcal{P}Y \end{array}$$

Being the fully general possible-world semantics, neighborhood semantics subsumes other possible-world semantics as special cases. Kripke semantics can be regarded as one such case. Given a Kripke frame (X, R), define $\mathcal{N} : X \to \mathcal{P}\mathcal{P}X$ so that

$$S \in \mathcal{N}(w) \iff \bar{R}(w) \subseteq S. \qquad (9.35)$$

Then $\mathcal{N}$ is such that:

(9.36) Each $\mathcal{N}(w)$ has the smallest element $\bar{R}(w)$,

and (9.31) coincides with (9.16). This indeed embeds **Krip**$_{\text{monot}}$ and **Krip**$_{\text{bdd}}$ fully into the categories of neighborhood frames with continuous maps and with open maps. The duality **Krip**$_{\text{bdd}} \simeq$ **CABAO**^{op} readily extends, too.

9.3.2 Topological semantics and its epistemic reading

Another important subcase of neighborhood semantics is *topological semantics* for modal logic S4, which is given by *topological spaces*. A topological space is a set X equipped with a topology—a family $\mathcal{O}X \subseteq \mathcal{P}X$ of subsets of X that satisfies the following axioms:

(9.37) $\mathcal{O}X$ is closed under finite intersection; in particular, the empty intersection X lies in $\mathcal{O}X$.

(9.38) $\mathcal{O}X$ is closed under arbitrary union; in particular, the empty union $\varnothing$ lies in $\mathcal{O}X$.

Sets $U \in \mathcal{O}X$ are called *open sets* of the space $(X, \mathcal{O}X)$; their complements $X \setminus U$ are called *closed*. Axiom (9.38) guarantees that every $S \subseteq X$ has the largest open subset, viz.

$$\text{int}(S) = \bigcup_{U \in \mathcal{O}X, U \subseteq S} U,$$

called the *interior* of S. This gives the interior operation $\text{int} : \mathcal{P}X \to \mathcal{P}X$, and topological semantics consists in interpreting $\Box$ with $[\![\Box]\!] = \text{int}$. Its logic is then S4, which is given by (M), (N$_{\text{ax}}$), (C), (T), and (S4).

Indeed, given a topological space $(X, \mathcal{O}X)$, define a neighborhood frame $(X, \mathcal{N})$ by

$$S \in \mathcal{N}(w) \iff w \in U \subseteq S \text{ for some } U \in \mathcal{O}X \quad (9.39)$$

(i.e., $S \in \mathcal{N}(w)$ are neighborhoods of w in the original sense in topology); its $\Box : \mathcal{P}X \to \mathcal{P}X$ of $(X, \mathcal{N})$ coincides with the interior operation int of $(X, \mathcal{O}X)$. In addition, between topological spaces, the continuity (9.33) and openness (9.34) of a function $f : X \to Y$ boil down to the usual definition:

(9.40) Continuous if $U \in \mathcal{O}Y$ implies that $f^{-1}[U] \in \mathcal{O}X$.

(9.41) Open if it is continuous and $V \in \mathcal{O}X$ implies that $f[V] \in \mathcal{O}Y$.

So the topological spaces are fully ebmedded into the neighborhood frames as those validating S4. We write **Sp** for the category of topological spaces and continuous maps.

The axioms of S4 are often used to axiomatize epistemic logic, in which $\Box \varphi$ is read roughly as "the cognizer α knows that φ". In fact, an insight from both formal epistemology and computer science is that a certain kind of epistemic inquiry exhibits topological structure. When a cognizer verifies or refutes hypotheses by a finite amount of observation, the family of hypotheses or propositions shows a structure in which the *observable* or *verifiable* propositions are the open sets $U \in \mathcal{O}X$ of a topological space $(X, \mathcal{O}X)$, and the *refutable* propositions are the closed sets, i.e. complements $X \setminus U$ of open sets U.[21] Using this to interpret $\Box$ and $\Diamond$ with (9.39), we have the following:[22]

(9.42) $w \in [\![\Box \varphi]\!]$ iff $U \subseteq [\![\varphi]\!]$ for some $U \in \mathcal{O}X$, i.e. iff φ can be verified by observation to be true at w. So we read $\Box \varphi$ as "φ is verifiably true".

---

[21] This idea is in Schulte and Juhl (1996); see ch. 4 of Kelly (1996) as well. Abramsky (1991) and Vickers (1996) lay out the idea that the open sets are the observabile propositions, by reading the axioms (9.37) and (9.38) of topology directly as epistemic principles. In a similar approach, Parikh et al. (2007) derives a kind of epistemic logic from an epistemic interpretation of topology, though not the logic of interior operation.

[22] A similar epistemic interpretation using (9.42)–(9.43) but with a much weaker structure in place of topological spaces $\mathcal{O}X$ is found in "evidence logic" (van Benthem et al., 2012), in which $\Box \varphi$ means that the cognizer has evidence to believe that φ.

(9.43) $w \in [\![\Diamond \varphi]\!]$ iff $[\![\varphi]\!] \cap U \neq \varnothing$ for every $U \in \mathcal{O}X$, i.e. iff φ cannot be refuted by observation to be false at w. So we read $\Diamond \varphi$ as "φ is *not* refutably false".

One may note that this introduces another, contigent notion of verifiability and refutability. While φ may be verified to be true (or refuted to be false) in some cases, in other cases it may not be verifiable even though true (or not refutable though false). In fact, the first notion of non-contingent verifiability and refutability can be characterized as follows. A proposition φ is

- verifiable iff it is verifiably true whenever true, i.e. that $\varphi \vdash \Box \varphi$ is valid.
- refutable iff it is refutably false whenever false, i.e. that $\neg \varphi \vdash \Box \neg \varphi$ is valid.

9.3.3 Gödel Translation

Topology provides semantics for verifiability as a modality, but it can also provide semantics for the logic of the family of verifiable propositions. A perspective in the spirit of syntax–semantics duality illuminates, in a unifying fashion, the close connection both between the two semantics and between their logics.

Recall from Section 9.1.2 that the denotational presentation of semantics uses homomorphisms $[\![-]\!] : A \to \mathcal{P}X$. Analyzing the classical version (9.9), we can see two key ideas:

(i) We allow any subset S in the family $\mathcal{P}X$ to be the interpretation $[\![\varphi]\!]$ of a proposition $\varphi \in A$.
(ii) The family $\mathcal{P}X$ comes with the Boolean structure $\cap, X \setminus -, \ldots, \subseteq$; we use it to interpret $\wedge, \neg, \ldots, \vdash$ of A.

In general, however, X may have more structure than merely being a set; then, instead of (9.3.3), one may use that structure to distinguish a subfamily $B \subseteq \mathcal{P}X$ of subsets of X as "admissible propositions" and to restrict the image of $[\![-]\!]$ to those subsets, so $[\![-]\!] : A \to B$. And then, instead of (9.3.3), B comes with its own algebraic structure for interpreting $\wedge, \neg, \ldots, \vdash$ of A.

Topological semantics for *intuitionistic logic* (see van Dalen, 2001) is a salient example. Given a topological space $(X, \mathcal{O}X)$, its axioms (9.37) and (9.38) make $\mathcal{O}X \subseteq \mathcal{P}X$ a Heyting algebra. ($\mathcal{O}X$ inherits $\wedge, \vee, \top, \bot,$ and $\leq$ from $\mathcal{P}X$, but neither $\Rightarrow$ nor $\neg$, as we will discuss shortly.) Then, deeming a subset or proposition admissible iff it is open, the following replaces (9.9).

(9.44) $[\![-]\!] : A \to \mathcal{O}X$ is an order-preserving homomorphism, interpreting $\wedge, \neg, \ldots,$ $\vdash$ of A with the Heyting structure on $\mathcal{O}X$.

This provides topological semantics for intuitionistic logic. If we use the verifiability interpretation of topology, this semantics, by taking $\mathcal{O}X$ instead of $\mathcal{P}X$, captures intuitionistic logic as the logic of verifiable propositions.[23]

---

[23] As another example, take a Hilbert space modelling a quantum system, and let X be the set of its states, and $\mathcal{L}X \subseteq \mathcal{P}X$ be the family of (subsets of X corresponding to) closed linear subspaces. Restricting $\mathcal{P}X$ to

Thus topology provides two semantics, one for modal logic and the other for intuitionistic logic; a categorical perspective helps to compare them in a manner that is simple but also powerful. Since $\mathcal{O}X$ is the image of the interior operation, let us write int : $\mathcal{P}X \twoheadrightarrow \mathcal{O}X$, which is epic, with the inclusion map inc : $\mathcal{O}X \hookrightarrow \mathcal{P}X$, which is monic, and $\square = $ inc $\circ$ int : $\mathcal{P}X \to \mathcal{P}X$, as in

$$\square \circlearrowright \mathcal{P}X \underset{\text{int}}{\overset{\text{inc}}{\underset{\longtwoheadrightarrow}{\longleftarrow\!\!\bot\!\!}}} \mathcal{O}X$$

and then int $\circ$ inc $= 1_{\mathcal{O}X}$ means $\square \circ \square = \square$. Moreover, inc $\dashv$ int, because by the definition of int(S)—the largest open subset of S—every $U \in \mathcal{O}X$ and $S \in \mathcal{P}X$ have inc(U) $\subseteq$ S iff U $\subseteq$ int(S). This means that $\square$ is the comonad of the adjunction inc $\dashv$ int.[24] Now inc and its associated comonad $\square$ encapsulate the comparison between the modal logic of $\mathcal{P}X$ and the intuitionistic logic of $\mathcal{O}X$.[25]

The map inc : $\mathcal{O}X \hookrightarrow \mathcal{P}X$ embeds verifiable propositions into all propositions. This does not, however, preserve all the Heyting structure of $\mathcal{O}X$. On the one hand, the axioms (9.37) and (9.38) mean that finite meets and all joins of $\mathcal{O}X$ coincide with those of $\mathcal{P}X$, so that inc preserves them. On the other hand, $\Rightarrow_{\mathcal{O}X}$ and $\neg_{\mathcal{O}X}$ of $\mathcal{O}X$ cannot coincide with $\Rightarrow_{\mathcal{P}X}$ and $\neg_{\mathcal{P}X}$ of $\mathcal{P}X$, since, e.g., $\mathcal{O}X$ is not closed under $\neg_{\mathcal{P}X} = X \setminus -$. Still, $\Rightarrow_{\mathcal{O}X} = $ int $\circ \Rightarrow_{\mathcal{P}X} \circ$ inc^2 and $\neg_{\mathcal{O}X} = $ int $\circ \neg_{\mathcal{P}X} \circ$ inc. Thus, inc is a kind of homomorphism interpreting $\top, \wedge, \bigvee, \Rightarrow_{\mathcal{O}X}, \neg_{\mathcal{O}X}$ with $\top, \wedge, \bigvee, \square \circ \Rightarrow_{\mathcal{P}X}, \square \circ \neg_{\mathcal{P}X}$. We can express this as inc($\wedge$) = $\wedge$, inc($\Rightarrow_{\mathcal{O}X}$) = $\square \circ \Rightarrow_{\mathcal{P}X}$, etc., as in

$$\begin{array}{ccc} \mathcal{P}X & \xleftarrow{\text{inc}} & \mathcal{O}X \\ \text{inc}(\circledast) \uparrow & \not\!\! & \uparrow \circledast \\ (\mathcal{P}X)^I & \xleftarrow{\text{inc}^I} & (\mathcal{O}X)^I \end{array} \qquad (9.45)$$

This is indeed the essence of the Gödel translation of intuitionistic logic into S4.[26] It translates intuitionistic φ as modal inc(φ) using inductive clauses

- inc($\circledast(\varphi_1, \ldots, \varphi_n)$) = inc($\circledast$)(inc($\varphi_1$), $\ldots$, inc(φ_n)),

$\mathcal{L}X$ is to deem a subset or proposition admissible iff it concerns an observable property of the system. Then $\mathcal{L}X$ forms an orthomodular lattice, and homomorphisms to $\mathcal{L}X$ provide semantics for quantum logic—the logic of quantum-observable propositions.

[24] Extending the quantum example in footnote 23, write Rwv if the states $w, v \in X$ are *not* orthogonal to each other. Then R is symmetric and so $\Diamond_R \dashv \square_R$ (see the discussion in Section 9.2.2 around where (B) was introduced), and $\mathcal{L}X$ is exactly the family of fixed points of the monad $\square_R \circ \Diamond_R$. See Bergfeld et al. (2015).

[25] Another example of modality that is a comonad is the "exponential modality" in linear logic; see Blute and Scott (2004). Also see Abramsky (1993) for a translation of intuitionistic logic into intuitionistic linear logic using the exponential modality.

[26] See sect. 3.9 of Chagrov and Zakharyaschev (1997).

which is just (9.45). And the "faithfulness" of the translation consists in the following.

Fact 9.1 For any composition $\circledast$ of $\top, \wedge, \bigvee, \Rightarrow_{\mathcal{O}X}, \neg_{\mathcal{O}X}$, we have

$$\circledast = \top_{\mathcal{O}X} \iff \text{inc}(\circledast) \circ \square^I = \top_{\mathcal{P}X},$$

where $\top_A : A^I \to A :: (\varphi_i)_{i \in I} \mapsto X$ for $A = \mathcal{O}X, \mathcal{P}X$.[27]

Proof. Any composition $\circledast$ of $\top, \wedge, \bigvee, \Rightarrow_{\mathcal{O}X}, \neg_{\mathcal{O}X}$ satisfies (9.45), so

$$\text{inc}(\circledast) \circ \square^I = \text{inc}(\circledast) \circ \text{inc}^I \circ \text{int}^I = \text{inc} \circ \circledast \circ \text{int}^I,$$

where

$$\circledast = \top_{\mathcal{O}X} \iff \text{inc} \circ \circledast \circ \text{int}^I = \text{inc} \circ \top_{\mathcal{O}X} \circ \text{int}^I = \top_{\mathcal{P}X}$$

since inc is monic and int epic. □

The upshot is this: the categorical perspective, that interpretations are homomorphisms, demonstrates that the Gödel translation of the logic of $\mathcal{O}X$ into the logic of $(\mathcal{P}X, \square)$, on the syntactic side, and the structural map inc : $\mathcal{O}X \hookrightarrow \mathcal{P}X$ (with the comonad $\square$), on the semantic side, are really the same thing.

9.3.4 Stone duality

We have observed how to embed topological semantics for intuitionistic logic into that for modal logic. This is to elucidate the relation between the two semantics from the standpoint of possible worlds: we take worlds, i.e. points of a topological space, as basic entities, and show how verifiable propositions arise as specially structured (viz. open) sets of worlds. Yet the categorical idea of syntax–semantics duality also provides an opposite standpoint, showing how worlds arise from the Heyting algebra of verifiable propositions as basic entities.

As seen in Section 9.3.3, $\mathcal{O}X$ is not a Heyting subalgebra of $\mathcal{P}X$, since the inclusion inc: $\mathcal{O}X \hookrightarrow \mathcal{P}X$ fails to preserve $\Rightarrow$ and $\neg$. But it preserves $\top, \wedge$, and all $\bigvee$, so let us focus on this structure: a *frame* (not to be confused with Kripke or neighborhood frames) is a lattice that has all $\bigvee$ and satisfies the "infinite distributive law"

$$\varphi \wedge \bigvee_{i \in I} \psi_i \leqslant \bigvee_{i \in I} (\varphi \wedge \psi_i). \tag{9.46}$$

Both $\mathcal{O}X$ and $\mathcal{P}X$ are frames, and inc a homomorphism of frames. We write **Frm** for the category of frames and frame homomorphisms. We should note that the frame structure determines the Heyting structure uniquely: each frame A has a unique (complete) Heyting algebra, the underlying frame of which is A.[28]

---

[27] $\square^I$ before $\text{inc}(\circledast)$ captures the base clause of the translation, that $\text{inc}(p) = \square p$ for $p \in V$.
[28] In short, any frame is a complete Heyting algebra. Nevertheless, neither infinitary $\bigwedge, \Rightarrow$, nor $\neg$ is part of the frame structure; i.e. frame homomorphisms are not required to preserve them.

From the "interpretations as homomorphisms" perspective, frame homomorphisms are interpretations of a logic with $\top$, $\wedge$, and all $\bigvee$, replacing (9.44) with

(9.47) $[\![-]\!] : A \to \mathcal{O}X$ is an order-preserving homomorphism interpreting $\top$, $\wedge$, $\bigvee$, $\vdash$ of A with the frame structure on $\mathcal{O}X$.

For a frame A, this means

$$[\![-]\!] \in \mathrm{Hom}_{\mathrm{Frm}}(A, \mathcal{O}X). \tag{9.48}$$

This is indeed part of syntax–semantics duality. Rewriting (9.47) in terms of $\mathbb{T} : X \to \mathcal{P}A$ as well as $\vDash$ and $[\![-]\!]$, we obtain (9.6), (9.8), and the conditions for $\top$ and infinitary $\bigvee$. These mean that $\mathbb{T}(w) : A \to 2$ is a frame homomorphism, or that $\mathbb{T} : X \to \mathrm{Pt}\,A$ for

$$\mathrm{Pt}\,A = \mathrm{Hom}_{\mathrm{Frm}}(A, 2) \subseteq \mathcal{P}A.$$

A frame homomorphism $u : A \to 2$, i.e. $u \in \mathrm{Pt}\,A$, is called a *point* of A, and a corresponding subset $u \subseteq A$ is called a "completely prime filter" of A.

Indeed, declare a set of points $U \subseteq \mathrm{Pt}\,A$ open iff U has the form $\{u \in \mathrm{Pt}\,A \mid u(\varphi) = 1\}$ for some $\varphi \in A$, i.e. iff U is the set of points at which φ is true; then (9.48) corresponds exactly to

$$\mathbb{T} \in \mathrm{Hom}_{\mathrm{Sp}}(X, \mathrm{Pt}\,A), \tag{9.49}$$

resulting in a syntax–semantics dual adjunction:[29]

$$\mathrm{Sp} \underset{\mathrm{Pt}\,=\,\mathrm{Hom}_{\mathrm{Frm}}(-, 2)}{\overset{\mathcal{O}}{\rightleftarrows}} \mathrm{Frm}^{\mathrm{op}}.$$

This is not a duality, but it restricts to one: the spaces of the form $\mathrm{Pt}\,A$ are called "sober", the frames of the form $\mathcal{O}X$ are called "spatial", and they are dual to each other.

Let us review this duality using the idea in (9.12) and (9.13), the instances of which in this case are

(i) Given the topological space X of worlds, its open sets $U \in \mathcal{O}X$ are the (admissible) propositions.
(ii) Given the frame A of propositions, its points $u \in \mathrm{Pt}\,A$ are the worlds.

In (i), the points, or worlds, of the space X are given as basic entities, and they form the frame $\mathcal{P}X$ of all propositions. The functor $\mathcal{O}$ extracts from this a subframe $\mathcal{O}X$ of admissible propositions. An instance of this is the idea in Section 9.3.2 of using epistemic structure on X to determine the family $\mathcal{O}X$ of verifiable propositions. On the other hand, in (ii), it is the family A of propositions that is given as basic. The

---

[29] $\mathcal{O}$ is a functor $\mathcal{O} : \mathrm{Sp}^{\mathrm{op}} \to \mathrm{Frm}$ because the continuity of $f : X \to Y$ means that $f^{-1} : \mathcal{P}Y \to \mathcal{P}X$ restricts to $\mathrm{Hom}_{\mathrm{Sp}}(f, 2) = f^{-1} : \mathcal{O}Y \to \mathcal{O}X$. In fact, similarly to $\mathcal{P} = \mathrm{Hom}_{\mathrm{Sets}}(-, 2)$, $\mathcal{O}$ is representable: $\mathcal{O} = \mathrm{Hom}_{\mathrm{Sp}}(-, 2)$ for the "Sierpiński space" 2, i.e. 2 with three open sets, 2, $\{1\}$, and $\varnothing$.

functor Pt then yields worlds and their topological space Pt A, as if A is just the family of verifiable propositions on Pt A, whereas, by the Gödel translation, we can regard points $w \in \text{Pt } A$ as classical worlds.

This duality is therefore important to our understanding of modal logic, both technically and conceptually. Technically, it keeps logic and semantics side by side, and provides a unified perspective on the structures, properties, and assumptions of logic and semantics. Yet, conceptually, it illuminates the way in which the classical, ontological picture of $\mathcal{P}X$ and the intuitionistic, epistemic picture of $\mathcal{O}X$ give rise to each other. Indeed, this duality will be all the more important once we extend the setting from propositional to quantified logic.

9.4 Autonomy of Substructure

Before moving on to quantified logic, we should observe a few concepts. They may appear to have nothing to do with quantified logic, but they will prove crucial in the semantic considerations of quantified modal logic.

9.4.1 Impossible worlds

When we apply modal logic to express philosophical concepts or issues, there are theorems or rules that Kripke semantics validates but that should not be valid for the sake of the application. Neighborhood semantics may invalidate them, which may make it more suitable for the application. For example, when we read $\Box \varphi$ as "it is more likely than not that φ", the axiom (C) should be invalid; since Kripke semantics manages to validate it, we need neighborhood semantics to model this reading of $\Box$.

Another potential example is the "problem of logical omniscience" in epistemic logic and doxastic logic (Hintikka, 1962, see also Meyer, 2001): When $\Box \varphi$ is read as "α knows that φ" (the same applies to "believes" in place of "knows"), the axiom (K) and the rule (N) mean that α knows every logical consequence of what she knows—in particular, (N) means that α knows every logical truth—no matter how hard its derivation may be. In many contexts of philosophical reflection on knowledge, we are interested in cognizers without such deductive power, so neither (K) nor (N) should be valid; Kripke semantics validates them, but neighborhood semantics does not.[30]

Neighborhood semantics, however, does not make the problem go away, since its rule (E) results in a variant of logical omniscience. If α knows that φ, she must know any ψ logically equivalent to φ. For instance let ψ be a logical truth that is hard to prove. Certainly, neighborhood semantics invalidates (N) and so $\Box \psi$ can fail—but then, by (E), $\Box \varphi$ must also fail for all the other logical truths φ, even ones as trivial as

---

[30] Logical omniscience may pose no problem when we are interested in, say, "knowability" than "knowing". For example, the use of Kripke semantics despite logical omniscience is often defended by taking $\Box \varphi$ as expressing "implicit knowledge"—the reading of $\Box \varphi$ as "α has semantic information that entails φ" makes logical omniscience unproblematic and indeed part of definition; see e.g. Hintikka (1962); van Benthem (2011). The interpretations laid out in Section 9.3.2 are cases of this type, too. This, of course, does not make the fact go away that it takes other semantics to model actual, "explicit" knowledge.

$p \to p$. In other words, neighborhood semantics can only model either cognizers who know all logical truths or ones who know none. This would hardly be a good solution to the problem.

Thus we need a semantics in which (E) fails. Yet, as explained in Section 9.3.1, (E) expresses the near triviality that, because $[\![\Box]\!]$ is a function, if $[\![\varphi]\!] = [\![\psi]\!]$ then $[\![\Box]\!][\![\varphi]\!] = [\![\Box]\!][\![\psi]\!]$. So (E) is a direct consequence of the following two core ideas behind the setup (9.15) of not just spatial but also algebraic semantics:

(i) We interpret sentences φ with $[\![\varphi]\!] \subseteq X$ and $\Box$ with a function $[\![\Box]\!]$ on $\mathcal{P}X$.
(ii) $\varphi \vdash \psi$ means $[\![\varphi]\!] \subseteq [\![\psi]\!]$ in $\mathcal{P}X$.

To invalidate (E), therefore, we need to rethink and break this combination. And that is precisely what the concept of "impossible world" (Hintikka, 1975) achieves.

The basic idea goes like this. Let X be a set of worlds, but we say only a subset $N \subseteq X$ of worlds are "logically possible" or "normal", and the others in $X \setminus N$ "impossible". As before, X is equipped with some structure for interpreting $\Box$; an accessibility relation $R : X \twoheadrightarrow X$ will serve our purpose. Then we use $\vDash \subseteq A \times X$ to interpret sentences in A, but we only assume that the classical constraints (9.6)–(9.8) apply to normal worlds $w \in N$. So, in terms of $[\![-]\!] : A \to \mathcal{P}X$, we have, e.g., the following, where we write $[\![\varphi]\!]_N$ for $[\![\varphi]\!] \cap N$:

- $[\![\varphi \wedge \psi]\!]_N = [\![\varphi]\!]_N \cap [\![\psi]\!]_N$;
- $[\![\neg \varphi]\!]_N = N \setminus [\![\varphi]\!]_N$; and
- If $\varphi \vdash \psi$ for $\varphi, \psi \in A$, then $[\![\varphi]\!]_N \subseteq [\![\psi]\!]_N$.

Thus the upshot is that we keep core idea (i) but replace (ii) with

(iii) $\varphi \vdash \psi$ means $[\![\varphi]\!]_N \subseteq [\![\psi]\!]_N$ in $\mathcal{P}N$.

Then (E) can fail because, if $\varphi \dashv\vdash \psi$, it means $[\![\varphi]\!]_N = [\![\psi]\!]_N$ and not $[\![\varphi]\!] = [\![\psi]\!]$. In short, the logically possible worlds N give a logically coherent description of a logically incoherent cognizer.

A toy model in Figure 9.1 comprises four worlds, normal w_i and impossible v_i.

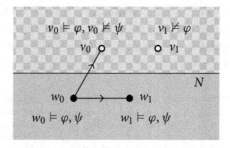

Figure 9.1 Impossible worlds $X \setminus N$ and logical truths φ, ψ.

Logical truths φ and ψ are both true at $w_i \in N$, so $[\![\varphi]\!]_N = [\![\psi]\!]_N = N$, i.e. $\top \vdash \varphi, \psi$; however, they may fail at $v_i \notin N$, as indicated in Figure 9.1 (it does not matter if $v_1 \vDash \psi$). The arrows indicate the worlds accessible from w_0, i.e. the states of affairs which the cognizer α at the state w_0 believes may be the case. Crucially, Rw_0v_0, so α at w_0 believes ψ may fail, i.e. $w_0 \nvDash \Box\psi$; in contrast, Rw_0u for no u with $u \nvDash \varphi$, so α at w_0 knows that φ, i.e. $w_0 \vDash \Box\varphi$.

This review of impossible-world semantics aims not so much at solving the problem of logical omniscience (indeed, logical omniscience will not weigh heavily in the remainder of this chapter) as at introducing the concept of impossible world. One may not quite find this concept central to modal logic and possible-world semantics. Nevertheless, some categorical reflection on both impossible worlds and quantified modal logic will reveal that the idea and structure of impossible worlds constitute a major point of conceptual and formal divergence among frameworks for quantified modal logic.

9.4.2 Closure and irrelevance

In Section 9.2.4, we observed how bounded morphisms $f : X \to Y$ of Kripke frames "preserved satisfaction", from a structural viewpoint using the diagram (9.29); this readily generalizes to neighborhood semantics by the same diagram in (9.34). In this section, we consider the special case in which the "satisfaction-preserving" f is an inclusion map; then $X \subseteq Y$ forms a special kind of substructure of Y, which we may call "autonomous". For (non)example, in fact, it is essential in the impossible-world semantics of Section 9.4.1 that the normal worlds $N \subseteq X$ are *not* autonomous in this sense.

Let $i : S \hookrightarrow X$ be an inclusion map. As mentioned in Section 9.2.4, given a Kripke frame (X, R_X) on X, its subframe (S, R_S) on S is defined as R_Swv iff R_Xwv. Or, given a neighborhood frame $(X, \mathcal{N}_X)$, its subframe $(S, \mathcal{N}_S)$ is defined by $\mathcal{N}_S(w) = \{i^{-1}[U] \mid U \in \mathcal{N}_X(w)\}$ (note that $i^{-1}[U] = U \cap S$). Then R_S is the largest relation on S making i monotone, i.e. satisfying the first square below and by duality the third; $\mathcal{N}_S$ is the smallest neighborhood map on S making i continuous, i.e. satisfying the second and third squares. (These are just squares from (9.26), (9.28), (9.33) rearranged and relabeled.)

$$\begin{array}{ccc}
X \xrightarrow{R_X} X & X \xrightarrow{\mathcal{N}_X} \mathcal{PP}X & \mathcal{P}X \xrightarrow{\Box_X} \mathcal{P}X \\
i \uparrow \quad \gtrsim \quad \uparrow i & i \uparrow \quad \swarrow \quad \uparrow \mathcal{PP}i & i^{-1} \downarrow \quad \nearrow \quad \downarrow i^{-1} \\
S \xrightarrow{R_S} S & S \xrightarrow{\mathcal{N}_S} \mathcal{PP}S & \mathcal{P}S \xrightarrow{\Box_S} \mathcal{P}S
\end{array}$$

(9.50)

Now, these squares may or may not commute; i.e., i may or may not be a bounded morphism of Kripke frames or an open map of neighborhood frames. In the Kripke

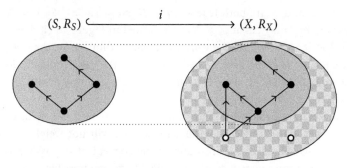

Figure 9.2 A Kripke subframe closed under accessibility.

case, this depends on whether S is "closed" under R_X—i.e.

$$
\begin{array}{ccc}
\mathcal{P}X & \xrightarrow{\Box_X} & \mathcal{P}X \\
{\scriptstyle i^{-1}}\downarrow & /\!\!/ & \downarrow{\scriptstyle i^{-1}} \\
\mathcal{P}S & \xrightarrow[\Box_S]{} & \mathcal{P}S
\end{array}
\qquad
\begin{array}{c}
[\![\varphi]\!]_X \longmapsto [\![\Box\varphi]\!]_X \\
\downarrow \qquad\qquad\qquad \downarrow \\
[\![\varphi]\!]_X \cap S \longmapsto \Box_S([\![\varphi]\!]_X \cap S) = [\![\Box\varphi]\!]_X \cap S
\end{array}
\tag{9.51}
$$

commutes iff $w \in S$ and $R_X wv$ imply that $v \in S$.[31] Figure 9.2 shows an example to help see the point of this condition. S is closed under R_X because, from each $w \in S$, there is no arrow going out of S; some arrows may come into S from outside, but they do not matter. Compare this to Figure 9.1 in which an arrow goes out of $N \subseteq X$.

In fact, S being closed under R—no arrow going out of S—means that whatever holds outside S is irrelevant to what holds inside. We can read this from (9.6)–(9.7), etc., and (9.16)–(9.17), in which the truth conditions of $w \vDash \varphi$ for $w \in S$ never refer to $v \notin S$, but this is indeed precisely expressed by (9.51). As we observed in Section 9.2.4, (9.51) gives

- $[\![-]\!]_S = i^{-1}[\![-]\!]_X$ when $[\![-]\!]_{S,V} = i^{-1}[\![-]\!]_{X,V}$ for the set V of propositional variables.

In a diagram, the example in (9.30) has the following; note (again) that $i^{-1}[-] = - \cap S$. (Note also that we use classical semantics for Boolean connectives; X contains no impossible worlds.)

---

[31] With a valuation $[\![-]\!]_{S,V} = [\![-]\!]_{X,V} \cap S$ added, a submodel $(S, R_S, [\![-]\!]_S)$ of a Kripke model $(X, R_X, [\![-]\!]_X)$ satisfying this condition is called a "generated submodel"; see Blackburn et al. (2001).

$$
\begin{array}{ccccc}
(\llbracket p\rrbracket_X, \llbracket q\rrbracket_X) & \xrightarrow{\llbracket\neg\rrbracket_X \times \llbracket\Box\rrbracket_X} & (\llbracket\neg p\rrbracket_X, \llbracket\Box q\rrbracket_X) & \xrightarrow{\llbracket\vee\rrbracket_X} & \llbracket\neg p \vee \Box q\rrbracket_X \\
{\scriptstyle i^{-1} \times i^{-1}}\downarrow & \mathbin{/\!\!/} & {\scriptstyle i^{-1} \times i^{-1}}\downarrow & \mathbin{/\!\!/} & \downarrow{\scriptstyle i^{-1}} \qquad (9.52) \\
(\llbracket p\rrbracket_S, \llbracket q\rrbracket_S) & \xrightarrow[\llbracket\neg\rrbracket_S \times \llbracket\Box\rrbracket_S]{} & (\llbracket\neg p\rrbracket_S, \llbracket\Box q\rrbracket_S) & \xrightarrow[\llbracket\vee\rrbracket_S]{} & \llbracket\neg p \vee \Box q\rrbracket_S.
\end{array}
$$

This states the irrelevance of $X \setminus S$ to S in the following sense. In evaluating at which worlds in S the sentence φ (e.g. $\neg p \vee \Box q$ above) is true, i.e. in computing $\llbracket\varphi\rrbracket_S$,

(i) One may first compute $\llbracket\varphi\rrbracket_X$ and then ignore $X \setminus S$ (by taking $\llbracket\varphi\rrbracket_S = \llbracket\varphi\rrbracket_X \cap S$)—in (9.52), the path from the top-left corner to the bottom-right through the top-right;

(ii) Yet she may also ignore $X \setminus S$ from the outset, see which propositional variables are true at which worlds in S (by taking $\llbracket-\rrbracket_{S,V} = \llbracket-\rrbracket_{X,V} \cap S$), and compute $\llbracket\varphi\rrbracket_S$ from that by treating S as a model on its own right—the path through the bottom-left corner.

Diagram (9.52) then means that (i) and (ii) yield the same result—i.e. that whatever holds outside S can be ignored entirely as in (ii), since it makes no difference to what holds within S. The crucial aspect is the following:

(9.53) There exists a function $f : \mathcal{P}S \to \mathcal{P}S$ that makes (9.51) commute in place of $\Box_S$ and thereby gives us the path of (ii) equivalent to that of (i).

One might think it is always possible to find some such f as in (9.53). It is not. A nonexample is given by impossible worlds in Section 9.4.1, with $S = N$. It is essential there that the normal worlds N are *not* closed under R, i.e. that impossible worlds $X \setminus N$ are *relevant* to N. Otherwise, i.e. if N were closed and $X \setminus N$ irrelevant to N, it would mean that the cognizer knows the impossible worlds to be impossible, making her logically omniscient. Or, in terms of diagrams, it would mean that R_N gives such $f = \Box_N$ as in (9.53); then it validates the rule (E), resulting in the logical-equivalence variant of omniscience. For, such f would give

$$\llbracket\Box\varphi\rrbracket_X \cap N \stackrel{(9.51)}{=} f(\llbracket\varphi\rrbracket_X \cap N) = f(\llbracket\psi\rrbracket_X \cap N) \stackrel{(9.51)}{=} \llbracket\Box\psi\rrbracket_X \cap N$$

whenever $\llbracket\varphi\rrbracket_X \cap N = \llbracket\psi\rrbracket_X \cap N$; or, contrapositively, if $\llbracket\Box\varphi\rrbracket_X \cap N \neq \llbracket\Box\psi\rrbracket_X \cap N$ whereas $\llbracket\varphi\rrbracket_X \cap N = \llbracket\psi\rrbracket_X \cap N$, there can be no such f. Indeed, (9.53) is equivalent to the validity of the rule (E).

9.4.3 Autonomy: definition and examples

From the discussion in Section 9.4.2, let us extract the core idea in general terms as in Definitions 9.1 and 9.2.

Definition 9.1 Given an inclusion map $i : S \hookrightarrow X$ and a function $f : (\mathcal{P}X)^n \to (\mathcal{P}X)^m$, we call a function $g : (\mathcal{P}S)^n \to (\mathcal{P}S)^m$ a *restriction* of f to S if it makes the square in

$$
\begin{array}{ccc}
X & (\mathcal{P}X)^n \xrightarrow{\ f\ } (\mathcal{P}X)^m \\
\uparrow i & (i^{-1})^n \downarrow \quad /\!/ \quad \downarrow (i^{-1})^m \\
S & (\mathcal{P}S)^n \xrightarrow{\ g\ } (\mathcal{P}S)^m
\end{array}
$$

commute, and we say that f is *restrictable* to S if it has a restriction to S. Note that a restriction, if it exists, is unique since $(i^{-1})^n$ is epic, so we write f_S for the restriction of f to S.

Fact 9.2 Given an inclusion map $i : S \hookrightarrow X$ and a function $f : (\mathcal{P}X)^n \to (\mathcal{P}X)^m$, write

$$f_S^\forall = (i^{-1})^m \circ f \circ \forall_i^n : (\mathcal{P}S)^n \to (\mathcal{P}S)^m,$$
$$f_S^\exists = (i^{-1})^m \circ f \circ \exists_i^n : (\mathcal{P}S)^n \to (\mathcal{P}S)^m$$

as candidates for g in Definition 9.1. Then

$$f_S^\exists \circ (i^{-1})^n \leq (i^{-1})^m \circ f \leq f_S^\forall \circ (i^{-1})^n,$$

and moreover the following are equivalent:

- f is restrictable to S;
- $f_S^\forall$ is the restriction of f to S, i.e. $f_S^\forall \circ (i^{-1})^n \leq (i^{-1})^m \circ f$;
- $f_S^\exists$ is the restriction of f to S, i.e. $(i^{-1})^m \circ f \leq f_S^\exists \circ (i^{-1})^n$; and
- $f_S^\forall = f_S^\exists$.

For instance, the submodels (S, R_S) of a model (X, R_X) correspond dually to their Kripke necessities $\Box_S = (\Box_X)_S^\forall = i^{-1} \circ \Box_X \circ \forall_i$.[32] Then Fact 9.2 means that, although $\Box_S$ may fail to be the restriction of $\Box_X$ to S, it is a good candidate in the sense that if it fails to be the restriction of $\Box_X$ then no $g : \mathcal{P}S \to \mathcal{P}S$ is and so $\Box_X$ is not restrictable.

---

[32] This fact underlies, e.g. the "reduction axiom" for $\Box$, i.e. $[!\varphi]\Box\psi \dashv\vdash \varphi \Rightarrow \Box[!\varphi]\psi$, in "public announcement logic" (see van Benthem, 2011), in which the operator $[!\varphi]$, "if it is publicly observed that φ then ...", is interpreted as

$$[\![[!\varphi]\psi]\!]_X = [\![\varphi]\!]_X [\![\Rightarrow]\!]_X [\![\psi]\!]_S = \forall_i [\![\psi]\!]_S$$

for $S = [\![\varphi]\!]_X$ and $i : S \hookrightarrow X$.

Note that the irrelevance argument in Section 9.4.2 involved not just the restrictability of $\Box_X$ but also that of $[\![\neg]\!]_X$, $[\![\vee]\!]_X$, and all the other operations on $\mathcal{P}X$ used in the interpretation. Therefore we enter the following:

Definition 9.2 Fix a signature τ. Given a τ-algebra $A = (\mathcal{P}X, f, \dots)$ on $\mathcal{P}X$, or a space on X associated dually with A, we say that a subset $S \subseteq X$ is *autonomous* from X if every f is restrictable to S.

The upshot is that the irrelevance of $X \setminus S$ to S consists in the autonomy of S.

A useful special case of autonomy is found in a *disjoint union*. Suppose a set X is a disjoint union

$$X = \sum_{i \in I} X_i$$

of a family $(X_i)_{i \in I}$ of sets. Then a space on X is a disjoint union of spaces on the partitions X_i iff each X_i is autonomous. Note that duality turns a disjoint union into a product,

$$\mathcal{P}X = \mathcal{P}(\sum_{i \in I} X_i) \cong \prod_{i \in I} \mathcal{P}X_i$$

$$S \longmapsto (S \cap X_i)_{i \in I}$$

$$\sum_{i \in I} S_i \longleftarrow\!\!\!\shortmid (S_i)_{i \in I}$$

and more generally

$$(\mathcal{P}X)^n = (\mathcal{P}(\sum_{i \in I} X_i))^n \cong (\prod_{i \in I} \mathcal{P}X_i)^n \cong \prod_{i \in I} (\mathcal{P}X_i)^n.$$

So the restrictability of f to X_i as in the square above is equivalent to f being a product map:

$$f = \prod_{i \in I} f_{X_i} : \prod_{i \in I} (\mathcal{P}X_i)^n \to \prod_{i \in I} (\mathcal{P}X_i)^m :: (S_{1,i}, \dots, S_{n,i})_{i \in I} \mapsto (f_{X_i}(S_{1,i}, \dots, S_{n,i}))_{i \in I}.$$

(9.54)

Observe moreover that (9.54) can be rewritten as

$$f(\sum_{i \in I} S_{1,i}, \dots, \sum_{i \in I} S_{n,i}) = \sum_{i \in I} f_{X_i}(S_{1,i}, \dots, S_{n,i}), \qquad (9.55)$$

i.e. f distributing over $\sum_{i \in I}$ into f_{X_i}.

An instance of this that will help us later is the *extensionality* of operations in the following sense. As we used earlier without stating explicitly, Boolean operations on $\mathcal{P}X$ are restrictable to any $S \subseteq X$. In particular, they are restrictable to each $w \in X$; e.g. $\cap$ is restrictable to each $w \in X$ and makes the following first square commute, but this is just the same square as the second:

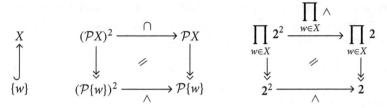

We can call an operation $f : (\mathcal{P}X)^n \to (\mathcal{P}X)^m$ extensional iff it is restrictable to each world $w \in X$, since it means that $f(S_1, \ldots, S_n) \cap \{w\}$, the output at w, depends solely on $S_1 \cap \{w\}, \ldots, S_n \cap \{w\}$, the inputs at w, ignoring all the other worlds $v \neq w$; i.e. it means that, with $f = [\![\circledast]\!]$, if two sentences φ_1 and φ_2 have the same truth values at w, they can be substituted for each other in $\circledast(\cdots \varphi_i \cdots)$ without altering its truth value at w. An example of non-extensional operation is, of course, modal $\square$ and $\diamond$. In propositional logic, this sense of extensionality coincides with truth-functionality; yet our definition of extensionality extends straightforwardly to first-order logic—we will use it in Section 9.6.3. The essential idea is that, when each world is autonomous, then the entire (many-world) model is just the disjoint union of the single-world models, so that the interpretation of sentences can be given "componentwise" at each world.

For a first-order extension, Definition 9.1 of restrictability needs to accommodate more general types of operations, but how to extend the definition should be obvious. For instance, similarly to the closedness of Kripke subframes, we have the following:

Fact 9.3 A subset $S \subseteq X$ is closed under a function $f : X^n \to X$ iff $f^{-1} : \mathcal{P}X \to \mathcal{P}(X^n)$ is restrictable to S.

9.5 A Categorical Approach to the First Order

Modal logicians have devoted the overwhelming majority of their inquiries to propositional modal logic and achieved a great advancement. In contrast, the subfield of quantified modal logic has been arguably much less successful. Philosophical logicians—most notably Carnap, Kripke, and David Lewis—have proposed semantics for quantified modal logic, but frameworks seem to keep ramifying rather than to converge. This is probably because building a system and semantics of quantified modal logic involves too many choices of technical and conceptual parameters, and perhaps because the field is lacking in a good methodology for tackling these choices in a unifying manner. The remainder of this chapter illustrates how the essential use of category theory helps this situation, both mathematically and philosophically. In this section, we only discuss the non-modal part of the semantics; yet this part already contains points of divergence regarding how to treat individuals, the satisfaction relation, and quantification, and our categorical approach enables us to analyze them structurally.

9.5.1 Syntax as functors

We need to review how first-order (and single-sorted) classical logic is treated in categorical terms.[33] In Section 9.1.1, we saw how the syntax of propositional logic gave rise to a (pre)algebra of sentences. In this subsection, we review how this idea extends to first-order logic.

Whereas the syntax of propositional logic has sentences, that of quantified logic has *formulas in contexts*: A *context* is an ordered (finite) tuple of (individual) variables *that are all distinct*. A formula φ can be "in" a context $(x_1, \ldots, x_n)$ if every free variable of φ is among $x_1, \ldots, x_n$. Then we write

$$x_1, \ldots, x_n \mid \varphi$$

for the formula-in-context, i.e. φ as regarded as an n-ary predicate. It is meant to express an n-ary property, so that an n-tuple $(a_1, \ldots, a_n)$ of individuals (that may *not* be all distinct) may or may not satisfy φ, with each a_i in place of x_i. For example, an ordered pair (a, b) satisfies $(x, y \mid x \text{ loves } y)$ iff a loves b. Note the following:

(9.56) The order of $x_1, \ldots, x_n$ matters. (a, b) satisfying $(y, x \mid x \text{ loves } y)$ means that b loves a, but not that a loves b.

(9.57) φ can be in $(x_1, \ldots, x_n)$ even if not all of $x_1, \ldots, x_n$ occur freely in φ. For example, "x loves y" can be in (x, y, z), so that (a, b, c) satisfies $(x, y, z \mid x \text{ loves } y)$ iff a loves b.

We also put terms (individual variables and constants, function symbols, and their composites) in contexts: a term t can be in $(x_1, \ldots, x_n)$ if every free variable of t is among $x_1, \ldots, x_n$, and we write

$$x_1, \ldots, x_n \mid t$$

for the term-in-context. It is meant to express an n-ary function that takes $x_1, \ldots, x_n$ as inputs and outputs t. And (9.56)–(9.57) apply again.

In classical propositional logic, sentences form a single Boolean (pre)algebra. In contrast, in first-order logic, we consider a family of algebras of formulas-in-contexts, one algebra for each (type of) context. This enables us to capture syntactic operations structurally, as maps between those algebras. To formulate this, although we only treat a single-sorted language in this chapter, it is still useful to write T for that single basic sort. Then a context $(x_1, \ldots, x_n)$ has type T^n, and a term-in-context $(x_1, \ldots, x_n \mid t)$ has type

$$(x_1, \ldots, x_n \mid t) : T^n \longrightarrow T$$
$$\text{``}(x_1, \ldots, x_n) \longmapsto t\text{''}$$

---

[33] See e.g. Pitts (2000) for a thorough exposition.

(be cautioned that the second line, in quotes, is merely a figurative heuristic of what is meant to be expressed, and not part of the formalism). One can also think of tuples of terms $(x_1, \ldots, x_n \mid t_1, \ldots, t_m) : T^n \to T^m$; this includes projections

$$p = (x_1, \ldots, x_n, y \mid x_1, \ldots, x_n) : T^{n+1} \longrightarrow T^n \qquad (9.58)$$
$$\text{``}(x_1, \ldots, x_n, y) \longmapsto (x_1, \ldots, x_n)\text{''}.$$

In effect, we are taking a category $\mathcal{C}_T$ with finite products, of types and terms. In addition, it is helpful to write formulas-in-contexts as "propositional functions",

$$(x_1, \ldots, x_n \mid \varphi) : T^n \to \text{Prop},$$

although in this chapter we do *not* count Prop as an object of $\mathcal{C}_T$. In the following, we may write $\bar{x}$ or $\bar{a}$ for tuples of variables $(x_1, \ldots, x_n)$ and of individuals $(a_1, \ldots, a_n)$.

Now, within the same context, a family of formulas are closed under Boolean operations; e.g. $(\bar{x} \mid \varphi)$ and $(\bar{x} \mid \psi)$ yield a conjunction $(\bar{x} \mid \varphi \wedge \psi)$. Thus, for each context $\bar{x} : T^n$, formulas-in-contexts form a Boolean (pre)algebra $P(T^n)$.[34] In fact, we identify formulas-in-contexts up to relabelling of (both free and bound) variables; e.g.:

- $(x, y \mid Rxy) = (y, x \mid Ryx)$, and
- $(x \mid \exists y. Rxy) = (x \mid \exists z. Rxz)$.

Syntactic operations then constitute functions between such algebras. For example, by (9.57), we can extend contexts by adding vacuous variables, which gives homomorphisms as follows:

$$P(T^n) \xrightarrow{P(p)} P(T^{n+1}) \qquad (9.59)$$
$$(\bar{x} \mid \varphi) \longmapsto (\bar{x}, y \mid \varphi).$$

This indeed corresponds dually to the projection $p = (\bar{x}, y \mid \bar{x}) : T^{n+1} \to T^n$ in (9.58), because (9.59) can be regarded as the precomposition with p as follows:

$$(\bar{x}, y \mid \varphi) : T^{n+1} \xrightarrow{(\bar{x}, y \mid \bar{x})} T^n \xrightarrow{(\bar{x} \mid \varphi)} \text{Prop}$$
$$\text{``}(\bar{x}, y) \longmapsto \bar{x} \longmapsto \varphi\text{''}.$$

Substitution of a term t for a free variable z is also a precomposition:[35]

---

[34] There is also an approach to quantification in which the family of all formulas forms a single algebra; see e.g. Scott (2008).

[35] For $\varphi[t/z]$ to make sense, t needs to be free for z in φ, but otherwise we can suitably relabel the bound variables in φ.

$$
\begin{array}{ccc}
P(T^{n+1}) & & (\bar{x}, z \mid \varphi) \\
{\scriptstyle P(\bar{x} \mid \bar{x}, t)} \downarrow & & \downarrow \\
P(T^n) & & (\bar{x} \mid \varphi[t/z])
\end{array}
\qquad
\begin{array}{c}
T^{n+1} \xrightarrow{(\bar{x}, z \mid \varphi)} \text{Prop} \\
{\scriptstyle (\bar{x} \mid \bar{x}, t)} \uparrow \; \nearrow {\scriptstyle (\bar{x} \mid \varphi[t/z])} \\
T^n
\end{array}
$$

(9.60)

We can do this for all terms or tuples thereof, so P forms a contravariant functor from $\mathcal{C}_T$ to **BA** (or **BpA**).

Furthermore, for every arrow f of $\mathcal{C}_T$, $P(f)$ has both left and right adjoints.[36] Quantification consists in monotone maps

$$
\begin{array}{c}
P(T^{n+1}) \xrightarrow{\exists y} P(T^n) \\
(\bar{x}, y \mid \psi) \longmapsto (\bar{x} \mid \exists y. \psi)
\end{array}
\qquad
\begin{array}{c}
P(T^{n+1}) \xrightarrow{\forall y} P(T^n) \\
(\bar{x}, y \mid \psi) \longmapsto (\bar{x} \mid \forall y. \psi) \cdot
\end{array}
$$

These are indeed left and right adjoints to $P(p)$ of (9.59), i.e. $\exists y \dashv P(p) \dashv \forall y$, according to the two-way rules of quantifiers:[37]

$$
\frac{\psi \vdash \varphi}{\exists y. \psi \vdash \varphi} \; (y \text{ not free in } \varphi)
\qquad
\frac{\varphi \vdash \psi}{\varphi \vdash \forall y. \psi} \; (y \text{ not free in } \varphi).
$$

Here the top lines are in the context $(\bar{x}, y)$, and the bottom in $\bar{x}$:

$$
\begin{array}{ccccccc}
P(T^{n+1}) & & (\bar{x}, y \mid \psi) & & (\bar{x}, y \mid \varphi) & & (\bar{x}, y \mid \psi) \\
\exists y \downarrow & \dashv & \downarrow & P(p) \uparrow & \uparrow & \dashv & \forall y \downarrow \\
P(T^n) & & (\bar{x} \mid \exists y. \psi) & & (\bar{x} \mid \varphi) & & (\bar{x} \mid \forall y. \psi) \cdot
\end{array}
$$

So $(\bar{x} \mid \varphi) \in P(T^n)$ expresses the "eigenvariable condition" that y does not occur freely in φ. As another example, the rule

$$
\frac{\psi \wedge y = z \vdash \varphi}{\psi \vdash \varphi[y/z]} \; (z \text{ not free in } \psi)
$$

---

[36] $P : \mathcal{C}_T^{\text{op}} \to$ **BA** assigns a Boolean homomorphism to each $P(f)$, but the adjoints to $P(f)$ do not have to be Boolean homomorphisms; they are only required to be monotone maps. For example, the left adjoint $\exists y \dashv P(f)$ preserves order and all joins, but not meets.

[37] The reader may be more familiar with the following version of the laws,

$$
\psi \vdash \exists y. \psi \qquad \frac{\psi \vdash \varphi}{\exists y. \psi \vdash \varphi} \; (y \text{ not free in } \varphi) \qquad \forall y. \psi \vdash \psi \qquad \frac{\varphi \vdash \psi}{\varphi \vdash \forall y. \psi} \; (y \text{ not free in } \varphi)
$$

but it is easy to see that the two formulations are equivalent.

of equality[38] means the following adjunction, where $\Delta = (x \mid x, x) : T \to T^2$:

$$
\begin{array}{ccc}
(\bar{x},y,z \mid \psi \wedge y = z) & (\bar{x},y,z \mid \varphi) \quad P(T^{n+2}) & \\
\uparrow \qquad \dashv \qquad \Big\uparrow P(1_{T^n} \times \Delta) & & \\
(\bar{x},y \mid \psi) & (\bar{x},y \mid \varphi[y/z]) \quad P(T^{n+1}) & \\
\end{array}
\qquad
\begin{array}{c}
(\bar{x},y,z \mid \varphi) \\
T^{n+2} \longrightarrow \text{Prop} \\
1_{T^n} \times \Delta = \uparrow \quad \nearrow \\
(\bar{x},y \mid \bar{x},y,y) \quad \quad (\bar{x},y \mid \varphi[y/z]). \\
T^{n+1}
\end{array}
$$

9.5.2 Semantics as natural transformations

One insight from Sections 9.1.2 and 9.1.3 is that, in the semantics of propositional logic, when we use a set X to interpret an algebra A of sentences, its dual $\mathcal{P}X$ is another algebra in the same category as A, so that a semantic interpretation of sentences consists in a homomorphism $[\![-]\!] : A \to \mathcal{P}X$. This extends naturally to first-order logic.

The structure P of algebras in Section 9.5.1 that arose from the syntax of first-order logic is an instance of *hyperdoctrine* (Lawvere, 1969): i.e. a contravariant functor F from $\mathcal{C}_T$ to a category of posets (or preorders) such that, for every arrow f in $\mathcal{C}_T$, $F(f)$ has left and right adjoints, subject to certain constraints coming from the syntax.[39] As an example of these constraints, when t is free for y in φ, the syntax identifies

- $(\exists y. \psi)[t/z]$, first quantifying ψ and then substituting t, and
- $\exists y(\psi[t/z])$, first substituting t and then quantifying the result,

as the same formula, and therefore requires the "Beck-Chevalley condition", i.e. that

$$
\begin{array}{ccc}
P(T^{n+2}) & \xrightarrow{P(\bar{x},y \mid \bar{x},y,t)} & P(T^{n+1}) \\
\exists y \downarrow & \simeq & \downarrow \exists y \\
P(T^{n+1}) & \xrightarrow{P(\bar{x} \mid \bar{x},t)} & P(T^n)
\end{array}
\qquad
\begin{array}{c}
(\bar{x},y,z \mid \psi) \longmapsto (\bar{x},y \mid \psi[t/z]) \\
\downarrow \qquad\qquad\qquad\qquad \downarrow \\
(\bar{x},z \mid \exists y. \psi) \longmapsto (\bar{x} \mid (\exists y. \psi)[t/z]) = (\bar{x} \mid \exists y(\psi[t/z]))
\end{array}
$$

(9.61)

commute; and similarly for $\forall$. Elements of the poset $F(T^n)$ are called predicates or attributes of type T^n. The insight from Section 9.1 then extends to first-order logic because, while we can use sets and functions to interpret $\mathcal{C}_T$, they give rise dually to another hyperdoctrine; then an interpretation of formulas consists in a homomorphism between hyperdoctrines, i.e. a natural transformation.

We first review what may be called a "single world" case. We interpret the sort T with a nonempty set $[\![T]\!] = D$. We use this as a domain of individuals, so that D^n is the set of n-tuples of individuals of type T^n. Moreover, to function symbols $f : T^n \to T$ (including constants with $n = 0$), we assign functions $[\![f]\!] : D^n \to D$. Then the interpretation extends to all the terms-in-contexts $(\bar{x} \mid \bar{t}) : T^n \to T^m$,

---

[38] It is easy to see that this rule is equivalent to the axioms $\top \vdash x = x$ and $\psi[x/y] \wedge x = y \vdash \psi$.
[39] There is a more general definition of hyperdoctrine; see Lawvere (1969); Pitts (2000).

with $[\![\bar{x} \mid \bar{t}]\!] : D^n \to D^m$. In sum, we take a functor $[\![-]\!] : \mathcal{C}_T \to \mathbf{Sets}$ perserving (finite) products. This is the spatial structure for the semantics.

Now, on the algebraic side, we interpret a formula-in-context $(\bar{x} \mid \varphi)$ by assigning a subset

$$[\![\bar{x} \mid \varphi]\!] \subseteq D^n.$$

We regard this as the "extension" of φ, i.e. the set of n-tuples $\bar{a} \in D^n$ of individuals that satisfy φ with a_i in place of x_i. We may write $\bar{a} \vDash_{\bar{x}} \varphi$ for "$a_1, \ldots, a_n$ satisfy φ with a_i for x_i", so

$$\bar{a} \vDash_{\bar{x}} \varphi \iff \bar{a} \in [\![\bar{x} \mid \varphi]\!]. \tag{9.62}$$

$[\![\bar{x} \mid \varphi]\!]$ are elements of the Boolean algebra $\mathcal{P}(D^n)$, whose Boolean structure interprets Boolean connectives with the straightforward extension of (9.6)–(9.8); e.g.

$$(9.63) \quad \left\{ \begin{array}{c} \bar{a} \vDash_{\bar{x}} \varphi \wedge \psi \\ \bar{a} \in [\![\bar{x} \mid \varphi \wedge \psi]\!] \end{array} \right\} \text{ iff } \left\{ \begin{array}{c} \bar{a} \vDash_{\bar{x}} \varphi \text{ and } \bar{a} \vDash_{\bar{x}} \psi \\ \bar{a} \in [\![\bar{x} \mid \varphi]\!], [\![\bar{x} \mid \psi]\!] \end{array} \right\}; \text{ so } [\![\bar{x} \mid \varphi \wedge \psi]\!] = [\![\bar{x} \mid \varphi]\!] \cap [\![\bar{x} \mid \psi]\!].$$

And the interpretation validates $\varphi \vdash \psi$ iff $[\![\bar{x} \mid \varphi]\!] \subseteq [\![\bar{x} \mid \psi]\!]$ (whenever the latter makes sense). Note that this setup only involves a single world: since $D^0 = \{w\}$ for w the "0-tuple" (i.e. the empty sequence), $\mathcal{P}(D^0) = 2$, and for (closed) sentences φ (9.62) amounts to

$$w \vDash \varphi \iff w \in [\![\varphi]\!], \qquad [\![\varphi]\!] = \begin{cases} \{w\} = 1 & \text{if } w \vDash \varphi, \\ \varnothing = 0 & \text{otherwise.} \end{cases}$$

Thus, in this interpretation the satisfaction of a (closed) sentence only concerns the single world w.

On the other hand, given a term-in-context $(\bar{x} \mid \bar{t}) : T^n \to T^m$, the precomposition with it (e.g. extension of contexts, substitution of terms, etc.) is interpreted by the inverse-image map

$$[\![\bar{x} \mid \bar{t}]\!]^{-1} : \mathcal{P}(D^m) \to \mathcal{P}(D^n)$$

of the interpretation $[\![\bar{x} \mid \bar{t}]\!] : D^n \to D^m$ of the term. For example, a function symbol $f : T^m \to T$ has

$$[\![\bar{x}, \bar{y} \mid \bar{x}, \bar{y}, f(\bar{y})]\!]^{-1} : [\![\bar{x}, \bar{y}, z \mid \varphi]\!] \mapsto [\![\bar{x}, \bar{y} \mid \varphi[f(\bar{y})/z]]\!] \tag{9.64}$$

because

$(\bar{a}, \bar{b}) \in [\![\bar{x}, \bar{y} \mid \varphi[f(\bar{y})/z]]\!] \iff [\![\bar{x}, \bar{y} \mid \bar{x}, \bar{y}, f(\bar{y})]\!](\bar{a}, \bar{b}) = (\bar{a}, \bar{b}, [\![f]\!](\bar{b})) \in [\![\bar{x}, \bar{y}, z \mid \varphi]\!]$;

therefore the diagrams in (9.60) for substitution of a term are preserved:

$$\begin{array}{ccc} \mathcal{P}(D^{n+1}) & [\![\bar{x},z \mid \varphi]\!] \\ [\![\bar{x} \mid \bar{x},t]\!]^{-1} \Big\downarrow & \Big\downarrow \\ \mathcal{P}(D^n) & [\![\bar{x} \mid \varphi[t/z]]\!] \end{array} \qquad \begin{array}{ccc} (\bar{a},[\![t]\!](\bar{a})) & D^{n+1} \\ [\![\bar{x} \mid \bar{x},t]\!] \Big\uparrow & \Big\uparrow \\ \bar{a} & D^n \end{array} \qquad \begin{array}{c} [\![\bar{x},z \mid \varphi]\!] \\ \searrow \qquad 2 \\ \nearrow \\ [\![\bar{x} \mid \varphi[t/z]]\!] \end{array}.$$
(9.65)

Note that inverse-image maps g^{-1} are complete homomorphisms and so with left and right adjoints $\exists_g \dashv g^{-1} \dashv \forall_g$. In this way, the dual $\mathcal{P} \circ [\![-]\!]$ of the structure $[\![-]\!]$ is a hyperdoctrine assigning

$$[\![\bar{x} \mid \varphi]\!] \in \mathcal{P}[\![T^n]\!], \qquad \mathcal{P}[\![\bar{x} \mid t]\!] : \mathcal{P}[\![T^m]\!] \to \mathcal{P}[\![T^n]\!].$$

It may be helpful to observe how quantification is interpreted. We mentioned this in (9.20) and thereafter; in particular, recall the definition of $\forall_R$, $\exists_R$, as well as that $\exists_f \dashv f^{-1} \dashv \forall_f$ for a function f. Then, for the projection

$$p = [\![\bar{x},y \mid \bar{x}]\!] : D^{n+1} \to D^n :: (\bar{a},b) \mapsto \bar{a},$$

the truth conditions for $\forall$ and $\exists$ are

(9.66) $\bar{a} \in [\![\bar{x} \mid \forall y.\varphi]\!]$ iff $(\bar{a},b) \in [\![\bar{x},y \mid \varphi]\!]$ for all $b \in D$, i.e. iff

- $(b_1,\ldots,b_{n+1}) \in [\![\bar{x},y \mid \varphi]\!]$ for all $(b_1,\ldots,b_{n+1}) \in D^{n+1}$ such that $p(b_1,\ldots,b_{n+1}) = \bar{a}$; so

$$[\![\bar{x} \mid \forall y.\varphi]\!] = \forall_p [\![\bar{x},y \mid \varphi]\!].$$

(9.67) $\bar{a} \in [\![\bar{x} \mid \exists y.\varphi]\!]$ iff $(\bar{a},b) \in [\![\bar{x},y \mid \varphi]\!]$ for some $b \in D$, i.e. iff

- $(b_1,\ldots,b_{n+1}) \in [\![\bar{x},y \mid \varphi]\!]$ for some $(b_1,\ldots,b_{n+1}) \in D^{n+1}$ such that $p(b_1,\ldots,b_{n+1}) = \bar{a}$; so

$$[\![\bar{x} \mid \exists y.\varphi]\!] = \exists_p [\![\bar{x},y \mid \varphi]\!].$$

Thus $[\![\exists y]\!] \dashv [\![\bar{x},y \mid \bar{x}]\!]^{-1} \dashv [\![\forall y]\!]$, as is supposed to be in a hyperdoctrine. Moreover, $[\![-]\!]$ preserves the commutative square (9.61) to

$$\begin{array}{ccc} \mathcal{P}[\![T^{n+2}]\!] & \xrightarrow{[\![\bar{x},y \mid \bar{x},y,t]\!]^{-1}} & \mathcal{P}[\![T^{n+1}]\!] \\ [\![\exists y]\!] \Big\downarrow & \mathrel{/\!/} & \Big\downarrow [\![\exists y]\!] \\ \mathcal{P}[\![T^{n+1}]\!] & \xrightarrow[{[\![\bar{x} \mid \bar{x},t]\!]^{-1}}]{} & \mathcal{P}[\![T^n]\!] \end{array} \qquad \begin{array}{ccc} [\![\bar{x},y,z \mid \psi]\!] & \longmapsto & [\![\bar{x},y \mid \psi[t/z]]\!] \\ \Big\downarrow & & \Big\downarrow \\ [\![\bar{x},z \mid \exists y.\psi]\!] & \longmapsto & [\![\bar{x} \mid \exists y.\psi[t/z]]\!] \end{array}.$$

Similarly, for $[\![-]\!]$ to give a hyperdoctrine, it needs to have the following adjunction for the diagonal map $\Delta = [\![x \mid x,x]\!] : D \to D^2 :: a \mapsto (a,a)$:

CATEGORIES AND MODALITIES   199

$$
\begin{array}{ccc}
\mathcal{P}D & & \mathcal{P}(D^2) \\
[\![x \mid \psi]\!] & \xrightarrow{\ \exists_\Delta\ } & [\![x,y \mid \psi \wedge x = y]\!] \\
[\![x \mid \varphi[x/y]]\!] & \xleftarrow[\ \Delta^{-1}\ ]{\bot} & [\![x,y \mid \varphi]\!]
\end{array}
$$

Hence, in particular,

$$[\![x,y \mid x = y]\!] = \exists_\Delta [\![x \mid \top]\!] = \exists_\Delta(D) = \{\,(a,a) \in D^2 \mid a \in D\,\},$$

so

(9.68)
$$\left\{\begin{array}{c}(\bar{a},b_1,b_2) \vDash_{\bar{x},y,z} y = z \\ (\bar{a},b_1,b_2) \in [\![\bar{x},y,z \mid y = z]\!]\end{array}\right\} \text{ iff } b_1 = b_2.$$

Using these and other operations, we can build up the interpretations $[\![\bar{x} \mid \varphi]\!]$ of all the formulas from the interpretations $[\![R]\!] \subseteq D^n$ of basic relation symbols.

Lastly, while in each context $\bar{x}$ the interpretation of formulas

$$[\![-]\!] : P(T^n) \to \mathcal{P}(D^n) :: (\bar{x} \mid \varphi) \mapsto [\![\bar{x} \mid \varphi]\!]$$

is a Boolean homomorphism, (9.64) means that

$$
\begin{array}{ccc}
P(T^{n+m+1}) & \xrightarrow{P(\bar{x},\bar{y} \mid \bar{x},\bar{y},f(\bar{y}))} & P(T^{n+m}) \\
{\scriptstyle [\![-]\!]} \downarrow & \ \!\!/\!\!/ & \downarrow {\scriptstyle [\![-]\!]} \\
\mathcal{P}(D^{n+m+1}) & \xrightarrow[{[\![\bar{x},\bar{y} \mid \bar{x},\bar{y},f(\bar{y})]\!]^{-1}}]{} & \mathcal{P}(D^{n+m})
\end{array}
\qquad
\begin{array}{ccc}
(\bar{x},\bar{y},z \mid \varphi) & \longmapsto & (\bar{x},\bar{y} \mid \varphi[f(\bar{y})/z]) \\
\downarrow & & \downarrow \\
[\![\bar{x},\bar{y},z \mid \varphi]\!] & \longmapsto & [\![\bar{x},\bar{y} \mid \varphi[f(\bar{y})/z]]\!]
\end{array}
$$

commutes; this type of commutation holds in general, so that the interpretation of formulas is a natural transformation $[\![-]\!] : P \to \mathcal{P} \circ [\![-]\!]$ (with abuse of notation).

This is the categorical formulation for the interpretation of first-order classical logic in a setup involving a single world w. For modal logic, however, we need to use a set X of worlds. Formally speaking, this can be done by replacing **Sets** above with the category **Sets**/X of sets over X, but to lay this out conceptually—in a manner relevant to the ideas by Kripke and by David Lewis, as we will in Section 9.6—we should first understand how to treat "non-existing" individuals.

9.5.3 Free logic

In Section 9.5.2, we used the notion of

(i) Domain of individuals, the set of individuals that may or may not satisfy given properties.

We then interpreted quantifiers in such a way that (9.5.3) happens to be the

(ii) Domain of quantification, the set of individuals that can be instances of quantified statements.

These two notions, however, may not coincide, once we consider an individual a that may satisfy properties but that is "non-existing" in the sense that a satisfying φ fails to imply "there is something that is φ"—this makes (ii) a proper subset of (i). Then we have a kind of *free logic* (see Lambert, 2001; and Scott, 2008). One may or may not accept such an individual as a philosophically legitimate notion; yet we have at least a technical reason to consider it for the sake of semantics of quantified modal logic, since we will technically need to consider whether an individual satisfies properties at a world even if it does not exist there.

We modify the classical semantics in Section 9.5.2 as follows. In addition to the domain D of individuals (i), we assign some (nonempty) domain $E \subseteq D$ of quantification (ii), and replace D with E in (9.66)–(9.67); so we have the following for all $\bar{a} \in D^n$ (and not just $\bar{a} \in E^n$):

(9.69) $\bar{a} \in [\![\bar{x} \mid \forall y.\varphi]\!]$ iff $(\bar{a}, b) \in [\![\bar{x}, y \mid \varphi]\!]$ for all $b \in E$.

(9.70) $\bar{a} \in [\![\bar{x} \mid \exists y.\varphi]\!]$ iff $(\bar{a}, b) \in [\![\bar{x}, y \mid \varphi]\!]$ for some $b \in E$.

That is, $\exists y.\varphi$ is true iff φ is true of some existing individual and not just any individual; $\forall y.\varphi$ is true iff φ is true of all the existing individuals though not necessarily all the individuals. Hence

$[\![\bar{x} \mid \exists y.\varphi]\!] = \exists_p((D^n \times E)[\![\wedge]\!][\![\bar{x}, y \mid \varphi]\!])$ $[\![\bar{x} \mid \forall y.\varphi]\!] = \forall_p((D^n \times E)[\![\Rightarrow]\!][\![\bar{x}, y \mid \varphi]\!])$
$\quad = \exists_p \circ \exists_i \circ i^{-1}[\![\bar{x}, y \mid \varphi]\!],$ $\quad = \forall_p \circ \forall_i \circ i^{-1}[\![\bar{x}, y \mid \varphi]\!],$
$[\![\exists y]\!] = \exists_p \circ \exists_i \circ i^{-1},$ $[\![\forall y]\!] = \forall_p \circ \forall_i \circ i^{-1}$

for the projection $p : D^{n+1} \to D^n :: (\bar{a}, b) \mapsto \bar{a}$ and other maps in

$$D^n \times E \xhookrightarrow{i} D^{n+1} \xtwoheadrightarrow{p} D^n$$
$$(\bar{a}, b) \mapsto (\bar{a}, b) \mapsto \bar{a}$$

$$\mathcal{P}(D^n \times E) \xleftarrow[\substack{\exists_i \\ \dashv \\ i^{-1} \\ \dashv \\ \forall_i}]{} \mathcal{P}(D^{n+1}) \xleftarrow[\substack{\exists_p \\ \dashv \\ p^{-1} \\ \dashv \\ \forall_p}]{} \mathcal{P}(D^n).$$

(It is then easy to check that $E = [\![x \mid \exists y.x = y]\!]$.) In this modified semantics, various classical laws of quantifiers fail, most crucially $\varphi \vdash \exists y.\varphi$ and $\forall y.\varphi \vdash \varphi$.

Yet the concepts of restrictability and autonomy (Definitions 9.1 and 9.2, with obvious generalization) give a certain way of finding classical logic within the interpretation. $[\![\forall y]\!]$ and $[\![\exists y]\!]$ are restrictable to E, as well as all the Boolean operations; so, by Fact 9.3, we have the following:

Fact 9.4 Let $[\![-]\!]$ be an interpretation of free logic with domains D of individuals and E of quantification. Suppose E is closed under $[\![f]\!]$ for every function symbol f, or one may even assume that the language has no function symbol. Then E is autonomous.

Hence the interpretation of every formula can be restricted to E; e.g.

$$\begin{array}{ccc}
([\![x,y \mid R_1xy]\!], [\![y,z \mid R_2yz]\!]) & \xrightarrow{[\![\exists y]\!] \circ [\![\wedge]\!] \circ (p_1^{-1} \times p_2^{-1})} & [\![x,z \mid \exists y(R_1xy \wedge R_2yz)]\!] \\
\downarrow & & \downarrow \\
([\![x,y \mid R_1xy]\!]_E, [\![y,z \mid R_2yz]\!]_E) & \xrightarrow[{[\![\exists y]\!]_E \circ [\![\wedge]\!]_E \circ (p_{1,E}^{-1} \times p_{2,E}^{-1})}]{} & [\![x,z \mid \exists y(R_1xy \wedge R_2yz)]\!]_E .
\end{array}$$
(9.71)

This means the following. A logician may admit non-existing individuals $D \setminus E$ in her model. Nevertheless, if she is only concerned with the validity in the existing individuals E, then what properties non-existing individuals may satisfy is irrelevant to what holds in E, so she can treat the substructure E as her model from the outset. (This breaks down if E is not closed under some interpretation $[\![f]\!]$ of a function, since $[\![f]\!]$ then lets us refer to a non-existing individual from within E and makes it relevant to existing individuals—or, more precisely, by Fact 9.3.)

The logician indeed obtains classical logic as the logic of E, since E is a classical structure in which the domains of individuals and of quantification coincide. More precisely,

(9.72)   When the language has no function symbol, the logic of the validity in domains of quantification is exactly classical quantified logic.

Fact 9.4 also entails the following, because $D^0 = E^0$ means that for (closed) sentences the vertical arrow on the right in (9.71) is the identity map.

(9.73)   Let φ be a (closed) sentence containing no function symbol. Then φ is valid in the semantics of free logic above iff it is valid in classical semantics.

This plays a role in some formulation of semantics for quantified modal logic, such as Kripke's.

9.6 First-Order Logic over Possible Worlds

9.6.1 What satisfaction is about

Let us now discuss how to extend the single-world interpretation in Section 9.5 to one involving a set X of worlds. This subsection and next deal with the spatial side of the matter, i.e. what type of structure should replace $[\![-]\!] : \mathcal{C}_T \to$ **Sets** of Section 9.5. As we announced there, it is $[\![-]\!] : \mathcal{C}_T \to$ **Sets**$/X$; we will reach this conclusion by analyzing two ideas, one by Kripke (1963) and the other by David Lewis (1968).

Kripke and Lewis both use a domain D of individuals as well as a set X of worlds, but they understand elements of D differently. In Kripke's interpretation, an individual $a \in D$ can exist in different worlds $w, v \in X$, and it makes sense to think of world-individual pairs $(w, a), (v, a) \in X \times D$ as "a at w" and "a at v" and to say that they are "the same individual in different worlds". Within the formal semantic structure, this idea of "transworld identity" is expressed solely by the identity of a.

In contrast, in Lewis's *counterpart theory* (1968), individuals $b \in D$ are individuated "within worlds": Each $b \in D$ is associated with a unique world $\pi(a) \in X$ in which it lives; so we never call $b_1, b_2 \in D$ identical if $\pi(b_1) \neq \pi(b_2)$—just the same way world-individual pairs $(w, a_1), (v, a_2) \in X \times D$ in Kripke's setup cannot be identical *as pairs* if $w \neq v$. As this comparison suggests, and as Lewis points out,[40] we can regard Kripke's $X \times D$ as a special case of Lewis's D, but, as Lewis also points out, his D is more general. He replaces Kripke's notion of transworld identity with the more general concept of *counterpart*, so that, even if $b_1, b_2 \in D$ fail to be identical since $\pi(b_1) \neq \pi(b_2)$, b_2 may still be the "counterpart of b_1 in $\pi(b_2)$". (We will discuss the idea of counterpart more in Section 9.7.)

Kripke defines a satisfaction relation between an n-ary formula φ, or, more precisely, a formula-in-context $(\bar{x} \mid \varphi) \in P(T^n)$, and an $(n+1)$-tuple $(w, \bar{a}) \in X \times D^n$. The latter is "an n-tuple $\bar{a}$ of individuals at a world w", so its satisfying φ means that "$a_1, \ldots, a_n$ satisfy φ at w, with a_i for x_i"—this results in the following modification of (9.62):

$$(w, \bar{a}) \vDash_{\bar{x}} \varphi \iff (w, \bar{a}) \in [\![\bar{x} \mid \varphi]\!]. \tag{9.74}$$

So $[\![\bar{x} \mid \varphi]\!] \subseteq X \times D^n$. This is to replace $[\![T^n]\!] = D^n$ of Section 9.5.2 with $[\![T^n]\!] = X \times D^n$. This includes the case of $n = 0$, in which $[\![T^0]\!] = X$.

Lewis provides truth conditions for the expression "$\varphi[\bar{b}/\bar{x}]$ holds in world w", so we can read this as a satisfaction relation $(w, \bar{b}) \vDash_{\bar{x}} \varphi$ between an n-ary formula φ and an $(n+1)$-tuple $(w, \bar{b})$,[41] and use (9.74) again. Nevertheless, it is not clear here what type $(w, \bar{b})$ should have, since in Lewis's formalism the expression "$\varphi[\bar{b}/\bar{x}]$ holds in w" is well-typed for

(i)   All $(w, \bar{b}) \in (X \cup D)^{n+1}$.[42]

For example, when $w \notin X$ is not a world, the formalism deems $(w, \bar{b}) \vDash_{\bar{x}} \forall y. \psi$ trivially the case rather than ill-typed. But this—"everything is ψ in w because w is not a world"—is clearly a formal artifact and not a conceptually meaningful feature of

---

[40] Lewis (1968), 115.

[41] Lewis does not explicitly use the notion of satisfaction relation; he formulates his semantics in terms of a translation of a quantified modal language into a non-modal language describing his models (this is why $(w, \bar{b}) \vDash_{\bar{x}} \varphi$ makes sense for any tuple $(w, \bar{b})$ of objects in the models). The inductive clauses of this translation, however, can be read as truth conditions.

[42] In fact, the model may have a set Y of objects larger than $X \cup D$, in which case the expression makes sense for all $(w, \bar{b}) \in Y^{n+1}$. We can, however, ignore objects not in $X \cup D$, by the same autonomy argument as in the following.

the semantics. So we may rule out the artificial cases of $w \notin X$ and $b_i \notin D$ and restrict $\vDash$ to

(ii) $(w, \bar{b}) \in X \times D^n$.

Yet one may find that this still includes a puzzling case—viz. $(w, \bar{b})$ such that not all b_i live in w. The puzzlement may be more striking in the special case mentioned above where an individual $b \in D$ in Lewis's sense happens to be "a at $\pi(b)$" from Kripke's setup; then, with $w \neq \pi(b)$, what does it mean to say

(∗) $(w, b) \vDash_x \varphi$, i.e. "a at $\pi(b)$ satisfies φ in $w \neq \pi(b)$"?

If one finds this more artificial than conceptual, she may choose to restrict $\vDash$ further to

(iii) $(w, \bar{b}) \in X \times D^n$ such that all b_i live in w, i.e. $\pi(b_1) = \cdots = \pi(b_n) = w$.

In fact, regardless of what one thinks of (∗), in Lewis's semantics the set of (iii) turns out autonomous, as we will see at the end of Section 9.7.3.[43] Therefore, if one is only interested in which (closed) sentences are true in which worlds—e.g. Lewis's formalism concerns how to model which sentences[44] are true in a world designated as "actual"—it does not matter whether (∗) is true, false, or ill-typed; we can restrict the model to (iii) and the semantics gives the same answer.[45] Hence we conclude that the set of (iii) is $[\![T^n]\!]$ in Lewis's semantics. For $n = 0$, it gives $[\![T^0]\!] = X$.

9.6.2 Semantic structures

We also need to discuss what type of structure $[\![\bar{x} \mid \bar{t}]\!]$ should interpret terms t and tuples thereof. Terms are not discussed in Kripke (1963) or Lewis (1968), but the canonical method of categorical logic turns out to provide a formalism that is conceptually reasonable.

We should first observe that, in both Kripke's and Lewis's semantics, $[\![T^n]\!]$ is the n-ary product of $[\![T]\!]$ in the category $\mathbf{Sets}/X$ of sets over X, so that $[\![-]\!] : \mathcal{C}_T \to \mathbf{Sets}/X$ is a functor preserving (finite) products. In Kripke's case, we have sets and projections

$$[\![T^n]\!] = X \times D^n,$$
$$p^n : X \times D^n \to X :: (w, \bar{a}) \mapsto w.$$

Then it is easy to see that p^n is the n-ary product of p in $\mathbf{Sets}/X$. In Lewis's case, we have shown

---

[43] The set of (ii) is also autonomous, so tuples $(w, \bar{b})$ with $w \notin X$ or $b_i \notin D$ can safely be ignored.
[44] Lewis does treat formulas with free variables, but reads them as universally quantified.
[45] Lewis takes the position that (∗) makes sense at least in some cases. For example, if φ is a basic unary predicate, then whether b satisfies φ is a matter independent of worlds, rather than holding in some worlds and not in others; so, in this sense, if b satisfies φ then it can be said to satisfy φ in every world. Nevertheless, the point of the autonomy of the set of (iii) is that Lewis's semantics is independent of this position of his.

$$\llbracket T^n \rrbracket = X \times_X D \times_X \cdots \times_X D = \{\, (w, \bar{b}) \in X \times D^n \mid w = \pi(b_1) = \cdots = \pi(b_n) \,\},$$
$$p^n : X \times_X D \times_X \cdots \times_X D \to X :: (w, \bar{b}) \mapsto w.$$

Then, again, p^n is the n-ary product of p in **Sets**/X. In both cases, the 0-ary product in **Sets**/X is $\llbracket T^0 \rrbracket = X$ with $p^0 = 1_X : X \to X$.

In fact, in Lewis's setup, we have the function $\pi : D \to X$ assigning to each individual $b \in D$ the unique world $\pi(b) \in X$ in which it lives, and this "residence map" π is an object of **Sets**/X isomorphic to $\llbracket T \rrbracket = X \times_X D$. Indeed, since $1_X : X \to X$ is terminal in **Sets**/X, each $\llbracket T^n \rrbracket = X \times_X D^n_X$ is isomorphic to the n-ary product of π in **Sets**/X,

$$D^n_X = D \times_X \cdots \times_X D = \{\, \bar{b} \in D^n \mid \pi(b_1) = \cdots = \pi(b_n) \,\} \subseteq D^n,$$
$$\pi^n : D \times_X \cdots \times_X D \to X :: \bar{b} \mapsto \pi(b_i),$$

called the n-ary *fibered product* of π over X. (The 0-ary product is again $\pi^0 = 1_X : X \to X$.) We can therefore equivalently use $\llbracket T^n \rrbracket = D^n_X$ instead of $X \times_X D^n_X$. This means replacing (9.74) with "$b_1, \ldots, b_n$ satisfy φ, with b_i for x_i",

$$\bar{b} \vDash_{\bar{x}} \varphi \iff \bar{b} \in \llbracket \bar{x} \mid \varphi \rrbracket, \tag{9.75}$$

where it is assumed that $\pi(b_1) = \cdots = \pi(b_n)$; the isomorphism means that (9.75) carries the same information as (9.74) since "at $\pi(b_i)$" can be recovered uniquely. The use of D^n_X for Lewis's semantics also shows that the subsumption of Kripke's $X \times D$ as a special case of Lewis's D extends naturally to tuples of individuals: Kripke's $(w, \bar{a}) \in X \times D^n = \llbracket T^n \rrbracket$, or an n-tuple ("a_1 at w", ..., "a_n at w"), is a special case of Lewis's n-tuple $\bar{b} \in D^n_X = \llbracket T^n \rrbracket$.

Another crucial fact is the equivalence of categories **Sets**/$X \simeq$ **Sets**^{X}. Let $\pi : D \to X$ be a function, i.e. an object of **Sets**/X. It can be described equivalently as a family of sets indexed by X, as follows: for each w, we call its inverse image

$$D_w = \pi^{-1}[\{w\}]$$

the *fiber* of D over w; Figure 9.3 depicts fibers D_w as line segments over $w \in X$. This gives a family $(D_w)_{w \in X}$ of sets indexed by $w \in X$—i.e. a functor $D_- : X \to$ **Sets** from the discrete category on X, as in Figure 9.3. Since the fibers D_w partition D, taking their disjoint union gives back

$$D = \sum_{w \in X} D_w,$$

with $\pi : D \to X$ sending $a \in D_w$ to w. In short, π is a *bundle* of fibers. A map f of **Sets**/X is then a bundle of functions $(f_w)_{w \in X}$ between fibers

$$(f : D \to E) = \left(\sum_{w \in X} (f_w : D_w \to E_w) : \sum_{w \in X} D_w \to \sum_{w \in X} E_w \right),$$

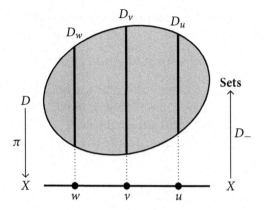

Figure 9.3 A function π as a bundle of fibers D_w.

since the commutative triangle on the left below means that f maps $b \in D_w$ over w to $f(b) \in E_w$ over w; thus f is a map of $\mathbf{Sets}^X$:

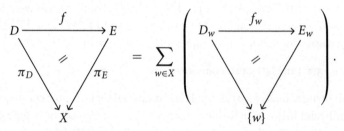

This enables us to regard the many-world structure $[\![-]\!] : \mathcal{C}_T \to \mathbf{Sets}/X$ as a bundle of single-world structures $[\![-]\!]_w : \mathcal{C}_T \to \mathbf{Sets}$. This is abstractly because

$$[\![-]\!] : \mathcal{C}_T \to (X \to \mathbf{Sets})$$
$$[\![-]\!]_- : X \to (\mathcal{C}_T \to \mathbf{Sets})$$

are the same thing, but let us spell this out in Kripke's and Lewis's terms.

First, here is how the idea goes for $[\![T^n]\!]$. For Lewis, $[\![T]\!]$ is given by his residence map $\pi : D \to X$. Its fiber $[\![T]\!]_w = D_w$ over $w \in X$ is the set of individuals that live in world w. So, its n-ary cartesian product $([\![T]\!]_w)^n = D_w^n$ is the set of n-tuples of individuals that live in world w. But this is the fiber

$$D_w^n = \{\bar{b} \in D_X^n \mid \pi(b_1) = \cdots = \pi(b_n) = w\} = (\pi^n)^{-1}[\{w\}]$$

of $\pi^n : [\![T^n]\!] = D_X^n \to X$ over w; therefore

$$[\![T^n]\!]_w = ([\![T]\!]_w)^n.$$

The same goes in Kripke's setup, where $[\![T]\!]$ has the projection $p : X \times D \to X$. Its fiber $[\![T]\!]_w$ over $w \in X$ is $p^{-1}(\{w\}) = \{w\} \times D \cong D$, i.e. the domain of individuals a

(perhaps regarded as "a at w"). Its n-ary cartesian product $(\llbracket T \rrbracket_w)^n = D^n$ is the set of n-tuples of individuals (perhaps with "at w"), but this is just the fiber

$$\llbracket T^n \rrbracket_w = (p^n)^{-1}[\{w\}] = \{w\} \times D^n$$

of $p^n : \llbracket T^n \rrbracket = X \times D^n \to X$ over w. In either case, each world w has a cartesian product $\llbracket T^n \rrbracket_w = (\llbracket T \rrbracket_w)^n$ as its domain of tuples, and the product $\llbracket T^n \rrbracket$ of $\llbracket T \rrbracket$ over X is recovered as the bundle of such cartesian products:

$$\llbracket T^n \rrbracket = \sum_{w \in X} \llbracket T^n \rrbracket_w = \sum_{w \in X} (\llbracket T \rrbracket_w)^n,$$

$$D_X^n = \sum_{w \in X} D_w^n,$$

$$X \times D^n = \sum_{w \in X} (\{w\} \times D^n) = \sum_{w \in X} D^n.$$

This extends to the interpretation of terms as well. Given that $\llbracket T^n \rrbracket$ are products of $\llbracket T \rrbracket$ in **Sets**/X, categorical wisdom says that, for things to work nicely, interpretations $\llbracket \bar{x} \mid \bar{t} \rrbracket$ of terms must be maps $\llbracket \bar{x} \mid t \rrbracket : \llbracket T^n \rrbracket \to \llbracket T^m \rrbracket$ of **Sets**/X. This technical advice in fact makes conceptual sense, too. In Lewis's setup, $\llbracket \bar{x} \mid t \rrbracket : \llbracket T^n \rrbracket \to \llbracket T^m \rrbracket$ in **Sets**/X means the constraint that it maps a tuple in D_w^n to a tuple in D_w^m. For example,

$$\llbracket x, y \mid \text{the last-born common ancestor of } x \text{ and } y \rrbracket : D_X^2 \to D$$

takes a pair of individuals b_1, b_2 as inputs and outputs their last-born common ancestor b_3. The constraint in question then means that b_1, b_2, and b_3 must be in the same world $\pi(b_i)$. This follows directly from our concluding stipulation in Section 9.6.1 that

$$\llbracket x, y, z \mid z \text{ is the last-born common ancestor of } x \text{ and } y \rrbracket \subseteq \llbracket T^3 \rrbracket = D_X^3$$

should be a property of triples of individuals from the same world. Thus, $\llbracket f \rrbracket$ is just a bundle of functions $\llbracket f \rrbracket_w$ on the domain D_w of each world $w \in X$. This applies to Kripke's setup, too:

$$\llbracket f \rrbracket : X \times D^n \to X \times D :: (w, \bar{a}) \mapsto (w, \llbracket f \rrbracket_w(\bar{a}))$$

is a bundle of functions $\llbracket f \rrbracket_w : D^n \to D$ that are all on the fixed domain D but that may behave differently in different worlds $w \in X$.

In sum, a many-world structure $\llbracket - \rrbracket$ is a bundle of fibers $\llbracket - \rrbracket_w$, each of which looks just like a single-world model of Section 9.5.

9.6.3 Interpretations of quantification

We move on to the algebraic side of the semantics, i.e. how to interpret formulas of quantified non-modal logic in **Sets**/X. The core idea is that a many-world model $\llbracket - \rrbracket$ is a disjoint union of an X-indexed family of single-world models $\llbracket - \rrbracket_w$, while we consider semantics whose non-modal part is extensional (in the sense we laid out in Section 9.4.3); therefore, using our idea of extensionality, the interpretation of a formula in the entire model is just an X-tuple of the single-world interpretations,

$$\mathcal{P}[\![T^n]\!] = \mathcal{P}(\sum_{w \in X} [\![T^n]\!]_w) \cong \prod_{w \in X} \mathcal{P}[\![T^n]\!]_w$$
$$[\![\bar{x} \mid \varphi]\!] \longmapsto ([\![\bar{x} \mid \varphi]\!]_w)_{w \in X} \quad (9.76)$$
$$\sum_{w \in X} [\![\bar{x} \mid \varphi]\!]_w \longleftarrow\!\!\mid ([\![\bar{x} \mid \varphi]\!]_w)_{w \in X}$$

as illustrated in Figure 9.4.

Applying this idea, we can interpret extensional operators by product maps. By straightforwardly adapting the discussion of disjoint union and extensionality in Section 9.4.3, an operation

$$f : (\mathcal{P}[\![T^\ell]\!])^n \to (\mathcal{P}[\![T^k]\!])^m$$

is extensional iff there is an X-indexed family $(f_w)_{w \in X}$ of operations

$$f_w : (\mathcal{P}[\![T^\ell]\!]_w)^n \to (\mathcal{P}[\![T^k]\!]_w)^m,$$

the product of which is f in the sense of (9.77), seen later—or, equivalently, into which f distributes over $\sum_{w \in X}$ in the sense of (9.78):

$$f = \prod_{w \in X} f_w : \prod_{w \in X} (\mathcal{P}[\![T^\ell]\!]_w)^n \to \prod_{w \in X} (\mathcal{P}[\![T^k]\!]_w)^m :: (S_{1,w}, \ldots, S_{n,w})_{w \in X} \mapsto$$
$$(f_w(S_{1,w}, \ldots, S_{n,w}))_{w \in X}, \quad (9.77)$$
$$f(\sum_{w \in X} S_{1,w}, \ldots \sum_{w \in X} S_{n,w}) = \sum_{w \in X} f_w(S_{1,w}, \ldots, S_{n,w}). \quad (9.78)$$

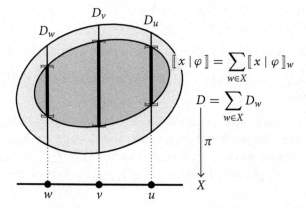

Figure 9.4 An interpretation $[\![-]\!]$ on a bundle π.

Thus, to interpret extensional ⊛, we can let $[\![⊛]\!]$ on the entire many-world model be the product of $[\![⊛]\!]_w$.

This is almost trivial when interpreting Boolean connectives. All the Boolean operations on $\mathcal{P}[\![T^n]\!]$ distributes over $\sum_{w \in X}$, so they can interpret the Boolean connectives just the same way they do in single-world models, using e.g. (9.6)–(9.8) (with types modified suitably). More significantly, the bundle-and-product idea works for quantifiers as well. In the classical case for instance, the projection

$$p = [\![\bar{x}, y \mid \bar{x}]\!] : D_X^{n+1} \to D_X^n :: (\bar{a}, b) \mapsto \bar{a}$$

over X (we use Lewis's $[\![T^n]\!] = D_X^n$ as an example) is a bundle of projections

$$p_w = [\![\bar{x}, y \mid \bar{x}]\!]_w : D_w^{n+1} \to D_w^n :: (\bar{a}, b) \mapsto \bar{a};$$

so, by duality,

$$\mathcal{P}(D_X^{n+1}) \cong \prod_{w \in X} \mathcal{P}(D_w^{n+1}) \xleftarrow[\forall_p]{\exists_p \;\;\; \dashv \;\;\; p^{-1} \;\;\; \dashv} \mathcal{P}(D_X^n) \cong \prod_{w \in X} \mathcal{P}(D_w^n)$$

are the products of $[\![\exists y]\!]_w = \exists_{p_w} \dashv p_w^{-1} \dashv [\![\forall y]\!]_w = \forall_{p_w}$. Therefore we can keep the categorical interpretation of quantifiers we observed in Section 9.5.2, i.e.

$$[\![\exists y]\!] = \exists_p, \qquad\qquad [\![\forall y]\!] = \forall_p,$$
$$[\![\bar{x} \mid \exists y.\varphi]\!] = \exists_p [\![\bar{x}, y \mid \varphi]\!], \qquad [\![\bar{x} \mid \forall y.\varphi]\!] = \forall_p [\![\bar{x}, y \mid \varphi]\!].$$

Spelling these out as truth conditions, we have the following adaptations of (9.66)–(9.67):

(9.79) $\bar{a} \in [\![\bar{x} \mid \forall y.\varphi]\!]$ iff $(\bar{a}, b) \in [\![\bar{x}, y \mid \varphi]\!]$ for all $b \in D_{\pi(a_i)}$, i.e. iff

- $(b_1, \ldots, b_{n+1}) \in [\![\bar{x}, y \mid \varphi]\!]$ for all $(b_1, \ldots, b_{n+1}) \in D_X^{n+1}$ such that $p(b_1, \ldots, b_{n+1}) = \bar{a}$.

(9.80) $\bar{a} \in [\![\bar{x} \mid \exists y.\varphi]\!]$ iff $(\bar{a}, b) \in [\![\bar{x}, y \mid \varphi]\!]$ for some $b \in D_{\pi(a_i)}$, i.e. iff

- $(b_1, \ldots, b_{n+1}) \in [\![\bar{x}, y \mid \varphi]\!]$ for some $(b_1, \ldots, b_{n+1}) \in D_X^{n+1}$ such that $p(b_1, \ldots, b_{n+1}) = \bar{a}$.

And these agree with Lewis's semantics. His idea, as expressed by (9.79)–(9.80), is to interpret quantifiers classically and extensionally at each world $w \in X$, using D_w as the domain of quantification at $w \in W$. Then the bundle $[\![-]\!] = \sum_{w \in X} [\![-]\!]_w$ of models of classical first-order logic, with the domain $D = \sum_{w \in X} D_w$, forms a model of classical first-order logic.

Kripke also interprets quantifiers extensionally, but uses free as opposed to classical logic. His $[\![T^n]\!] = X \times D^n = \sum_{w \in X} D^n$, or $[\![T^n]\!]_w = D^n$, means that D is the domain of individuals at every world. Yet he introduces a map $E_- : X \to \mathcal{P}D$ to use each $E_w \subseteq D$ as the domain of quantification at w; i.e. $E_w \subseteq D$ is the set of individuals that exist in w. This gives the following adaptation of (9.69)–(9.70) to the type of satisfaction in (9.74):

(9.81) $(w, \bar{a}) \in [\![\bar{x} \mid \forall y. \varphi]\!]$ iff $(w, \bar{a}, b) \in [\![\bar{x}, y \mid \varphi]\!]$ for all $b \in E_w$.

(9.82) $(w, \bar{a}) \in [\![\bar{x} \mid \exists y. \varphi]\!]$ iff $(w, \bar{a}, b) \in [\![\bar{x}, y \mid \varphi]\!]$ for some $b \in E_w$.

These give

$$[\![\bar{x} \mid \exists y. \varphi]\!] = \exists_p \circ \exists_i \circ i^{-1}[\![\bar{x}, y \mid \varphi]\!], \quad [\![\bar{x} \mid \forall y. \varphi]\!] = \forall_p \circ \forall_i \circ i^{-1}[\![\bar{x}, y \mid \varphi]\!],$$
$$[\![\exists y]\!] = \exists_p \circ \exists_i \circ i^{-1}, \quad\quad\quad\quad\quad [\![\forall y]\!] = \forall_p \circ \forall_i \circ i^{-1}$$

for the following inclusion i and projection p:

$$\sum_{w \in X}(D^n \times E_w) \xhookrightarrow{\;\;i\;\;} \sum_{w \in X} D^{n+1} \xrightarrow{\;\;p\;\;} \sum_{w \in X} D^n\,.$$
$$(w, \bar{a}, b) \longmapsto (w, \bar{a}, b) \longmapsto (w, \bar{a})$$

$[\![\forall y]\!]$ and $[\![\exists y]\!]$ thus given are, as an argument similar to that in the previous paragraph shows, the products of $[\![\forall y]\!]_w$ and $[\![\exists y]\!]_w$ from Section 9.5.3.

The upshot is that, in Kripke's setup, each $[\![-]\!]_w$ is a single-world model of free logic, with the domain E_w of quantification. Fact 9.4 and its corollary (9.72) then state that E_w is a model of classical logic autonomous from $[\![-]\!]_w$, as far as non-modal logic is concerned. Now consider the transpose $E \subseteq X \times D$ of $E_- : X \to \mathcal{P}D$, i.e. the set of world-individual pairs (w, a) such that a exists in w. It can also be written as a bundle $E = \sum_{w \in X} E_w$ of fibers E_w. Therefore the autonomy of each fiber carries over to the bundle:

Fact 9.5 Let $[\![-]\!]$ be an interpretation of free logic over X as above. Suppose that E is closed under $[\![f]\!]$ for every function symbol f or simply that the language has no function symbol. Then E is an autonomous interpretation of classical first-order logic.

One may compare this situation to Lewis's again: Kripke's $X \times D$ is a special case of Lewis's D as a domain of individuals; yet Kripke's $E \subseteq X \times D$ is a special case of Lewis's D as a domain of quantification that forms a model of classical logic.

9.7 First-Order Modal Logic

We are finally ready to discuss interpretations of first-order modal logic. The categorical approach makes it possible to provide general semantics in a systematic fashion.[46] Yet a general semantics itself is more of a means than an end of this section. Our primary goal is to show, by taking Kripke's and Lewis's as examples, how categorical methods—including the characterization of several semantics of seemingly different makeup as special cases of the same general setup—can help philosophical logicians to analyze technical and conceptual assumptions of existing frameworks for first-order modal logic.

9.7.1 A family of spaces for a family of algebras

As we saw in Sections 9.2 and 9.3, propositional modal logic and its possible-world semantics consist in equipping Boolean algebras with modal operators. On the other hand, one of the lessons of Sections 9.5 and 9.6 is that we should use a family of Boolean algebras, $[\![T^n]\!]$ for each type T^n of context, to interpret first-order logic. Formally speaking, it is a natural consequence of these two ideas to equip each algebra in this family with a modal operator, and then the spatial side of the story runs in parallel by duality.

In propositional modal logic, we take a propositional language and add operators $\Box, \Diamond$ to it; this means, since the sentences form an algebra A, that we add $\Box, \Diamond : A \to A$. Similarly, take a first-order language and add $\Box, \Diamond$; this means, since the formulas-in-context form a hyperdoctrine $P : \mathcal{C}_T^{\mathrm{op}} \to \mathbf{BA}$, to add

$$\Box_n, \Diamond_n : P(T^n) \to P(T^n) :: (\bar{x} \mid \varphi) \mapsto (\bar{x} \mid \Box\varphi), (\bar{x} \mid \Diamond\varphi)$$

for each type T^n of context.

Correspondingly, an interpretation $[\![-]\!]$ of P that assigns a set $[\![T^n]\!]$ to each T^n therefore needs operations

$$\Box_n, \Diamond_n : \mathcal{P}[\![T^n]\!] \to \mathcal{P}[\![T^n]\!] :: [\![\bar{x} \mid \varphi]\!] \mapsto [\![\bar{x} \mid \Box\varphi]\!], [\![\bar{x} \mid \Diamond\varphi]\!]$$

for each T^n. This further means that, by duality, each $[\![T^n]\!]$ has a spatial structure, most generally a neighborhood system $\mathcal{N}_n : [\![T^n]\!] \to \mathcal{PP}[\![T^n]\!]$, that corresponds to $\Box_n, \Diamond_n$. This is why $[\![-]\!] : \mathcal{C}_T \to \mathbf{Sets}/X$ should assign to each type T^n some space (and not just set) $[\![T^n]\!]$, including $[\![T^0]\!] = X$.

---

[46] See e.g. Kracht and Kutz (2007); Braüner and Ghilardi (2007).

This goal can be achieved by taking a category **X** of spaces in place of **Sets** in $[\![-]\!]$: $\mathcal{C}_T \to \mathbf{Sets}/X$.[47] Note, however, that it is not straightforward at all what kind of maps **X** should have. The structure on the spaces $[\![T^n]\!]$ or the algebras $\mathcal{P}[\![T^n]\!]$ is there to capture

   (i) The behavior of propositional connectives, $\wedge$, $\neg$, ..., and $\Box$, $\Diamond$, operating on formulas in the same context.

In contrast, the maps $[\![\bar{x} \mid \bar{t}]\!] : [\![T^n]\!] \to [\![T^m]\!]$ preserve (part of) that structure in order to capture

   (ii) The logical relationship between formulas from different contexts—i.e. the behavior of quantification, extension of contexts, substitution of terms, and other syntactic operations.

And (ii) involves much more difficult issues than (i) as to what behavior we want or need. Indeed, even the syntax of the language depends on these issues: e.g. whereas the two ways of obtaining $\neg\varphi[t/y]$ from φ, i.e. $(\neg\varphi)[t/y]$ and $\neg(\varphi[t/y])$, yield the same result in the syntax (also recall (9.61) in Section 9.5.2), it is a significant and recalcitrant question whether we have to (or want to) similarly take $\Box(\varphi[t/y])$ and $(\Box\varphi)[t/y]$ as the same formula or to distinguish them:

$$\begin{array}{ccc} P(T^{n+1}) \xrightarrow{P(\bar{x}\mid\bar{x},t)} P(T^n) & \quad & (\bar{x},y \mid \varphi) \longmapsto (\bar{x} \mid \varphi[t/y]) \\ \Box \downarrow \qquad\qquad\qquad \downarrow \Box & & \downarrow \qquad\qquad\qquad\qquad \downarrow \\ P(T^{n+1}) \xrightarrow[P(\bar{x}\mid\bar{x},t)]{} P(T^n) & & (\bar{x},y \mid \Box\varphi) \mapsto (\bar{x} \mid (\Box\varphi)[t/y]) \stackrel{?}{=} (\bar{x} \mid \Box(\varphi[t/y])). \end{array}$$

Such syntactical issues, in fact, affect logical laws that appear to be within the simpler territory (i) (we will see an example in Section 9.7.3).

Issues involved in (ii) are not merely mathematical, but have philosophical imports. Therefore categorical methods alone may not provide solutions to these issues. Categorical methods do, however, connect these issues to properties of maps of **X** in a conceptually direct and powerful manner. It is a strong virtue of categorical methods to always keep spatial (and ontological) ideas and algebraic (and logical) ones side by side. This facilitates both mathematical and philosophical analyses of the interaction between logic, semantics, and ontology for first-order modal logic. The remainder of this section shows several case studies of this virtue.

---

[47] For instance, the semantics of Awodey and Kishida (2008) takes the category **X** of topological spaces and local homeomorphisms; Kishida (2011) takes the category **X** of neighborhood frames and "local isomorphisms". One can also replace **Sets**/X with a more general category, such as the one of presheaves on some algebra associated to X, but it can often be regarded as having $\mathbf{Sets}/X \simeq \mathbf{Sets}^X$ as an underlying structure. See also Goldblatt (1979); Ghilardi and Meloni (1988); Ghilardi (1989); Makkai and Reyes (1995); Kracht and Kutz (2002, 2007); Braüner and Ghilardi (2007); Gabbay et al. (2009); Awodey et al. (2014).

9.7.2 Kripke's semantics

The setup in Section 9.7.1 involves spaces $[\![T^n]\!]$ for all $n \in \mathbb{N}$; this means that, if we are to use accessibility relations to interpret $\Box$ and $\Diamond$, we use relations $R_n : [\![T^n]\!] \twoheadrightarrow [\![T^n]\!]$ for all $n \in \mathbb{N}$. This may appear to be a rather different setup from more standard ideas in semantics for quantified modal logic. Yet the point of the discussion in Section 9.7.1 is that as long as one interprets quantified modal logic with subsets $[\![\bar{x} \mid \varphi]\!] \subseteq [\![T^n]\!]$—or equivalently with a satisfaction relation $\vDash \subseteq [\![T^n]\!] \times P(T^n)$—she is committed to using, whether explicitly or implicitly, $\mathbb{N}$-many operators $\Box, \Diamond : \mathcal{P}[\![T^n]\!] \to \mathcal{P}[\![T^n]\!]$ and by duality $\mathbb{N}$-many spaces T^n. In this subsection and next, we observe what kind of spaces $[\![T^n]\!]$ are working implicitly in Kripke's and Lewis's semantics. Due to the syntactic complication mentioned in Section 9.7.1, we assume in this subsection through Section 9.7.4 that the language has no function symbols including individual constants.

We first deal with Kripke's semantics. Since we assume the language has no function symbols, the consideration we will lay out shortly guarantees that the syntax can have $(\Box \varphi)[x/y] = \Box(\varphi[x/y])$ in this subsection (though not in the subsequent ones). Now let $[\![-]\!]$ be a many-world structure in Kripke's setup in Section 9.6.3—i.e. a bundle of models $([\![-]\!]_w)_{w \in X}$ for free logic. His idea is to equip the set X of worlds with an accessibility relation $R_X : X \twoheadrightarrow X$ and then to modify his propositional interpretation (9.16)–(9.17) simply by attaching a single tuple $\bar{a}$ of individuals:

(9.83) $(w, \bar{a}) \in [\![\bar{x} \mid \Box \varphi]\!]$ iff $(v, \bar{a}) \in [\![\bar{x} \mid \varphi]\!]$ for all $v \in X$ such that $R_X w v$.

(9.84) $(w, \bar{a}) \in [\![\bar{x} \mid \Diamond \varphi]\!]$ iff $(v, \bar{a}) \in [\![\bar{x} \mid \varphi]\!]$ for some $v \in X$ such that $R_X w v$.

Note that $\bar{a}$ is the only tuple of individuals occurring above, either in the left-hand side or in the right. The point is the role of "transworld-identity": we may read (9.83) as

- The tuple $\bar{a}$ satisfies φ necessarily at w iff *it* satisfies φ at every v accessible from w,

using "it", which is to deem $(w, \bar{a})$ and $(v, \bar{a})$ "the same thing in different worlds".

This idea of transworld-identity is indeed reflected formally on the spaces $[\![T^n]\!] = X \times D^n$. The conditions (9.83)–(9.84) define $\Box_n, \Diamond_n : \mathcal{P}(X \times D^n) \to \mathcal{P}(X \times D^n)$ for each $X \times D^n$, but R_X in (9.83)–(9.84) is not a relation on $X \times D^n$. Instead, the relation R_n on $X \times D^n$ that is dual to $\Box_n, \Diamond_n$ must satisfy the following:

(9.85) $(w, \bar{a}) \in \Box_n(S)$ iff $(v, \bar{b}) \in S$ for all $(v, \bar{b}) \in X \times D_X$ such that $R_n(w, \bar{a})(v, \bar{b})$.

(9.86) $(w, \bar{a}) \in \Diamond_n(S)$ iff $(v, \bar{b}) \in S$ for some $(v, \bar{b}) \in X \times D_X$ such that $R_n(w, \bar{a})(v, \bar{b})$.

Comparing these to (9.83)–(9.84) shows that R_n must be the following, where the "$\bar{a} = \bar{b}$" part states that "$\bar{a}$ at w" and "$\bar{b}$ at v" are transworld-identical:

$$R_n(w, \bar{a})(v, \bar{b}) \iff R_X w v \text{ and } \bar{a} = \bar{b}. \tag{9.87}$$

Thus, Kripke implicitly uses a family of Kripke frames $(X \times D^n, R_n)_{n \in \mathbb{N}}$. Then the semantics has the logic K sound (though it does not follow so immediately as it may appear; cf. Lewis's case in Section 9.7.3).

Since $[\![-]\!]$ is a structure for free logic, classical logic fails to be sound in Kripke's semantics. Nevertheless, the corollary (9.73) of Fact 9.4 implies, by an extensionality-and-bundle argument (cf. Fact 9.5), that Kripke's semantics validates classical logic when restricted to (closed) sentences. Taking advantage of this, Kripke axiomatizes the logic of his semantics with the constraint that only closed sentences can be asserted.

We should mention a nice consequence of (9.87), viz., that it makes $[\![f]\!]$ a bounded morphism (9.27), with (9.29) commuting, for every map f in $\mathcal{C}_T$. (This breaks down if the language has function symbols; we currently assume it does not.) For example for the diagonal map

$$\Delta = [\![x \mid x, x]\!] : X \times D \to X \times D^2 :: (w, a) \mapsto (w, a, a)$$

the following commutes:

$$\begin{array}{ccc}
\mathcal{P}(X \times D^2) & \xrightarrow{[\![x \mid x,x]\!]^{-1}} & \mathcal{P}(X \times D) \\
\Box \downarrow & & \downarrow \Box \\
\mathcal{P}(X \times D^2) & \xrightarrow[{[\![x \mid x,x]\!]^{-1}}]{} & \mathcal{P}(X \times D)
\end{array} \qquad \begin{array}{c} [\![x, y \mid \varphi]\!] \longmapsto [\![x \mid \varphi[x/y]]\!] \\ \downarrow \qquad\qquad \downarrow \\ [\![x, y \mid \Box\varphi]\!] \longmapsto [\![x \mid (\Box\varphi)[x/y]]\!] = [\![x \mid \Box(\varphi[x/y])]\!] \end{array}$$

(9.88)

This is why we can take $(\Box\varphi)[x/y] = \Box(\varphi[x/y])$. This plays a crucial role in e.g. the following derivation of $x = y \vdash \Box(x = y)$, in which we make subsitution explicit to expose the role of $\Box(\varphi[x/y]) = (\Box\varphi)[x/y]$ (we also use the rule on $=$ and the rule (N)):

$$\dfrac{\dfrac{\dfrac{x = y \vdash x = y}{\top \vdash (x = y)[x/y]}}{\top \vdash \Box((x = y)[x/y])} \quad \dfrac{\dfrac{}{(\Box(x = y))[x/y] \vdash (\Box(x = y))[x/y]}}{(\Box(x = y))[x/y] \wedge x = y \vdash \Box(x = y)}}{x = y \vdash \Box(x = y)}$$

9.7.3 Lewis's semantics

We move on to review Lewis's semantics in terms of his counterpart theory.[48] In the remainder of this section, the syntax needs to make the substitution of terms an explicit basic operator and to distinguish $\Box(\varphi[t/x])$ and $(\Box\varphi)[t/x]$.

Now let $[\![-]\!]$ be a many-world structure in Lewis's setup in Section 9.6.3. To interpret $\Box, \Diamond$ and provide $\Box_n, \Diamond_n : \mathcal{P}(D_X^n) \to \mathcal{P}(D_X^n)$, he introduces a *counterpart* relation

---

[48] See Kracht and Kutz (2007) and Braüner and Ghilardi (2007) for more detailed reviews of how the counterpart-theoretic semantics can be formulated categorically.

$R_D : D \twoheadrightarrow D$, for which we read $R_D ab$ as "b is a counterpart of a". Then his idea is as follows, the first for a unary φ and the second for a general n-ary φ; replace "every" with "some" for "possibly" in place of "necessarily":

- a satisfies φ necessarily (at $\pi(a)$) iff, for every world w, every counterpart of a in w satisfies φ (at w).
- $\bar{a}$ satisfies φ necessarily (at $\pi(a_i)$) iff, for every world w, every tuple of respective counterparts of $\bar{a}$ in w satisfies φ (at w).

Formally, suppose $\bar{x} = (x_1, \ldots, x_n)$ are *all the* free variables in φ; then

(9.89) $\bar{a} \in [\![\,\bar{x} \mid \Box\varphi\,]\!]$ iff $\bar{b} \in [\![\,\bar{x} \mid \varphi\,]\!]$ for all $\bar{b} \in D_X^n$ such that $R_D a_i b_i$ for all i with $1 \leqslant i \leqslant n$.

(9.90) $\bar{a} \in [\![\,\bar{x} \mid \Diamond\varphi\,]\!]$ iff $\bar{b} \in [\![\,\bar{x} \mid \varphi\,]\!]$ for some $\bar{b} \in D_X^n$ such that $R_D a_i b_i$ for all i with $1 \leqslant i \leqslant n$.

These include the case of $n = 0$, in which 0-tuples $w, v, \ldots \in D_X^0 = X$ satisfy

- $w \in [\![\Box\varphi]\!]$ iff $v \in [\![\varphi]\!]$ for all $v \in X$, i.e., iff $[\![\varphi]\!] = X$; and
- $w \in [\![\Diamond\varphi]\!]$ iff $v \in [\![\varphi]\!]$ for some $v \in X$, i.e., iff $[\![\varphi]\!] \neq \varnothing$.

Thus, Lewis implicitly uses the following relations R_n on D_X^n, including $R_0 = X \times X$:

$$R_n \bar{a}\bar{b} \iff R_D a_i b_i \text{ for all } i \text{ with } 1 \leqslant i \leqslant n. \tag{9.91}$$

We should note the following regarding (9.91) and the assumption on $\bar{x}$ in (9.89)–(9.90). Unlike Kripke's (9.87), Lewis's (9.91) fails to make the diagonal map Δ a bounded morphism, although it makes Δ monotone (9.26); i.e. (9.28) holds but that (9.29) may not. Also, a projection

$$p = [\![\,\bar{x}, y \mid \bar{x}\,]\!] : D_X^{n+1} \to D_X^n :: (\bar{a}, b) \mapsto \bar{a}$$

is monotone but not a bounded morphism, so the following "$\leqslant$" holds but the square may fail to commute in the way (9.88) does, and similarly for $\Diamond$ (with "$\geqslant$" in place of "$\leqslant$"):

$$
\begin{array}{ccc}
\mathcal{P}(D_X^n) \xrightarrow{[\![\,\bar{x},y \mid \bar{x}\,]\!]^{-1}} \mathcal{P}(D_X^{n+1}) & \quad & [\![\,\bar{x} \mid \varphi\,]\!] \longmapsto [\![\,\bar{x},y \mid \varphi\,]\!] \\
\Box_n \downarrow \quad \leqslant \quad \downarrow \Box_{n+1} & & \downarrow \qquad\qquad \downarrow \\
\mathcal{P}(D_X^n) \xrightarrow[{[\![\,\bar{x},y \mid \bar{x}\,]\!]^{-1}}]{} \mathcal{P}(D_X^{n+1}) & & [\![\,\bar{x} \mid \Box\varphi\,]\!] \mapsto [\![\,\bar{x},y \mid \Box\varphi\,]\!] \subseteq \Box_{n+1}[\![\,\bar{x},y \mid \varphi\,]\!].
\end{array}
$$
(9.92)

This is why, following Lewis's idea, we assume all of $\bar{x}$ to occur freely in φ in (9.89)–(9.90). Given a formula φ, $[\![\,\bar{x}, y \mid \Box\varphi\,]\!]$ must be well-defined uniquely from $[\![\,\bar{x} \mid \varphi\,]\!]$, but the failure of (9.92) to commute means that we must choose which of the two paths defines it. Lewis's assumption in effect defines

$$[\![\bar{x}, \bar{y} \mid \Box\varphi]\!] = [\![\bar{x} \mid \bar{x}, \bar{y}]\!]^{-1} \Box_n [\![\bar{x} \mid \varphi]\!] \text{ for the set } \bar{x} \text{ of free variables in } \varphi. \qquad (9.93)$$

This, however, has significant ramifications to the modal logic of the semantics. Probably the most striking is that the axiom (K) fails in Lewis's semantics. To analyze the invalidity, decompose K into (M), (N_{ax}), and (C); then, even though each $\Box_n$, the Kripke necessity of R_n, is a monotone map preserving all meets, the semantics fails (M) while validating (N_{ax}) and (C). To see this, let x be the free variable in φ, let ψ be closed, and suppose $[\![x \mid \varphi]\!] \subseteq [\![x \mid \psi]\!] = \pi^{-1}[\![\psi]\!]$; then the monotonicity of $\Box_n$ entails (i) below, but (M) requires (ii) by (9.93), whereas (9.92) means (ii) may be properly stronger than (i).

$$[\![x \mid \Box\varphi]\!] = \Box_1 [\![x \mid \varphi]\!] \quad \begin{matrix} \text{(i)} & \Box_1 \circ \pi^{-1}[\![\psi]\!] \\ \swarrow & \\ & \text{\textbackslash\textbackslash/ (9.92)} \\ \nwarrow & \\ \text{(ii)} & [\![x \mid \Box\psi]\!] = \pi^{-1} \circ \Box_0 [\![\psi]\!]. \end{matrix}$$

Thus (M) needs replacing with

$$\frac{\varphi \vdash \psi}{\Box\varphi \vdash \Box\psi} \text{ (every free variable in } \varphi \text{ occurs freely in } \psi\text{)}. \qquad (M_L)$$

In contrast, Kripke's (9.87) makes each projection p a bounded morphism, and therefore we can interpret $\Box$ uniformly, in the sense that $\Box_n$ applied to $[\![\bar{x} \mid \varphi]\!]$ always yields $\Box_n [\![\bar{x} \mid \varphi]\!] = [\![\bar{x} \mid \Box\varphi]\!]$ regardless of the arity of φ. In Section 9.7.5, we will discuss more on such squares as (9.88) and (9.92).

The following is also worth observing: although we take $[\![T^n]\!] = D_X^n$, let us assume $[\![T^n]\!] = X \times D^n$ instead as if in Kripke's setup, and read (9.89)–(9.90) accordingly. Even then, however, $\Box$ and $\Diamond$ they define are restrictable to D_X^n (with the inclusion $i : D_X^n \cong X \times_X D_X^n \hookrightarrow X \times D^n :: \bar{b} \mapsto (\pi(b_i), \bar{b})$. This is because Lewis defines $R_\Box ab$, "b is a counterpart of a in w", as implying that $\pi(b) = w$, "b is in w". Therefore (9.91) implies that each D_X^n is closed under R_n, so, as shown in Section 9.4.2, each $\Box_n$ is restrictable to D. On the other hand, all the other relevant operations, including $[\![\forall y]\!]$ and $[\![\exists y]\!]$, are restrictable to D_X^n—one can show this by Fact 9.4 and an extensionality-and-bundle argument, but it really is the same fact as Fact 9.5. So D_X^n is autonomous, a fact that played an essential role in concluding $[\![T^n]\!] = D_X^n$ in Section 9.6.1.

9.7.4 Converse Barcan formula

In Sections 9.7.2 and 9.7.3, we showed how Kripke's and Lewis's semantics could be regarded as special cases of the framework mentioned in Section 9.7.1. We should stress the categorical character of this. Different frameworks of semantics may implement their interpretations with semantic structures $[\![T^n]\!]$, $[\![t]\!]$ of conceptually and set-theoretically different makeup. The point of our categorical approach—say, subsuming Kripke's or Lewis's semantics into a more general one—is not quite to translate one makeup to another. It is rather to analyze and dissect each makeup and to bring to light the essential aspects of the semantic structure. Categorical methods make this possible by identifying structural properties (e.g. that $[\![T^n]\!]$ is a product of $[\![T]\!]$ in a certain category) and connecting those aspects of the semantic structure precisely to the algebraic structure of the resulting or desired logic (e.g. that the well-definedness of this syntactic operation consists in that property of a certain map).

In this subsection, we lay out an example of such use of categorical methods, by analyzing a theorem called the *converse Barcan formula*:

$$\Box \forall x. \varphi \vdash \forall x. \Box \varphi. \tag{CBF}$$

This can be derived using the classical rules of $\forall$ and the rule (M) of modal logic:

$$\frac{\forall x. \varphi \vdash \varphi}{\frac{\Box \forall x. \varphi \vdash \Box \varphi}{\Box \forall x. \varphi \vdash \forall x. \Box \varphi}}$$

In fact, (M_L) is enough for this derivation, and therefore Lewis's semantics validates (CBF). On the other hand, Kripke's semantics does not, since its quantifier logic is free and not classical, invalidating the first step of the derivation. There is, however, a well-known fact characterizing when (CBF) is valid in Kripke's semantics.

Fact 9.6 An interpretation $[\![-]\!]$ in Kripke's semantics validates (CBF) iff it *has an increasing domain* in the sense that

(9.94) $R_X wv$ implies that $E_w \subseteq E_v$.

Due to this result, it has become a sort of orthodoxy of philosophical logicians that the converse Barcan formula is about increasing domains. We, on the other hand, propose revising this orthodoxy using formal concepts and tools we have introduced so far. Specifically, we show the following:

(9.95)   In Kripke's semantics, increasing domains are characterized by the autonomy of the domain E of quantification.

(9.96)   In a more general setup, where the notion of increasing domain may no longer make sense, it is the autonomy of E that characterizes (CBF).

Therefore, as we argue, the converse Barcan formula is really about the autonomy, and the connection to increasing domains, Fact 9.6, is merely a derivative, local fact about Kripke's specific semantics.

Let us observe (9.95) first. (9.87) rewrites (9.94) as (i) below, but it is indeed equivalent to (ii) for all $n \in \mathbb{N}$.

(i) $(w, a) \in E$ and $R_1(w, a)(v, b)$ implies that $(v, b) \in E$; i.e. E is closed under R_1.

(ii) $(w, \bar{a}) \in E_X^n$ and $R_n(w, \bar{a})(v, \bar{b})$ implies that $(v, \bar{b}) \in E_X^n$; i.e. E_X^n is closed under R_n.

Therefore $[\![-]\!]$ has an increasing domain iff all $\Box_n$ are restrictable to E. By Fact 9.5, therefore, we have the following:

Fact 9.7 Let $[\![-]\!]$ be an interpretation $[\![-]\!]$ in Kripke's semantics. Since the language has no function symbol, E is autonomous iff $[\![-]\!]$ has an increasing domain; in this case E models classical quantified modal logic.

Let us now show (9.96). We take a free-logic interpretation $[\![-]\!]$ of quantified modal logic, from the general setting of Section 9.7.1; we only assume the following conditions, which are all satisfied by Kripke's, Lewis's, and many more semantics (though most are classical and not free).[49]

(iii) $[\![-]\!]$ has a domain $E \subseteq [\![T]\!]$ of quantification, as in Kripke's setup, so that

$$[\![\forall y]\!] = \forall_p \circ \forall_i \circ i^{-1}$$

for each projection $p : [\![T^{n+1}]\!] \to [\![T^n]\!]$ and inclusion $i : E_X^{n+1} \hookrightarrow [\![T^{n+1}]\!]$.

(iv) $[\![-]\!]$ has a family $\Box_n : \mathcal{P}[\![T^n]\!] \to \mathcal{P}[\![T^n]\!]$ that interprets $\Box \varphi$ with (9.93) as in Lewis's case.

(v) Projections p are continuous in the sense of $p^{-1} \circ \Box_n \leqslant \Box_{n+1} \circ p^{-1}$ as in (9.92).

(vi) Each $\Box_n$ is monotone and preserves binary meets, validating (M_L) and (C) by (v).

(vii) $\Box_{n+1}(S) \subseteq p^{-1} \circ \Box_n(\top_n)$ for any $S \subseteq [\![T^{n+1}]\!]$ and $\top_n = [\![T^n]\!]$. This means that $[\![-]\!]$ validates $\Box \psi \vdash \Box \top(\bar{x})$ for any ψ which can be in the context $(\bar{x}, y)$ and a tautology $\top(\bar{x})$ whose set of free variables is $\bar{x}$. This is trivially the case in semantics validating the axiom (N_{ax}).[50]

In particular, we do *not* assume $\Box_n$ are Kripke necessities corresponding to accessibility relations. Therefore the notion of increasing domain may not make sense in $[\![-]\!]$, whereas the restrictability of $\Box_n$ to E does. Now, when $(\bar{x}, y)$ are the free variables in φ, (iii)-(iv) imply that $[\![-]\!]$ validates (CBF) iff

$$\Box_n \circ \forall_p \circ \forall_i \circ i^{-1} \leqslant \forall_p \circ \forall_i \circ i^{-1} \circ \Box_{n+1}. \tag{9.97}$$

Then we have the following:

Fact 9.8 If $[\![-]\!]$ satisfies (iii)-(viii), then (9.97) holds iff $\Box_{n+1}$ is restrictable to E. So, since the language has no function symbol, $[\![-]\!]$ validates (CBF) iff E is autonomous.

---

[49] For example, various presheaf and sheaf semantics in Goldblatt (1979); Ghilardi and Meloni (1988); Ghilardi (1989); Reyes (1991); Makkai and Reyes (1995); Awodey and Kishida (2008); Gabbay et al. (2009); Kishida (2011); Awodey et al. (2014). Neighborhood semantics of Arló-Costa and Pacuit (2006) in its full generality does not satisfy (vi)-(viii), but it has subcases that do.

[50] (N_{ax}) is invalid in e.g. the neighborhood semantics in Kishida (2011), which nonetheless validates $\Box \psi \vdash \Box \top(\bar{x})$.

9.7.5 More from the categorical perspective

The analysis of the converse Barcan formula in Section 9.7.4 is just one instance of the assistance that categorical methods offer in clarifying logical, semantic, and ontological assumptions and their relationship. There are many more ways in which category theory can help. One is to connect the continuity (9.33) and openness (9.34) of maps between spaces in the semantics to the preservation of algebraic structures by homomorphisms, which in turn expresses syntactical and axiomatic properties of the logic. We already saw a bit of this in Section 9.7.3 regarding the commutation of (9.88) and (9.92).

In (9.93), p^{-1} of the projection p expands context, so its commuting with $\Box$, the commutation of (9.92), means that $\Box$ behaves uniformly across all contexts (recall the discussion of (9.93)). On the other hand, in (9.88), Δ^{-1} of the diagonal map Δ duplicates a variable by subsitution, so Δ^{-1} needs to commute with $\Box$, making (9.88) commute, for the syntax to identify $(\Box\varphi)[x/y] = \Box(\varphi[x/y])$. In general, substitution of terms $[t/z]$ as in (9.60) is interpreted by the inverse-image operation $[\![t]\!]^{-1}$ of $[\![t]\!]$, as in (9.65) (with 2 replaced by $\mathcal{P}X = \sum_{w\in X} 2$ in the case of **Sets**/X). Therefore its commuting with $\Box$,

$$\begin{array}{ccc}
\mathcal{P}[\![T^{n+1}]\!] & \xrightarrow{[\![\bar{x}\mid x,t]\!]^{-1}} & \mathcal{P}[\![T^n]\!] \\
{\scriptstyle \Box_{n+1}}\downarrow & \!\!\!\!/\!\!\!/ & \downarrow{\scriptstyle \Box_n} \\
\mathcal{P}[\![T^{n+1}]\!] & \xrightarrow[{[\![\bar{x}\mid x,t]\!]^{-1}}]{} & \mathcal{P}[\![T^n]\!]
\end{array}
\qquad
\begin{array}{c}
[\![\bar{x},y\mid \varphi]\!] \longmapsto [\![\bar{x}\mid \varphi[t/z]]\!] \\
\downarrow \qquad\qquad\qquad \downarrow \\
[\![\bar{x},y\mid \Box\varphi]\!] \mapsto [\![\bar{x}\mid (\Box\varphi)[t/z]]\!] = [\![\bar{x}\mid \Box(\varphi[t/z])]\!]
\end{array}$$

(9.98)

is necesary for the syntax to identify $(\Box\varphi)[t/z] = \Box(\varphi[t/z])$.

As mentioned in Section 9.7.3, for Lewis's accessibility relations given by (9.91), projections p and the diagonal Δ fail to be bounded morphisms, and so do $[\![t]\!]$. On the other hand, as mentioned in Section 9.7.2, Kripke's (9.87) makes p and Δ bounded morphisms. Nevertheless, $[\![t]\!]$ may not be bounded morphisms—indeed, we can characterize bounded morphisms in Kripke's setup as follows: a function $f : X \times D^n \to X \times D$ over X is a bounded morphism with respect to (9.87) iff

(i) $R_X wv$ implies that $f_w(\bar{a}) = f_v(\bar{a})$ for all $\bar{a} \in D^n$.

(Recall the notation f_w from the last part of Section 9.6.2.) Therefore, if $[\![f]\!]$ is a bounded morphism, $f(\bar{x}) = y \vdash \Box(f(\bar{x}) = y)$ is valid (cf. the discussion at the end of Section 9.7.2); i.e. y can be the f of x only if it is necessarily so. One may well find this is too strong a requirement, preventing reasonable applications of first-order modal logic. What the categorical perspective provides by connecting (i) and (9.98) is a simple yet powerful argument that (i) and the syntac condition $(\Box\varphi)[t/z] = \Box(\varphi[t/z])$ are two sides of the same coin in Kripke's setup.

One may choose to keep the syntax clean and accept all the syntactic identifications mentioned earlier. Then the categorical perspective tells her that not just projections p and the diagonal map Δ but also all the interpretations $[\![f]\!]$ of function symbols, need to be bounded morphisms—or, more generally, open maps. This means, for Kripke frames, that the residence map π is a "Kripke sheaf" (Goldblatt, 1979; see also Gabbay et al., 2009, Kishida 2011). Or, for topological spaces, π is a local homeomorphism, or a sheaf over X (see Awodey and Kishida, 2008); i.e. take the category LH of spaces and local homeomorphisms, replace **Sets** with LH in **Sets**/X, and obtain LH/X, which is equivalent to the cateogry **Sh**(X) of sheaves over the space X.

In fact, the sheaf setting has more conceptual merits. One is that, since **Sets**/X and **Sh**(X) both form toposes, we can apply categorical logic in toposes to readily obtain a sheaf semantics for higher-order typed modal logic (Awodey et al., 2014). Another merit is that sheaves extend the duality perspective of propositional logic to the first and higher orders. For instance, as Awodey and Forssell (2013) showed, sheaves can be used to extend the Stone duality to first-order logic. Similarly, for modal logic, sheaves extend the Gödel comparison of Boolean and Heyting algebras we saw in Section 9.3.3 to a geometric morphism between the classical universe **Sets**/X over the space X of possible worlds and the constructive universe **Sh**(X) over X (see Braüner and Ghilardi, 2007; Awodey and Kishida, 2008). This comparison of the two universes (together with the equivalence between sheaves and étale bundles) can also be put somewhat more conceptually as follows, directly extending the comparison at the propositional level we saw in Section 9.3.4: From the classical point of view, the constructive ontology is given by constructive intensions. On the other hand, from the constructive point of view, the classical ontology arises by providing more and more objects as certain filters.

Relatedly, it is worth noting that there is another modality for toposes, called "geometric modality", and that it serves as a tool for comparing the toposes of sheaves and of presheaves over a site. Also, similarly to the case of toposes, modal logic can also be found in type theory, with a modality operating not only on propositions but on all types. The understanding of modal logic as a tool for comparing logics plays a prominent role here, because a modality compares the logic of an entire universe with that of a (reflective) subuniverse. A salient example in homotopy type theory (cf. Shulman's chapter, Chapter 3) is "propositional truncation", which compares the logic of all types with the "traditional" logic of sets and propositions.

9.8 Conclusion

Category theory provides a powerful methodology for the semantic modeling and philosophical analysis of both propositional and quantified modal logic. This should be useful to any logician who wants to model something in a modal-logical setting. It also helps any philosopher to analyze and dissect the conceptual and formal makeup of a given semantics. It is moreover essential if the logician or the philosopher believes that the simplicity, generality, applicability, or conceptual clarity of her formalism constitutes a criterion for the success of her endeavor. Furthermore, modal logic is not just a tool of modeling; it arises naturally in many contexts of the foundations of logic and mathematics, as soon as one considers phenomena in these contexts from a structural, categorical perspective. These are reasons why philosophers and logicians should take seriously the interaction between categories and modalities.

References

Abramsky, S. (1991). Domain theory in logical form. *Annals of Pure and Applied Logic* 51, 1–77.
Abramsky, S. (1993). Computational interpretations of linear logic. *Theoretical Computer Science* 111, 3–57.
Adámek, J., Herrlich, H., and Strecker, G. E. (1990). *Abstract and Concrete Categories: The Joy of Cats*. Wiley, New York.
Adams, R. M. (1974). Theories of actuality. *Noûs* 8, 211–31.
Arló-Costa, H., and Pacuit, E. (2006). First-order classical modal logic. *Studia Logica* 84, 171–210.
Awodey, S., and Forssell, H. (2013). First-order logical duality. *Annals of Pure and Applied Logic* 164, 319–48.
Awodey, S., and Kishida, K. (2008). Topology and modality: the topological interpretation of first-order modal logic. *Review of Symbolic Logic* 1, 146–66.
Awodey, S., Kishida, K., and Kotzsch, H.-C. (2014). Topos semantics for higher-order modal logic. *Logique et Analyse* 228, 591–636.
Bergfeld, J. M., Kishida, K., Sack, J., and Zhong, S. (2015). Duality for the logic of quantum actions. *Studia Logica* 103, 781–805.
Blackburn, P., de Rijke, M., and Venema, Y. (2001). *Modal Logic*. Cambridge University Press, Cambridge, UK.
Blute, R., and Scott, P. (2004). Category theory for linear logicians, in T. Ehrhard, J.-Y. Girard, P. Ruet and P. Scott (eds), *Linear Logic in Computer Science*, 3–64. Cambridge University Press, Cambridge, UK.
Bräuner, T., and Ghilardi, S. (2007). First-order modal logic, in P. Blackburn, J. van Benthem and F. Wolter (eds), *Handbook of Modal Logic*, 546–620. Elsevier, London.
Carboni, A., and Walters, R. F. C. (1987). Cartesian bicategories i. *Journal of Pure and Applied Algebra* 49, 11–32.
Chagrov, A., and Zakharyaschev, M. (1997). *Modal Logic*. Clarendon Press, Oxford.
Chellas, B. F. (1980). *Modal Logic: An Introduction*. Cambridge University Press, Cambridge, UK.
Conradie, W., and Palmigiano, A. (2012). Algorithmic correspondence and canonicity for distributive modal logic. *Annals of Pure and Applied Logic* 163, 338–76.
Davey, B. A., and Priestley, H. A. (2002). *Introduction to Lattices and Order*, second edn. Cambridge University Press, Cambridge, UK.
Freyd, P. J., and Scedrov, A. (1990). *Categories, Allegories*. North-Holland, Amsterdam.
Gabbay, D. M., Shehtman, V., and Skvortsov, D. (2009). *Quantification in Nonclassical Logic*, Vol. 1. Elsevier, London.
Ghilardi, S. (1989). Presheaf semantics and independence results for some non classical first order logics. *Archive for Mathematical Logic* 29, 125–36.

Ghilardi, S., and Meloni, G. (1988). Modal and tense predicate logic: models in presheaves and categorical conceptualization, in F. Borceux (ed.), *Categorical Algebra and its Applications*, 130–42. Springer, London.
Goldblatt, R. (1979). *Topoi: The Categorial Analysis of Logic*. North-Holland, Amsterdam.
Goldblatt, R. (2006). Mathematical modal logic: a view of its evolution, in D. M. Gabbay and J. Woods (eds), *Handbook of the History of Logic*, Volume 7: *Logic and the Modalities in the Twentieth Century*, 1–98. Elsevier, London.
Goré, R., and Tiu, A. (2007). Classical modal display logic in the calculus of structures and minimal cut-free deep inference calculi for s5. *Journal of Logic and Computation* 17, 767–94.
Hansen, H. H., Kupke, C., and Leal, R. A. (2014). Strong completeness for iteration-free coalgebraic dynamic logics, in J. Diaz, I. Lanese, and D. Sangiorgi (eds), *Theoretical Computer Science*, 281–95. Springer, London.
Hansen, H. H., Kupke, C., and Pacuit, E. (2009). Neighbourhood structures: bisimilarity and basic model theory. *Logical Methods in Computer Science* 5, 1–38.
Harel, D., Kozen, D., and Tiuryn, J. (2000). *Dynamic Logic*. MIT Press, Cambridge, MA.
Hermida, C. (2011). A categorical outlook on relational modalities and simulations. *Information and Computation* 209, 1505–17.
Hintikka, J. (1962). *Knowledge and Belief: An Introduction to the Logic of the Two Notions*. Cornell University Press, Cambridge, UK.
Hintikka, J. (1975). Impossible possible worlds vindicated. *Journal of Philosophical Logic* 4, 475–84.
Hughes, G. E., and Cresswell, M. J. (1996). *A New Introduction to Modal Logic*. Routledge, London.
Jacobs, B. (2015). A recipe for state-and-effect triangles, in L. S. Moss and P. Sobociński (eds), *Sixth Conference on Algebra and Coalgebra in Computer Science (CALCO 2015)*, 116–29. Schloss Dagstuhl, Leibniz-Zentrum für Informatik.
Johnstone, P. T. (1982). *Stone Spaces*. Cambridge University Press, Cambridge, UK.
Johnstone, P. T. (2002). *Sketches of an Elephant: A Topos Theory Compendium*, Vol. 1. Clarendon Press, Oxford.
Kelly, K. T. (1996). *The Logic of Reliable Inquiry*. Oxford University Press, Oxford.
Kishida, K. (2011). Neighborhood-sheaf semantics for first-order modal logic. *Electronic Notes in Theoretical Computer Science* 278, 129–43.
Kracht, M., and Kutz, O. (2002). The semantics of modal predicate logic i: Counterpart-frames, in F. Wolter, H. Wansing, M. de Rijke and M. Zakharyaschev (eds), *Advances in Modal Logic*, Vol. 3, 299–320. World Scientific, Singapore.
Kracht, M., and Kutz, O. (2007). Logically possible worlds and counterpart semantics for modal logic, in D. Jacquette (ed.), *Philosophy of Logic*, 943–95. Elsevier, London.
Kripke, S. (1963). Semantical considerations on modal logic. *Acta Philosophica Fennica* 16, 83–94.
Lambert, K. (2001). Free logics, in L. Goble (ed.), *The Blackwell Guide to Philosophical Logic*, 258–79. Blackwell Publishers, Oxford.
Lawvere, F. W. (1969). Adjointness in foundations. *Dialectica* 23, 281–96.
Lemmon, E. J., and Scott, D. S. (1977). *The 'Lemmon Notes': An Introduction to Modal Logic*. Basil Blackwell, Oxford.
Lewis, C. I. (1918). *A Survey of Symbolic Logic*. University of California Press, Los Angeles.

Lewis, D. (1968). Counterpart theory and quantified modal logic. *Journal of Philosophy* 65, 113–26.

Makkai, M., and Reyes, G. E. (1995). Completeness results for intuitionistic and modal logic in a categorical setting. *Annals of Pure and Applied Logic* 72, 25–101.

Meyer, J.-J. C. (2001). Epistemic logic, in L. Goble (ed.), *The Blackwell Guide to Philosophical Logic*, 183–202. Blackwell Publishers, Oxford.

Montague, R. (1968). Pragmatics, in R. Klibansky (ed.), *Contemporary Philosophy: A Survey*, 102–22. La Nuova Italia Editrice, Florence.

Montague, R. (1969). On the nature of certain philosophical entites. *The Monist* 53, 159–94.

Moss, L. S. (1999). Coalgebraic logic. *Annals of Pure and Applied Logic* 96, 277–317.

Parikh, R., Moss, L. S., and Steinsvold, C. (2007). Topology and epistemic logic, in M. Aiello, I. Pratt-Hartmann, and J. van Benthem (eds), *Handbook of Spatial Logics*, 299–341. Springer, New York.

Pitts, A. M. (2000). Categorical logic, in S. Abramsky, D. M. Gabbay, and T. S. E. Maibaum (eds), *Handbook of Logic in Computer Science*, Volume 5: *Algebraic and Logical Structures*, 39–128. Oxford University Press, Oxford.

Pratt, V. R. (1995). The stone gamut: a coordinatization of mathematics, in *Proceedings of Tenth Annual IEEE Symposium on Logic in Computer Science*, 444–54. IEEE, New York.

Reyes, G. E. (1991). A topos-theoretic approach to reference and modality. *Notre Dame Journal of Formal Logic* 32, 359–91.

Sadrzadeh, M., and Dyckhoff, R. (2009). Positive logic with adjoint modalities: proof theory, semantics and reasoning about information. *Electronic Notes in Theoretical Computer Science* 249, 451–70.

Schulte, O., and Juhl, C. (1996). Topology as epistemology. *The Monist* 79, 141–7.

Scott, D. S. (1970). Advice in modal logic, in K. Lambert (ed.), *Philosophical Problems in Logic*, 143–73. Reidel, Dordrecht.

Scott, D. S. (2008). The algebraic interpretation of quantifiers: intuitionistic and classical, in A. Ehrenfeucht, V. W. Marek, and M. Srebrny (eds), *Andrzej Mostowski and Foundational Studies*, 289–312. IOS Press, Amsterdam.

Segerberg, K. (1971). *An Essay in Classical Modal Logic*. University of Uppsala.

Thomason, S. K. (1975). Categories of frames for modal logic. *Journal of Symbolic Logic* 40, 439–42.

van Benthem, J. (2011). *Logical Dynamics of Information and Interaction*. Cambridge University Press, Cambridge, UK.

van Benthem, J., Fernández-Duque, D., and Pacuit, E. (2012). Evidence logic: a new look at neighborhood structures, in T. Bolander, T. Braüner, S. Ghilardi, and L. Moss (eds), *Advances in Modal Logic*, Vol. 9, 97–118. College Publications, London.

van Dalen, D. (2001). Intuitionistic logic, in L. Goble (ed.), *The Blackwell Guide to Philosophical Logic*, 224–57. Blackwell Publishers, Oxford.

Venema, Y. (2001). Temporal logic, in L. Goble (ed.), *The Blackwell Guide to Philosophical Logic*, 203–23. Blackwell Publishers, Oxford.

Venema, Y. (2007). Algebras and coalgebras, in P. Blackburn, J. van Benthem, and F. Wolter (eds), *Handbook of Modal Logic*, 331–426. Elsevier, London.

Vickers, S. (1996). *Topology via Logic*. Cambridge University Press, Cambridge, UK.

von Karger, B. (1998). Temporal algebra. *Mathematical Structures in Computer Science* 8, 277–320.

10

Proof Theory of the Cut Rule

J. R. B. Cockett and R. A. G. Seely

10.1 Introduction

The cut rule is a very basic component of any sequent-style presentation of a logic. This essay starts by describing the categorical proof theory of the cut rule in a calculus which allows sequents to have many formulas on the left but only one on the right of the turnstile. We shall assume a minimum of structural rules and connectives: in fact, we shall start with none. We will then show how logical features can be added to this proof theoretic substrate in a modular fashion. The categorical semantics of the proof theory of this modest starting point, assuming just the cut rule, lies in multicategories. We shall refer to the resulting logic as **multi-logic**[1] to emphasize this connection.

The list of formulas to the left of the turnstile are separated by commas. This comma may be "represented" by a logical connective called "tensor", written $\otimes$. We may regard this connective as a primitive conjunction which lacks the usual structural rules of weakening and contraction. When this connective is present it is usual to also "represent" the empty list of formulas with a constant called the "tensor unit", written $\top$, which may be regard as a primitive "true". Of course, truth and falsity in these logics is not the central issue; rather the main interest is how proofs in these logics behave. When these "representing" connectives are assumed to be present we call the result $\otimes$-multi-logic. Significantly, the categorical semantics of the proof theory of a $\otimes$-multi-logic lies in the doctrine of monoidal categories.

Our next step is to consider logics whose sequents have many formulas on both the left and right of the turnstile. Again we assume no structural rules and no connectives, and start with just the cut rule, adapted, however, to this two-sided setting. The resulting logic then has its categorical semantics in polycategories[2] and consequently we shall refer to it as **poly-logic**.

---

[1] The "multi-" prefix indicates that one can have a list of formulas to the left of the turnstile but only one formula to the right.

[2] The "poly-" prefix indicates that one can have a list of formulas both to the left and to the right of the turnstile.

Again we may add connectives to represent the commas. This time, however, we need two different connectives: one for the commas on the left, given by "tensor", and one for the commas on the right, given by a "par", written $\oplus$. We may regard the latter as a primitive disjunction which lacks the usual structural rules. Both the tensor and par connectives have units: the unit for the par, written $\bot$, may be regarded as a primitive "false". The categorical semantics of the proof theory of this very minimal logic with connectives then lies in the doctrine of linearly distributive categories.

In the presence of negation, this logic is precisely the multiplicative fragment of linear logic.[3] It is possible to reduce a two-sided sequent to a one-sided sequent by moving formulas on the left of the turnstile to the right while negating them. This meant that, in the development of linear logic, the behaviour of $\otimes\oplus$-poly-logic could be—and was—avoided. In particular, this meant that the categorical semantics of two-sided proof systems was also avoided. Thus, the fundamental importance to these systems of the natural transformation

$$\delta: A \otimes (B \oplus C) \longrightarrow (A \otimes B) \oplus C,$$

called a linear distribution, was overlooked.

A linear distribution may be viewed as a "tensorial strength" in which the object A is pushed into the structure $B \oplus C$. Dually, it may be viewed as a "tensorial costrength" in which the object $A \otimes B$ is the structure which remains after pulling out C. These linear notions of strength pervade the features of poly-logics, and thus in particular of linear logic, and provide a unifying structure for them.

It is important to note that, even in this very basic $\otimes\oplus$-poly-logic, the behaviour of the—so called—"multiplicative" units (i.e. the unit of the tensor, $\top$, and the unit of the par, $\bot$) is very subtle. Indeed, it is the behaviour of the units at this very basic level that makes deciding the equality of proofs difficult. Exactly how difficult was an open problem until recently. For multiplicative linear logic with units, deciding equality is PSPACE complete (Heijltjes and Houston, 2014), though without units the problem is linear.

The full proof theory of linear logic can be built in a modular fashion from the basic semantics of linearly distributive categories. Negation, that is the requirement that every object have a complement, is fundamental to linear logic. For a linearly distributive category, having a complement is a property rather than extra structure.[4] This means in a linearly distributive category either an object has a negation or it

---

[3] Readers should note that we use a notation different from that of Girard (1987), specifically using $\oplus$ for his par $\invamp$; also we use $\top$ as the unit for $\otimes$, $\bot$ as the unit for $\oplus$ (or $\invamp$). We use $\times, +, 1, 0$ for the additives, as categorically they are product and coproduct, terminal and initial objects.

[4] The key point here is that a property is something possessed or not possessed by the object under discussion, not something imposed from "outside". For example, a set may have many different group structures imposed on it (so group structure is "structure" imposed on the set), but a group may or may not be Abelian—this is not extra structure, but a property of the group. You can make a set into a group, but you cannot make a group into an Abelian group (unless it was so already).

does not. It is, furthermore, possible to "complete" a linearly distributive category by formally adding negations. A linearly distributive category in which every object has a complement is precisely a ∗-autonomous category and ∗-autonomous categories provide precisely the categorical semantics for the—so-called—"multiplicative fragment" of linear logic.

In the posetal case, the fact that complements are a property is exactly the observation that in a distributive lattice either an element has a complement or it does not. Of course, as is well known, a distributive lattice in which every object has a complement is a Boolean algebra. It is worth remarking that the fact that a distributive lattice can always be embedded in a Boolean algebra has not made these lattices a lesser area of study. Quite the converse is true: the theory of distributive lattices has hugely enriched the development and understanding of ordered structures. The reason and motivation for studying more general structures, such as linearly distributive categories, in the context of linear logic is exactly analogous: it enriches and broadens our understanding of the "linear" world.

To arrive at the full structure of linear logic, after negation we require the presence of the "additives", that is categorical products and coproducts, and the "exponential" modalities: ! "of course" and ? "why not". Note that having additives is, again, a property rather than structure: either a linearly distributive category has additives or it does not. On the other hand, the exponentials are structure. Semantically they are provided by functors which are appropriate to the categorical doctrine: these are called linear functors and they are actually pairs of functors (e.g. product and coproduct for the additives, ! and ? for the exponentials) which satisfy certain coherence requirements (taking the form of equations). For additives, this coherence structure amounts to the requirement that the tensor distributes over the coproduct and that the "par" distributes over the product. For the exponentials, they must form a linear functor pair which supports duplication. The fact that all the structure of linear logic—including the multiplicatives—may be described in terms of linear functors is one of the remarkable insights gained from the categorical view of its proof theory.

The techniques for studying the proof theoretic structure of fragments of linear logic are also useful in analyzing other categorical—and bicategorical—structures involving monoidal structure. A key tool, introduced by Jean-Yves Girard (1987), was a graphical representation of proofs, "proof nets", to represent the formal derivations or "proofs" of linear logic. The use of graphical languages has now become ubiquitous. The circuit diagrams we shall use here are a graphical representation of *circuits* which have their origin as a term logic for monoidal categories. Circuits are much more generally applicable than Girard's proof nets and provide a bridge between geometric and graphical intuitions (Joyal and Street, 1991). They are, on the one hand, a formal mathematical language but crucially, at the same time they have an intuitive graphical representation.

As graphical languages have become an almost indispensable tool for visualizing linear logic proofs and, more generally, maps in monoidal categories, we take the time here to describe how "circuits" are formalized and we illustrate their use. A major benefit of using circuits is that they make coherence requirements (i.e. which diagrams must commute) very natural. For example, the coherence requirements of linear functors (see Section 10.6.1) are somewhat overwhelming when presented "algebraically", but when seen graphically are very natural.

The use of graphical techniques is also illustrated in our resolution of the coherence problem for ∗-autonomous categories. Here circuits are used to construct free linearly distributive categories. Morphisms in these categories are given by circuits, modulo certain equivalences which are generated by graph rewrites. The rewriting system, which is a reduction–expansion system with equalities, is analyzed to produce a notion of normal form, and this allows us to derive a procedure for not only determining the existence of morphisms (between given objects) but also for determining the equality of morphisms (between the same objects).

Some of our techniques and perspectives may seem to lie outside what is traditionally regarded as "proof theory", but they are firmly rooted in the proof theoretic traditions which follow Gentzen's natural deduction. Our approach will be moderately informal, aiming to give the essential ideas involved so the reader may more easily read the technical papers which may be found online.

We wish to dedicate this exposition to the memory of Joachim (Jim) Lambek. Jim Lambek started the field of categorical proof theory with papers in the 1950s and 1960s, and he has been an inspiration for so much of our own work for the past several decades. As a person, Jim will be missed by all his friends and colleagues; his work will remain a vital influence in all the fields in which he worked.

10.1.1 Prerequisites

This essay is not entirely self-contained. In particular, we shall assume some familiarity with basics of category theory and of formal logic (especially fragments of linear logic), and with the connections between these (often referred to as the "Curry–Howard isomorphism"). More specifically, the main such assumptions involve the following topics. A refresher on many of these may be found in other chapters in this volume, in the references to this essay, and in standard references.

> **Basic notions of categories.** Included in this is the definition of a category (consisting of objects and morphisms or arrows between them, with structure characterizing identity morphisms and composition of morphisms).
> **Sequent calculus.** The reader should be familiar with sequent calculus presentations of logics, specifically how a logic is generated by basic (logical) operations, axioms for these, and deduction rules which specify how the operations operate.
> **Categorical proof theory.** The "equivalence" between objects of a category and well-formed formulas of a logic, and between morphisms of a category and derivations (usually modulo an equivalence relation) will be basic to this essay.[5] This

---

[5] In view of the "Curry–Howard isomorphism", this "equivalence" also extends to the types and terms of a type theory, as may be seen in other chapters in this volume.

"equivalence" also applies to other categorical structures such as multicategories and polycategories. Our presentation of this equivalence will be "high level", rather than in terms of explicit details. So, for example, when we consider monoidal categories (Section 10.3.1), we regard an object $A \otimes B$ and a "symmetry map" $A \otimes B \to B \otimes A$ as a logical formula, rather like a weak notion of conjunction $A \wedge B$, together with a logical entailment $A \wedge B \vdash B \wedge A$ (think of propositional logic, where $A \wedge B$ is logically equivalent to $B \wedge A$). Conversely, given a logical theory, one can construct a category from its formulas and derivations. And so a suitable notion of equivalence may be established between the logical and categorical notions; the techniques of one type of structure may be applied to the other type, giving new techniques for the study of each. The details are not necessary for a first reading of the essay, but will be useful for a deeper understanding. Such details may be found in many of the references provided.

Graphical representation of logical derivations. There are many ways to denote proofs in a logic. The reader is probably familiar with several, such as Hilbert-style axioms and deduction rules, sequent calculus, natural deduction, and combinators, and may even be familiar with others, for example using the style of programming languages. In this chapter, we shall present a graphical notation for representing proofs in some simple logics (essentially fragments of linear logic (Girard, 1987)). The point of such a notation is that it is particularly well adapted to resolving the types of questions we have about the categorical (and so logical) structures involved, particularly coherence questions (When are two maps equal? When are two derivations equivalent?), better adapted, in fact, than other presentations of the structure. For example, by using a logical presentation of $*$-autonomous categories, we are able to give a procedure for determining equality of maps (an open question at the time of our paper (Blute et al., 1996b)).

We shall be fairly explicit how derivations may be represented by graphs, but it must be said that our approach is hardly the only one. There are many variations in such graphical representations. An excellent survey of many used for monoidal categories may be found in (Selinger, 2010). The main point to stress here is that graphs, called "circuits" in this paper, are used to represent maps in various structured categories, such as monoidal categories, and so (by the "Curry-Howard isomorphism") also derivations (formal proofs) in various logics. However, if these are to be maps in a category with structure, then we must be sure that appropriate equalities of maps hold, and furthermore, if the graphs are also derivations in a logic, such equalities of maps must be coherent with appropriate equivalences of derivations. One reason the use of graphical representations has been so successful is that not only are these equations simply handled, but many such equations appear "for free" (such as the categorical axioms), and moreover different graphical representations have different virtues in this regard—one's preference for one over another usually depends on exactly which is most convenient for the purpose at hand.

Consider the symmetry transformation we mentioned earlier in the context of monoidal categories: $A \otimes B \to B \otimes A$. It is usual in such a situation to require that applying such a symmetry map twice $A \otimes B \to B \otimes A \to A \otimes B$ should equal the identity map on $A \otimes B$. Graphically this amounts to a standard "string rewrite" in our circuit calculus. In a real sense, this is applying Descartes' connection between geometry and algebra to logic (and proof theory) and category theory *via* several devices, including term logics and graphical calculii.

This connection goes both ways. Prawitz (1971) wanted to give a notion of "equivalence of proofs", in terms of natural rewriting rules on derivations (in natural deduction). These turned out to be the rules needed to capture the corresponding categorical structure. For example, for the $\wedge, \Rightarrow$ fragment of intuitionistic logic, Prawitz' rewriting rules are just what was needed to capture Cartesian closed categories.[6] Likewise the "string rewrites" needed for logical structure are geometrically natural.

10.2 Circuits and the Basic Proof Theory of the Cut Rule: Multicategories

We shall start with the most basic logical system involving a cut rule, namely logic with only the cut rule, and its representation in multicategories. The proof theory consists of sequents with a list of formulas on the left of the turnstile, and just one formula on the right (i.e. many premises, one conclusion). For logicians, we remark that this means we shall dispense with the usual structural rules of contraction, thinning, and exchange. So the only logical axiom of the system is the identity axiom, and the only other rule is the cut rule. It is useful to name sequents. Using Lambek's conventions, these are the deduction rules:

$$\frac{}{1_A : A \vdash A}\ ax \qquad \frac{f : \Gamma \vdash A \quad g : \Gamma_1, A, \Gamma_2 \vdash B}{g\langle \Gamma_1, f, \Gamma_2 \rangle : \Gamma_1, \Gamma, \Gamma_2 \vdash B}\ cut.$$

When less precision does not cause ambiguity, we shall abbreviate the conclusion of the cut rule as $g\langle f \rangle$, Γ_1 and Γ_2 being understood.

In addition to the identity axioms, we shall allow a set of sequents $f : \Gamma \vdash A$ to be used as a starting point for generating proofs: these are often referred to as "non-logical axioms". The proof theory embodied by the two axioms above together with a specified set of non-logical axioms we refer to as a **multi-logic**.

We shall represent sequents graphically with "circuits" whose nodes (shown as boxes) represent sequents, and whose edges or "wires" represent the formulas making

---

[6] The situation can be more complicated, and certainly depends on the presentation of the logic. For example, Zucker (1974) showed that the natural equivalences differ if intuitionist propositional logic is presented with the sequent calculus compared to natural deduction (but also, see Urban, 2014). Seely (1979, 1987) showed that natural deduction directly gives the categorical equivalences for conjunction, but not for disjunction—some extra permutation equivalences are needed for the latter. In this chapter, we shall see that for a simple substructural logic (the "multiplicative" fragment of linear logic), although the "tensor" and "par", which represent a sort of conjunction and disjunction, have a simple equivalence structure, their units are much subtler. The graphical rewrites for the tensor and par are very obvious, but those for the units are certainly not, and have been reinvented several times since our presentation (Blute et al., 1996b; Koh and Ong, 1999; Lamarche and Straßburger, 2004; Hughes, 2012). It is our view that the closest representation of the "essence" of a logical system is its categorical presentation.

up the sequent. The premises of the sequent correspond to the wires entering the box from above, while the conclusion is the wire leaving the box below. So, for example, given non-logical axioms f and g (as above), the cut rule constructs $g\langle\Gamma_1, f, \Gamma_2\rangle$, represented graphically as follows:

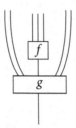

A is the type of the output of the top sequent (box) and is connected by a wire of that type to an input of the bottom sequent (box), B is the type of the final output. An axiom sequent is simply represented by the wire corresponding to the formula.

The categorical proof theory of such a logic is Lambek's multicategories (Lambek, 1969). A multicategory consists of a set of objects and a set of multimorphisms. Each multimorphism has a domain consisting of a list of objects and a codomain consisting of a single object. Each object has an identity multimorphism, whose codomain is the object itself, and whose domain is the singleton list consisting of just that object. Composition is cut, as described earlier. Appropriate equivalences must be imposed: there are two identity axioms (for "pre-composition" and for "post-composition" of a multimorphism by an identity morphism), and two associativity axioms (when cut is done "vertically" or "horizontally"). Using notation similar to that above, if we are given $f: \Gamma \to A, g: \Gamma_1, A, \Gamma_2 \to B, h: \Gamma_3, B, \Gamma_4 \to C$, then we require $h\langle g\rangle\langle f\rangle = h\langle g\langle f\rangle\rangle$. Also, given $f: \Gamma \to A, g: \Gamma_1 \to B, h: \Gamma_3, A, \Gamma_4, B, \Gamma_5 \to C$, we require $h\langle g\rangle\langle f\rangle = h\langle f\rangle\langle g\rangle$. A nice summary may be found in Lambek (1989). This structure is more intuitively presented by the circuit diagrams. The two identity axioms, pictorially represented, merely amount to noticing that extending a wire with an identical (i.e. an identity) wire does not change the circuit. The two associativity axioms merely assert that the two "obvious" ways to "compose" two diagrams produce the same circuit. To illustrate this, here are the two circuits resulting from the two associativity axioms.

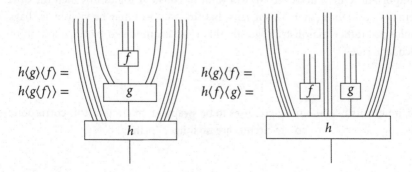

The appropriate notion of morphism of multicategories, called multifunctor, preserves the composition and the identities. This corresponds to an interpretation of one multi-logic into another.

A multi-logic is determined by its presentation as propositions and non-logical axioms: these determine a "multi-graph". However, given a multi-graph, that is, a set of objects and a set of "multi-arrows" (arrows whose domain is a list of objects, and whose codomain is a single object), it is clear how we can construct a multicategory which is generated by the multi-graph. One closes the multi-arrows under composition (equivalently cut), and one factors out by the equivalences required of a multicategory. Thus, given a presentation of a multi-logic, we may generate a multicategory. It is clear the construction outlined above will provide the free multicategory generated by the multi-graph corresponding to the non-logical axioms. This is the categorical semantics of the multi-logic.

Indeed, we have an equivalence of categories (2-categories in fact) between basic cut logics (generated by non-logical axioms with proof identifications) and multicategories (based on non-logical axioms). Thus we have constructed the following "triangle of doctrines", where the two-headed arrows represent equivalences of categories:

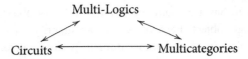

Next we shall add connectives to this logic, and explain the corresponding categorical notions, features, and circuits.

10.3 ⊗-Multi-logic, Representable Multicategories, and Monoidal Categories

We now pass to a simple categorical structure, the free monoidal category generated by some objects and morphisms, in order to understand the effect of introducing connectives into a multi-logic. In effect, we shall be "representing" the commas in the sequents of multi-logic by connectives.

In our discussion of multi-logic and multicategories, we allowed our sequents to have lists of premises. This means that the order in which the premises occurred is important. Often, however, we will want to consider logics in which the order of premises does not matter. This means that the premises may be viewed as "bags", or "multisets" rather than lists. Logically this is accommodated by the addition of the "exchange rule":

$$\frac{\Gamma_1, A, B, \Gamma_2 \vdash C}{\Gamma_1, B, A, \Gamma_2 \vdash C} \; exch,$$

which permits neighbouring premises to be swapped. In circuits, this corresponds to allowing wires to cross so the circuits are no longer "planar".

Often one refers to a (multi-)logic in which the order of premises matters as a "non-commutative" logic and one in which the exchange law is present as a "commutative" logic. On the categorical side one refers to multicategories in which there is no exchange rule as being "non-symmetric" and those in which the exchange rule (crossing wires) is allowed as being "symmetric".

10.3.1 Monoidal categories

We begin by fixing notation, recalling the definition of a monoidal category.

Definition 10.3.1 (Monoidal categories) *A monoidal category $\langle \mathbf{C}, \otimes, \top \rangle$ consists of a category $\mathbf{C}$ with an associative bifunctor (a "tensor") $\otimes$ with a unit $\top$. If the tensor is symmetric, we shall refer to $\mathbf{C}$ as a symmetric monoidal category.*

To say the tensor has a unit, and is associative means that we have the natural isomorphisms

$$u^R_\otimes : A \otimes \top \to A \quad u^L_\otimes : \top \otimes A \to A \quad a_\otimes : (A \otimes B) \otimes C \to A \otimes (B \otimes C),$$

which must satisfy the following coherence equations (expressed as commuting diagrams):

$$\begin{array}{c}
(A \otimes \top) \otimes B \xrightarrow{a_\otimes} A \otimes (\top \otimes B) \\
{}_{u^R_\otimes \otimes 1} \searrow \quad \swarrow {}_{1 \otimes u^L_\otimes} \\
A \otimes B
\end{array}$$

$$\begin{array}{c}
((A \otimes B) \otimes C) \otimes D \xrightarrow{a_\otimes \otimes 1} (A \otimes (B \otimes C)) \otimes D \\
{}_{a_\otimes} \downarrow \qquad\qquad\qquad\qquad \downarrow {}_{a_\otimes} \\
(A \otimes B) \otimes (C \otimes D) \qquad A \otimes ((B \otimes C) \otimes D) \\
{}_{a_\otimes} \searrow \qquad\qquad \swarrow {}_{1 \otimes a_\otimes} \\
A \otimes (B \otimes (C \otimes D))
\end{array}$$

The tensor is symmetric when there is, in addition, the natural isomorphism

$$c_\otimes : A \otimes B \to B \otimes A,$$

which must satisfy the coherence requirements:

$$\begin{array}{c}
A \otimes B \xrightarrow{c_\otimes} B \otimes A \\
\parallel \qquad \downarrow {}_{c_\otimes} \\
A \otimes B
\end{array}$$

$$\begin{array}{c}
(A \otimes B) \otimes C \xrightarrow{a_\otimes} A \otimes (B \otimes C) \\
{}_{c_\otimes \otimes 1} \downarrow \qquad\qquad\qquad \downarrow {}_{c_\otimes} \\
(B \otimes A) \otimes C \qquad\qquad (B \otimes C) \otimes A \\
{}_{a_\otimes} \downarrow \qquad\qquad\qquad \downarrow {}_{a_\otimes} \\
B \otimes (A \otimes C) \xrightarrow{1 \otimes c_\otimes} B \otimes (C \otimes A)
\end{array}$$

One may think of the tensor as representing a (weak) notion of conjunction ("and"), but this is a conjunction without the structural rules of contraction and thinning, and in the non-symmetric case, without exchange as well.

10.3.2 Sequent calculus and circuits for monoidal categories

The sequent calculus presentation of the logic of monoidal categories adds to basic cut logic the tensor $\otimes$ as a logical operator, together with a unit $\top$ for the tensor (generating well-founded formulas in the usual way, so that if A, B are well-founded formulas, so is $A \otimes B$, as are atomic formulas and $\top$), as well as the rules

$$(\otimes R) \frac{\Gamma \vdash A \quad \Delta \vdash B}{\Gamma, \Delta \vdash A \otimes B} \qquad (\otimes L) \frac{\Gamma_1, A, B, \Gamma_2 \vdash C}{\Gamma_1, A \otimes B, \Gamma_2 \vdash C}$$

$$(TR) \frac{}{\vdash \top} \qquad (TL) \frac{\Gamma_1, \Gamma_2 \vdash C}{\Gamma_1, \top, \Gamma_2 \vdash C}$$

Corresponding to these rules, we enrich our circuits with nodes for the tensor and its unit: we have "tensor introduction" and "tensor elimination" nodes, "unit introduction" and "unit elimination" nodes, and any non-logical axioms we may have assumed (which now may involve composite types or formulas involving tensor). The introduction and elimination nodes look like this:

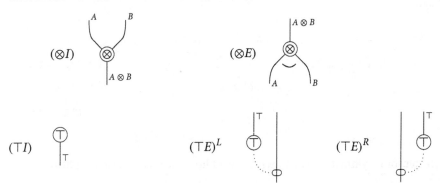

The (optional) small arc drawn below the $(\otimes E)$ node indicates that this node is of a different nature from its $(\otimes I)$ cousin—in part, this reflects that the introduction node represents a valid deduction $A, B \vdash A \otimes B$ of the logic, whereas the elimination node does not (as $A \otimes B \vdash A, B$ is not a valid multi-sequent). We call the elimination node a "switching link". It provides a way to replace two premises A, B with the single premise $A \otimes B$, as given by the $(\otimes L)$ rule. We have placed a restriction on non-logical components, that they have only one output wire, but that restriction could be replaced with a restriction that the output wires of a non-logical component be switching. Using the introduction link, one can tensor together output wires so as to have a single output wire in their place. Note that in the present monoidal case, any node that has several output wires must be switching (at the output wires), and no node has switching wires at the input. The idea of "switching" will be particularly relevant when we consider poly-logics, as there which wires are "switching" becomes a matter of greater subtlety and importance.

The curved "lasso"-like wires, called thinning links, used in the unit nodes are rather different from the other components, especially in the more general cases soon to be considered, and so are denoted with dotted lines. The reader can consider the loop at the end of the lasso as a movable node and the lasso itself as a wire. The unit nodes correspond to (derivable) sequents $\vdash \top$, $\top, A \vdash A$, and $A, \top \vdash A$.

10.3.2.1 REPRESENTABILITY

Sequents $A \vdash B$ are derivable in this sequent calculus if and only if there is a corresponding circuit with one input wire of type A, and one output wire of type B. Any valid circuit can be "represented" by such a one-in-one-out circuit by tensoring all the inputs wires together and tensoring all the output wires together. These one-in-one-out sequents/circuits then correspond to morphisms in a monoidal category.

Thinking in terms of natural deduction, the sequent rules induce bijective correspondences indicated by these "rules":

$$\dfrac{\Gamma_1, A \otimes B, \Gamma_2 \vdash C}{\Gamma_1, A, B, \Gamma_2 \vdash C} \qquad \dfrac{\Gamma_1, \top, \Gamma_2 \vdash C}{\Gamma_1, \Gamma_2 \vdash C}$$

This sets up a natural bijection between proofs and multi-arrows in a "representable multicategory" (Hermida, 2000). Multi-arrows then correspond bijectively to one-in-one-out sequents, and thus to maps in a tensor category. To show that these do correspond properly, the simplest route is via the circuits.

10.3.3 Circuit rewrites

Of course, merely having circuits is just the start of the matter. We want circuits to not only correspond to proofs in the logic but also to have the structure of a monoidal category. In the present simplified case, any circuit with just one input and one output always represents a morphism in a monoidal category. In general, we should expect there to be a more complicated "correctness" condition on circuits which corresponds to how proofs in the logic (or maps in the categorical doctrine) are constructed. Shortly, we shall meet such a correctness condition for poly-logics.

A question, arising not only from categorical considerations, but also from logic (Prawitz, 1971), is: When are two morphisms or proofs equivalent? This is in some sense *the* fundamental question for the logics we are discussing. It turns out that circuits are a very convenient tool for resolving this question. We shall now focus on this question.

There is an equivalence relation on circuits, generated by the following rewrites which we present as a reduction–expansion system modulo equalities.

Reductions:

(There is a mirror image rewrite for the unit, with the unit edge and nodes on the other side of the A edge.)

Expansions:

(Again, there is a mirror image rewrite for the unit, with the thinning edge on the other side of the unit edge and node.)

In addition to these rewrites, there are also a number of equivalences, see Figure 10.1, which must be imposed to account for the unit isomorphisms. These basic rewiring moves (Figure 10.1) may be summarized by a result, originally proved by Todd Trimble in his PhD dissertation (and later published in (Blute et al., 1996b)), which says that a thinning link may be moved to any position in its "empire", which essentially means from its initial position to any wire which is connected to that position whatever the switch settings, where the thinning link itself is not used for that connection (details in Blute et al., 1996b). The basic moves of Figure 10.1 give more "atomic" moves which generate the larger "empire" moves. An example may be seen in Figure 10.5, where, for instance, the first step moves a thinning wire on the far left up over the top of the circuit and down to the right-hand side of the circuit. The next rewiring step then moves a thinning link top right down along the wires to a position bottom right (that move is possible, for each setting of the switches of the bottom two $\oplus$ links, because of the thinning wires which allow two paths around each $\oplus$ link). And so on . . .

At the basic monoidal level, these equivalences actually allow the unit lasso to be moved onto *any* wire—provided a circuit is produced (for example, there should be no cycles). For this reason, when dealing with this level of logic the thinning links will often be omitted. However, we shall shortly see why thinning links cannot be omitted in the setting with more structure we shall consider. With that setting in mind, we have

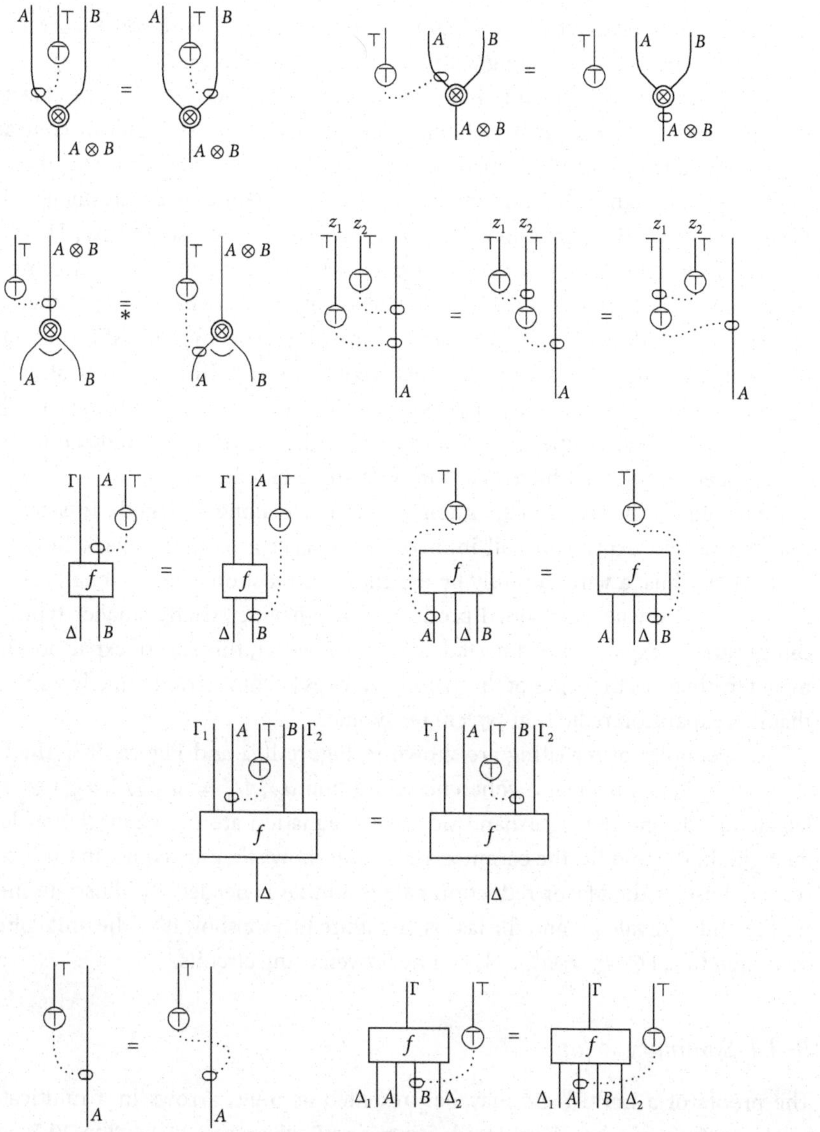

PROOF THEORY OF THE CUT RULE 235

Figure 10.1 Unit rewirings

also included some diagrams with many outputs, which cannot occur in the monoidal setting. The reader should imagine a single output in those cases for now.

10.3.3.1 NORMAL FORMS

Given the system of reductions and expansions, modulo equations, as described earlier, we can show that the reduction and expansion rewrites terminate, that there is a Church–Rosser theorem, modulo equations, and so there is a notion of expanded normal form, again modulo equations, for proof circuits. Essentially, the shape of a normal form involves tensor elimination steps (at the top), with some rearrangement of wires (in the middle), ending with tensor introductions (at the bottom).

In the present case, all this is not too difficult to see. First, notice that the reductions always remove material and so must certainly terminate. In contrast, the expansions always introduce material as the idea is that they "express" the type of the wire they expand. One can imagine repeatedly expanding a wire along its length. This means that the expansions do not terminate in the usual sense. However, in an expansion/reduction system, expansions which can be immediately removed by reduction (we call these *reducible* expansions) have no net effect on the system. Thus, an expansion/reduction system works by reducing terms to reduced form and using irreducible expansions to move between reduced forms (this is a sort of annealing process). This means one only applies expansion rules to reduced terms and, having applied the expansion, one immediately reduces the result. An expanded normal form is a reduced circuit for which all expansions are reducible.

Returning to the case of expanding a wire twice along its length, it is easy to see that the second expansion will immediately trigger a reduction back to the single expansion. Thus, a wire can only be expanded at most once on its length. A wire of composite type when expanded produces wires having strictly smaller types which can in turn be expanded in a nested fashion. However, this sort of expansion nesting must terminate as the types of the produced wires become strictly smaller. This shows that this expansion reduction system terminates.

Two examples of rewriting are shown in Figure 10.2 and Figure 10.3. The former shows Mac Lane's pentagon coherence condition $a_\otimes; a_\otimes = a_\otimes \otimes 1; a_\otimes; 1 \otimes a_\otimes$: the left-hand side and the right-hand side of the equation are shown on the far left and far right. In the middle, the common reduction shows they are equal. In this example, repeated use of the tensor reduction rule is all that is needed. To illustrate the need for the unit equivalence and the lassos, in Figure 10.3 we show how the unit coherence condition $(a_\otimes; 1 \otimes u_\otimes^L = u_\otimes^R \otimes 1)$ can be proven using circuits.

10.3.4 Summary so far

The proofs of a multi-logic may be presented as multi-arrows in a multicategory. These, in turn can be represented using circuit diagrams. In Section 10.5, we will provide a more formal treatment of circuits. Representable multicategories are monoidal categories and these correspond to $\otimes$-multi-logics. $\otimes$-multi-logics have a cut elimination theorem. Viewed as a rewriting system on circuits this becomes an expansion/reduction rewriting system which allows one to decide equality of proofs. Finally, the equivalence classes of circuits (or of derivations) are morphisms of a category of circuits, and this is the free (symmetric) monoidal category (over a generating multigraph of components).

We have already seen that circuits form a (symmetric) monoidal category where the objects are formulas and morphisms are equivalence classes of circuits. That it is the

PROOF THEORY OF THE CUT RULE   237

free one, such is the force of Mac Lane's coherence theorem; a proof of these claims (in the linearly distributive context) may be found in Blute et al. (1996b); see also Cockett and Seely (1997a) and Schneck (1999). This may be summarized by the following conceptual diagram (which actually represents 2-equivalences between appropriate chosen 2-categories):

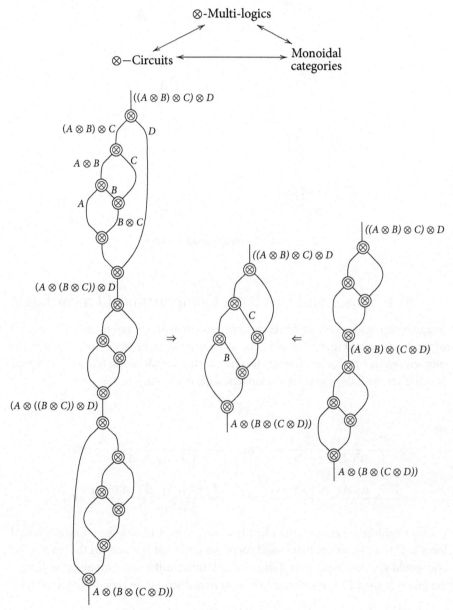

Figure 10.2 Pentagon coherence condition

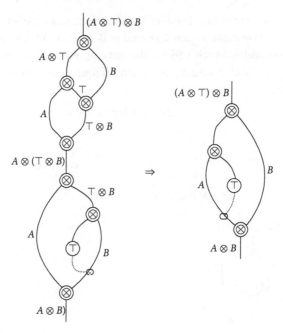

Figure 10.3 Unit coherence condition

10.4 Tensor and Par: Basic Components of Linear Logic

Imagine setting up a poly-logic: its sequents permit many (or no) premises on the left of the turnstile and many (or no) conclusions on the right of the turnstile. For the moment assume no commutativity: this means the cut rule has to take several forms, which gives the following presentation as a sequent calculus.

$$\frac{}{A \vdash A} \; id$$

$$\frac{\Gamma \vdash A \quad \Gamma_1, A, \Gamma_2 \vdash \Delta}{\Gamma_1, \Gamma, \Gamma_2 \vdash \Delta} \; cut_1 \qquad \frac{\Gamma \vdash \Delta_1, A, \Delta_2 \quad A \vdash \Delta}{\Gamma \vdash \Delta_1, \Delta, \Delta_2} \; cut_2$$

$$\frac{\Gamma_1 \vdash \Delta_1, A \quad A, \Gamma_2 \vdash \Delta_2}{\Gamma_1, \Gamma_2 \vdash \Delta_1, \Delta_2} \; cut_3 \qquad \frac{\Gamma_1 \vdash A, \Delta_1 \quad \Gamma_2, A \vdash \Delta_2}{\Gamma_2, \Gamma_1 \vdash \Delta_2, \Delta_1} \; cut_4.$$

One might have expected the identity axiom $A \vdash A$ to have had a more general form as $\Gamma \vdash \Gamma$. However, that would correspond to what is known as the "mix" rule, and would give the logic quite a distinct and different flavour. Among other things, the mix rule would allow a circuit to have several disconnected components; without mix, circuits must be connected. The circuits corresponding to the four cut rules are drawn below. Note that for simplicity, we have represented lists of formulas, such as Γ, as well as single formulas, such as A, by single wires. In each case, the only wires that *must* correspond to single formulas are those that join the two boxes being cut.

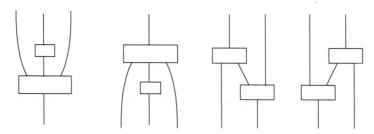

The reader should notice that any circuit inductively built from these rules, that is a **poly-circuit**, not only must be connected but also cannot have any cycles (i.e. must be acyclic). This, in fact, is precisely the *correctness criterion* for poly-circuits.

This circuit calculus is the basis for the categorical structure of a polycategory, as discussed for example in Cockett and Seely (1997b), Cockett et al. (2003). In fact, circuits over an arbitrary poly-graph, with their natural notion of equivalence and satisfying the correctness criterion, form the free polycategory over that poly-graph.

10.4.1 $\otimes\oplus$-Poly-logic

To capture the appropriate categorical notion, we must add appropriate tensor structures so that the "commas" of poly-logic, on both sides of the poly-sequents, are represented. The commas on the left of the turnstile are interpreted differently from those on the right. Thus, there are two connectives: $\otimes$ ("tensor") for commas on the left and $\oplus$ ("par") for commas on the right. The behaviour of these connectives is determined by requiring that polycategory be representable in the sense that there are the following bijective correspondences between poly-maps:

$$\frac{\Gamma_1, \Gamma_2 \vdash \Delta}{\Gamma_1, \top, \Gamma_2 \vdash \Delta} \qquad \frac{\Gamma \vdash \Delta_1, \Delta_2}{\Gamma \vdash \Delta_1, \bot, \Delta_2}$$

$$\frac{\Gamma_1, X, Y, \Gamma_2 \vdash \Delta}{\Gamma_1, X \otimes Y, \Gamma_2 \vdash \Delta} \qquad \frac{\Gamma \vdash \Delta_1, X, Y, \Delta_2}{\Gamma \vdash \Delta_1, X \oplus Y, \Delta_2}$$

Translating this into a sequent calculus presentation gives the sequent calculus presentation of $\otimes\oplus$-**poly-logic**. This consists of the cut rules of poly-logic, as seen earlier at the beginning of section 10.4, together with the following logical rules governing these new connectives:

$$\frac{\Gamma_1, \Gamma_2 \vdash \Delta}{\Gamma_1, \top, \Gamma_2 \vdash \Delta} \; (\top L) \qquad\qquad \frac{}{\vdash \top} \; (\top R)$$

$$\frac{}{\bot \vdash} \; (\bot L) \qquad\qquad \frac{\Gamma \vdash \Delta_1, \Delta_2}{\Gamma \vdash \Delta_1, \bot, \Delta_2} \; (\bot R)$$

$$\frac{\Gamma_1, X, Y, \Gamma_2 \vdash \Delta}{\Gamma_1, X \otimes Y, \Gamma_2 \vdash \Delta} \; (\otimes L) \qquad\qquad \frac{\Gamma_1 \vdash \Delta_1, X \quad \Gamma_2 \vdash Y, \Delta_2}{\Gamma_1, \Gamma_2 \vdash \Delta_1, X \otimes Y, \Delta_2} \; (\otimes R)$$

$$\frac{\Gamma_1, X \vdash \Delta_1 \quad Y, \Gamma_2 \vdash \Delta_2}{\Gamma_1, X \oplus Y, \Gamma_2 \vdash \Delta_1, \Delta_2} \; (\oplus L) \qquad\qquad \frac{\Gamma \vdash \Delta_1, X, Y, \Delta_2}{\Gamma \vdash \Delta_1, X \oplus Y, \Delta_2} \; (\oplus R).$$

The circuits that correspond to these rules are induced by the following basic nodes: the familiar tensor and tensor unit nodes from the tensor circuits, and dual nodes for the par and par unit $\bot$:

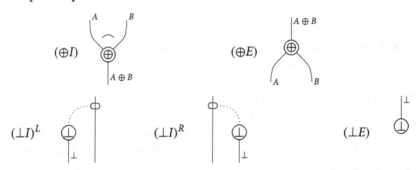

Using these components any proof of $\otimes \oplus$-poly-logic may be represented by a circuit. However, not every circuit made of these components represents a proof! Thus, to guarantee that a circuit does represent a proof, there is an additional correctness criterion which must be satisfied. One way to express this correctness criterion, due to Girard, is by a "switching condition": a circuit satisfies this condition if, whichever way one sets the "switch" in each switching link, the circuit remains connected and acyclic. To set the switch in a switching link one disconnects one of the two wires between which the "switch" links. A circuit satisfying this correctness criterion (and thus represents a proof in $\otimes \oplus$-poly-logic) is a $\otimes \oplus$-circuit.

Note that using this criterion one can demonstrate that a circuit *does not* satisfy the correctness criterion by a non-deterministic polynomial time (NP) algorithm. This algorithm works by guessing a configuration of the switches and proving the circuit with that set of disconnections is either cyclic or disconnected. Providing a configuration of switches to witness the *incorrectness* of a circuit is a very effective way of showing the invalidity of a circuit. So, the problem of determining correctness is a co-NP problem. To show that a circuit satisfies the correctness criterion using this approach one must potentially try all possible switching configurations ... and there are exponentially many of these. Algorithmically this would be somewhat disastrous!

Fortunately, Danos and Regnier (1989) describes a linear time algorithm for the correctness criterion which is based on more directly checking that a circuit represents

10.4.2 Why are thinning links necessary?

A curiosity is the apparent lack of symmetry between unit introduction and elimination links. Logically they correspond to the (bijective) correspondences we saw earlier. We might have expected the $\top$ elimination link ($\top E$) and the $\bot$ introduction link ($\bot I$) to be without a lasso. But this simply does not work! As this is a rather crucial aspect of our circuit calculus, some discussion of this is in order.

We shall want circuit identities corresponding to the equivalence of the following proof (which uses a cut on the left $\top$) and the identity axiom:

$$\frac{\vdash \top \quad \dfrac{\top \vdash \top}{\top, \top \vdash \top}}{\top \vdash \top}$$

(and dually for $\bot$). If we had let the ($\top E$) link be without lasso, as suggested earlier, this identity would become

[diagram: two $\top$ links stacked vertically with a wire connecting them, equals a single $\top$ wire]

This will not do, however—the lack of a thinning link here is fatal to the coherence questions which concern us. To see why, consider the following simple example which compares the identity with the par twist map applied to the tensor unit "par"ed with itself:

[diagram: two circuits involving $T \oplus T$, the left showing an identity and the right showing a twist, labeled "and" between them]

Given the above identity without thinning links these would both be equivalent to the same net as expanding the wires of type $\top$ would lose the twist because the circuit becomes disconnected.

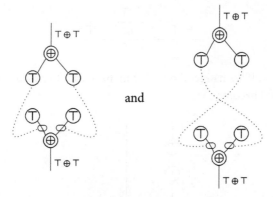

The twist and the identity on $\top \oplus \top$ are not equivalent as morphisms.[7] The point is that with thinning links we can at least distinguish these maps as nets, as we see next:

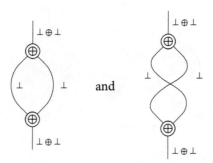

so there is hope that we can arrange for them to be inequivalent (and indeed they are). Note, however, the different behaviour of the units: if we replace $\top$ with $\bot$, then these nets do correspond to equivalent derivations, since $\bot \oplus \bot$ is isomorphic to $\bot$ and the identity is the same as the twist on $\bot$. Thus, these two nets when we expand the $\bot$ identity wires in the same manner must be equivalent:

---

[7] A remark for "experts": an example of a linearly distributive category where this is the case is Chu(Set, 2). This is more easily seen considering the dual $\bot \otimes \bot$. The unit $\bot$ is the tuple $\langle 2, 1 \rangle$ (with the obvious map $2 \times 1 \to 2$), and $\bot \otimes \bot = \langle 2 \times 2, \emptyset \rangle$ (with the empty map). It's clear that the twist and the identity are not equal.

To make these circuits equivalent it is clear that we must be able to rewire the thinning links in some manner, but equally not all rewirings can be permissible, so as to keep the distinctions seen with the two maps $\top \oplus \top \to \top \oplus \top$ above. It is not too surprising that rewirings will be required. Thinning links merely indicate a point at which a unit (or counit) has been introduced by thinning and there is considerable inessential choice going on here. For example, consider the three sequent calculus derivations of the sequent $A, \top, B \to A \otimes B$ obtained by thinning in each of the possible places (these clearly should be equivalent):

$$\frac{\frac{A \to A \quad B \to B}{A, B \to A \otimes B}}{A, \top, B \to A \otimes B} \quad = \quad \frac{\frac{A \to A}{A, \top \to A} \quad B \to B}{A, \top, B \to A \otimes B} \quad = \quad \frac{A \to A \quad \frac{B \to B}{\top, B \to B}}{A, \top, B \to A \otimes B}$$

As circuits, these are just the $(\otimes I)$ node with a $(\top E)$ link attached to the three possible links. It turns out that the allowable rewirings are essentially those from Figure 10.1, their obvious "duals" involving $\bot$ and $\oplus$, and a few involving interactions between these two structures (Blute et al., 1996b).

10.4.3 Linearly distributive categories

To introduce linearly distributive categories, consider the following conceptual diagram, recalling the connection among tensor circuit diagrams, multicategories, and monoidal categories. This states the analogous connection between the various structures for two tensors should also hold. The intention behind the definition of a linearly distributive category is, thus, that it should be viewed as the (one-in-one-out) maps of a "representable polycategory":

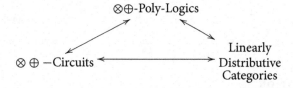

A linearly distributive category has two tensors: the "tensor" $(\otimes, a_\otimes, u^L_\otimes, u^R_\otimes)$, and the "par" $(\oplus, a_\oplus, u^L_\oplus, u^R_\oplus)$ both satisfying the usual coherences of a tensor. The interaction of these tensors is mediated (in the non-symmetric case) by two natural linear distribution maps:

$$\delta^L: A \otimes (B \oplus C) \to (A \otimes B) \oplus C \qquad \delta_R: (B \oplus C) \otimes A \to B \oplus (C \otimes A).$$

These must satisfy a number of coherences which, categorically, may be seen as (linear) strength coherences. Before discussing these, however, it is worth understanding the manner in which the linear distributions arise from the interaction of having representation for the commas and the behaviour of the cut. The following derivation of δ^L demonstrates this interaction:

$$\dfrac{\dfrac{\overline{B\oplus C \vdash B\oplus C}\ \text{id}}{B\oplus C \vdash B,C}\ \oplus R \quad \dfrac{\overline{A\otimes B \vdash A\otimes B}\ \text{id}}{A,B \vdash A\otimes B}\ \otimes L}{\dfrac{A, B\oplus C \vdash A\otimes B, C}{A\otimes (B\oplus C) \vdash (A\otimes B)\oplus C}\ \otimes L, \oplus R.}\ \text{Cut}$$

The definition of a linearly distributive category is subject to a number of symmetries; these arise from reversing the tensor ($A\otimes B \mapsto B\otimes A$), reversing the par ($A\oplus B \mapsto B\oplus A$), and reversing the arrows themselves while simultaneously swapping tensor and par (thus, $\delta^L \colon A\otimes(B\oplus C) \to (A\otimes B)\oplus C$ becomes $\delta^R \colon (A\oplus B)\otimes C \to A\oplus(B\otimes C)$). We present three coherence diagrams which are complete in the sense that together with these symmetries they can generate all the coherences for (non-symmetric) linearly distributive categories:

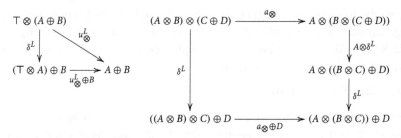

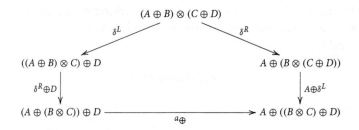

10.4.3.1 NEGATION

A key ingredient in the original account of linear logic, which is missing in linearly distributive categories, is negation. So just what do we need to obtain negation in a linearly distributive category? First, we need a function (on objects), which we shall denote $A \mapsto A^\perp$. In the non-symmetric case we shall need two such object functions $A \mapsto A^\perp$ and $A \mapsto {}^\perp A$. For simplicity, we shall outline the symmetric case here, where ${}^\perp A = A^\perp$. Also we need two parametrized families of maps[8]

$$A\otimes A^\perp \xrightarrow{\gamma_A} \perp \qquad \top \xrightarrow{\tau_A} A^\perp \oplus A,$$

which satisfy the coherence conditions

---

[8] We do not assume any naturality for these maps. In fact, they turn out to be dinatural transformations.

$$
\begin{array}{ccc}
A \otimes \top \xrightarrow{A \otimes \tau_A} A \otimes (A^\perp \oplus A) & \quad & \top \otimes A^\perp \xrightarrow{\tau_A \otimes A^\perp} (A^\perp \oplus A) \otimes A^\perp \\
\Big\downarrow u_\otimes^R \quad \quad \Big\downarrow \delta^L & & \Big\downarrow u_\otimes^L \quad \quad \Big\downarrow \delta^R \\
\quad \quad (A \otimes A^\perp) \oplus A & & \quad \quad A^\perp \oplus (A \otimes A^\perp) \\
\Big\downarrow \quad \quad \Big\downarrow \gamma_A \oplus A & & \Big\downarrow \quad \quad \Big\downarrow A^\perp \oplus \gamma_A \\
A \xleftarrow{u_\oplus^L} \perp \oplus A & & A^\perp \xleftarrow{u_\oplus^R} A^\perp \oplus \perp
\end{array}
$$

As circuits, this is as follows. The links are represented as "bends":

(γ) [bend diagram with A, $A^\perp$ and $\neg$]   (τ) [bend diagram with $A^\perp$, A and $\neg$]

And the equivalences (rewrites) are these:

($\neg$ Reduction)   [diagram] $\Longrightarrow$ $\Big| A$

($\neg$ Expansion) $A^\perp \Big|$ $\Longrightarrow$ [diagram]

Of course, the hope is that the function $A \mapsto A^\perp$ is a contravariant functor, and the families of maps are dinatural transformations. This is indeed the case. In fact, the category of $\otimes\oplus$-circuits with negation (in this sense) generated from a poly-graph is the free $*$-autonomous category generated by said poly-graph. The 2-category of linearly distributive categories with negation and linear functors (which we shall discuss shortly) is equivalent to the category of $*$-autonomous categories with monoidal functors. Moreover, there is a conservative extension result, stating that the extension of the tensor–par fragment of linear logic to full multiplicative linear logic (which includes negation) is conservative. More precisely, the functor from the category of linearly distributive categories to the category of $*$-autonomous categories extends to an adjunction (the right adjoint being the forgetful functor), whose unit is full and faithful, and whose counit is an equivalence (Blute et al., 1996b).

One consequence of this is that we now have a good circuit calculus for $*$-autonomous categories—the categorical doctrine corresponding to multiplicative fragment of linear logic—and this allows us to give a decision procedure for equality of maps in these categories. Bear in mind that this decision procedure necessarily involves a search as this is a PSPACE complete problem (Heijltjes and Houston, 2014) and so is not very efficient! As an example of this at work, here is a classic problem (not completely solved until Blute et al. (1996b)), usually called the triple-dual diagram:

$$\begin{array}{ccc} & ((A \multimap I) \multimap I) \multimap I & \\ & \nearrow^{1} \searrow^{k_A \multimap 1} & \\ ((A \multimap I) \multimap I) \multimap I & \xleftarrow{ k_{A \multimap 1} } & (A \multimap I) \end{array}$$

(using $k_A : A \to ((A \multimap I) \multimap I)$, the exponential transpose of evaluation.)

In ∗-autonomous categories, (or monoidal closed categories), this diagram generally does not commute. This is easy to see if I is not a unit. If I is a unit, then the diagram does commute if $I = \bot$, generally does not commute if $I = \top$, but does commute if $A = I = \top$.

We note that such instances of this diagram in fact can be done in the linearly distributive context, if we define the internal hom $A \multimap B$ as $A^\bot \oplus B$, replace I with a unit and $I^\bot$ with the other unit, and replace the negation links with appropriate derived rules corresponding to the (iso)morphisms $\top \otimes \bot \to \bot$ and $\top \to \bot \oplus \top$. Then we translate the composite $k_{A \multimap 1}; k_A \multimap 1$ into a proof net: the left side of Figure 10.4 is a step on the way to its expanded normal form (we write B for $A^\bot$ to prepare for the version of the net that may be constructed in the linearly distributive context). The right side is a similar step in calculating the expanded normal form of the identity circuit.

In the circuits of Figure 10.4, if I were not a unit, these would be the expanded normal forms, and clearly these nets are not the same. An old idea due to Lambek

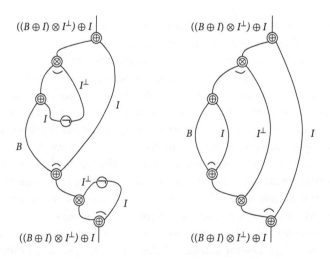

Figure 10.4 Two sides of the triple-dual triangle

(1969) may be seen here: the "generality" of the first net is clearly a derivation of the sequent $((B \oplus C) \otimes C^\perp) \oplus D \to ((B \oplus D) \otimes E^\perp) \oplus E$, whereas the generality of the second is $((B \oplus C) \otimes D) \oplus E \to ((B \oplus C) \otimes D) \oplus E$. This is no surprise; it is exactly what one would expect if I was not a unit. Next consider the case if I is a unit.

If $I=\top$, say, then the nodes at I and $I^\perp$ must be expanded, since in expanded normal form, each occurrence of a unit (recall $\top^\perp = \perp$) must either come from or go to a null node. This in effect transforms several of the edges in the graphs above into thinning links. We leave it as an exercise to show that in this case no rewiring is possible, and hence the diagram does not commute. And similarly that the rewiring may be done if $I = \perp$, so that diagram does commute.

But now consider the case where $A=I=\top$ (where the diagram commutes). We must show how to rewire the net corresponding to the compound morphism to give the identity. This is shown in Figure 10.5: the point here is that with $A = \top$ we have an extra unit and thinning link (corresponding to the wire/thinning link for $\perp$ replacing B at the left), which has a possible rewiring. Although it is not immediately obvious why this thinning link should be rewired, doing so makes other rewirings possible,

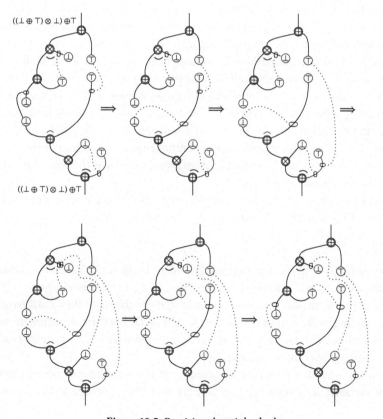

Figure 10.5 Rewiring the triple-dual

and once the required rewirings are done, the initial rewiring is reversed to finish with the expanded normal form of the identity map.

10.5 Circuits

While the circuits diagrams provide a convenient representation of morphisms or proofs, they have a formal underpinning which is also of interest. Although it is tempting to conflate the two notions, a circuit diagram is, in fact, a diagrammatic representation of something more fundamental, which we call a *circuit*. Circuits provide a term logic for monoidal settings. Such term logics have a distinguished pedigree, going back to Einstein's summation notation for vector calculations,[9] and including diagram notations by Feynman and by Penrose, for example. Their notations were more concretely attached to the specifics of the vector space contexts for which they were intended. Our circuits—which are derived not only from Girard's proof nets but also from the work of Joyal and Street (1991)—are intended to be more general. Vector space manipulations are a canonical example of tensor category manipulations, so that there is a direct ancestry is not a surprise. However, unlike our predecessors, including Joyal and Street, we explicitly intended our notation to be a term logic, and in particular we applied it to solve the coherence problems associated with linearly distributive categories. Despite this, the term logic was initially invented to facilitate calculations in monoidal categories, and so we shall return to this more straightforward application in the present essay. To extend these ideas to the full linearly distributive case involves adding the structures which we have described earlier using—the more user friendly but equivalent representation—circuit diagrams.

We start by making precise the notion of a **typed circuit**. To build a typed circuit one needs a set of **types**, $\mathcal{T}$, and a set of **components**, $\mathcal{C}$. Each component $f \in \mathcal{C}$ has a signature $\text{sig}(f) = (\alpha, \beta)$, a pair of lists of types, where α is the type of **input ports** and β the type of **output ports**.

To obtain a **primitive circuit expression** one attaches to a component two lists of variables. Thus, if $\text{sig}(h) = ([A, B], [C, D])$, then we may write

$$h^{x_1, x_2}_{y_1, y_2}$$

where the variables in the superscripted list are the **input variables**: each variable must have the correct type for its corresponding port so $x_1: A$ (*viz* x_1 is of type A) and $x_2: B$. The subscripted variable list contains the output variables and they must have types corresponding to the output ports, so $y_1: C$, and $y_2: D$. The variable names in each list must be distinct. The resulting (primitive) circuit expression has a list of input variables $[x_1, x_2]$ and a list of output variables $[y_1, y_2]$.

In the more familiar term logic associated with algebraic theories one does not have output variables. To emphasize that they are something peculiar to this "monoidal"

---

[9] We thank Gordon Plotkin for bringing this to our attention.

PROOF THEORY OF THE CUT RULE 249

term logic, Lambek referred to them as *covariables*. While we shall not adopt this terminology here, we shall discover that the term does convey their intent.

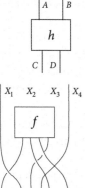

A primitive circuit presented as a circuit diagram is just a box with a number of (typed) input and output ports. The input wires in this diagram represent the list of typed input variables and the output wires represent the list of typed output variables.

One "plugs" (primitive) circuit expressions together to form new circuit expressions by juxtaposition, just as one attaches two circuit diagrams together:

$$f^{x_2,x_3}_{y_1,z_1,y_5,z_2}; g^{x_1,z_2,x_4,z_1}_{y_2,y_3,y_4}$$

The output variables of the first component which are common to the input variables of the second component become bound in this juxtaposition and indicate how the components are connected. An output variable when it is bound in this juxtaposition is bound to a unique input variable, or in Lambek's terminology, covariables bind to unique variables. To perform a legal juxtaposition the unbound input variables must be distinct and, similarly, the unbound output variables must be distinct. A variable clash occurs when this requirement is violated. One can always rename variables to avoid variable clashes.

When one avoids variable clashes, the juxtaposition operation is associative. Furthermore, when a juxtaposition does not cause any output variable to became bound to any input variables, one can exchange the order of the juxtaposition.

Note that we have allowed the wires representing the bindings of z_1 and z_2 to "cross", and indeed to access x_1 and x_4 as inputs also requires crossings. Allowing wires to cross in this manner corresponds to having commutativity of the underlying logic. To obtain a non-symmetric or *planar* juxtaposition, we would have to properly treat the inputs and outputs as lists of variables instead of viewing them as bags (or multi-sets). This would require that we alter the criteria for juxtaposition (the details are explicitly given in Blute et al. (1996b)).

A (non-planar) circuit expression C can be **abstracted** by indexing it by a non-repeating list of input and output variables. This is written

$$\left\langle C \mid^{x_1,\ldots,x_n}_{y_1,\ldots,y_m} \right\rangle$$

Furthermore, one can indicate the types of the input and output wires by the notation

$$\left\langle C \mid^{x_1:T_1,\ldots,x_n:T_n}_{y_1:T'_1,\ldots,y_m:T'_m} \right\rangle$$

An abstraction must be *closed* in the sense that all the free input variables of C occur in the abstracting input variable list and all the free output variables of C occur in the abstracting output list. Furthermore, any variable in the abstracting input list which is not a free input of C must occur in the abstracting output list and, similarly, any variable in the abstracting output list which is not a free output of C must occur in the abstracting input list.

In particular, we can use this technique of abstracting to isolate a wire (or many wires) as $\langle \emptyset \mid_{x:T}^{x:T} \rangle$, where $\emptyset$ is the empty circuit and the unit for juxtaposition. This is to be regarded as the "identity map" on the type T. The ability to abstract (and the existence of an empty circuit) are important when we consider how to form categories from circuits.

When a circuit expression is abstracted in this fashion all the wire names become bound. Externally an abstraction presents only a list of typed input ports and a list of typed output ports. This permits an abstracted circuit expression to be used as if it were a primitive component. An abstraction used as a component is equivalent to the circuit obtained by removing the abstraction with a substitution of wires *outward* with a renaming of the bound internal wires away from the external wires so as to avoid variable clashes. To see why the variables of the abstraction are used to substitute the external wires it suffices to consider the use of the "identity" abstraction mentioned earlier (or indeed any abstraction with "straight-through" wires):

$$\langle\langle \emptyset \mid_x^x \rangle_z^y \mid_z^y \rangle \Longrightarrow \langle \emptyset \mid_x^x \rangle$$

The operation of removing an abstraction we call abstraction **dissipation**; it is analogous to a β-reduction. The reverse operation is to **coalesce** an abstraction. These operations become particularly important when we consider how one adds rules of surgery, as discussed in the following. In the non-symmetric case, a **planar abstraction** must also preserve the order of the wires.

We may now define the notion of a (non-planar) circuit based on a set of components:

Definition 10.5.1 (Non-planar circuits)

(i) C-**circuit expressions** *are generated by the following:*
- *The empty circuit, $\emptyset$, is a circuit expression.*
- *If c_1 and c_2 are circuit expressions which can be juxtapositioned (with no variable clash), then $c_1; c_2$ is a circuit expression.*
- *If $f \in C$ is a component with $\mathrm{sig}(f) = (\alpha, \beta)$ and V is a non-repeating wire list with type α and W is a non-repeating wire list with type β, then f_W^V is a circuit expression.*
- *If F is an abstracted circuit with signature $\mathrm{sig}(F) = (\alpha, \beta)$ and V is a non-repeating wire list with type α and W is a non-repeating wire list with type β, then F_W^V is a circuit expression.*

(ii) *A **circuit** is an abstracted circuit expression.*

One circuit expression (and by inference circuit) is equivalent to another precisely when one can obtain the second from the first by a series of the following operations:

- Juxtaposition reassociation (with possible bound variable renaming to avoid clashes), $c_1; (c_2; c_3) = (c_1; c_2); c_3$,
- Empty circuit elimination and introduction, $c; \emptyset = c = \emptyset; c$,
- Exchanging non-interacting circuits, $c_1; c_2 = c_2; c_1$,
- Renaming of bound variables,
- Abstraction coalescing and dissipating.

The fact that circuit equivalence under these operations is decidable is immediately obvious when one presents them graphically. Indeed, while it is nice to have a syntax for circuits it is very much more natural and intuitive to simply draw them!

The C-circuits, besides permitting these standard manipulations, can also admit arbitrary additional identities. These take the form of equalities, $c_1 = c_2$, between (closed) abstracted circuits with the same signature. To use such an identity in a circuit, it is necessary to be able to coalesce one of the sides, say c_1 (up to α-conversion) within the circuit. Once this has been done one can replace c_1 with c_2 and dissipate the abstraction. Diagrammatically this corresponds to a surgical operation of cutting out the left-hand side and replacing it with the right-hand side. Accordingly, such additional identities are often referred to as *rules of surgery*. The circuit reductions and expansions we saw earlier are examples of such rules of surgery.

10.5.1 $\otimes$-Circuits: a term logic for monoidal categories

The basic components required to provide a circuit-based term logic for monoidal categories are as follows:

$(\otimes_I)_{A \otimes B}^{A,B}$ $\qquad\qquad$ $\otimes$-introduction

$(\otimes_E)_{A,B}^{A \otimes B}$ $\qquad\qquad$ $\otimes$-elimination

$(\top_I)_\top$ $\qquad\qquad$ unit introduction

$(\top_E^R)_A^{A,\top}$ $\qquad\qquad$ unit right elimination (thinning)

$(\top_E^L)_A^{\top,A}$ $\qquad\qquad$ unit left elimination (thinning).

The rules of surgery providing the reduction system for $\otimes$-multi-logic are expressed as follows:

$$\left\langle (\otimes_I)_z^{x_1,x_2}; (\otimes_E)_{y_1,y_2}^{z} \,\Big|\, {}^{x_1:A,x_2:B}_{y_1:A,y_2:B} \right\rangle \Rightarrow \left\{ {}^{x_1:A,x_2:B}_{x_1:A,x_2:B} \right\} \tag{10.1}$$

$$\left\langle (\top_I)_z; (\top_E^L)_{x_2}^{z,x_1} \,\Big|\, {}^{x_1:A}_{x_2:A} \right\rangle \Rightarrow \left\{ {}^{x:A}_{x:A} \right\} \tag{10.2}$$

$$\left\langle (\top_I)_z; (\top_E^R)_{x_2}^{x_1,z} \,\Big|\, {}^{x_1:A}_{x_2:A} \right\rangle \Rightarrow \left\{ {}^{x:A}_{x:A} \right\} \tag{10.3}$$

The rules of surgery providing the expansion rules for $\otimes$-multi-logic are next. Recall these should be thought of as expressing the type of the wire:

$$\left\langle \begin{vmatrix} z:A\otimes B \\ z:A\otimes B \end{vmatrix} \right\rangle \Rightarrow \left\langle (\otimes_E)^z_{z_1,z_2}; (\otimes_I)^{z_1,z_2}_z \begin{vmatrix} z:A\otimes B \\ z:A\otimes B \end{vmatrix} \right\rangle \qquad (10.4)$$

$$\left\langle \begin{vmatrix} x:\top \\ x:\top \end{vmatrix} \right\rangle \Rightarrow \left\langle (\top_I)_z; (\top^L_E)^{z,x_1}_{x_2} \begin{vmatrix} x_1:\top \\ x_2:\top \end{vmatrix} \right\rangle \qquad (10.5)$$

$$\left\langle \begin{vmatrix} x:\top \\ x:\top \end{vmatrix} \right\rangle \Rightarrow \left\langle (\top_I)_z; (\top^R_E)^{x_1,z}_{x_2} \begin{vmatrix} x_1:\top \\ x_2:\top \end{vmatrix} \right\rangle \qquad (10.6)$$

We shall leave as an exercise for the reader the translation of the unit rewirings into this term calculus—graphically they were given in Figure 10.1. For example, the first one may be written thus:

$$\left\langle (\top^R_E)^{x,z}_x; (\otimes_I)^{x,y}_w \begin{vmatrix} x:A,z:\top,y:B \\ w:A\otimes B \end{vmatrix} \right\rangle = \left\langle (\top^L_E)^{z,y}_y; (\otimes_I)^{x,y}_w \begin{vmatrix} x:A,z:\top,y:B \\ w:A\otimes B \end{vmatrix} \right\rangle$$

As an illustration of the term calculus at work, consider the following example which is the coherence condition for the tensor unit as shown in Figure 10.3. We shall write the variables $x_1, x_2, \ldots$ as simply $1, 2, \ldots$, numbering the wires, top to bottom, left to right. In this way, the topmost link in the left-hand diagram is $(\otimes_E)^1_{2,3}$ (wire 1 comes into the $(\otimes E)$ link, and wires 2, 3 leave it, 2 on the left, 3 on the right, so that 1 refers to a wire of type $(A \otimes \top) \otimes B$, 2 to a wire of type $A \otimes \top$, and 3 to a wire of type B). Here are the details of the rewriting showing the coalescing, surgery, and dissipation steps:

$$\left\langle (\otimes_E)^1_{2,3}; (\otimes_E)^2_{4,5}; (\otimes_I)^{5,3}_6; (\otimes_I)^{4,6}_7; (\otimes_E)^7_{8,9}; (\otimes_E)^9_{10,11}; (\top^L_E)^{10,11}_{11}; (\otimes_I)^{8,11}_{12} \ |^1_{12} \right\rangle$$

$$= \left\langle (\otimes_E)^1_{2,3}; (\otimes_E)^2_{4,5}; (\otimes_I)^{5,3}_6; \left\langle (\otimes_I)^{4,6}_7; (\otimes_E)^7_{8,9} \ |^{4,6}_{8,9} \right\rangle^{4,7}_{8,9}; (\otimes_E)^9_{10,11}; (\top^L_E)^{10,11}_{11}; (\otimes_I)^{8,11}_{12} \ |^1_{12} \right\rangle$$

$$\Rightarrow \left\langle (\otimes_E)^1_{2,3}; (\otimes_E)^2_{4,5}; (\otimes_I)^{5,3}_6; \left\langle |^{4,6}_{4,6} \right\rangle^{4,6}_{8,9}; (\otimes_E)^9_{10,11}; (\top^L_E)^{10,11}_{11}; (\otimes_I)^{8,11}_{12} \ |^1_{12} \right\rangle (\otimes\text{-reduction})$$

$$= \left\langle (\otimes_E)^1_{2,3}; (\otimes_E)^2_{4,5}; (\otimes_I)^{5,3}_6; (\otimes_E)^6_{10,11}; (\top^L_E)^{10,11}_{11}; (\otimes_I)^{4,11}_{12} \ |^1_{12} \right\rangle$$

$$= \left\langle (\otimes_E)^1_{2,3}; (\otimes_E)^2_{4,5}; \left\langle |\ (\otimes_I)^{5,3}_6; (\otimes_E)^6_{10,11} \ |^{5,3}_{10,11} \right\rangle^{5,3}_{10,11}; (\top^L_E)^{10,11}_{11}; (\otimes_I)^{4,11}_{12} \ |^1_{12} \right\rangle$$

$$\Rightarrow \left\langle (\otimes_E)^1_{2,3}; (\otimes_E)^2_{4,5}; \left\langle |^{5,3}_{5,3} \right\rangle^{5,3}_{10,11}; (\top^L_E)^{10,11}_{11}; (\otimes_I)^{4,11}_{12} \ |^1_{12} \right\rangle (\otimes\text{-reduction})$$

$$= \left\langle (\otimes_E)^1_{2,3}; (\otimes_E)^2_{4,5}; (\top^L_E)^{5,3}_{11}; (\otimes_I)^{4,11}_{12} \ |^1_{12} \right\rangle$$

$$= \left\langle (\otimes_E)^1_{2,3}; (\otimes_E)^2_{4,5} \left\langle (\top^L_E)^{z,y}_y; (\otimes_I)^{x,y}_w \ |^{x,z:\top,y}_w \right\rangle^{4,5,3}_{12} \ |^1_{12} \right\rangle \text{ (coalescing)}$$

$$= \left\langle (\otimes_E)^1_{2,3}; (\otimes_E)^2_{4,5} \left\langle (\top^R_E)^{x,z}_x; (\otimes_I)^{x,y}_w \ |^{x:A,z:\top,y:B}_{w:A\otimes B} \right\rangle^{4,5,3}_{12} \ |^1_{12} \right\rangle \text{ (surgery: tensor unit axiom)}$$

$$= \left\langle (\otimes_E)^1_{2,y}; (\otimes_E)^2_{x,z} (\top^R_E)^{x,z}_x; (\otimes_I)^{x,y}_w \ |^1_w \right\rangle \text{ (dissipation)}$$

$$= \left\langle (\otimes_E)^1_{2,3}; (\otimes_E)^2_{4,5} (\top^R_E)^{4,5}_4; (\otimes_I)^{4,3}_6 \ |^1_6 \right\rangle \text{ (renaming).}$$

This amply illustrates why it is so attractive to work with the diagrammatic representation! However, it is important to know that under the hood of circuit diagrams there is a fully formal (if somewhat verbose) notion of circuits which, for example, could be implemented on a computer.

10.6 Functor Boxes

The use of circuit diagrams—and similar graphical tools using similar but different conventions—is very widespread, and includes applications to systems developed for handling quantum computing (dagger categories) and for fixpoints and feedback (traced monoidal categories). They are even a useful device in understanding the categorical proof theory of classical logic. It is well known that the usual Lambek-style approach collapses to posetal proof theory, so one cannot distinguish between proofs of the same sequents. However, interesting non-posetal proof theory for classical logic may be obtained by a construction on top of the proof theoretic substrate provided by linearly distributive categories or ∗-autonomous categories (Lamarche and Straßburger, 2005, 2006; Führmann and Pym, 2007).

For the category theorist, one question is always paramount: what are the morphisms? In the present context, what would be the suitable functors between linearly distributive categories? What logical structure would be suitable for handling interpretations from one poly-logic to another? And specifically, one is led to questions such as how would one represent modal operators (or other operators, for that matter) in poly logic? How would we adapt circuits to this purpose?

The answer we developed, in Cockett and Seely (1999), was a description of circuits for structured functors, and indeed, is sufficient to account for a variety of logics, including a simple linear modal logic (Blute et al., 2002). The basic idea is similar to the proof boxes of Girard (1987) for the exponentials ! and ?, as described in (Blute et al., 1996a). But for more general functors, a slightly different approach was needed, which we shall sketch here. Full details are available in Cockett and Seely (1999).

The first question to address is how to handle functors, indeed, "Why do functors at all?" (i.e. "Why boxes?"). We shall see an example at the end of this section: modal logic. The first use of boxes was by Girard (1987) for the exponentials ! and ?, which were called modalities from the beginning. They were necessary to be able to interpret intuitionistic logic in linear logic, and in fact were really the reason for his initial development of linear logic. More traditional modal logic would also seem to require functors for □ and ◇ (necessity and possibility). But we must handle these at the level of derivations (as well as formulas), so in effect we need to be able to "apply" the modalities to morphisms as well as to formulas. And the simplest way (or so it seems!) is to simply take a subgraph corresponding to a derivation, and replace it with one corresponding to the image of that derivation under the modality. And that is easily handled with the boxes we shall describe in the following. Interestingly, this notion was independently discovered by Melliès (2006).

Figure 10.6 Simple and monoidal functor boxes

We start with an ordinary functor $F: \mathbf{C} \to \mathbf{D}$. Given a morphism $f: A \to B$ in $\mathbf{C}$, represented as a component with input wire of type A and output wire of type B, the corresponding morphism $F(f): F(A) \to F(B)$ in $\mathbf{D}$ is represented by simply "boxing" the component, as shown in Figure 10.6.

Note that the box bears a label with the name of the functor. These functor boxes have one input and one output. If the component f is a poly-map, then it is necessary to tensor the inputs and par the outputs to obtain a one-in-one-out map before the functor box can be applied. We shall relax this condition soon in discussing monoidal and linear functors. The half oval through which the wire leaves the box is called the "principle port". This is not really essential here, but we include it for comparison with the structured boxes that will be described next: at that point its role will become clear. Note also the typing changes the box imposes on a wire as it passes into or out of a box.

There are two obvious rewrites to express functoriality: an "expansion" which takes an identity wire of type $F(A)$ and replaces it with an identity wire of type A which is then "boxed", and a "reduction" which "merges" two functor boxes, one of which directly "feeds" into the next (box "eats" box).

Now we consider the situation where there is some structure on both categories and functor. Suppose first that the categories are monoidal categories, and the functor is also monoidal, meaning that preservation of the tensor is lax. The functor F is monoidal if there are natural transformations $m_\otimes : F(A) \otimes F(B) \to F(A \otimes B)$ and $m_\top : \top \to F(\top)$ satisfying the equations

$$u^L_\otimes = m_\top \otimes 1; m_\otimes; F(u^L_\otimes) : \top \otimes F(A) \to F(A)$$
$$a_\otimes; 1 \otimes m_\otimes; m_\otimes = m_\otimes \otimes 1; m_\otimes; F(a_\otimes)$$
$$: (F(A) \otimes F(B)) \otimes F(C) \to F(A \otimes (B \otimes C)),$$

and in the symmetric case, the next equation as well:

$$m_\otimes; F(c_\otimes) = c_\otimes; m_\otimes \quad : \quad F(A) \otimes F(B) \to F(B \otimes A).$$

We will soon also want the dual notion (for the dual par $\oplus$): a functor G is comonoidal if there are natural transformations $n_\oplus : G(A \oplus B) \to G(A) \oplus G(B)$ and $n_\bot : G(\bot) \to \bot$ satisfying equations dual to those above.

To capture the effect in circuits of requiring a functor to be monoidal, we modify the effect of the "functor box", as shown in Figure 10.6.

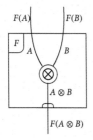

Note that for F monoidal, we may relax the supposition that the boxed subgraph is "one-in-one-out" to allow multi-arrows which have many input wires (but still just one output wire). One might expect that we would have to add components representing the two natural transformations $m_\otimes, m_\top$ that are necessary for F to be monoidal. However, it is an easy exercise to show that these can be induced by the formation rule for monoidal functor boxes: $m_\top$ is the case where f is the ($\top\ I$) node (no inputs and one output $\top$), and $m_\otimes$ is shown left.

There are reduction and expansion rewrites for these boxes (and one for handling "twists" permitted by symmetry). The necessary reduction rewrite is shown in Figure 10.7 (we refer to this saying one box "eats" the other). And the "expansion" rule and the "twist" rule are shown in Figure 10.8.

We have already indicated what the nets are for $m_\otimes, m_\top$. It is a fairly straightforward exercise to show that the equations are consequences of the net rewrites given earlier, and that the rewrites correspond to commutative diagrams, if F is monoidal. (An exercise we leave for the reader(!)—but recall that the details may be found in Cockett and Seely (1999).) As an example, Figure 10.9 shows that the equation dealing with "reassociation" is true for any F whose functor boxes satisfy the circuit rewrites we have given so far. So this circuit syntax is indeed sound and complete for monoidal functors. For comonoidal functors, we just use a dual syntax, with the corresponding rewrites. Note then that for comonoidal functors, the principle port will be at the top of the box (this is the role of the principle port, to distinguish monoidal functors from comonoidal ones).

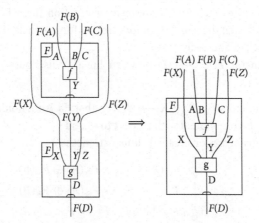

Figure 10.7 Box-eats-box rule

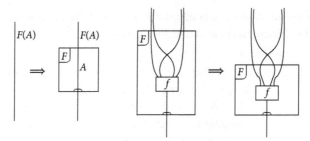

Figure 10.8 Expansion and twist rules

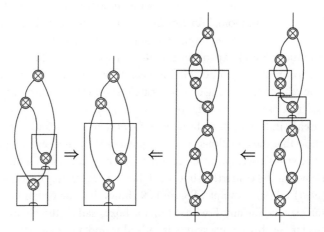

Figure 10.9 Functor boxes are monoidal: reassociation

10.6.1 Linear functors

To handle linear logic, it turns out that the suitable notion of functor is what we call a "linear functor": this is really a pair of functors related by a shadow of duality. This duality becomes explicit in the presence of negation (i.e. when the category is ∗-autonomous), for then the pair of functors are de Morgan duals. This is discussed in more detail in Cockett and Seely (1999).

So, a linear functor $F: \mathbf{C} \to \mathbf{D}$ between linearly distributive categories $\mathbf{C}, \mathbf{D}$, consists of:

1. a pair of functors $F_\otimes, F_\oplus : \mathbf{X} \to \mathbf{Y}$ such that $F_\otimes$ is monoidal with respect to $\otimes$, and $F_\oplus$ is comonoidal with respect to $\oplus$ and
2. natural transformations (called "linear strengths")

$$\nu^R_\otimes : F_\otimes(A \oplus B) \to F_\oplus(A) \oplus F_\otimes(B)$$
$$\nu^L_\otimes : F_\otimes(A \oplus B) \to F_\otimes(A) \oplus F_\oplus(B)$$
$$\nu^R_\oplus : F_\otimes(A) \otimes F_\oplus(B) \to F_\oplus(A \otimes B)$$
$$\nu^L_\oplus : F_\oplus(A) \otimes F_\otimes(B) \to F_\oplus(A \otimes B)$$

satisfying coherence conditions corresponding to the requirements that the linear strengths are indeed strengths, and that the various transformations are compatible with each other. (These are listed explicitly in Cockett and Seely (1999).) A representative sample is given here—the rest are generated by the obvious dualities:

$$v_\otimes^R; n_\perp \oplus 1; u_\oplus^L = F_\otimes(u_\oplus^L)$$
$$F_\otimes(a_\oplus); v_\otimes^R; 1 \otimes v_\otimes^R = v_\otimes^R; n_\oplus \oplus 1; a_\oplus$$
$$F_\otimes(a_\oplus); v_\otimes^R; 1 \otimes v_\otimes^L = v_\otimes^L; v_\otimes^R \oplus 1; a_\oplus$$
$$1 \otimes v_\otimes^R; \delta_L^L; v_\oplus^R \oplus 1 = m_\otimes; F_\otimes(\delta_L^L); v_\otimes^R$$
$$1 \otimes v_\otimes^L; \delta_L^L; m_\otimes \oplus 1 = m_\otimes; F_\otimes(\delta_L^L); v_\otimes^L$$

In Cockett and Seely (1999) the definition of linear transformations, which are necessary to describe the 2-categorical structure of linear logic, is also given, but we shall not pursue that further here.

Now we extend the syntax of functor boxes to linear functors. In fact all that is necessary is to generalize the monoidal boxes to allow the boxed circuit to have arbitrarily many inputs and outputs. So for the monoidal component $F_\otimes$ of a linear functor F, the functor boxes will have the formation rule shown in Figure 10.10, and the comonoidal component will have the dual rule (just turn the page upside down). Please note the typing of this formation rule carefully: at the top of the box, the functor applied is the functor $F_\otimes$ associated with the box, but at the bottom, only the wire that leaves through the principal port gets an $F_\otimes$ attached to it; the other wires get the comonoidal $F_\oplus$ attached to them. (The dual situation applies for the $F_\oplus$ boxes.) This is the role of the principal port in our notation (and is similar to the notation used in Blute et al. (1996a) for the "exponential" or "modal" operators ! and ?). There may be only one principal port, though there may be arbitrarily many other ("auxiliary") ports.

It is then quite easy to represent the $v_\otimes^R$ map as a boxed ($\oplus E$) node—the right output wire of the node passes through the principal port. The three other linear strengths are given similarly: $v_\otimes^L$ is the ($\oplus E$) node boxed with a $F_\otimes$ box, the left output wire passing through the principal port. The two $v_\oplus$ maps are given by the ($\otimes I$) node boxed by

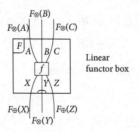

Figure 10.10 Linear functor box

the $F_\oplus$ box, with either the right or the left input port being the principal port, as appropriate.

Associated with these box formation rules are several rewrites. The expansion rewrite remains as before, but the reduction rewrite must be generalized to account for the more general f; this is done in the obvious fashion. Similarly, in the symmetric case we generalize the rewrites that move a "twist" outside a box. In fact, in the symmetric case it is convenient to regard the order of inputs/outputs as irrelevant, so that these rewrites are in fact equalities of circuits. In addition, we must account for the interaction between $F_\otimes$ and $F_\oplus$ boxes, which gives a series of rewrites that allow one box to "eat" another whenever a non-principal wire of one type of box becomes the principal wire of the dual type. We give an example of this in Figure 10.11, along with the other rules mentioned in this paragraph. The reader may generate the dual rules. Note that in this table we have illustrated circuits with crossings of wires; in the non-symmetric case, such crossings must not occur, so some wires must be absent from these rewrites. The rewrites dealing with pulling a "twist" out of a box are only relevant in the symmetric case of course. We have illustrated one, where the "twisted" wires are inputs; it is also possible that the "twisted" wires are outputs, and one may (or may not) be the wire through the principle port.

These rules are sound and complete; we must verify that any $F = (F_\otimes, F_\oplus)$ which allows such a calculus is indeed a linear functor, and conversely, that any linear functor allows such rewrites (Cockett and Seely, 1999). More interestingly, however, is what can be done with such functors (and their circuits). The key fact is that all the basic structure of linear logic (with the exception of negation) can be described in terms of linear functors: in a natural manner, tensor and par together form a linear functor, as do Cartesian product and coproduct. And not surprisingly, so do the exponentials ! and ?. If negation is added to the mix (so the category is $*$-autonomous, not merely linearly distributive), then each pair consists of de Morgan duals, and in that context, the pair really just amounts to the monoidal functor in the pair (the comonoidal functor just being its de Morgan dual).

There is a natural logic associated with a linear functor. A special case is illuminating. Consider a linear functor $F: \mathbf{X} \to \mathbf{X}$. Write $\Box$ for $F_\otimes$ and $\Diamond$ for $F_\oplus$. We shall not give a complete description of the logic one obtains from this (Blute et al., 2002), but here are some highlights:

$$v^L_\otimes: \Box (A \oplus B) \;\;\to\;\; \Box A \oplus \Diamond B$$
$$m_\otimes: \Box A \otimes \Box B \;\;\to\;\; \Box(A \otimes B).$$

In a classical setting, these would be equivalent to

$$\Box(A \Rightarrow B) \;\;\to\;\; (\Box A \Rightarrow \Box B)$$
$$\Box A \wedge \Box B \;\;\to\;\; \Box(A \wedge B),$$

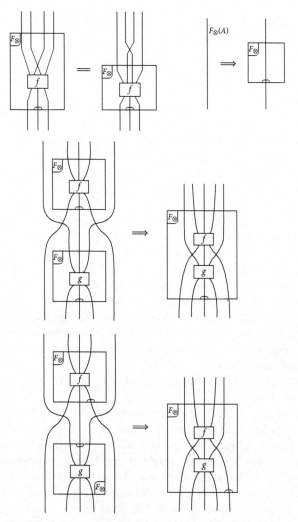

Figure 10.11 Some reduction and expansion rewrites for functor boxes

the first being "normality" of the logic, and the second being one half (the linear half!) of the standard isomorphism

$$\Box A \wedge \Box B \longleftrightarrow \Box(A \wedge B).$$

The following rule (and its dual cousins) holds in basic linear modal logic:

$$\frac{A_1, A_2, \cdots, A_m, B \vdash C_1, C_2, \cdots, C_n}{\Box A_1, \Box A_2, \cdots, \Box A_m, \Diamond B \vdash \Diamond C_1, \Diamond C_2, \cdots, \Diamond C_n}$$

This is a familiar rule in many modal situations; for example it occurs in the process calculus of Hennessy and Milner (1985).

References

Blute, R., Cockett, J.R.B., and Seely, R.A.G. (1996a). ! and ?—storage as tensorial strength. *Mathematical Structures in Computer Science* 6(04), 313–51.

Blute, R., Cockett, J.R.B., and Seely, R.A.G. (2002). The logic of linear functors. *Mathematical Structures in Computer Science* 12(4), 513–39.

Blute, R., Cockett, J.R.B., Seely, R.A.G., and Trimble, T. H. (1996b). Natural deduction and coherence for weakly distributive categories. *Journal of Pure and Applied Algebra* 113(3), 229–96.

Cockett, J.R.B., Koslowski, J., and Seely, R.A.G. (2000). Introduction to linear bicategories. *Mathematical Structures in Computer Science* 10(02), 165–203.

Cockett, J.R.B., Koslowski, J., and Seely, R.A.G. (2003). Morphisms and modules for poly-bicategories. *Theory and Applications of Categories* 11(2), 15–74.

Cockett, J.R.B., and Seely, R.A.G. (1997a). Proof theory for full intuitionistic linear logic, bilinear logic, and mix categories. *Theory and Applications of Categories* 3(5), 85–131.

Cockett, J.R.B., and Seely, R.A.G. (1997b). Weakly distributive categories. *Journal of Pure and Applied Algebra* 114(2), 133–73 (updated version available on http://www.math.mcgill.ca/rags).

Cockett, J.R.B., and Seely, R.A.G. (1999). Linearly distributive functors. *Journal of Pure and Applied Algebra* 143(1), 155–203.

Danos, V., and Regnier, L. (1989). The structure of multiplicatives. *Archive for Mathematical Logic* 28(3), 181–203.

Führmann, C., and Pym, D. (2007). On categorical models of classical logic and the geometry of interaction. *Mathematical Structures in Computer Science* 17(05), 957–1027.

Girard, J.-Y. (1987). Linear logic. *Theoretical Computer Science* 50(1), 1–101.

Heijltjes, W., and Houston, R. (2014). No proof nets for mll with units: proof equivalence in mll is pspace-complete, in *Proceedings of the Joint Meeting of the Twenty-Third EACSL Annual Conference on Computer Science Logic (CSL) and the Twenty-Ninth Annual ACM/IEEE Symposium on Logic in Computer Science (LICS)*, ACM, p. 50.

Hennessy, M., and Milner, R. (1985). Algebraic laws for nondeterminism and concurrency. *Journal of the ACM (JACM)* 32(1), 137–61.

Hermida, C. (2000). Representable multicategories. *Advances in Mathematics* 151(2), 164–225.

Hughes, D. J. (2012). Simple free ∗-autonomous categories and full coherence. *Journal of Pure and Applied Algebra* 216(11), 2386–410.

Joyal, A., and Street, R. (1991). The geometry of tensor calculus, I. *Advances in Mathematics* 88(1), 55–112.

Koh, T. W., and Ong, C. (1999). Explicit substitution internal languages for autonomous and ∗-autonomous categories. *Electronic Notes in Theoretical Computer Science* 29, 151.

Lamarche, F., and Straßburger, L. (2004). On proof nets for multiplicative linear logic with units, in *International Workshop on Computer Science Logic*, 145–59. Springer, Berlin.

Lamarche, F., and Straßburger, L. (2005). Constructing free boolean categories, in *20th Annual IEEE Symposium on Logic in Computer Science (LICS'05)*, 209–18. IEEE, Piscataway, NJ.

Lamarche, F., and Straßburger, L. (2006). From proof nets to the free∗-autonomous category. *arXiv preprint cs/0605054*.

Lambek, J. (1969). Deductive systems and categories II. standard constructions and closed categories, in *Category Theory, Homology Theory and Their Applications I*, 76–122. Springer, Berlin.

Lambek, J. (1989). Multicategories revisited, in *Categories in Computer Science and Logic: Proceedings of the AMS-IMS-SIAM Joint Summer Research Conference Held June 14–20, 1987*, Vol. 92, pp. 217–37. American Mathematical Society, Providence, RI.

Melliès, P.-A. (2006). Functorial boxes in string diagrams, in *International Workshop on Computer Science Logic*, 1–30. Springer, Berlin.

Prawitz, D. (1965). *Natural Deduction*. Almqvist and Wiksell, Stockholm, Sweden.

Prawitz, D. (1971). Ideas and results in proof theory. *Studies in Logic and the Foundations of Mathematics* 63, 235–307.

Schneck, R. R. (1999). Natural deduction and coherence for non-symmetric linearly distributive categories. *Theory and Applications of Categories* 6(9), 105–46.

Seely, R.A.G. (1979). Weak adjointness in proof theory, in *Applications of Sheaves*, 697–701. Springer, Berlin.

Seely, R.A.G. (1987). Modelling computations: a 2-categorical framework, in *Proc. Symposium on Logic in Computer Science*, 65–71. Computer Society of the IEEE, Washington, DC.

Seely, R.A.G. (1989). Linear logic, ∗-autonomous categories and cofree algebras. *Categories in Computer Science and Logic* 92, 371–82.

Selinger, P. (2010). A survey of graphical languages for monoidal categories, in *New Structures for Physics*, 289–355. Springer, Berlin.

Urban, C. (2014). Revisiting Zucker's work on the correspondence between cut-elimination and normalisation, in *Advances in Natural Deduction*, 31–50. Springer, Berlin.

Zucker, J. (1974). The correspondence between cut-elimination and normalization. *Annals of Mathematical Logic* 7(1), 1–112.

11

Contextuality: At the Borders of Paradox

Samson Abramsky

11.1 Introduction

Logical consistency is usually regarded as a minimal requirement for scientific theories, or indeed for rational thought in general. Paraconsistent logics (Priest, 2002) aim at constraining logical inference, to prevent inconsistency leading to triviality.

However, a richer, more nuanced view of consistency is forced on us if we are to make sense of one of the key features of quantum physics, namely **contextuality**.

The phenomenon of contextuality is manifested in classic No-Go theorems such as the Kochen–Specker paradox (Kochen and Specker, 1967); and in a particular form, also appears in Bell's theorem (Bell, 1964), and other non-locality results such as the Hardy paradox (Hardy, 1993). The close relationship of these results to issues of consistency is suggested by the very fact that terms such as "Kochen–Specker paradox" and "Hardy paradox" are standardly used.

At the same time, these arguments are empirically grounded. In fact, a number of experiments to test for contextuality have already been performed (Bartosik et al., 2009; Kirchmair et al., 2009; Zu et al., 2012). Moreover, it has been argued that contextuality can be seen as an essential **resource** for quantum advantage in computation and other information-processing tasks (Raussendorf, 2013; Howard et al., 2014).

What, then, is the essence of contextuality? In broad terms, we propose to describe it as follows:

Contextuality arises where we have a family of data which is **locally consistent**, but **globally inconsistent**.

In more precise terms, suppose that we have a family of data $\{D_c\}_{c \in \mathsf{Cont}}$, where each D_c describes the data which is observed in the context c. We can think of these contexts as ranging over various experimental or observational scenarios Cont. This data is locally consistent in the sense that, for all contexts c and d, D_c and D_d agree on their overlap; that is, they give consistent information on those features which are common to both contexts.

Figure 11.1 Penrose tribar

However, this data is globally inconsistent if there is no global description D_g of **all** the features which can be observed in **any** context, which is consistent with all the local data D_c, as c ranges over the set of possible contexts Cont.

An immediate impression of how this situation might arise is given by impossible figures such as the Penrose tribar (Penrose, 1992), shown in Figure 11.1.

If we take each leg of the tribar, and the way each pair of adjacent legs are joined to each other, this gives a family of locally consistent data, where consistency here refers to realizability as a solid object in 3-space. However, the figure as a whole is inconsistent in this sense. We will see more significant examples arising from quantum mechanics shortly.

Why should we regard this phenomenon of contextuality as disturbing? Consider the following situation. Certain fundamental physical quantities are being measured, e.g. electron spin or photon polarization. Some of these physical quantities can be measured together. In a famous example, Alice and Bob are spacelike separated, and a pair of particles are prepared at some source, and one is sent to Alice, and one to Bob. Then they can each measure the spin of their particle in a given direction. Each such measurement has two possible outcomes, spin up or down in the given direction. We shall refer to this choice of direction as a **measurement setting**. Now imagine this procedure being repeated, with Alice and Bob able to make different choices of measurement setting on different repetitions. This gives rise to a family of local data, the statistics of the outcomes they observe for their measurements. This set of local data is moreover locally consistent, in the sense that Alice's observed statistics for the outcome for various measurement settings is independent of Bob's choice of measurements, and vice versa. However, let us suppose that this data is globally inconsistent: there is no way of explaining what Alice and Bob see in terms of some joint distribution on the outcomes for all possible measurement settings.

The import of Bell's theorem and related results is that exactly this situation is predicted by quantum mechanics. Moreover, these predictions have been extensively confirmed by experiment, with several loophole-free Bell tests having been published in 2015 (Giustina et al., 2015; Hensen et al., 2015; Shalm et al., 2015).

What this is saying is that the fundamental physical quantities we are measuring, according to our most accurate and well-confirmed physical theory, cannot be taken to have objective values which are independent of the context in which they are being measured.

In the light of this discussion, we can elucidate the connection between contextuality and inconsistency and paradox. There is no inconsistency in what we can actually observe directly—no conflict between logic and experience. The reason for this is that we cannot observe all the possible features of the system at the same time. In terms of the Alice–Bob scenario, they must each choose one of their measurement settings each time they perform a measurement. Thus, we can never observe directly what the values for the other measurement settings are, or "would have been". In fact, we must conclude that they have **no well-defined values**.

The key ingedient of quantum mechanics which enables the possibility of contextual phenomena, while still allowing a consistent description of our actual empirical observations, is the presence of **incompatible observables**: variables that cannot be measured jointly. In fact, we can say that the experimental verification of contextual phenomena shows that **any** theory capable of predicting what we actually observe must incorporate this idea of incompatible observables.

The compatible families of observables, those which **can** be measured together, provide the empirically accessible windows onto the behaviour of microphysical systems. These windows yield local, and locally consistent, information. However, in general there may be no way of piecing this local data together into a global, context-independent description. This marks a sharp difference with classical physics, with consequences which are still a matter for foundational debate.

The phenomenon of contextuality thus leaves quantum mechanics, and indeed any empirically adequate "post-quantum" theory, on the borders of paradox, without actually crossing those borders. This delicate balance is not only conceptually challenging, but appears to be deeply implicated in the **quantum advantage**: the possibility apparently offered by the use of quantum resources to perform various information-processing tasks better than can be achieved classically.

11.2 Basic Formalization

We shall now show how our intuitive account of contextuality can be formalized, using tools from category theory. This will provide the basis for a mathematical theory of contextuality. It can also, perhaps, serve as a case study for how conceptual discussions can be turned into precise mathematics using categories.

We recall the basic ingredients of our informal discussion:

- Notions of **context** of observation or measurement, and of **features** of the system being observed or measured. Certain features, but in general not all, can be observed in each context. Those features which can be observed in the same context are deemed **compatible**.
- A collection of observations will give rise to a family of data $\{D_c\}$ describing the information yielded by these observations, ranging over the contexts.

- We say that such a family of data is **locally consistent** if for each pair of contexts c, d, the data D_c, D_d yields consistent descriptions of the features which are common to c and d.
- We say that the family of data is **globally consistent** if there is some data description D_g giving information on all the features which can appear in any context, and which is consistent with the local data, in the sense that each local description D_c can be recovered by restricting D_g to the features which can be observed in the context c. If there is no such global description, then we say that the family of data $\{D_c\}$ is **contextual**.

To formalize these ideas, we fix sets Cont, Feat of contexts and features, together with a function

$$\Phi : \mathsf{Cont} \longrightarrow \mathcal{P}(\mathsf{Feat}),$$

which for each context c gives the set of features which can be observed in c.

We shall make the assumption that this function is injective, ruling out the possibility of distinct contexts with exactly the same associated features. This is not essential, but simplifies the notation, and loses little by way of examples. We can then **identify** contexts with their associated set of features, so we have $\mathsf{Cont} \subseteq \mathcal{P}(\mathsf{Feat})$. We moreover assume that this family is closed under subsets: if $C \subseteq D \in \mathsf{Cont}$, then $C \in \mathsf{Cont}$. This says that if a set of features is compatible, and can be observed jointly, so is any subset.

To model the idea that there is a set of possible data descriptions which can arise from performing measurements or observations on the features in a context, we assume that there is a map

$$P : \mathsf{Cont} \longrightarrow \mathsf{Set},$$

where $P(C)$ is the set of possible data for the context C.

We now come to a crucial point. In order to define both local and global consistency, we need to make precise the idea that the information from a larger context, involving more features, can be cut down or restricted to information on a smaller one. That is, when $C \subseteq D$, $C, D \in \mathsf{Cont}$, we require a function

$$\rho_C^D : P(D) \longrightarrow P(C).$$

We call such a function a **restriction map**. We require that these functions satisfy the obvious conditions:

- If $C \subseteq D \subseteq E$, then $\rho_C^D \circ \rho_D^E = \rho_C^E$
- $\rho_C^C = \mathrm{id}_{P(C)}$.

This says that P is a functor,

$$P : \mathsf{Cont}^{\mathrm{op}} \longrightarrow \mathsf{Set},$$

where Cont is the poset *qua* category of subsets under inclusion. Such a functor is called a **presheaf**.

In fact, it will be convenient to assume that P is defined on arbitrary subsets of features, compatible or not, so that P is a presheaf

$$P : \mathcal{P}(\mathsf{Feat})^{\mathrm{op}} \longrightarrow \mathsf{Set}.$$

This will allow us to formulate global consistency.

11.2.1 Notation

It will be convenient to write $d|_C := \rho_C^D(d)$ when $d \in P(D)$ and $C \subseteq D$. In this notation, the functoriality conditions become

$$(d|_D)|_C = d|_C, \qquad d|_C = d.$$

We now have exactly the tools we need to define local and global consistency.

A family of data is now a family $\{d_C\}_{C \in \mathsf{Cont}}$, where $d_C \in P(C)$. Such a family is locally consistent if whenever $C \subseteq D$, then $d_C = (d_D)|_C$. Equivalently, for any $C, D \in \mathsf{Cont}$, $d_C|_{C \cap D} = d_D|_{C \cap D}$. This formalizes the idea that the data from the two contexts yields consistent information with respect to their overlap, i.e. their common features.

The family is globally consistent if there is some $d_g \in P(\mathsf{Feat})$, global data on the whole set of features, such that, for all contexts C, $d_g|_C = d_C$; that is, the local data can be recovered from the global description. It is the absence of such a global description which is the signature of contextuality.

These notions of local and global consistency have significance in the setting of sheaf theory (Mac Lane and Moerdijk, 1992; Kashiwara and Schapira, 2005), which is precisely the mathematics of the passage between local and global descriptions, and the obstructions to such a passage which may arise. If we think of the family of subsets Cont as an "open cover", in the discrete topology $\mathcal{P}(\mathsf{Feat})$, then a locally consistent family $\{d_C\}_{C \in \mathsf{Cont}}$ is exactly a **compatible family** in the terminology of sheaf theory. The condition for this family to be globally consistent is exactly that is satisfies the **gluing condition**. The condition for the presheaf P to be a sheaf is exactly that for every open cover, every compatible family satisfies the gluing condition in a unique fashion. Thus, contextuality arises from obstructions to instances of the sheaf condition.

We will return to this link to the topological language of sheaf theory shortly.

11.3 Example

We shall now examine an Alice–Bob scenario, of the kind discussed in Section 11.1, in detail. The scenario is depicted in Figure 11.2. Alice can choose measurement settings a_1 or a_2, while Bob can choose b_1 or b_2. Alice can measure her part of the system with her chosen setting, and observe the outcome 0 or 1. Bob can perform the same operations with respect to his part of the system. They send the outcomes to a common target.

CONTEXTUALITY 267

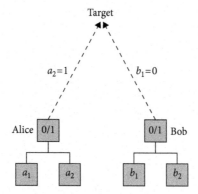

Figure 11.2 Alice–Bob scenario

A	B	(0, 0)	(1, 0)	(0, 1)	(1, 1)
a_1	b_1	1/2	0	0	1/2
a_1	b_2	3/8	1/8	1/8	3/8
a_2	b_1	3/8	1/8	1/8	3/8
a_2	b_2	1/8	3/8	3/8	1/8

Figure 11.3 The Bell table

We now suppose that Alice and Bob perform repeated rounds of these operations. On different rounds, they may make different choices of which measurement settings to use, and they may observe different outcomes for a given choice of setting. The target can compile statistics for this series of data, and infer probability distributions on the outcomes. The probability table in Figure 11.3 records the result of such a process.

Consider for example the cell at row 2, column 3 of the table. This corresponds to the following event:

- Alice chooses measurement setting a_1 and observes the outcome 0.
- Bob chooses measurement setting b_2 and observes the outcome 1.

This event has the probability 1/8, conditioned on Alice's choice of a_1 and Bob's choice of b_2.

Each row of the table specifies a probability distribution on the possible joint outcomes, conditioned on the indicated choice of settings by Alice and Bob.

We can now ask:

> How can such an observational scenario be realized?

The obvious classical mechanism we can propose to explain these observations is depicted in Figure 11.4.

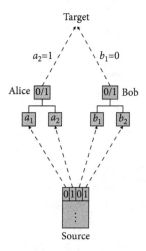

Figure 11.4 A Source

We postulate a **source** which on each round chooses outcomes for each of the possible measurement settings a_1, a_2, b_1, b_2. Alice and Bob will then observe the values which have been chosen by the source. We can suppose that this source is itself randomized, and chooses the outcomes according to some probability distribution P on the set of 2^4 possible assignments.

We can now ask the question: is there any distribution P which would give rise to the table specified in Figure 11.3?

11.3.1 Important note

A key observation is that, in order for this question to be non-trivial, we must assume that the choices of measurement settings made by Alice and Bob are **independent** of the source.[1] If the source could determine which measurements are to be performed on each round, as well as their values, then it becomes a trivial matter to achieve **any** given probability distribution on the joint outcomes.

Under this assumption of independence, it becomes natural to think of this scenario as a kind of **correlation game**. The aim of the source is to achieve as high a degree of correlation between the outcomes of Alice and Bob as possible, whatever the choices made by Alice and Bob on each round.

---

[1] This translates formally into a conditional independence assumption, which we shall not spell out here; see e.g. Bell (1985), Brandenburger and Yanofsky (2008).

11.3.2 Logic rings a Bell

We shall now make a very elementary and apparently innocuous deduction in elementary logic and probability theory, which could easily be carried out by students in the first few weeks of a Probability 101 course.

Suppose we have propositional formulas $\varphi_1, \ldots, \varphi_N$. We suppose further that we can assign a probability p_i to each φ_i.

In particular, we have in the mind the situation where the boolean variables appearing in φ_i correspond to empirically testable quantities, such as the outcomes of measurements in our scenario; φ_i then expresses a condition on the outcomes of an experiment involving these quantities. The probabilities p_i are obtained from the statistics of these experiments.

Now suppose that these formulas are **not simultaneously satisfiable**. Then (e.g.)

$$\bigwedge_{i=1}^{N-1} \phi_i \to \neg \phi_N, \quad \text{or equivalently} \quad \phi_N \to \bigvee_{i=1}^{N-1} \neg \phi_i.$$

Using elementary probability theory, we can calculate

$$p_N \leq \text{Prob}(\bigvee_{i=1}^{N-1} \neg \phi_i) \leq \sum_{i=1}^{N-1} \text{Prob}(\neg \phi_i) = \sum_{i=1}^{N-1}(1 - p_i) = (N-1) - \sum_{i=1}^{N-1} p_i.$$

The first inequality is the monotonicity of probability, and the second is sub-additivity. Hence we obtain the inequality

$$\sum_{i=1}^{N} p_i \leq N - 1.$$

We shall refer to this as a **logical Bell inequality**, for reasons to be discussed later. Note that it hinges on a purely logical consistency condition.

11.3.3 Logical analysis of the Bell table

We return to the probability table from Figure 11.2.

	(0, 0)	(1, 0)	(0, 1)	(1, 1)
(a_1, b_1)	1/2	0	0	1/2
(a_1, b_2)	3/8	1/8	1/8	3/8
(a_2, b_1)	3/8	1/8	1/8	3/8
(a_2, b_2)	1/8	3/8	3/8	1/8

If we read 0 as true and 1 as false, the highlighted entries in each row of the table are represented by the following propositions:

$$\varphi_1 = (a_1 \wedge b_1) \vee (\neg a_1 \wedge \neg b_1) = a_1 \leftrightarrow b_1$$

$$\varphi_2 = (a_1 \wedge b_2) \vee (\neg a_1 \wedge \neg b_2) = a_1 \leftrightarrow b_2$$

$$\varphi_3 = (a_2 \wedge b_1) \vee (\neg a_2 \wedge \neg b_1) = a_2 \leftrightarrow b_1$$

$$\varphi_4 = (\neg a_2 \wedge b_2) \vee (a_2 \wedge \neg b_2) = a_2 \oplus b_2.$$

The events on first three rows are the correlated outcomes; the fourth is anticorrelated. These propositions are easily seen to be jointly unsatisfiable. Indeed, starting with φ_4, we can replace a_2 with b_1 using φ_3, b_1 with a_1 using φ_1, and a_1 with b_2 using φ_2, to obtain $b_2 \oplus b_2$, which is obviously unsatisfiable.

It follows that our logical Bell inequality should apply, yielding the inequality

$$\sum_{i=1}^{4} p_i \leq 3.$$

However, we see from the table that $p_1 = 1$, $p_i = 6/8$ for $i = 2, 3, 4$. Hence the table yields a violation of the Bell inequality by $1/4$.

This rules out the possibility of giving an explanation for the observational behaviour described by the table in terms of a classical source. We might then conclude that such behaviour simply cannot be realized. However, **in the presence of quantum resources, this is no longer the case.** More specifically, if we use the Bell state $(|\uparrow\uparrow\rangle + |\downarrow\downarrow\rangle)/\sqrt{2}$, with Alice and Bob performing 1-qubit local measurements corresponding to directions in the XY-plane of the Bloch sphere, at relative angle $\pi/3$, then this behaviour is **physically realizable** according to the predictions of quantum mechanics—our most highly confirmed physical theory.[2]

More broadly, we can say that this shows that quantum mechanics predicts correlations which exceed those which can be achieved by any classical mechanism. This is the content of **Bell's theorem** (Bell, 1964), a famous result in the foundations of quantum mechanics, and in many ways the starting point for the whole field of quantum information. Moreover, these predictions have been confirmed by many experiments (Aspect et al., 1982; Aspect, 1999).

The logical Bell inequality we used to derive this result is taken from Abramsky and Hardy (2012). Bell inequalities are a central technique in quantum information and foundations. In Abramsky and Hardy (2012) it is shown that **every** Bell inequality (i.e. every inequality satisfied by the "local polytope") is equivalent to a logical Bell inequality, based on purely logical consistency conditions.

---

[2] For further details on this, see e.g. Abramsky (2015).

11.4 More on Formalization

We note that the example in the previous section, which is representative of those studied in quantum contextuality and non-locality, has a specific form which can be reflected naturally in our mathematical formalism.

In particular, the "features" take the form of a set X of variables, which can be measured or observed. Thus, we take Feat $= X$. The set Cont consists of compatible sets of variables, those which can be measured jointly. The result of measuring a compatible set of variables C is a **joint outcome** in $\prod_{x \in C} O_x$, where O_x is the set of possible outcomes or values for the variable $x \in X$. To simplify notation, we shall assume there is a single set of outcomes O for all variables, so a joint outcome for C is an element of O^C—an assignment of a value in O to each $x \in C$. By repeatedly observing joint outcomes for the variables in a context C, we obtain statistics, from which we can infer a probability distribution $d \in \mathsf{Prob}(O^C)$, where $\mathsf{Prob}(X)$ is the set of probability distributions on a set X.[3] So the presheaf P has the specific form $P(C) = \mathsf{Prob}(O^C)$. What about the restriction maps?

Note, first, that the assignment $C \mapsto O^C$ has a natural contravariant functorial action (it is a restriction of the contravariant hom functor). If $C \subseteq D$, then restriction is just function restriction to a subset of the domain:

$$\rho_C^D : O^D \longrightarrow O^C, \qquad \rho_C^D(f) = f|_C.$$

Moreover, the assignment $X \mapsto \mathsf{Prob}(X)$ can be extended to a (covariant) functor $\mathcal{D}$: Set $\longrightarrow$ Set: $\mathcal{D}(X) = \mathsf{Prob}(X)$, and given $f : X \longrightarrow Y$, $\mathcal{D}(f) : \mathcal{D}(X) \longrightarrow \mathcal{D}(Y)$ is the push-forward of probability measures along f:

$$\mathcal{D}(f)(d)(U) = d(f^{-1}(U)).$$

This functor extends to a monad on Set; the discrete version of the Giry monad (Giry, 1982).

Composing these two functors, we can define a presheaf $P : \mathcal{P}(X)^{\mathrm{op}} \longrightarrow$ Set, with $P(C) = \mathcal{D}(O^C)$, and if $C \subseteq D$, $d \in \mathcal{D}(O^D)$:

$$\rho_C^D(d)(U) = d(\{s \in O^D \mid s|_C \in U\}).$$

Note that if $C \subseteq D$, we can write D as a disjoint union $D = C \sqcup C'$, and $O^D = O^C \times O^{C'}$. Restriction of $s \in O^D$ to C is projection onto the first factor of this product. Thus, restriction of a distribution is **marginalization**.

A compatible family of data for this presheaf is a family $\{d_C\}_{C \in \mathsf{Cont}}$ of probability distributions, $d_C \in \mathcal{D}(O^C)$. Local consistency, i.e. compatibility of the family, is the condition that for all $C, D \in \mathsf{Cont}$,

$$d_C|_{C \cap D} = d_D|_{C \cap D}.$$

---

[3] To avoid measure-theoretic technicalities, we shall assume that we are dealing with discrete distributions. In fact, in all the examples we shall consider, the sets X and O will be finite.

This says that the distributions on O^C and O^D have the same marginals on their overlap, i.e. the common factor $O^{C \cap D}$.

Global consistency is exactly the condition that there is a **joint distribution** $d \in \mathcal{D}(O^X)$ from which the distributions d_C can be recovered by marginalization.

We refer again to the Alice–Bob table in Figure 11.3. We anatomize this table according to our formal scheme. The set of variables is $X = \{a_1, a_2, b_1, b_2\}$. The set of contexts, indexing the rows of the table, is

$$\mathsf{Cont} = \{\{a_1, b_1\}, \quad \{a_2, b_1\}, \quad \{a_1, b_2\}, \quad \{a_2, b_2\}\}.$$

These are the compatible sets of measurements. The set of outcomes is $O = \{0, 1\}$. The columns are indexed by the possible joint outcomes of the measurements. Thus, for example the matrix entry at row (a_2, b_1) and column $(0, 1)$ indicates the **event**

$$\{a_2 \mapsto 0, \ b_1 \mapsto 1\}.$$

The set of events relative to a context C is the set of functions O^C. Each row of the table, indexed by context C, specifies a distribution $d_C \in \mathcal{D}(O^C)$. Thus, the set of rows of the table is the family of data $\{d_C\}_{C \in \mathsf{Cont}}$. One can check that this family is locally consistent, i.e. compatible. In this example, this is exactly the No-Signalling principle (Ghirardi et al., 1980; Popescu and Rohrlich, 1994), a fundamental physical principle which is satisfied by quantum mechanics. Suppose that $C = \{a, b\}$, and $C' = \{a, b'\}$, where a is a variable measured by Alice, while b and b' are variables measured by Bob. Then under relativistic constraints, Bob's choice of measurement—b or b'—should not be able to affect the distribution Alice observes on the outcomes from her measurement of a. This is captured by the No-Signalling principle, which says that the distribution on $\{a\} = \{a, b\} \cap \{a, b'\}$ is the same whether we marginalize from the distribution e_C, or the distribution $e_{C'}$. This is exactly compatibility.

Thus, we can see that this table, and the others like it which are standardly studied in non-locality and contextuality theory within quantum information and foundations, falls exactly within the scope of our definition.

At the same time, we can regard **any** instance of our general definition of a family of data as a generalized probability table; compatibility becomes a generalized form of No-Signalling (which, suitably formulated, can be shown to hold in this generality in quantum mechanics (Abramsky and Brandenburger, 2011)); and we have a very general probabilistic concept of contextuality.

It might appear from our discussion thus far that contextuality is inherently linked to probabilistic behaviour. However, this is not the case. In fact, our initial formalization in Section 11.2 was considerably more general; and even within the current more specialized version, we can gain a much wider perspective by generalizing the notion of distribution. We recall first that a probability distribution of finite support on a set X can be specified as a function

$$d : X \longrightarrow \mathbb{R}_{\geq 0},$$

where $\mathbb{R}_{\geq 0}$ is the set of non-negative reals, satisfying the normalization condition

$$\sum_{x \in X} d(x) = 1.$$

This condition guarantees that the range of the function lies within the unit interval $[0, 1]$. The finite support condition means that d is zero on all but a finite subset of X. The probability assigned to an event $E \subseteq X$ is then given by

$$d(E) = \sum_{x \in E} d(x).$$

This is easily generalized by replacing $\mathbb{R}_{\geq 0}$ by an arbitrary **commutative semiring**, which is an algebraic structure $(R, +, 0, \cdot, 1)$, where $(R, +, 0)$ and $(R, \cdot, 1)$ are commutative monoids satisfying the distributive law

$$a \cdot (b + c) = a \cdot b + a \cdot c.$$

Examples include the non-negative reals $\mathbb{R}_{\geq 0}$ with the usual addition and multiplication, and the booleans $\mathbb{B} = \{0, 1\}$ with disjunction and conjunction playing the rôles of addition and multiplication, respectively.

We can now define a functor $\mathcal{D}_R$ of R-distributions, parameterized by a commutative semiring R. Given a set X, $\mathcal{D}_R(X)$ is the set of R-distributions of finite support. The functorial action is defined exactly as we did for $\mathcal{D} = \mathcal{D}_{\mathbb{R}_{\geq 0}}$. In the boolean case, $\mathbb{B}$-distributions on X correspond to non-empty finite subsets of X. In this boolean case, we have a notion of **possibilistic contextuality**, where we have replaced probabilities by boolean values, corresponding to possible or impossible. Note that there is a homomorphism of semirings from $\mathbb{R}_{\geq 0}$ to $\mathbb{B}$, which sends positive probabilities to 1 (possible), and 0 to 0 (impossible). This lifts to a map on distributions, which sends a probability distribution to its **support**. This in turn sends generalized probability tables $\{d_C\}_{C \in \text{Cont}}$ to possibility tables. We refer to this induced map as the **possibilistic collapse**.

11.4.1 Measurement scenarios and empirical models

To understand how contextuality behaves across this possibilistic collapse, it will be useful to introduce some terminology. The combinatorial shape of the situations we are considering is determined by the following data:

- The set X of variables,
- The set Cont of contexts, i.e. the compatible sets of variables, and
- The set of outcomes O.

Accordingly, we shall call (X, Cont, O) a **measurement scenario**. Once we have fixed a measurement scenario $\Sigma = (X, \text{Cont}, O)$, and a semiring R, we have the notion of a compatible family of R-distributions $\{e_C\}_{C \in \text{Cont}}$, where $e_C \in \mathcal{D}_R(O^C)$. We will call such compatible families **empirical models**, and write $\text{EM}(\Sigma, R)$ for the set of empirical models over the scenario Σ and the semiring R. We refer to **probabilistic empirical models** for $R = \mathbb{R}_{\geq 0}$, and **possibilistic empirical models** for $R = \mathbb{B}$.

Given an empirical model $e \in \mathsf{EM}(\Sigma, R)$, we call a distribution $d_g \in \mathcal{D}_R(O^X)$ such that, for all $C \in \mathsf{Cont}$, $d_g|_C = e_C$, a **global section** for e. The existence of such a global section is precisely our formulation of global consistency. Thus

$$\boxed{e \text{ is contextual if and only if it has no global section}}$$

The term "empirical model" relates to the fact that this is purely observational data, without prejudice as to what mechanism or physical theory lies behind it. In the terminology of quantum information, it is a **device-independent** notion (Barrett et al., 2005). It is a key point that **contextuality is in the data**. It does not presuppose quantum mechanics, or any other specific theory.

A homomorphism of semirings $h : R \longrightarrow S$ induces a natural transformation $\bar{h}$ from the presheaf of R-valued distributions to the presheaf of S-valued distributions. In particular, if $h : \mathbb{R}_{\geq 0} \longrightarrow \mathbb{B}$ is the unique semiring homomorphism from the positive reals to the booleans, then $\bar{h}$ is the possibilistic collapse. Given a global section d_g for an empirical model $e \in EM(\Sigma, R)$, it is easy to see that $\bar{h}(d_g)$ is a global section for $\bar{h}(e)$. Thus, we have the following result.

Proposition 11.4.1 *If $\bar{h}(e)$ is contextual, then so is e. In particular, if the possibilistic collapse of a probabilistic empirical model e is contextual, then e is contextual.*

This is a strict one-way implication. It is strictly harder to be possibilistically contextual than probabilistically contextual, as we shall now see.

11.5 Example: The Hardy Paradox

Consider the table in Figure 11.5.

This table depicts the same kind of scenario we considered previously. However, the entries are now either 0 or 1. The idea is that a 1 entry represents a positive probability. Thus, we are distinguishing only between **possible** (positive probability) and

	(0,0)	(0,1)	(1,0)	(1,1)
(a_1, b_1)	1			
(a_1, b_2)	0			
(a_2, b_1)	0			
(a_2, b_2)				0

Figure 11.5 The Hardy Paradox

impossible (zero probability). In other words, the rows correspond to the **supports** of some (otherwise unspecified) probability distributions. Moreover, only four entries of the table are filled in. Our claim is that just from these four entries, referring only to the supports, we can deduce that the behaviour recorded in this table is contextual. Moreover, this behaviour can again be realized in quantum mechanics, yielding a stronger form of Bell's theorem, due to Lucien Hardy (1993).[4]

11.5.1 What do "observables" observe?

Classically, we would take the view that physical observables directly reflect properties of the physical system we are observing. These are objective properties of the system, which are independent of our choice of which measurements to perform—of our **measurement context**. More precisely, this would say that for each possible state of the system, there is a function λ which for each measurement m specifies an outcome $\lambda(m)$, **independently of which other measurements may be performed**. This point of view is called **non-contextuality**, and may seem self-evident. However, this view is **impossible to sustain** in the light of our **actual observations of (micro)-physical reality**.

Consider once again the Hardy table depicted in Figure 11.5. Suppose there is a function λ which accounts for the possibility of Alice observing value 0 for a_1 and Bob observing 0 for b_1, as asserted by the entry in the top left position in the table. Then this function λ must satisfy

$$\lambda : a_1 \mapsto 0, \quad b_1 \mapsto 0.$$

Now consider the value of λ at b_2. If $\lambda(b_2) = 0$, then this would imply that the event that a_1 has value 0 and b_2 has value 0 is possible. However, **this is precluded** by the 0 entry in the table for this event. The only other possibility is that $\lambda(b_2) = 1$. Reasoning similarly with respect to the joint values of a_2 and b_2, we conclude, using the bottom right entry in the table, that we must have $\lambda(a_2) = 0$. Thus, the only possibility for λ consistent with these entries is

$$\lambda : a_1 \mapsto 0, \quad a_2 \mapsto 0, \quad b_1 \mapsto 0, \quad b_2 \mapsto 1.$$

However, this would require the outcome $(0, 0)$ for measurements (a_2, b_1) to be possible, and this is **precluded** by the table.

We are thus forced to conclude that the Hardy models are contextual. Moreover, it is **possibilistically contextual**. By virtue of Proposition 11.4.1, we know that any probabilstic model with this support must be probabilistically contextual. On the other hand, if we consider the support of the table in Figure 11.3, it is easy to see that it is not possibilistically contextual. Thus, we see that possibilistic contextuality is strictly stronger than probabilistic contextuality.

---

[4] For a detailed discussion of realizations of the Bell and Hardy models in quantum mechanics, see sect. 7 of Abramsky (2013b). Further details on the Hardy construction can be found in a number of papers (Hardy, 1993; Mermin, 1994).

11.6 Strong Contextuality

Logical contextuality as exhibited by the Hardy paradox can be expressed in the following form: there is a local assignment (in the Hardy case, the assignment $a_1 \mapsto 0, b_1 \mapsto 0$) which is in the support, but which cannot be extended to a global assignment which is compatible with the support. This says that the support cannot be covered by the projections of global assignments. A stronger form of contextuality is when **no global assignments are consistent with the support at all**. Note that this stronger form does not hold for the Hardy paradox.

We pause to state this in more precise terms. Consider a possibilistic model $e \in \mathsf{EM}(\Sigma, \mathbb{B})$. For example, e may be the possibilistic collapse of a probabilistic model. We can consider the boolean distribution e_C as (the characteristic function of) a subset of O^C. In fact, the compatibility of e implies that it is a subpresheaf of $C \mapsto O^C$. We write $S_e(C) \subseteq O^C$ for this subpresheaf. A global assignment $g \in O^X$ is **consistent with** e if $g|_C \in S_e(C)$ for all $C \in \mathsf{Cont}$. We write $S_e(X)$ for the set of such global assignments. A **compatible family of local assignments for** e is a family $\{s_C\}_{C \in \mathsf{Cont}}$ with $s_C \in S_e(C)$, such that for all $C, D \in \mathsf{Cont}$, $s_C|_{C \cap D} = s_D|_{C \cap D}$.

Proposition 11.6.1 There is a bijective correspondence between $S_e(X)$ and compatible families of local assignments for e.

This follows directly from the observation that the presheaf $C \mapsto O^C$ is a sheaf, and hence S_e as a subpresheaf is separated.

We can now characterize possibilistic and strong contextuality in terms of the extendability of local assignments to global ones.

Proposition 11.6.2 Let $e \in \mathsf{EM}(\Sigma, \mathbb{B})$ be a possibilistic empirical model. The following are equivalent:

1. e is possibilistically contextual.
2. There exists a local assignment $s \in S_e(C)$ which cannot be extended to a compatible family of local assignments.
3. There exists a local assignment $s \in S_e(C)$ such that, for all $g \in S_e(X)$, $g|_C \neq s$.

Also, the following are equivalent:

1. e is strongly contextual.
2. No local assignment $s \in S_e(C)$ can be extended to a compatible family of local assignments.
3. $S_e(X) = \emptyset$.

Several much-studied constructions from the quantum information literature exemplify strong contextuality. An important example is the Popescu–Rohrlich (PR) box (Popescu and Rohrlich, 1994) shown in Figure 11.6.

This is a behaviour which satisfies the **no-signalling principle** (Ghirardi et al., 1980), meaning that the probability of Alice observing a particular outcome for her choice of

A	B	(0,0)	(1,0)	(0,1)	(1,1)
a_1	b_1	1	0	0	1
a_1	b_2	1	0	0	1
a_2	b_1	1	0	0	1
a_2	b_2	0	1	1	0

Figure 11.6 The PR Box

measurement (e.g. $a_1 = 0$) is independent of whether Bob chooses measurement b_1 or b_2; and vice versa. That is, Alice and Bob cannot signal to one another—the importance of this principle is that it enforces compatibility with relativistic constraints. However, despite satisfying the no-signalling principle, the PR box does not admit a quantum realization. Note that the full support of this model corresponds to the propositions used in showing the contextuality of the Bell table from Figure 11.3, and hence the fact that these propositions are not simultaneously satisfiable shows the strong contextuality of the model.

In fact, there is provably no bipartite quantum-realizable behaviour of this kind which is strongly contextual (Lal, 2011; Mansfield, 2014). However, as soon as we go to three or more parties, strong contextuality does arise from entangled quantum states. A notable example is provided by the GHZ states (Greenberger et al., 1990). Thus, we have a strict hierarchy of strengths of contextuality,

$$\text{probabilistic} < \text{possibilistic} < \text{strong}$$

exemplified in terms of well-known examples from the quantum foundations literature as follows:

$$\text{Bell} < \text{Hardy} < \text{GHZ}.$$

11.7 Visualizing Contextuality

The tables which have appeared in our examples can be displayed in a visually appealing way which makes the fibred topological structure apparent, and forms an intuitive bridge to the formal development of the sheaf-theoretic ideas.

First, we look at the Hardy table from Fig. 11.5, displayed as a "bundle diagram" on the left of Figure 11.7. Note that all unspecified entries of the Hardy table are set to 1.

What we see in this representation is the **base space** of the variables a_1, a_2, b_1, b_2. There is an edge between two variables when they can be measured together. The pairs of co-measurable variables correspond to the rows of the table. In terms of quantum theory, these correspond to pairs of **compatible observables**. Above each vertex is a **fibre** of those values which can be assigned to the variable—in this example, 0 and 1 in each fibre. There is an edge between values in adjacent fibres precisely when the

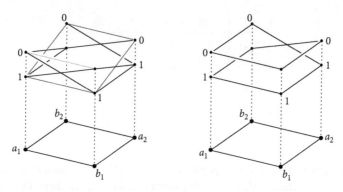

Figure 11.7 The Hardy table and the PR box as bundles

corresponding **joint outcome** is possible, i.e. has a 1 entry in the table. Thus there are three edges for each of the pairs $\{a_1, b_2\}$, $\{a_2, b_1\}$, and $\{a_2, b_2\}$.

Note that compatibility is expressed topologically by the fact that paths through the fibres can always be extended. That is, if we have an edge over $\{a, b\}$ then there must be an edge over $\{b, a'\}$ with a common value for b, and similarly an edge over $\{a, b'\}$ with a common value for a. This is a local condition.

A **global assignment** corresponds to a closed path traversing all the fibres exactly once. We call such a path **univocal** since it assigns a unique value to each variable. Note that there is such a path, which assigns 1100 to $a_1 b_1 a_2 b_2$; thus, the Hardy model is not strongly contextual. However, there is no such path which includes the edge 00 over $a_1 b_1$. This shows the possibilistic contextuality of the model.

Next, we consider the PR box displayed as a bundle on the right of Figure 11.7. In this case, the model is strongly contextual, and accordingly there is no univocal closed path. We can see that the PR box is a discrete version of a Möbius strip.

11.8 Contextuality, Logic, and Paradoxes

We return to our theme that contextuality arguments lie at the borders of paradox, but they do not cross those borders. We shall now show that a similar analysis can indeed be applied to some of the fundamental logical paradoxes.

11.8.1 Liar cycles

A Liar cycle of length N is a sequence of statements as shown in Figure 11.8.

For $N = 1$, this is the classic Liar sentence

$$S : S \text{ is false.}$$

These sentences contain two features which go beyond standard logic: references to other sentences, and a truth predicate. While it would be possible to make a more refined analysis directly modelling these features, we will not pursue this here,

$$S_1 : S_2 \text{ is true,}$$

$$S_2 : S_3 \text{ is true,}$$

$$\vdots$$

$$S_{N-1} : S_N \text{ is true,}$$

$$S_N : S_1 \text{ is false.}$$

Figure 11.8 The Liar cycle

noting that it has been argued extensively and rather compellingly in much of the recent literature on the paradoxes that the essential content is preserved by replacing statements with these features by **boolean equations** (Levchenkov, 2000; Wen, 2001; Cook, 2004; Walicki, 2009). For the Liar cycles, we introduce boolean variables $x_1, \ldots, x_n$, and consider the following equations:

$$x_1 = x_2, \quad \ldots, \quad x_{n-1} = x_n, \quad x_n = \neg x_1.$$

The "paradoxical" nature of the original statements is now captured by the inconsistency of these equations.

Note that we can regard each of these equations as fibred over the set of variables which occur in it:

$$\{x_1, x_2\} : \quad x_1 = x_2$$

$$\{x_2, x_3\} : \quad x_2 = x_3$$

$$\vdots$$

$$\{x_{n-1}, x_n\} : \quad x_{n-1} = x_n$$

$$\{x_n, x_1\} : \quad x_n = \neg x_1.$$

Any subset of up to $n - 1$ of these equations is consistent; while the whole set is inconsistent.

Up to rearrangement, the Liar cycle of length 4 corresponds exactly to the PR box. The usual reasoning to derive a contradiction from the Liar cycle corresponds precisely to the attempt to find a univocal path in the bundle diagram on the right of Figure 11.7. To relate the notations, we make the following correspondences between the variables of Figure 11.7 and those of the boolean equations:

$$x_1 \sim a_2, \quad x_2 \sim b_1, \quad x_3 \sim a_1, \quad x_4 \sim b_2.$$

Thus, we can read the equation $x_1 = x_2$ as "a_2 is correlated with b_1", and $x_4 = \neg x_1$ as "a_2 is anti-correlated with b_2".

Now suppose that we try to set a_2 to 1. Following the path in Figure 11.7 on the right leads to the following local propagation of values:

$$a_2 = 1 \leadsto b_1 = 1 \leadsto a_1 = 1 \leadsto b_2 = 1 \leadsto a_2 = 0$$
$$a_2 = 0 \leadsto b_1 = 0 \leadsto a_1 = 0 \leadsto b_2 = 0 \leadsto a_2 = 1.$$

The first half of the path corresponds to the usual derivation of a contradiction from the assumption that S_1 is true, and the second half to deriving a contradiction from the assumption that S_1 is false.

We have discussed a specific case here, but the analysis can be generalized to a large class of examples along the lines of Cook (2004) and Walicki (2009).

11.8.2 The Robinson Consistency Theorem

As a final remark on the connections between contextuality and logic, we consider a classic result, the Robinson Joint Consistency Theorem (Robinson, 1956). It is usually formulated in a first-order setting. The version we will use has the following form:

Theorem 11.8.1 (Robinson Joint Consistency Theorem) Let T_i be a theory over the language L_i, $i = 1, 2$. If there is no sentence ϕ in $L_1 \cap L_2$ with $T_1 \vdash \phi$ and $T_2 \vdash \neg\phi$, then $T_1 \cup T_2$ is consistent.

Thus, this theorem says that two compatible theories can be glued together. In this binary case, local consistency implies global consistency. Note, however, that an extension of the theorem beyond the binary case fails. That is, if we have three theories which are pairwise compatible, it need not be the case that they can be glued together consistently. A minimal counter-example is provided at the propositional level by the following "triangle":

$$T_1 = \{x_1 \leftrightarrow \neg x_2\}, \quad T_2 = \{x_2 \leftrightarrow \neg x_3\}, \quad T_3 = \{x_3 \leftrightarrow \neg x_1\}.$$

This example is well known in the quantum contextuality literature as the **Specker triangle** (Liang et al., 2011). Although not quantum realizable, it serves as a basic template for more complex examples which are. See Kochen and Specker (1967), Cabello et al. (1996), and Abramsky and Brandenburger (2011) for further details.

11.9 Perspectives and Further Directions

The aim of this article has been to give an exposition of the basic elements of the sheaf-theoretic approach to contextuality introduced by the present author and Adam Brandenburger in Abramsky and Brandenburger (2011), and subsequently developed extensively with a number of collaborators, including Rui Soares Barbosa, Shane Mansfield, Kohei Kishida, Ray Lal, Carmen Constantin, Nadish de Silva, Giovanni

Caru, Linde Wester, Lucien Hardy, Georg Gottlob, Phokion Kolaitis, and Mehrnoosh Sadrzadeh.

11.9.1 Some further developments in quantum information and foundations

- The sheaf-theoretic language allows a unified treatment of non-locality and contextuality, in which results such as Bell's theorem (Bell, 1964) and the Kochen–Specker theorem (Kochen and Specker, 1967) fit as instances of more general results concerning obstructions to global sections.
- A hierarchy of degrees of non-locality or contextuality is identified (Abramsky and Brandenburger, 2011). This explains and generalizes the notion of "inequality-free" or "probability-free" non-locality proofs, and makes a strong connection to logic (Abramsky, 2013b). This hierarchy is lifted to a novel classification of multipartite entangled states, leading to some striking new results concerning multipartite entanglement, which is currently poorly understood. In joint work with Carmen Constantin and Shenggang Ying, it is shown that, with certain bipartite exceptions, all entangled n-qubit states are "logically contextual", admitting a Hardy-style contextuality proof. Moreover, the local observables witnessing logical contextuality can be computed from the state (Abramsky et al., 2016).
- The obstructions to global sections witnessing contextuality are characterized in terms of sheaf cohomology in joint work with Shane Mansfield and Rui Barbosa (Abramsky et al., 2012), and a range of examples are treated in this fashion. In later work with Shane Mansfield, Rui Barbosa, Kohei Kishida, and Ray Lal (Abramsky et al., 2015), the cohomological approach is carried further, and shown to apply to a very general class of "All-versus-Nothing" arguments for contextuality, including a large class of quantum examples arising from the stabilizer formalism.
- A striking connection between no-signalling models and global sections with signed measures ("negative probabilities") is established in joint work with Adam Brandenburger (Abramsky and Brandenburger, 2011). An operational interpretation of such negative probabilities, involving a signed version of the strong law of large numbers, has also been developed (Abramsky and Brandenburger, 2014).

11.9.2 Logical Bell inequalities

Bell inequalities are a central technique in quantum information. The discussion in Section 11.9.3 is based on joint work with Lucien Hardy (Abramsky and Hardy, 2012), in which a general notion of "logical Bell inequality", based on purely logical consistency conditions, is introduced, and it is shown that every Bell inequality (i.e. every inequality satisfied by the "local polytope") is equivalent to a logical Bell inequality. The notion is developed at the level of generality of the sheaf-theoretic framework

(Abramsky and Brandenburger, 2011), and hence applies to arbitrary contextuality scenarios, including multipartite Bell scenarios and Kochen–Specker configurations.

11.9.3 Contextual semantics in classical computation

The generality of the approach, in large part based on the use of category-theoretic tools, has been used to show how contextuality phenomena arise in classical computation.

- An isomorphism between the basic concepts of quantum contextuality and those of relational database theory is shown in Abramsky (2013a).
- Connections between non-locality and logic have been developed (Abramsky, 2013b). A number of natural complexity and decidability questions are raised in relation to non-locality.
- Our discussion of the Hardy paradox in Section 11.5 showed that the key issue was that a local section (assignment of values) could not be extended to a global one consistently with some constraints (the "support table"). This directly motivated some joint work with Georg Gottlob and Phokion Kolaitis (Abramsky et al., 2013), in which we studied a refined version of **constraint satisfaction**, dubbed "robust constraint satisfaction", in which one asks if a partial assignment of a given length can always be extended to a solution. The tractability boundary for this problem is delineated, and this is used to settle one of the complexity questions previously posed in Abramsky (2013b).
- Application of the contextual semantics framework to natural language semantics was initiated in joint work with Mehrnoosh Sadrzadeh (Abramsky and Sadrzadeh, 2014). In this paper, a basic part of the Discourse Representation Structure framework (Kamp and Reyle, 1993) is formulated as a presheaf, and the gluing of local sections into global ones is used to represent the resolution of anaphoric references.

Other related work includes Mansfield (2013); Barbosa (2014); Mansfield and Barbosa (2014); Barbosa (2015); Constantin (2015); de Silva (2015); Hyttinen et al. (2015); Abramsky et al. (2016); Kishida (2016); Raussendorf (2016).

11.9.4 Envoi

It has been suggested that complex systems dynamics can emerge most fruitfully at the "edge of chaos" (Langton, 1990; Waldrop, 1993). The range of contextual behaviours and arguments we have studied and shown to have common structure suggests that a rich field of phenomena in logic and information, closely linked to key issues in the foundations of physics, arise at the borders of paradox.

References

Abramsky, S. (2013a). Relational databases and Bell's theorem, in V. Tannen, L. Wong, L. Libkin, W. Fan, W. Tan and M. Fourman (eds), *In Search of Elegance in the Theory and Practice of Computation: Essays Dedicated to Peter Buneman*, 13–35. Springer, Berlin.

Abramsky, S. (2013b). Relational hidden variables and non-locality. *Studia Logica* 101(2), 411–52.

Abramsky, S. (2015). Contextual semantics: from quantum mechanics to logic, databases, constraints, and complexity, in E. Dzhafarov, S. Jordan, R. Zhang and V. Cervantes (eds), *Contextuality from Quantum Physics to Psychology*, 23–50. World Scientific, Singapore.

Abramsky, S., Barbosa, R. S., Kishida, K., Lal, R. and Mansfield, S. (2015). Contextuality, cohomology and paradox, in *24th EACSL Annual Conference on Computer Science Logic, CSL 2015, September 7–10, 2015, Berlin, Germany*, 211–28. http://dx.doi.org/10.4230/LIPIcs.CSL.2015.211

Abramsky, S., Barbosa, R. S., Kishida, K., Lal, R., and Mansfield, S. (2016). Possibilities determine the combinatorial structure of probability polytopes. *Journal of Mathematical Psychology* 74, 58–65.

Abramsky, S., and Brandenburger, A. (2011). The sheaf-theoretic structure of non-locality and contextuality. *New Journal of Physics* 13(11), 113036.

Abramsky, S., and Brandenburger, A. (2014). An operational interpretation of negative probabilities and no-signalling models, in F. van Breugel, E. Kashefi, C. Palamidessi and J. Rutten (eds), *Horizons of the Mind: A Tribute to Prakash Panagaden*, 59–75. Springer, Berlin.

Abramsky, S., Constantin, C. M., and Ying, S. (2016). Hardy is (almost) everywhere: nonlocality without inequalities for almost all entangled multipartite states. *Information and Computation* 250, 3–14.

Abramsky, S., Gottlob, G., and Kolaitis, P. G. (2013). Robust constraint satisfaction and local hidden variables in quantum mechanics, in F. Rossi (ed.), *Proceedings of the Twenty-Third IJCAI*, 440–6. AAAI Press, Palo Alto, CA.

Abramsky, S., and Hardy, L. (2012). Logical Bell inequalities. *Physical Review A* 85(6), 062114.

Abramsky, S., Mansfield, S., and Barbosa, R. (2012). The cohomology of non-locality and contextuality. *Electronic Proceedings in Theoretical Computer Science* 95, 1–15.

Abramsky, S., and Sadrzadeh, M. (2014). Semantic unification: a sheaf theoretic approach to natural language, in C. Casadio, B. Coecke, M. Moortgat and P. Scott (eds), *Categories and Types in Logic, Language, and Physics, A Festshrift for Jim Lambek*, Vol. 8222 of *Lecture Notes in Computer Science*, 1–13. Springer, Berlin.

Aspect, A. (1999). Bell's inequality test. *Nature* 398(6724), 189–90.

Aspect, A., Dalibard, J., and Roger, G. (1982). Experimental test of Bell's inequalities using time-varying analyzers. *Physical Review Letters* 49(25), 1804–7.

Barbosa, R. S. (2014). On monogamy of non-locality and macroscopic averages: examples and preliminary results. arXiv preprint arXiv:1412.8541.

Barbosa, R. S. (2015). Contextuality in quantum mechanics and beyond. DPhil thesis, University of Oxford.

Barrett, J., Hardy, L., and Kent, A. (2005). No signaling and quantum key distribution. *Physical Review Letters* 95(1), 010503.

Bartosik, H., Klepp, J., Schmitzer, C., Sponar, S., Cabello, A., Rauch, H., and Hasegawa, Y. (2009). Experimental test of quantum contextuality in neutron interferometry. *Physical Review Letters* 103(4), 40403.

Bell, J. (1985). An exchange on local beables. *Dialectica* 39, 85–96.

Bell, J. S. (1964). On the Einstein-Podolsky-Rosen paradox. *Physics* 1(3), 195–200.

Brandenburger, A., and Yanofsky, N. (2008). A classification of hidden-variable properties. *Journal of Physics A: Mathematical and Theoretical* 41, 425302.

Cabello, A., Estebaranz, J. M., and García-Alcaine, G. (1996). Bell-Kochen-Specker theorem. *Physical Letters A* 212(4), 183–7.

Constantin, C. M. (2015). Sheaf-theoretic methods in quantum mechanics and quantum information theory. arXiv preprint arXiv:1510.02561.

Cook, R. T. (2004). Patterns of paradox. *Journal of Symbolic Logic* 69(03), 767–74.

de Silva, N. (2015). Unifying frameworks for nonlocality and contextuality. arXiv preprint arXiv:1512.05048.

Ghirardi, G.-C., Rimini, A., and Weber, T. (1980). A general argument against superluminal transmission through the quantum mechanical measurement process. *Lettere Al Nuovo Cimento Series 2* 27(10), 293–8.

Giry, M. (1982). A categorical approach to probability theory, in *Categorical Aspects of Topology and Analysis*, 68–85. Springer, Berlin.

Giustina, M., Versteegh, M. A., Wengerowsky, S., Handsteiner, J., Hochrainer, A., Phelan, K., Steinlechner, F., Kofler, J., Larsson, J.-Å., Abellán, C., et al. (2015). Significant-loophole-free test of bells theorem with entangled photons. *Physical Review Letters* 115(25), 250401.

Greenberger, D. M., Horne, M. A., Shimony, A., and Zeilinger, A. (1990). Bell's theorem without inequalities. *American Journal of Physics* 58(12), 1131–43.

Hardy, L. (1993). Nonlocality for two particles without inequalities for almost all entangled states. *Physical Review Letters* 71(11), 1665–8.

Hensen, B., Bernien, H., Dréau, A., Reiserer, A., Kalb, N., Blok, M., Ruitenberg, J., Vermeulen, R., Schouten, R., Abellán, C., et al. (2015). Loophole-free bell inequality violation using electron spins separated by 1.3 kilometres. *Nature* 526(7575), 682–6.

Howard, M., Wallman, J., Veitch, V., and Emerson, J. (2014). Contextuality supplies the 'magic' for quantum computation. *Nature* 510(7505), 351–5.

Hyttinen, T., Paolini, G., and Väänänen, J. (2015). Quantum team logic and bells inequalities. *Review of Symbolic Logic* 8(04), 722–42.

Kamp, H., and Reyle, U. (1993). *From Discourse to Logic: Introduction to Model-Theoretic Semantics of Natural Language, Formal Logic and Discourse Representation Theory*. Springer, Berlin.

Kashiwara, M., and Schapira, P. (2005). *Categories and Sheaves*, Vol. 332. Springer Science & Business Media, Berlin.

Kirchmair, G., Zähringer, F., Gerritsma, R., Kleinmann, M., Gühne, O., Cabello, A., Blatt, R., and Roos, C. F. (2009). State-independent experimental test of quantum contextuality. *Nature* 460(7254), 494–7.

Kishida, K. (2016). Logic of local inference for contextuality in quantum physics and beyond. arXiv preprint arXiv:1605.08949.

Kochen, S., and Specker, E. P. (1967). The problem of hidden variables in quantum mechanics. *Journal of Mathematics and Mechanics* 17(1), 59–87.

Lal, R. (2011). A sheaf-theoretic approach to cluster states. Private communication.

Langton, C. G. (1990). Computation at the edge of chaos: phase transitions and emergent computation. *Physica D: Nonlinear Phenomena* 42(1), 12–37.

Levchenkov, V. (2000). Boolean equations with many unknowns. *Computational Mathematics and Modeling* 11(2), 143–53.

Liang, Y.-C., Spekkens, R. W., and Wiseman, H. M. (2011). Specker's parable of the overprotective seer. *Physics Reports* 506(1), 1–39.

Mac Lane, S., and Moerdijk, I. (1992). *Sheaves in Geometry and Logic*. Springer, Berlin.

Mansfield, S. (2013). The mathematical structure of non-locality and contextuality. PhD thesis, University of Oxford.

Mansfield, S. (2014). Completeness of Hardy non-locality: consequences & applications, in *Informal Proceedings of 11th International Workshop on Quantum Physics & Logic*.

Mansfield, S., and Barbosa, R. S. (2014). Extendability in the sheaf-theoretic approach: construction of Bell models from Kochen-Specker models. arXiv preprint arXiv:1402.4827.

Mermin, N. (1994). Quantum mysteries refined. *American Journal of Physics* 62, 880.

Penrose, R. (1992). On the cohomology of impossible figures. *Leonardo* 25(3/4), 245–7.

Popescu, S., and Rohrlich, D. (1994). Quantum nonlocality as an axiom. *Foundation of Physics* 24(3), 379–85.

Priest, G. (2002). Paraconsistent logic, in *Handbook of Philosophical Logic*, 287–393. Springer, Berlin.

Raussendorf, R. (2013). Contextuality in measurement-based quantum computation. *Physical Review A* 88, 022322. http://link.aps.org/doi/10.1103/PhysRevA.88.022322

Raussendorf, R. (2016). Cohomological framework for contextual quantum computations. arXiv preprint arXiv:1602.04155.

Robinson, A. (1956). A result on consistency and its application to the theory of definition. *Indagationes Mathematicae* 18(1), 47–58.

Shalm, L. K., Meyer-Scott, E., Christensen, B. G., Bierhorst, P., Wayne, M. A., Stevens, M. J., Gerrits, T., Glancy, S., Hamel, D. R., Allman, M. S. et al. (2015). Strong loophole-free test of local realism. *Physical Review Letters* 115(25), 250402.

Waldrop, M. M. (1993). *Complexity: The Emerging Science at the Edge of Order and Chaos*. Simon and Schuster, New York.

Walicki, M. (2009). Reference, paradoxes and truth. *Synthese* 171(1), 195–226.

Wen, L. (2001). Semantic paradoxes as equations. *Mathematical Intelligencer* 23(1), 43–8.

Zu, C., Wang, Y.-X., Deng, D.-L., Chang, X.-Y., Liu, K., Hou, P.-Y., Yang, H.-X., and Duan, L.-M. (2012). State-independent experimental test of quantum contextuality in an indivisible system. *Physical Review Letters* 109(15), 150401.

12
Categorical Quantum Mechanics I: Causal Quantum Processes

Bob Coecke and Aleks Kissinger

12.1 Introduction

This chapter is the first of a three-part overview on categorical quantum mechanics (CQM), an area of applied category theory that over the past twelve years or so has become increasingly prominent within physics, mathematics, and computer science, and even has spin-offs in other areas such as computational linguistics. Probably the most appealing feature of CQM is the use of diagrams, which are related to the usual Hilbert space model via symmetric monoidal categories and structures therein. However, we have written this overview in such a way that no prior knowledge on category theory is required. In fact, it can be seen as a first encounter with the relevant parts of category theory.

We start with boxes and wires, which together make up diagrams. The wires stand for systems, and the boxes stand for processes. Symmetric monoidal categories then arise when endowing diagrams with operations of sequential and parallel composition. There are very good reasons to start with diagrams, rather than with traditional category-theoretic axioms, one being that the set-theoretic underpinning of category theory invokes an additional level of bureaucracy, namely dealing with such details as the bracketing of expressions, which has no counterpart in the reality that one aims to describe. In other words, the traditional symbolic presentation of monoidal categories suffers from a substantial overhead as compared to its diagrammatic counterpart, to the extent that monoidal category theory itself becomes much simpler if one takes diagrams as a starting point. From this perspective, the role of the traditional presentation of monoidal categories is reduced to providing a bridge to standard mathematical models, for example, the presentation of quantum theory using Hilbert space.

Next we define very general *quantum types*. Despite this generality, we are able to derive important results that characterize the behaviour of quantum systems, most notably, the *no-broadcasting theorem* (Barnum et al., 1996). The diagrammatic formalism also makes it remarkably easy to assert compliance with the theory of relativity, namely, in terms of a *causality postulate* (Chiribella et al., 2010), which, in category-theoretic terms, simply boils down to the tensor unit being terminal (Coecke, 2014). Again, important results follow straightforwardly, such as *unitarity of evolution* and *Stinespring dilation*.

12.1.1 Other overviews, surveys and tutorials on CQM

This three-part overview has a big brother, namely, a forthcoming textbook (Coecke and Kissinger, 2016c) by the same authors. This three-part overview, which amounts to some 70 pages, strives to be self-contained, but the book (which is about 12 times that size) will provide a more comprehensive introduction suitable for students and researchers in a wide variety of disciplines and levels of experience.

There is another forthcoming book (Heunen and Vicary, 2016), but rather than providing a complete presentation of quantum theory from first principles, it puts many of the core aspects of CQM studied in earlier papers in one place, and also emphasizes how the structures used in CQM appear in other areas of mathematics.

A tutorial that provides a pedestrian introduction to the relevant category theory for CQM, and is complementary to this three-part one, is Coecke and Paquette (2011).

There have been shorter overviews on CQM before in the spirit of this one, e.g. Coecke (2005, 2009), but the material simply wasn't ripe enough at that time for a fully comprehensive presentation of quantum theory.

12.1.2 Comparison to previous versions of CQM

The most prominent difference between this overview and the way CQM has been presented in the past is the fact that we take diagrams as our starting point, rather than standard category-theoretic structures and consistently introduce all new ingredients of the formalism in those terms. While process ontology has played a motivational role throughout the entire CQM endeavour, see e.g. Coecke (2011), only now does every single ingredient emerge from this ontology. This idea to start with diagrams as a primitive notion, even when a symbolic alternative is available, has also been advocated by Hardy (2011).

As compared to the paper by Abramsky and Coecke that initiated CQM (Abramsky and Coecke, 2004), a now well-established difference is that we no longer rely on biproducts (a non-diagrammatic concept) for the description of classical types. They have been superseded by special processes called *spiders*, which allow one to express the characteristic processes associated with classical data (most importantly: 'copy' and 'delete') and reason about them diagrammatically (Coecke and Pavlovic, 2007; Coecke et al., 2010).

A more recent innovation is the central role now played by Chiribella, D'Ariano, and Perinotti's causality postulate (Chiribella et al., 2010). In the past, issues of normalization were mostly ignored in CQM. Only recently it became apparent that the type of normalization captured in the causality postulate is something structurally very fundamental to processes which are physically realizable. Notably, all the results in Section 12.5 of this chapter rely on it.

12.2 Process Theories

> *The art of progress is to preserve order amid change, and to preserve change amid order.*
>
> Alfred North Whitehead, *Process and Reality* (1929)

After his attempt to complete the set-theoretic foundations of mathematics in collaboration with Russell, Whitehead's venture into the natural sciences made him realize that the traditional timeless ontology of substances, and not in the least their static set-theoretic underpinning, does not suit natural phenomena. Instead, he claims, it is processes and their relationships which should underpin our understanding of those phenomena.

Can one turn this stance into a formal underpinning for natural sciences? Category theory is a big step in that direction, but falls short of shaking off its set-theoretic shackles.

12.2.1 Processes as diagrams

We shall use the term *process* to refer to anything that has zero or more inputs and zero or more outputs. For instance, the function

$$f : \mathbb{R} \times \mathbb{R} \to \mathbb{R} :: (x, y) \mapsto x^2 + y \tag{12.1}$$

is a process which takes two real numbers as input and produces one real number as output. We represent such a process as a *box* with some *wires* coming in the bottom to represent input systems and some wires coming out the top to represent output systems. For example, we could write the function (12.1) like this:

 (12.2)

The labels on wires are called *system-types* or simply *types*.

Similarly, a computer program is a process which takes some data (e.g. from memory) as input and produces some new data as output. For example, a program that sorts lists might look like this:

The following are also perfectly good processes:

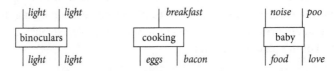

Clearly, the world around us is packed with processes!

We can *wire together* simple processes to make more complicated processes, which are described by *diagrams*:

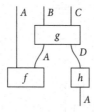

We form diagrams by plugging the outputs of some processes into the inputs of others. This is allowed only if the types of the output and the input match. For example, the following two processes:

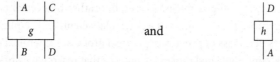

can be connected in some ways, but not in others, depending on the types of their wires:

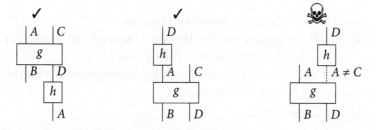

This restriction on which wirings are allowed is an essential part of the language of diagrams, in that it tells us when it makes sense to apply a process to a certain system and prevents occurrences like this:

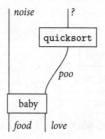

which probably wouldn't be very good for your computer!

One thing to note is we haven't yet been too careful to say what a diagram actually *is*. A complete description of a diagram consists of:

1. what boxes it contains, and
2. how those boxes are connected.

So the diagram refers to the 'drawing' of boxes and wires without the interpretation of the diagram as a process. However, it makes no reference to where boxes are written on the page. Hence, if two diagrams can be deformed into each other (without changing connections of course) then they are equal:

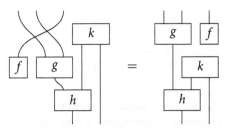

(12.3)

When we think about interpreting the boxes in a diagram as processes, we are usually not interested in all possible processes, but rather in a certain class of related processes. For example, practicioners of a particular scientific discipline will typically only study a particular class of processes: physical processes, chemical processes, biological processes, computational processes, etc. For that reason, we organize processes into *process theories*.

Definition 12.2.1 A *process theory* consists of:

(i) a collection T of *system-types* represented by wires,
(ii) a collection P of *processes* represented by boxes, where for each process in P the input types and output types are taken from T, and
(iii) a means of 'wiring processes together'; that is, an operation that interprets a diagram of processes in P as a process in P.

In particular, (iii) guarantees that

process theories are 'closed under wiring processes together',

since it is this operation that tells us what 'wiring processes together' means. In some cases this operation consists of literally plugging things together with physical wires, like in the theory of **electrical devices**:

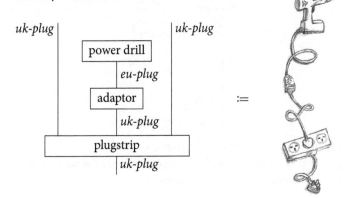

In other cases this will require some more work, and sometimes there is more than one obvious choice available. We shall see in the following that in traditional mathematical practice one typically breaks down 'wiring processes together' in two sub-operations: parallel composition and sequential composition of processes.

According to Definition 12.2.1 a process theory tells us how to *interpret* the boxes and wires in a diagram. But, crucially, by doing so it also tells us which diagrams consisting of those processes should be considered *equal*.

For example, suppose we define a simple process theory for **computer programs**, where the types are data-types (e.g. integers, booleans, lists) and the processes are computer programs. Then, consider a short program which takes a list as input and sorts it. It might be defined this way (don't worry if you can't read the code, neither can half of the authors):

$$\boxed{\text{quicksort}} \quad := \quad \begin{cases} \text{qs [] = []} \\ \text{qs (x :: xs) =} \\ \quad \text{qs [y | y <- xs; y < x] ++ [x] ++} \\ \quad \text{qs [y | y <- xs; y >= x]} \end{cases}$$

Wiring together programs means sending the output of one program to the input of another program. Taking two programs to be equal if they behave the same (disregarding some details like execution time, etc.), our process theory yields equations like this one:

$$\boxed{\text{quicksort}} \atop \boxed{\text{quicksort}} \quad = \quad \boxed{\text{quicksort}}$$

i.e. sorting a list twice has the same effect as sorting it once. Unlike Eq. (12.3), this is not just a matter of deforming one diagram into another, but actually represents a non-trivial equation between processes.

The reason we call a process theory a *theory* is that it comes with lots of such equations, and these equations are precisely what allows us to draw conclusions about the processes we are studying. Another thing one expects from a theory is that it makes predictions. In particular, one usually expects a theory to produce some numbers (e.g. probabilities) that can verified by experiments. Toward that end, we first identify two special kinds of processes:

- *States* are processes without any inputs. In operational terms they are 'preparation procedures'. We represent them as follows:

- *Effects* are processes without any outputs. We have borrowed this terminology from quantum theory, where effects play a key role. From now on we represent them as follows:

When composing a state and an effect a third special kind of process arises which neither has inputs nor outputs, called a *number*. Every process theory has at least one number, given by the 'empty diagram':

$$1 := \quad \boxed{}$$

We label this number '1' because combining the empty diagram with any other process yields the same process again.

It is natural to interpret the number arising from the composition of a state and an effect as the *probability* that, given the system is in that state, the effect happens:

$$\left. \begin{array}{r} \text{test} \left\{ \triangle_\pi \right. \\ \text{state} \left\{ \nabla_\psi \right. \end{array} \right\} \text{probability.}$$

We refer to this procedure as the *generalized Born rule*.

In the Introduction we announced that we would introduce category-theoretic definitions after we introduced the relevant diagrammatic notion. However, the simple notion of a diagram introduced above has a not-so-simple category-theoretic description. We will now first introduce a more sophisticated notion of a diagram, which actually has a simpler category-theoretic counterpart, and this will serve as a stepping stone to defining the category-theoretic counterpart to the diagrams considered above.

12.2.2 Circuit diagrams

Given that boxes represent processes, we can define two basic *composition operations* on processes with the interpretations

$$f \otimes g := \text{'process } f \text{ takes place \underline{while} process } g \text{ takes place'}$$
$$f \circ g := \text{'process } f \text{ takes place \underline{after} process } g \text{ takes place'}.$$

The *parallel composition* operation, written '$\otimes$', consists of placing a pair of diagrams side-by-side:

Any two diagrams can be composed in this manner, since placing diagrams side by side does not involve connecting anything. This reflects the fact that both processes are happening independently of each other.

This composition operation is associative:

$$\left(\;\boxed{f}\otimes\boxed{g}\;\right)\otimes\boxed{h}\;=\;\boxed{f}\;\boxed{g}\;\boxed{h}\;=\;\boxed{f}\otimes\left(\;\boxed{g}\otimes\boxed{h}\;\right) \quad (12.4)$$

and it has a unit, the empty diagram:

$$\boxed{f}\otimes\quad=\quad\otimes\boxed{f}\;=\;\boxed{f} \quad (12.5)$$

Parallel composition is defined for system-types as well. That is, for types A and B, we can form a new type $A \otimes B$, called the *joint system-type*:

$$A \otimes B \;:=\; A \quad B$$

There is also a special 'empty' system-type, symbolically denoted I, which is used to represent 'no inputs', 'no ouputs', or both.

The *sequential composition* operation, written '$\circ$', consists of connecting the outputs of one diagram to the inputs of another diagram:

$$\left(\;\begin{array}{c}E\;F\\ \boxed{g}\\ \boxed{f}\\ C\;D\end{array}\;\right)\circ\left(\;\begin{array}{c}C\;D\\ \boxed{a}\;\boxed{b}\\ A\;B\end{array}\;\right)\;=\;\begin{array}{c}E\;F\\ \boxed{g}\\ \boxed{f}\\ \boxed{a}\;\boxed{b}\\ A\;B\end{array}$$

In other words, the process on the right happens first, followed by the process on the left, taking the output of the first process as its input. Clearly not any pair of diagrams can be composed in this manner: the number and type of the inputs of the left process must match the number and type of the outputs of the right process.

The sequential composition operation is also associative:

$$\left(\;\boxed{h}\circ\boxed{g}\;\right)\circ\boxed{f}\;=\;\begin{array}{c}h\\ g\\ f\end{array}\;=\;\boxed{h}\circ\left(\;\boxed{g}\circ\boxed{f}\;\right) \quad (12.6)$$

and it also has a unit. This time, it's a plain wire of appropriate type:

$$\left| \quad \circ \quad \begin{array}{c} |B \\ \boxed{f} \\ |A \end{array} \right. = \begin{array}{c} |B \\ \boxed{f} \\ |A \end{array} \circ \left| \quad = \begin{array}{c} |B \\ \boxed{f} \\ |A \end{array} \right. \tag{12.7}$$

These two composition operations, due to their diagrammatic origin, also obey an *interchange law*. Since we have

$$\left(\boxed{g_1} \otimes \boxed{g_2} \right) \circ \left(\boxed{f_1} \otimes \boxed{f_2} \right) = \left(\boxed{g_1} \; \boxed{g_2} \right) \circ \left(\boxed{f_1} \; \boxed{f_2} \right) = \begin{array}{cc} \boxed{g_1} & \boxed{g_2} \\ \boxed{f_1} & \boxed{f_2} \end{array}$$

$$\left(\boxed{g_1} \circ \boxed{f_1} \right) \otimes \left(\boxed{g_2} \circ \boxed{f_2} \right) = \left(\begin{array}{c} \boxed{g_1} \\ \boxed{f_1} \end{array} \right) \otimes \left(\begin{array}{c} \boxed{g_2} \\ \boxed{f_2} \end{array} \right) = \begin{array}{cc} \boxed{g_1} & \boxed{g_2} \\ \boxed{f_1} & \boxed{f_2} \end{array}$$

it follows that

$$\left(\boxed{g_1} \otimes \boxed{g_2} \right) \circ \left(\boxed{f_1} \otimes \boxed{f_2} \right) = \left(\boxed{g_1} \circ \boxed{f_1} \right) \otimes \left(\boxed{g_2} \circ \boxed{f_2} \right) \tag{12.8}$$

Note that ∘ assumes there is some ordering on the input/output wires, and plugs them together 'in order'. We can then express different orders by means of two wires crossing over each other:

which as called a *swap*.

Definition 12.2.2 A diagram is a *circuit* if it can be constructed by composing boxes, including identities and swaps, by means of ⊗ and ∘.

Every diagram that we have seen so far in this chapter is in fact a circuit. Here is an example of the assembly of such a circuit:

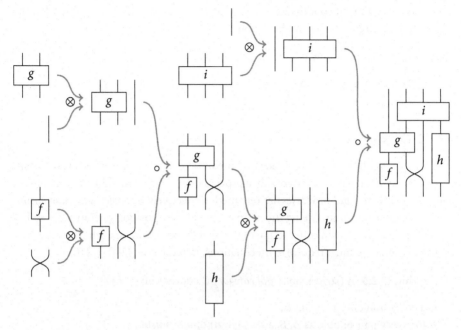

But not all diagrams are circuits. To understand which ones are, we provide an equivalent characterization that doesn't refer to the manner that one can build these diagrams, but in terms of a property that must be satisfied.

Definition 12.2.3 A *directed path* of wires is a list of wires $(w_1, w_2, \ldots, w_n)$ in a diagram such for all $i < n$, the wire w_i is an input to some box for which the wire w_{i+1} is an output. A *directed cycle* is a directed path that starts and ends at the same box.

An example of a directed path is shown in bold here:

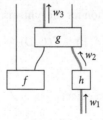

and an an example of a directed cycle is:

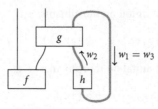

While diagrams only allow inputs to be connected to outputs, we indeed have done nothing (so far) to rule out the existence of directed cycles. This is precisely what the restricting to 'circuit diagrams' does for us:

Theorem 12.2.4 The following are equivalent:

- a diagram is a circuit, and,
- it contains no directed cycles.

12.2.3 Category-theoretic counterpart

We are now ready to provide a category-theoretic counterpart to diagrams. A *symmetric monoidal category* is essentially a process theory where all diagrams are circuit diagrams. Systems and processes are renamed, respectively, as *objects* and *morphisms*, and rather than diagrams being the main actor, the composition operations $\circ$ and $\otimes$ are taken as primitive. In order to guarantee that these operations behave as they did with diagrams, we must require extra equations, namely Eqs (12.4)–(12.8).

Definition 12.2.5 A (strict) *monoidal category* $\mathcal{C}$ consists of:

- a collection $\mathrm{ob}(\mathcal{C})$ of *objects*,
- for every pair of objects A, B, a set $\mathcal{C}(A, B)$ of *morphisms*,
- for every object A, a special *identity morphism* $1_A \in \mathcal{C}(A, A)$,
- a *sequential composition* operation for morphisms:

$$(- \circ -) : \mathcal{C}(B, C) \times \mathcal{C}(A, B) \to \mathcal{C}(A, C)$$

- a *parallel composition* operation for objects:

$$(- \otimes -) : \mathrm{ob}(\mathcal{C}) \times \mathrm{ob}(\mathcal{C}) \to \mathrm{ob}(\mathcal{C})$$

- a *unit object* $I \in \mathrm{ob}(\mathcal{C})$, and
- a *parallel composition* operation for morphisms:

$$(- \otimes -) : \mathcal{C}(A, B) \times \mathcal{C}(C, D) \to \mathcal{C}(A \otimes C, B \otimes D)$$

such that:

- $\otimes$ is associative and unital on objects:

$$(A \otimes B) \otimes C = A \otimes (B \otimes C) \qquad A \otimes I = A = I \otimes A$$

- $\otimes$ is associative and unital on morphisms:

$$(f \otimes g) \otimes h = f \otimes (g \otimes h) \qquad f \otimes 1_I = f = 1_I \otimes f$$

- $\circ$ is associative and unital on morphisms:

$$(h \circ g) \circ f = h \circ (g \circ f) \qquad 1_B \circ f = f = f \circ 1_A$$

- ⊗ and ∘ satisfy the *interchange law*:

$$(g_1 \otimes g_2) \circ (f_1 \otimes f_2) = (g_1 \circ f_1) \otimes (g_2 \circ f_2).$$

Note we often write $f : A \to B$ as shorthand for $f \in \mathcal{C}(A, B)$, which indicates that we are thinking of morphisms as maps of some kind.

Definition 12.2.6 A *symmetric monoidal category* (SMC) is a monoidal category with a swap morphism:

$$\sigma_{A,B} : A \otimes B \to B \otimes A$$

defined for all objects A, B, satisfying:

- $\sigma_{B,A} \circ \sigma_{A,B} = 1_{A \otimes B}$
- $(f \otimes g) \circ \sigma_{A,B} = \sigma_{B',A'} \circ (g \otimes f)$
- $\sigma_{A,I} = 1_A$
- $(1 \otimes \sigma_{A,C}) \circ (\sigma_{A,B} \otimes 1_C) = \sigma_{A,B \otimes C}$.

The diagrammatic counterparts to the first two of these equations are

which are simply diagram deformations. The remaining two equations are again tautologies in terms of diagrams:

It goes without saying that the category-theoretic definition is much more involved than its diagrammatic counterpart. And in fact, it gets worse when we drop the 'strict' from it, as explained in Coecke and Paquette (2011, § 3.4.4). Simply stated, dropping strictness allows for some of the defining equations not to hold on-the-nose, but are allowed some 'wiggle room'. However, making this precise requires a number of *natural isomorphisms*, which express 'how' one 'wiggles' from the left-hand side (LHS) to the right-hand side (RHS), together with a bunch of *coherence conditions* which force all the natural isomorphisms to fit together well. The reason why it does make sense (and is in fact necessary) to consider this very involved definition is that pretty much all set-theoretic structures which organize themselves into monoidal categories are in fact of the non-strict variety. However, there is a standard procedure for turning every non-strict monoidal category into an equivalent strict one, so considering only strict monoidal categories (as we shall do throughout these papers) yields no real loss of generality.

Example 12.2.7 The category FHilb, whose objects are finite-dimensional Hilbert spaces and whose morphisms, are linear maps forms a (non-strict) SMC. Sequential composition is just the usual composition of linear maps, parallel composition is tensor product, and the unit object is the 1-dimensional Hilbert space $\mathbb{C}$. The category Hilb of all Hilbert spaces and bounded linear maps also forms an SMC. However, it does not form a compact closed category, which we'll define in the next section, so in categorical quantum mechanics, we typically focus on FHilb.

The precise connection between SMCs and circuit diagrams is the following:

Theorem 12.2.8 Any circuit diagram can be interpreted as a morphism in an SMC, and two morphisms are equal according to the axioms of an SMC if and only if their circuit diagrams are equal.

We won't say that much about the categories that correspond to arbitrary diagrams, since they will be superseded by the notions introduced in the next section. The main point is that one must adjoin an additional operation, called *trace*, for every triple A, B, C of objects:

$$\text{tr}_{B,C}^A : \mathcal{C}(A \otimes B, A \otimes C) \to \mathcal{C}(B, C).$$

This trace obeys a bunch of axioms that guarantee the following counterpart diagrammatic behaviour:

$$\text{tr}_{B,C}^A ::$$

While such a trace does play a role in quantum theory, we will soon see that it arises as a derived concept in a simpler type of category.

12.2.4 Reference and further reading

The use of diagrams started with Penrose's diagrammatic calculus for abstract tensor systems (Penrose, 1971). The proof that abstract tensor systems characterize the free traced symmetric monoidal category was given by Kissinger (2014).

Monoidal categories are due to Benabou (1963), with their modern formulation—and the fact that any monoidal category is equivalent to a strict monoidal category—being worked out by Mac Lane (1963). The connection between circuit diagrams and symmetric monoidal categories was established by Joyal and Street (1991), where they are referred to as 'progressive diagrams'.

12.3 String Diagrams

> When two systems, of which we know the states by their respective representatives, enter into temporary physical interaction due to known forces between them, and when after a time of mutual influence the systems separate again, then they can no longer be described in the same way as before, viz. by endowing each of them with a representative of its own. I would not call that one but rather the characteristic trait of quantum mechanics, the one that enforces its entire departure from classical lines of thought.
>
> Erwin Schrödinger (1935)

By 1935, Schrödinger had already realized that the biggest gulf between quantum theory and our classical ways of thinking was really that, when it comes to quantum systems, the whole is more than the sum of its parts. In the classical world, for instance, it is possible to totally describe the state of two systems—say . . . two objects sitting on a table—by first totally describing the state of the first object then totally describing the state of the second object. This is a fundamental property one expects of a classical, or *separable* (or, in category-theory lingo: *Cartesian*) universe. However, as Schrödinger points out, there exist states predicted by quantum theory (and observed in the lab!) which do not obey this 'obvious' law about the physical world. Schrödinger called this new, totally non-classical phenomenon *verschränkung*, which later became translated to the dominant scientific language as *entanglement*.

In contrast to this great insight, it took physicists many years to actually exploit this property and reveal the features of the quantum world that are direct consequences of it. Most strikingly, it took physicists some 60 years to discover what is probably the most direct consequence: *quantum teleportation*.

12.3.1 Non-separability

Definition 12.3.1 *String diagrams* can be defined equivalently as:

- diagrams consisting of boxes and wires, where we additionally allow inputs to be connected to inputs and outputs to be connected to outputs, for example:

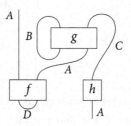

- circuit diagrams that contain a special state and a special effect:

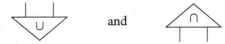

  and

for each type for which we have:

$$(12.9)$$

We can relate these two equivalent definitions by writing the cup state and cap effect from the second definition as cup- and cap-shaped pieces of wire:

so that the equations in (12.9) become

$$(12.10)$$

Hence they are reduced to simple wire deformations:

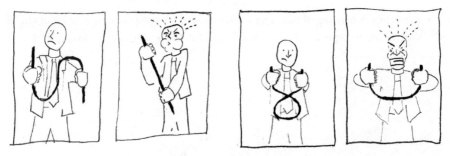

String diagrams capture what Schrödinger described as the characteristic trait of quantum mechanics: the existence of non-separable states. Visualizing cups and caps as wires captures this intuition that the two systems involved cannot be separated from each other (since, of course, we are not allowed to cut wires). We can state this more formally as follows:

Proposition 12.3.2 If a theory is described by string diagrams, and all two-system states ψ are $\otimes$-separable, i.e. there exist states ψ_1 and ψ_2 such that

$$\psi = \psi_1 \; \psi_2$$

then every process f is $\circ$-separable; i.e. it can be written as

$$f = \psi \circ \pi$$

for some effect π and some state ψ.

Proof. By assumption, the cup-state is $\otimes$-separable:

$$\cup = \psi_1 \; \psi_2$$

So, for any process f we have:

$$f = \cap\!\cup f = \cap f \psi_1 \psi_2 = \psi_2 f\!\cap\!\psi_1 = \phi\,\pi$$

□

What is this telling us? First, we should note that all processes being $\circ$-separable is a *bad thing*. It is the same as saying all processes in the theory simply throw away the input and produce a constant output (possibly multiplied by some number depending on the input). In other words, nothing ever happens! As this is an absurd condition for any 'reasonable' process theory, all bipartite (i.e. two-system) states cannot be $\otimes$-separable.

In such reasonable theories, string diagrams guarantee that the collection of two-system states is just as rich as the processes involving the same types:

Proposition 12.3.3 Any theory described by string diagrams has *process–state duality*. That is, for any two types A and B we have a 1-to-1 correspondence between processes and states of the form

$$f : A \to B \quad \longleftrightarrow \quad \psi : A \otimes B$$

which is realized by

12.3.2 Traces and transposes

For a process f with one of its inputs having the same type as one of its outputs, the *partial trace* with respect to that input/output pair is

$$\mathrm{tr}_{B,C}^{A}\left(\begin{array}{c} \\ f \\ \end{array}\right) := $$

The total trace, or simply the *trace*, is the special case where $B = C = I$. A typical property of the trace is *cyclicity*, which now comes just from the observation that—since only connectivity matters—the following two diagrams are the same:

The *transpose* of a process f is the process

or equivalently, by (12.10),

Then, clearly if we transpose twice, we get back where we started:

In other words, transposition is an *involution*. And now comes the really cool bit. First, we deform our boxes a bit:

$$\boxed{f} \rightsquigarrow \boxed{f/}$$

Now, if we express the transpose of f as a box labelled 'f', but rotated 180°:

$$\boxed{/f} := \smile\!\boxed{f/}\!\frown$$

the definition of a transpose gets built in to this '180° rotation'-notation:

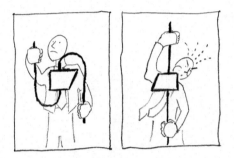

From this, it also follows that

$$\boxed{f/}\smile \;=\; \smile\boxed{/f} \qquad \boxed{f/}\frown \;=\; \frown\boxed{/f}$$

So we can slide boxes across cups and caps just like beads on a wire:

$$\boxed{\tau}\smile \;=\; \smile\boxed{\tau}\;=\;\smile\boxed{\tau} \qquad \boxed{\tau}\frown \;=\; \frown\boxed{\tau}\;=\;\frown\boxed{\tau}$$

Now, let's play a game. Here is the challenge: Aleks and Bob are far apart, Aleks possesses a system in a state ψ, and Bob needs this state. Suppose they also share another state, namely the cup-state. So, we have this situation:

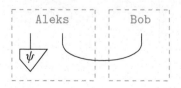

Starting from this arrangement, is there something Aleks and Bob can do in the '?'-marked regions indicated below, that results in Bob obtaining ψ?

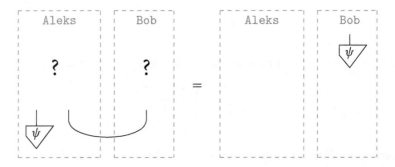

Here is a simple solution:

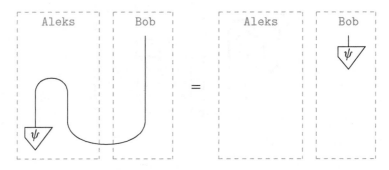

If you the reader came up with this solution yourself, then you on your own did in a few seconds what all the physicists in the world failed to do between 1930 and 1990: discover quantum teleportation!

12.3.3 Adjoints and unitarity

There is a slight caveat in Section 12.3.2's discussion of teleportation. While the cap is a process that Aleks can attempt to 'do', he'll need a bit of luck to succeed. This is because the cap process is not *causal* (a concept we'll soon meet), so it can only be realized with some probability strictly less than 1. As a consequence, Aleks might not get the effect he wants (the cap), but some other effect $\phi_1, \phi_2, \ldots$ which, by process–state duality, we can represent as a cap with some 'unwanted' box:

which we'll called the *error*. This box obstructs Bob's direct access to the state he so much desires:

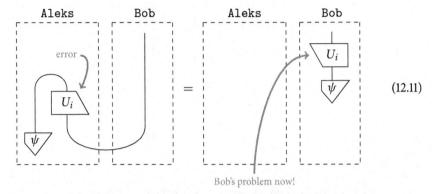

Bob's problem now!

(12.11)

So, all is lost? Of course not! Bob just needs to figure out how to 'undo' this error by means of some process.

Before going there, we need to enrich our language of string diagrams a bit. In addition to rotating boxes 180°, we will now also reflect them vertically:

$$\begin{array}{c} \vert B \\ \boxed{f} \\ \vert A \end{array} \;\mapsto\; \begin{array}{c} \vert A \\ \boxed{f} \\ \vert B \end{array}$$

and refer to this reflected box as the *adjoint*. Like transposition, this is an involution, and it can be applied to entire diagrams in the obvious way:

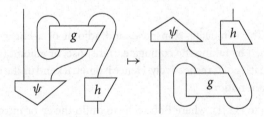

Rotating by 180° degrees then reflecting vertically is the same as just reflecting horizontally:

$$\boxed{f} \;\mapsto\; \boxed{f} \;\mapsto\; \boxed{f}$$

This horizontal reflection, obtained by composing the transpose and the adjoint, is called the *conjugate*. So all together, boxes now come in quartets:

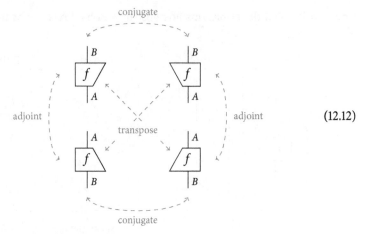
(12.12)

Having adjoints around enables us to make the following definition:

Definition 12.3.4 A process U an *isometry* if we have

$$\begin{array}{c} U \\ U \end{array} = A \qquad (12.13)$$

and it is *unitary* if its adjoint is also an isometry.

Each of these names may sound familiar to some readers, which is of course no coincidence:

Example 12.3.5 The conjugate, transpose, and adjoint of a linear map f is a new linear map obtained by taking the conjugate, transpose, or adjoint (a.k.a. conjugate-tranpose) of the matrix of f, respectively. Hence isometries and unitaries of linear maps are just the standard notions.

We now return to (12.11), where Bob seeks to undo the error introduced by Aleks' process. In the case where U is an isometry (or better yet, a unitary), Bob simply needs to apply its adjoint to undo it:

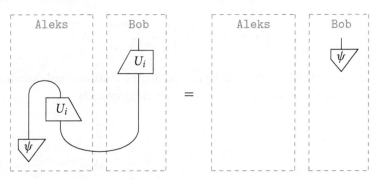

12.3.4 Adjoints and connectedness

We haven't said much about adjoints, just that they have to preserve diagrams. In fact, the ontology of the adjoint and corresponding postulates are still the subject of ongoing research. In the following we will make use of one particular additional condition that one may want adjoints to satisfy. This extra condition comes from taking the notion that an adjoint really is a reflection seriously.

Suppose we have a ∘-non-separable process. One could then imagine that it has some internal structure, say a collection of tubes or machines connecting some inputs to outputs:

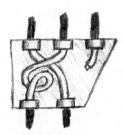

If we now compose this process with its adjoint, i.e. its vertical reflection, then these internal connections match up:

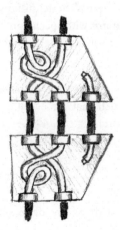

so one expects the resulting process also to be ∘-non-separable, that is,

$$\left(\exists \psi, \phi : \; \boxed{f} \;=\; \begin{array}{c}\phi\\ \psi\end{array}\right) \iff \left(\exists \psi', \phi' : \; \begin{array}{c}f\\ f\end{array} \;=\; \begin{array}{c}\phi'\\ \psi'\end{array}\right) \quad (12.14)$$

Indeed this assumption holds for our main example:

Example 12.3.6 For linear maps, ○-separable means rank-1. It is a well-known fact from linear algebra that the rank of

is the same as the rank of f. Hence (12.14) is satisfied.

While this assumption makes sense visually, it can fail in surprising places:

Example 12.3.7 Consider a process theory whose processes are relations $R \subseteq A \times B$ where ○ is the usual composition of relations, ⊗ is the Cartesian product, and I is the 1-element set $\{*\}$. The adjoint $R^\dagger$ of a relation R is its converse. That is, $(b, a) \in R^\dagger$ if and only if $(a, b) \in R$. Then the following relation from the two-element set $\{0, 1\}$ to itself fails to satisfy (12.14):

$$\{(0,0), (0,1), (1,1)\} \subseteq \{0,1\} \times \{0,1\}$$

12.3.5 Category-theoretic counterpart

In Definition 12.3.1 we gave two equivalent definitions of string diagram, the latter of which consisting of circuit diagrams with special processes called cups and caps. The category-theoretic counterpart for string diagrams proceeds in pretty much the same way:

Definition 12.3.8 A *compact closed category* (with symmetric self-duality) is a symmetric monoidal category $\mathcal{C}$ such that for every object $A \in \mathrm{ob}(\mathcal{C})$, there exists morphisms $\epsilon_A \in \mathcal{C}(A \otimes A, I)$ and $\eta_A \in \mathcal{C}(I, A \otimes A)$ such that

$$(\epsilon_A \otimes 1_A) \circ (1_A \otimes \eta_A) = 1_A$$

$$\epsilon_A \circ \sigma_{A,A} = \epsilon_A$$
$$\sigma_{A,A} \circ \eta_A = \eta_A.$$

Compact closed categories satisfy an analogue to Theorem 12.2.8, but for string diagrams:

Theorem 12.3.9 Any string diagram can be interpreted as a morphism in a compact closed category, and two morphisms are equal according to the axioms of a compact closed category if and only if their string diagrams are equal.

Example 12.3.10 The category FHilb is compact closed. For a Hilbert space A, fix a basis $\{e_i\}_i$ in A then let

$$\eta_A : \mathbb{C} \to A \otimes A \qquad \text{and} \qquad \epsilon_A : A \otimes A \to \mathbb{C}$$

be linear maps defined as follows:

$$\eta_A(1) = \sum_i e_i \otimes e_i \qquad \epsilon_A(e_i \otimes e_j) = \begin{cases} 1 & \text{if } i = j \\ 0 & \text{otherwise.} \end{cases}$$

Using Dirac's 'bra-ket' notation, which is popular in the quantum computing literature, these maps can be written as

$$\eta_A = \sum_i |i\rangle \otimes |i\rangle \qquad \epsilon_A = \sum_i \langle i| \otimes \langle i|.$$

In category theory, closure refers to the fact that one can represent sets of morphisms $\mathcal{C}(A, B)$ as the states of another object, which we could denote as $A \Rightarrow B$. Compact closed categories have the very convenient feature that we can take

$$A \Rightarrow B := A \otimes B.$$

We saw this feature in Proposition 12.3.3, under the name 'process–state duality'.

Just as we simplified previously by restricting to *strict* monoidal categories, here we simplify from a more general notion of compact closed categories to symmetrically self-dual compact closed categories. The former only require that $\epsilon_A \in \mathcal{C}(A^* \otimes A, I)$ and $\eta_A \in \mathcal{C}(I, A \otimes A^*)$ exists for some object A^* (called the *dual* of A). Then, 'self-duality' means we can choose A^* to just be A again, and 'symmetric' means this choice gets along well with symmetries. This of course makes sense from a diagrammatic point of view (cf. the right-most equations in (12.10)), and as pointed out by Selinger (2010) is necessary for interpreting string diagrams as morphisms without ambiguity.

Remark 12.3.11 The definition of η_A and ϵ_A from Example 12.3.10 depend on the choice of basis. We can avoid this by dropping self-duality, in which case we take A^* to be the dual space and let

$$\epsilon_A(\xi \otimes a) = \xi(a).$$

This uniquely fixes ϵ_A (and therefore η_A) without making reference to a basis.

So to summarize, we have the following correspondences between diagrams and categories:

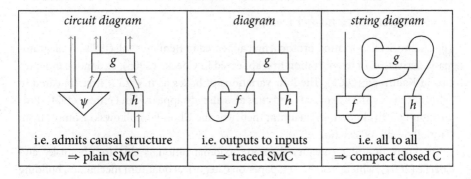

Notably, the more sophisticated (i.e. restrictive) kinds of diagrams correspond to the simplest kinds of SMCs, and vice versa.

Unsurprisingly, the category-theoretic definition of adjoints is again more involved than its diagrammatic counterpart, simply for the reason that we now must say carefully what 'preserving diagrams under reflection' means in terms of the language of SMCs.

Definition 12.3.12 For a strict SMC $\mathcal{C}$, a *dagger functor* assigns to each morphism $f : A \to B$ a new morphism $f^\dagger : B \to A$ such that

$$(f^\dagger)^\dagger = f \qquad (g \circ f)^\dagger = f^\dagger \circ g^\dagger \qquad (f \otimes g)^\dagger = f^\dagger \otimes g^\dagger \qquad \sigma_{AB}^\dagger = \sigma_{BA}.$$

In category-theoretic parlance, this is therefore a 'strict symmetric monoidal functor that is *identity-on-objects* and *involutive*'.

Definition 12.3.13 A strict dagger-symmetric monoidal category (†-SMC) is a strict SMC with a chosen dagger functor. For a *dagger-compact closed category* (with symmetric self-duality), we additionally assume $\eta_A^\dagger = \epsilon_A$.

Example 12.3.14 For FHilb, the dagger functor sends each linear map $f : A \to B$ to its linear-algebraic adjoint $f^\dagger : B \to A$, i.e. the unique linear map such that

$$\langle b|f(a)\rangle = \langle f^\dagger(b)|a\rangle$$

for all $a \in A, b \in B$.

There is a tight connection between equations between linear maps and equations between string diagrams. Namely, any equation between string diagrams involving linear maps $f, g, h, \ldots$ (and their adjoints) holds generically—i.e. for <u>all</u> linear maps $f, g, h, \ldots$—precisely when the diagrams themselves are equal. This is formally stated as a *completeness* theorem:

Theorem 12.3.15 FHilb is complete for string diagrams.

In other words, if we don't know anything about the linear maps in a diagram (i.e. we treat them as 'black boxes'), diagrammatic reasoning is already the best we can do.

12.3.6 Reference and further reading

The quantum teleportation protocol first appeared in Bennett et al. (1993). A diagrammatic derivation of teleportation first appeared in Coecke (2003), and independently, also in Kauffman (2005). The four variations of boxes as in (12.12) first appeared in Selinger (2007). The use of caps and cups also already appeared in Penrose (1971). Proposition 12.3.3 is known in quantum theory as the Choi–Jamiołkowski isomorphism (Jamiołkowski, 1972; Choi, 1975).

The corresponding category-theoretic axiomatization is due to Abramsky and Coecke (2004), which was the first paper on categorical quantum mechanics, building further on Kelly's compact closed categories (Kelly, 1972) by adjoining a dagger functor. Dagger compact categories had already appeared in Baez and Dolan (1995) as a special case of a more general construct in n-categories. Theorem 12.3.15 is due to Selinger (2011a). A comprehensive (at the time) survey of monoidal categories and their various graphical languages is given in Selinger (2011b).

12.4 Quantum Processes

I would like to make a confession which may seem immoral: I do not believe absolutely in Hilbert space any more.

John von Neumann, letter to Garrett Birkhoff (1935)

Let us summarize what we have seen thus far. In the first section, we introduced a general formalism to reason about interacting processes based on diagrams. In the following section, we saw how the simple assumption that the diagrams are string diagrams allows a kindergartner to derive quantum teleportation. And then we bumped into a caveat, which we claimed came from something called *causality*.

We claimed before that, rather than simply 'doing' a cap effect, Aleks must perform a non-deterministic processes which might instead yield a cap-effect with some error. But that's just a bunch of words. None of the ingredients coming from string diagrams can actually help us derive this fact.

There's also a second issue here. Before Bob can correct the error, he must know which error U_i happened. The only way this is possible is if Aleks picks up the phone and tells him. This requires distinguishing a phone call from a quantum system in our diagrams, i.e. diagrams will need to involve two kinds of types: classical types and quantum types, and their distinct behaviour should also be evident in diagrammatic terms. In fact, it is the diagrammatic formulation of quantum types that will allow us to define this (in)famous causality postulate which brought us here in the first place. Moreover, we will then be able to diagrammatically derive the fact that the cap-state cannot be invoked with certainty.

This section concerns specification of a very general kind of quantum type. Despite its generality, we will already be able to prove some highly non-trivial features of quantum systems, most notably, the no-broadcasting theorem. This sets the stage for exploring alternative models of quantum theory, which go beyond Hilbert spaces.

12.4.1 Quantum types

Back in Section 12.2.1, we interpreted a state meeting an effect as a probability, i.e. a positive number:

$$\left.\begin{array}{r}\text{test}\left\{\vcenter{\hbox{$\triangle\!\!\!\pi$}}\right. \\ \text{state}\left\{\vcenter{\hbox{$\nabla\!\!\!\psi$}}\right.\end{array}\right\}\ \text{probability.} \tag{12.15}$$

At that time, we didn't have a language rich enough to give a notion like 'positivity', but thanks to reflection (i.e. adjoints and conjugates) we now do. A complex number is positive precisely when it is the product of some number λ and its conjugate $\bar{\lambda}$. Being the composition of a number and its conjugate thus gives us a generalization of the condition of a number being positive:

$$\text{number} \quad \text{vs.} \quad \text{generalized positive number} \tag{12.16}$$

So, we could use expressions like the one on the right to compute probabilities, and indeed this is how it is typically done in quantum theory. However, there is a way to have our cake and eat it too. Namely, we can retain the simplicity of (12.15) while secretly ensuring the result is always a positive number.

The trick is to double everything. By pairing two wires, we make a new thicker kind of wire:

which gives us a *quantum* system type. We can then keep un-doubled types around for representing classical systems—something we'll use extensively in part II of this overview—obtaining this simple distinction:

$$\frac{\text{classical}}{\text{quantum}} = \frac{\text{single wire}}{\text{double wires}}$$

On these doubled wires, we can then build doubled boxes, which consist of the box itself, along with its conjugate:

When we apply this recipe to states and effects, we get a generalized positive number whenever the two meet:

Thus, the thing we called the generalized Born rule, applied to these new doubled processes, becomes what some will recognize as the usual Born rule from quantum theory.

Doubling preserves string diagrams:

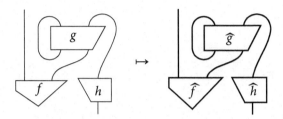

as long as we are a bit careful about where the wires go for boxes with many inputs/outputs:

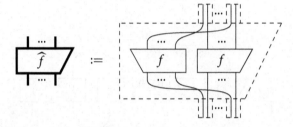

In particular, this gives us the following expression for the doubled cap:

$$\cup \;:=\; \widehat{\cup} \;=\; \cup$$

and similarly for the cup.

It also preserves all equations between string diagrams, so, for example, our derivation of teleportation carries over to the doubled world. What about the converse? Do equations in doubled diagrams carry over to their un-doubled counterparts? The answer turns out to be no, but luckily this is a feature, not a bug.

Proposition 12.4.1 In a theory described by string diagrams, if

$$\widehat{f} \;=\; \widehat{g} \tag{12.17}$$

then there exist numbers λ and μ, with $\widehat{\lambda} = \widehat{\mu}$, such that

$$\langle\lambda\rangle\ \boxed{f}\ =\ \langle\mu\rangle\ \boxed{g} \tag{12.18}$$

The converse also holds provided $\widehat{\lambda} = \widehat{\mu}$ is cancellable.

Proof. Let λ and μ be

$$\langle\lambda\rangle := \boxed{f}\ \boxed{f} \qquad \langle\mu\rangle := \boxed{g}\ \boxed{f}$$

Unfolding (12.17) yields

$$\boxed{f}\ \boxed{f}\ =\ \boxed{g}\ \boxed{g} \tag{12.19}$$

So

$$\langle\lambda\rangle = \boxed{f}\ \boxed{f}\ \boxed{f}\ \boxed{f}\ \overset{(12.19)}{=}\ \boxed{g}\ \boxed{f}\ \boxed{f}\ \boxed{g} = \langle\mu\rangle$$

and

$$\langle\lambda\rangle\ \boxed{f}\ =\ \boxed{f}\ \boxed{f}\ \boxed{f}\ \overset{(12.19)}{=}\ \boxed{g}\ \boxed{f}\ \boxed{g}\ =\ \langle\mu\rangle\ \boxed{g}$$

□

Typically, 'cancellable' means non-zero, and the condition that $\widehat{\lambda} = \widehat{\mu}$ means the absolute values of λ and μ coincide. In that case, the condition (12.18) is more commonly referred to as 'equal up to a (global) complex phase $\frac{\mu}{\lambda}$'. One of the uglier aspects of the standard quantum formalism is indeed that it contains redundant numbers of the form $\frac{\mu}{\lambda}$, which have no physical significance. Thus a nice side effect of doubling is precisely removing this redundancy.

12.4.2 Pure processes and discarding

A state ψ is called *normalized* if composing with its adjoint yields 1, i.e. the empty diagram:

$$\begin{array}{c}\psi\\\psi\end{array} = \Box$$

When we pass to the doubled world, we therefore have a way to throw away states $\widehat{\psi}$ arising from normalized states ψ. Simply connect the two halves together:

$$\cdots = \begin{array}{c}\psi\\\psi\end{array} = \Box$$

This new effect, which has no counterpart in the un-doubled underlying theory, is called *discarding*, and we denote it as follows:

$$\overline{\top} := \cdots$$

This then immediately gives rise to new boxes as well:

$$\widehat{f} := \begin{array}{c} f \quad f \end{array} = \begin{array}{c} f \quad f \end{array} \qquad (12.20)$$

which we call *impure*, or *mixed*, processes—for reasons that shall become clear in the follow-up paper to this one. These impure processes naturally occur, as the diagram indicates, when we are only considering part of a composite system, and ignoring (i.e. discarding) another part. In other words, we are considering systems that are in interaction with their *environment*. A *pure* process is then one that doesn't involve discarding, i.e. that doesn't have a wire connecting its two halves. Impure processes will typically be denoted by Φ, and and impure states by ρ, in contrast to the pure ones which carry a 'hat'.

Discarding multiple systems is the same as discarding one, bigger system. Hence we can put any process consisting of pure processes and discarding in the form of (12.20) by grouping all of the discarding processes together into a single effect:

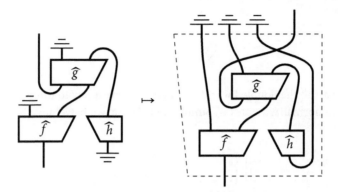

As a consequence:

Proposition 12.4.2 Any string diagram consisting of processes of the form (12.20) is again of that form. Hence they form a process theory.

Another consequence is that any diagram of pure processes and discarding can be put into a standard form involving a pure process and a single discard. In other words, any (possibly impure) process Φ has a *purification* $\hat{f}$:

$$\Phi = \hat{f} \qquad (12.21)$$

Condition (12.14) on adjoints now allows us establish an important connection between $\otimes$-separability and purity of the *reduced state* of a composite system, i.e. what remans if we discard part of it:

$$\rho$$

Proposition 12.4.3 Consider a theory that admits string diagrams and in which adjoints obey (12.14). If the reduced state of a bipartite state ρ is pure:

$$\rho = \phi \qquad (12.22)$$

then it $\otimes$-separates as follows:

$$\rho = \rho' \, \phi'$$

Proof. Writing ρ in the form (12.20):

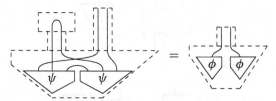 (12.23)

and substituting this into (12.22) we obtain

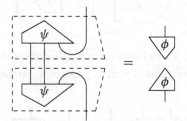

and hence

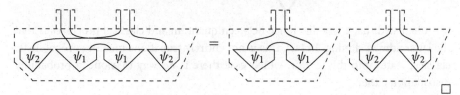

Deforming this equation via process–state duality and transposition, we get

Then, by (12.14) there exist ψ_1 and ψ_2 such that

hence

Plugging in to (12.23) indeed yields a $\otimes$-separable state of the required form:

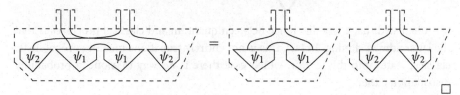

$\square$

By process–state duality this fact straightforwardly extends to processes:

Proposition 12.4.4 Consider a theory that admits string diagrams and in which adjoints obey (12.14). If a *reduced process* of a process Φ is pure:

$$\text{(diagram)} \tag{12.24}$$

then it $\otimes$-separates as follows:

$$\text{(diagram)} \tag{12.25}$$

Proof. Bend the wire in (12.24):

$$\text{(diagram)}$$

Treating the two rightmost wires above as a single system, this is the reduced state of a bipartite state. Since the reduced state is pure, by Proposition 12.4.3 it separates as follows:

$$\text{(diagram)}$$

Unbend the wire and we're done. □

12.4.3 No-broadcasting

A *cloning process* Δ is a process that takes any state as input and produces two copies of that state as output:

$$\text{(diagram)} \tag{12.26}$$

It is often said that a key difference between quantum and classical processes is that the latter admits cloning. In fact, this is not entirely true if one includes probabilistic classical states into the mix, in which case there is no way to 'clone' a probability distribution either.

However, what is possible classically is *broadcasting*. That is, there exists a process Δ such that, when a state ρ is fed in and either output is discarded, we are left with ρ:

$$\vcenter{\hbox{[diagram]}} = \vcenter{\hbox{[diagram]}} = \vcenter{\hbox{[diagram]}} \qquad (12.27)$$

It is easily seen diagrammatically that broadcasting is indeed a weaker notion than cloning:

$$\vcenter{\hbox{[diagram]}} \stackrel{(12.26)}{=} \vcenter{\hbox{[diagram]}} = \vcenter{\hbox{[diagram]}}$$

and similarly for discarding the other output. Rather than making explicit reference to the state ρ, we can also give the broadcasting equations (12.27) in a state less (or, if you want, point less) form:

$$\vcenter{\hbox{[diagram]}} \stackrel{(l)}{=} \vcenter{\hbox{[diagram]}} \stackrel{(r)}{=} \vcenter{\hbox{[diagram]}} \qquad (12.28)$$

Theorem 12.4.5 If in a theory described by string diagrams adjoints obey (12.14), then the theory obtained by doubling and adjoining discarding cannot have a broadcasting process.

Proof. By Eq. (12.28l) the reduced state of Δ is pure, namely a plain wire, so by Proposition 12.4.4 we have

$$\vcenter{\hbox{[diagram]}} = \vcenter{\hbox{[diagram]}} \qquad (12.29)$$

for some state ρ. Hence it follows that

$$\vcenter{\hbox{[diagram]}} \stackrel{(12.28r)}{=} \vcenter{\hbox{[diagram]}} \stackrel{(12.29)}{=} \vcenter{\hbox{[diagram]}}$$

Since the identity is o-separable, so is every other process involving that type, and hence the system must be trivial for Δ to exist. $\square$

12.4.4 Category-theoretic counterpart

Doubling and adding discarding has a categorical counterpart as well. Rather than building it up piecewise from doubled processes and discarding, the categorical construction just declares that all morphisms should be of the form (12.20):

Definition 12.4.6 For a compact closed category $\mathcal{C}$, we can form a new compact closed category $\text{CPM}[\mathcal{C}]$ with objects $\widehat{A}$ for every $A \in \text{ob}(\mathcal{C})$. The morphisms $f : \widehat{A} \to \widehat{B}$ are those morphisms $f \in \mathcal{C}(A \otimes A, B \otimes B)$ which are of the form

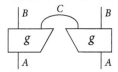

for some object C and morphism $g : A \to C \otimes B$.

One needs to do a bit of work to show that this is indeed a compact closed category. For instance, the parallel composition of $f : \widehat{A} \to \widehat{B}$ and $f' : \widehat{A'} \to \widehat{B'}$ should give something of the form

$$f \otimes f' \in \mathcal{C}((A \otimes A') \otimes (A \otimes A'), (B \otimes B') \otimes (B \otimes B')),$$

which involves some reshuffling of wires. We refer to references in the next section for details.

The acronym CPM refers to 'completely positive map', and indeed when you apply this construction to FHilb, you get the category whose morphisms are completely positive maps.

12.4.5 Reference and further reading

The doubling construction was introduced by Coecke (2007), including the proof of Proposition 12.4.1. Around the same time, the generalization to impure was introduced by Selinger (2007). The idea that this can be done by adding the discarding process was put forward in Coecke (2008).

The no-broadcasting theorem first appeared in Barnum et al. (1996), and our derivation from doubling and (12.14) is novel. Another 'generalized no-broadcasting theorem' is Barnum et al. (2007), which, rather than process theories, concerns generalized probabilistic theories.

12.5 Causality

> *A new scientific truth does not triumph by convincing its opponents and making them see the light, but rather because its opponents eventually die, and a new generation grows up that is familiar with it.*
>
> Max Planck (1936)

The beginning of the previous century saw two revolutions in physics: quantum theory and relativity. While the first one is a theory which associates probabilities to non-deterministic processes (usually) involving microscopic systems, the second concerns the geometry of space–time. Evidently, since there is only one reality, those two theories should not contradict each other. Amazingly, this compatibility can already be obtained within the generality of diagrams, provided there are discarding effects. If these discarding effects happen to arise from doubling as described in Section 12.4, then many more results follow.

12.5.1 Causal processes

Suppose we apply a process to some inputs, but then discard all of its outputs. Then the performance has gone to waste, and we could as well have simply discarded its inputs. In fact, this obvious assumption is quite vital for even being able to perform science. It allows one to 'discard' (i.e. ignore) everything that does not directly affect an experiment, such as stuff happening in some other galaxy. However innocent and/or obvious this principle sounds, it has many striking consequences, warranting an important-sounding name:

Definition 12.5.1 In a process theory where each system has a distinguished discarding effect, a process Φ is called *causal* if we have

$$\underset{\Phi}{\boxed{}}\!\!\bot \;=\; \bot \tag{12.30}$$

We call a theory causal if all its processes are.

Note that (12.30) should be read as discarding <u>all</u> of the outputs on the LHS results in discarding <u>all</u> of the inputs on the RHS. As a special case, states have no inputs, so nothing must be discarded on the RHS:

$$\underset{\rho}{\triangledown}\!\!\bot \;=\;$$

This is the usual condition for a (possibly impure) quantum state being normalized. Similarly, effects have no outputs, so nothing must be discarded on the LHS:

$$\underset{\rho}{\vartriangle} \;=\; \bot$$

Clearly, this is bad news for those who like variety:

Theorem 12.5.2 In a causal theory, each system-type admits only one effect: discarding.

In the case of two systems this means that

$$\bar{\top}\;\bar{\top}$$

is the only available effect. So now it should be clear that, if we restrict to causal effects, teleportation is no longer possible:

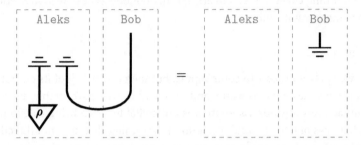

The way around this problem is to avoid Theorem 12.5.2 simply by having a classical output, corresponding to Aleks' outgoing phone call to Bob. As we will see in the follow-up paper (Coecke and Kissinger, 2016a), such a classical output is what enables one to accommodate the non-determinism alluded to in Section 12.3.3.

12.5.2 Evolution and Stinespring dilation

We now consider theories arising from doubling. In our first result, we will actually derive something that is typically assumed a priori in the standard formulation of quantum theory due to von Neumann (1932).

Theorem 12.5.3 In a theory obtained by doubling and adjoining discarding, the following are equivalent for any pure process $\widehat{U}$:

1. $\widehat{U}$ is causal:

$$\bar{\top} \circ \widehat{U} = \bar{\top}$$

2. U is an isometry:

$$U^\dagger \circ U = \mathbb{1}$$

3. and $\widehat{U}$ is an isometry:

$$\begin{array}{c}\widehat{U} \\ \widehat{U}\end{array} = \;\Big|$$

Proof. Unfolding the causality equation, we have

$$\overline{U\;U} = \cap$$

and we recover (12.13) simply by un-bending, so 1 ⇔ 2. 2 ⇔ 3 follows by Proposition 12.4.1. □

Corollary 12.5.4 Under the assumptions of the previous theorem, the following are equivalent for pure processes:

1. $\widehat{U}$ is causal and invertible, and
2. $\widehat{U}$ is unitary.

Dropping purity yields a standard result of quantum information theory:

Theorem 12.5.5 (Stinespring dilation) Under the assumptions of the previous theorem, for every causal process Φ there exists an isometry $\widehat{U}$ such that

$$\Phi = \widehat{U} \qquad (12.31)$$

Proof. We have (12.31) immediately if we let $\widehat{U}$ be the purification of Φ, as in (12.21). So it suffices to prove that $\widehat{U}$ is an isometry in (12.31). By causality of Φ, it follows that $\widehat{U}$ must also be causal:

$$\widehat{U} \stackrel{(12.31)}{=} \Phi \stackrel{(12.30)}{=} \;\overline{\top}$$

which, by Theorem 12.5.3, implies that $\widehat{U}$ is an isometry. □

12.5.3 No-signalling

In this section, we'll set doubling aside again and look at the general case of theories with discarding, and how diagrams in such a theory relate to *causal structures*. A causal structure is simply a directed graph (without cycles) that indicates some collection of events in space–time, where an edge from e_1 to e_2 indicates that e_1 could have an influence on e_2; i.e. e_2 is in the *causal future* of e_1. For example, the following causal structure

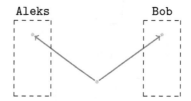

indicates that Aleks and Bob may have some shared history, but now they have moved far away from each other, so that they can no longer directly communicate. More precisely, they are so far away that the speed of light prohibits Aleks sending any kind of message to Bob and vice versa.

A process theory is said to be *non-signalling* if each process Φ in a diagram with a fixed causal structure can only have an influence on processes in the causal future of Φ. We claim that any causal process theory is non-signalling.

First, we note that we can associate a causal structure to any circuit diagram by associating boxes to events, and declaring a box Φ to be in the causal future of another box Ψ if an output of Ψ is wired to an input of Φ:

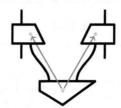

Note that we can freely add some extra input/output wires to account for the fact that Aleks and Bob can interact with their own processes locally:

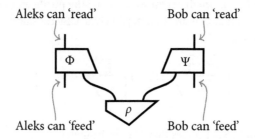

However, in this situation, non-signalling dictates that, since Bob does not have access to the output of Aleks' process (as this would require Aleks sending a message faster than the speed of light), Aleks shouldn't be able to influence Bob via his input. To see this is the case, let's see things from Bob's perspective by discarding Aleks' output:

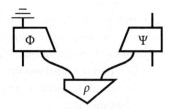

and see if Bob can learn anything about Aleks' input from his own input–output pair. By causality we have

and hence it follows that

So from Bob's perspective, his input–output pair is ∘-separated from Aleks' input. Thus no signalling from Aleks to Bob can take place. By symmetry it also follows that Bob cannot signal to Aleks.

Theorem 12.5.6 If a process theory has a discarding process for each type and it satisfies causality, then it is non-signalling.

12.5.4 Category-theoretic counterpart

In contrast to some of the previous definitions, one usually encounters the following definition in the first lesson of a course on category theory:

Definition 12.5.7 An object T is called *terminal* if for every object A, there exists a unique morphism $! : A \to T$.

It's a quick exercise to show that, if a terminal object exists, it is unique (up to isomorphism). A monoidal category is called *causal* if the monoidal unit I is a terminal object, which is the same as saying that every morphism in the category satisfies the causality postulate.

12.5.5 Reference and further reading

The causality postulate was only recently identified as a core principle of quantum theory, by Chiribella, D'Ariano, and Perinotti (Chiribella et al., 2010), as one of a series of axioms from which quantum theory was reconstructed.

Stinespring dilation first appeared within the context of C*-algebras (Stinespring, 1955). Our derivation of unitarity and Stinespring dilation from causality is new.

The first proof of the non-signalling theorem for quantum theory can be found in Ghirardi et al. (1980). The derivation of non-signalling from the causality principle is taken from Coecke (2014). A similar result is also in Fritz (2014), but in those papers more structure is used in order to establish this fact.

12.6 What Comes Next

Simply by stating that a quantum type arises from doubling we derived many typical features of quantum theory. The motivation for doubling arose from distinguishing quantum types from classical types. However, we haven't said anything yet about classical types, nor how quantum and classical types interact. Having a grip on this is essential to understanding concepts such as mixing, measurement, and entanglement, and will enable us to give fully comprehensive descriptions of several quantum protocols as diagrams. This is the content of the follow-up paper (Coecke and Kissinger, 2016a).

In the final paper (Coecke and Kissinger, 2016b) we discuss the important quantum theoretical notion of complementarity, as well as a strengthening thereof, which, among many other things, will enable us to derive quantum non-locality. The corresponding category-theoretic notions to classicality and complementarity will involve certain kinds of algebraic structures within a monoidal category, namely certain Frobenius algebras and Hopf-algebras.

References

Abramsky, S., and Coecke, B. (2004). A categorical semantics of quantum protocols, in *Proceedings of the 19th Annual IEEE Symposium on Logic in Computer Science (LICS)*, 415–25. arXiv:quant-ph/0402130.

Baez, J. C., and Dolan, J. (1995). Higher-dimensional algebra and topological quantum field theory. *Journal of Mathematical Physics* 36, 6073. arXiv:q-alg/9503002.

Barnum, H., Barrett, J., Leifer, M., and Wilce, A. (2007). A generalized no-broadcasting theorem, *Physical Review Letters* 99(24), 240501.

Barnum, H., Caves, C. M., Fuchs, C. A., Jozsa, R., and Schumacher, B. (1996). Noncommuting mixed states cannot be broadcast. *Physical Review Letters* 76, 2818.

Benabou, J. (1963). Categories avec multiplication. *Comptes rendus des seances de l'Academie des sciences. Paris* 256, 1887–90.

Bennett, C. H., Brassard, G., Crepeau, C., Jozsa, R., Peres, A., and Wootters, W. K. (1993). Teleporting an unknown quantum state via dual classical and Einstein-Podolsky-Rosen channels. *Physical Review Letters* 70(13), 1895–9.

Chiribella, G., D'Ariano, G. M., and Perinotti, P. (2010). Probabilistic theories with purification. *Physical Review A* 81(6), 062348.

Choi, M.-D. (1975). Completely positive linear maps on complex matrices. *Linear Algebra and Its Applications* 10, 285–90.

Coecke, B. (2003). The logic of entanglement: an invitation. Technical Report RR-03-12, Department of Computer Science, Oxford University. http://www.cs.ox.ac.uk/files/933/RR-03-12.ps

Coecke, B. (2005). Kindergarten quantum mechanics, in A. Khrennikov (ed.), *Quantum Theory: Reconsiderations of the Foundations III*, 81–98. AIP Press, Melville, NY. arXiv:quant-ph/0510032.

Coecke, B. (2007). De-linearizing linearity: projective quantum axiomatics from strong compact closure. *Electronic Notes in Theoretical Computer Science* 170, 49–72. arXiv:quant-ph/0506134.

Coecke, B. (2008). Axiomatic description of mixed states from Selinger's CPM-construction. *Electronic Notes in Theoretical Computer Science* 210, 3–13.

Coecke, B. (2009). Quantum picturalism. *Contemporary Physics* 51, 59–83. arXiv:0908.1787.

Coecke, B. (2011). A universe of processes and some of its guises, in H. Halvorson (ed.), *Deep Beauty: Understanding the Quantum World through Mathematical Innovation*, 129–86. Cambridge University Press, Cambridge, UK. arXiv:1009.3786.

Coecke, B. (2014). Terminality implies non-signalling. arXiv:1405.3681.

Coecke, B., and Kissinger, A. (2016a). Categorical quantum mechanics II: classical-quantum interaction. *International Journal of Quantum Information* 14, 1640020. arXiv:1605.08617 [quant-ph].

Coecke, B., and Kissinger, A. (2016b). Categorical quantum mechanics III: strong complementarity. In preparation.

Coecke, B., and Kissinger, A. (2016c). *Picturing Quantum Processes: A First Course in Quantum Theory and Diagrammatic Reasoning*. Cambridge University Press, Cambridge, UK.

Coecke, B., and Paquette, É. O. (2011). Categories for the practicing physicist, in B. Coecke (ed.), *New Structures for Physics, Lecture Notes in Physics*, 167–271. Springer, Berlin. arXiv:0905.3010.

Coecke, B., Paquette, É. O., and Pavlović, D. (2010). Classical and quantum structuralism, in S. Gay and I. Mackie (eds), *Semantic Techniques in Quantum Computation*, 29–69. Cambridge University Press, Cambridge, UK. arXiv:0904.1997.

Coecke, B., and Pavlovic, D. (2007). Quantum measurements without sums, in G. Chen, L. Kauffman, and S. Lamonaco (eds), *Mathematics of Quantum Computing and Technology*, 567–604. Taylor and Francis, London. arXiv:quant-ph/0608035.

Fritz, T. (2014). Beyond Bell's theorem II: Scenarios with arbitrary causal structure. arXiv:1404.4812.

Ghirardi, G.-C., Rimini, A., and Weber, T. (1980). A general argument against superluminal transmission through the quantum mechanical measurement process. *Lettere al nuovo cimento* 27(10), 293–8.

Hardy, L. (2011). Foliable operational structures for general probabilistic theories, in H. Halvorson (ed.), *Deep Beauty: Understanding the Quantum World through Mathematical Innovation*, 409–42. Cambridge University Press, Cambridge, UK. arXiv:0912.4740.

Heunen, C., and Vicary, J. (2016). *Lectures on Categorical Quantum Mechanics*. Oxford University Press, Oxford.

Jamiołkowski, A. (1972). Linear transformations which preserve trace and positive semidefiniteness of operators. *Reports on Mathematical Physics* 3, 275–8.

Joyal, A., and Street, R. (1991). The geometry of tensor calculus I. *Advances in Mathematics* 88, 55–112.

Kauffman, L. H. (2005). Teleportation topology. *Optics and Spectroscopy* 99, 227-32.

Kelly, G. M. (1972). Many-variable functorial calculus I, in G. M. Kelly, M. Laplaza, G. Lewis, and S. M. Lane (eds), *Coherence in Categories*, Vol. 281 of Lecture Notes in Mathematics, pp. 66-105. Springer-Verlag, Berlin.

Kissinger, A. (2014). Abstract tensor systems as monoidal categories, in C. Casadio, B. Coecke, M. Moortgat, and P. Scott (eds), *Categories and Types in Logic, Language, and Physics: Festschrift on the Occasion of Jim Lambek's 90th Birthday*, Vol. 8222 of Lecture Notes in Computer Science. Springer, Berlin. arXiv:1308.3586.

Mac Lane, S. (1963). Natural associativity and commutativity. *Rice University Studies* 49(4), 28-46.

Penrose, R. (1971). Applications of negative dimensional tensors, in *Combinatorial Mathematics and its Applications*, 221-44. Academic Press, San Diego, CA.

Selinger, P. (2007). Dagger compact closed categories and completely positive maps. *Electronic Notes in Theoretical Computer Science* 170, 139-63.

Selinger, P. (2010). Autonomous categories in which $A \cong A^*$, in *7th Workshop on Quantum Physics and Logic (QPL 2010)*.

Selinger, P. (2011a). Finite dimensional Hilbert spaces are complete for dagger compact closed categories (extended abstract). *Electronic Notes in Theoretical Computer Science* 270(1), 113-19.

Selinger, P. (2011b). A survey of graphical languages for monoidal categories, in B. Coecke (ed.), *New Structures for Physics, Lecture Notes in Physics*, 275-337. Springer-Verlag, Berlin. arXiv:0908.3347.

Stinespring, W. F. (1955). Positive functions on C*-algebras. *Proceedings of the American Mathematical Society* 6(2), 211-16.

von Neumann, J. (1932). *Mathematische grundlagen der quantenmechanik*. Springer-Verlag, Berlin. Translation, *Mathematical Foundations of Quantum Mechanics*. Princeton University Press, Princeton, NJ, 1955.

13
Category Theory and the Foundations of Classical Space–Time Theories

James Owen Weatherall

13.1 Introduction

There are certain questions that arise in philosophy of science—including the philosophies of particular sciences—for which it would be useful to have a formal apparatus by which to represent a scientific theory abstractly.[1] To this end, in the context of a sustained critique of the so-called "semantic view" of theories, Halvorson (2012) has suggested that philosophers look to category theory for tools and inspiration. The proposal is that for some purposes, it is useful to think of a scientific theory as a collection of (mathematical) models—though not, as in the semantic view, a *bare* set of models, but rather a *structured* set of models, or more specifically, a *category* of models.[2]

Although the present chapter develops this idea, I do not defend it.[3] Instead, my goal is to show how the idea can be put to work, by reviewing how it has been fruitfully applied to a cluster of related issues concerning symmetry, structure, and equivalence in the context of classical (i.e. non-quantum) field theories. Contemporary interest in these issues can be traced back to Stein (1967) and Earman (1989b), who

---

[1] Of course, there is a long history of proposals aiming to do just this, beginning with Ramsey (1931); Carnap (1968 [1928]); and Russell (2007 [1927]). For overviews of this history and the current state of the field, see Lutz (2015) and Halvorson (2016), and references therein.

[2] In some cases, one might also consider topological structure (Fletcher, 2016); measure theoretic structure (Curiel, 2014); etc. The idea of using category theory to compare physical theories traces back to Mac Lane (1968), who suggested that the Legendre transform between Hamilton and Lagrangian mechanics may be understood as a natural transformation—a result connected with recent work on the structure of classical mechanics (North, 2009; Curiel, 2013; Barrett, 2015a). See also Lawvere and Schanuel (1986).

[3] A more direct defense can be found in Halvorson (Chapter 17, this volume). Still, I take what follows to have some probative value: in my view, fruitful application of the sort described here is the most compelling reason to develop a formal program in philosophy of science.

showed how seventeenth-century debates concerning substantivalism and relationism in Newtonian gravitation can (and should) be understood to be about whether a classical spacetime is endowed with certain geometrical structure.[4] Meanwhile, Earman and Norton (1987) and Stachel (1989) have argued that the infamous "hole argument" leads to puzzles about the structure of space-time in general relativity.[5] Recent debates concerning interpretations of Yang–Mills theory, and so-called "gauge theories" more generally, may be understood along similar lines (Belot, 1998, 2003; Healey, 2001, 2004, 2007; Arntzenius, 2012).[6]

I will begin by introducing the category theory that will appear in the sequel. Then I will show how this framework recaptures (old) intuitions about relative amounts of structure in classical space-time theories. This discussion will lead to classical electromagnetism and a discussion of "gauge" structure. Finally, I will consider more difficult issues arising in general relativity and Yang–Mills theory. In these last two cases, I will argue, the tools developed in the earlier parts of the chapter provide new traction on thorny interpretational issues of current interest.[7]

Before proceeding, let me flag a worry. Although some of the results I describe in the body of the chapter are non-trivial, the category theory I use is elementary and, arguably, appears only superficially. Meanwhile, there are areas of mathematics and mathematical physics closely related to the theories I discuss—synthetic differential geometry; higher gauge theory—where category theory plays a much deeper role.[8] But as I said above, my goal is to review some ways in which thinking of a physical theory as a category of models bears fruit for issues of antecedent philosophical interest. The role the category theory ends up playing is to regiment the discussion, providing the mathematical apparatus needed to make questions of theoretical structure and equivalence precise enough to settle. The fact that fairly simple ideas bear low-hanging fruit only provides further motivation for climbing into the higher branches.

---

[4] This particular thread continues: recently, Saunders (2013), Knox (2014), and Weatherall (2016c) have considered new arguments concerning just what structure is presupposed by Newton's *Principia*.

[5] For up-to-date overviews of the state of the art on the hole argument, see Norton (2011) and Pooley (2013).

[6] These issues also connect up with more general concerns in philosophy of science and metaphysics, concerning (1) the role of symmetry in guiding our attempts to extract metaphysical morals from our scientific (read: physical!) theories (Ismael and van Fraassen, 2003; Baker, 2010; Dasgupta, 2016) and (2) the relationship between the structural features of our theories and various forms of realism (Worrall, 1989; Ladyman and Ross, 2009; French, 2014).

[7] The material presented here draws on a number of recent papers by the author and collaborators, including Rosenstock et al. (2015); Rosenstock and Weatherall (2016); and Weatherall (2016a,d,e).

[8] For more on synthetic differential geometry, see Kock (2006); and for applications to physics, see Mac Lane (1968), Lawvere and Schanuel (1986), and, more recently, Reyes (2011). Higher gauge theory is described by Baez and Schreiber (2007).

13.2 Structure and Equivalence in Category Theory

There are many cases in which we want to say that one kind of mathematical object has more structure than another kind of mathematical object.[9] For instance, a topological space has more structure than a set. A Lie group has more structure than a smooth manifold. A ring has more structure than a group. And so on. In each of these cases, there is a sense in which the first sort of object—say, a topological space—results from taking an instance of the second sort—say, a set—and adding something more—in this case, a topology. In other cases, we want to say that two different kinds of mathematical object have the same amount of structure. For instance, given a Boolean algebra, I can construct a special kind of topological space, known as a Stone space, from which I can uniquely reconstruct the original Boolean algebra; and vice versa.

These sorts of relationships between mathematical objects are naturally captured in the language of category theory,[10] via the notion of a *forgetful functor*. For instance, there is a functor $F : \mathbf{Top} \to \mathbf{Set}$ from the category **Top**, whose objects are topological spaces and whose arrows are continuous maps, to the category **Set**, whose objects are sets and whose arrows are functions. This functor takes every topological space to its underlying set, and it takes every continuous function to its underlying function. We say this functor is forgetful because, intuitively speaking, it forgets something: namely the choice of topology on a given set.

The idea of a forgetful functor is made precise by a classification of functors due to Baez et al. (2004). This requires some machinery. A functor $F : \mathbf{C} \to \mathbf{D}$ is said to be *full* if for every pair of objects A, B of **C**, the map $F : \hom(A, B) \to \hom(F(A), F(B))$ induced by F is surjective, where $\hom(A, B)$ is the collection of arrows from A to B. Likewise, F is *faithful* if this induced map is injective for every such pair of objects. Finally, a functor is *essentially surjective* if for every object X of **D**, there exists some object A of **C** such that $F(A)$ is isomorphic to X.

If a functor is full, faithful, and essentially surjective, we will say that it forgets *nothing*. A functor $F : \mathbf{C} \to \mathbf{D}$ is full, faithful, and essentially surjective if and only if it is essentially invertible; i.e. there exists a functor $G : \mathbf{D} \to \mathbf{C}$ such that $G \circ F : \mathbf{C} \to \mathbf{C}$ is naturally isomorphic to $1_\mathbf{C}$, the identity functor on **C**, and $F \circ G : \mathbf{D} \to \mathbf{D}$ is naturally isomorphic to $1_\mathbf{D}$. (Note, then, that G is also essentially invertible, and thus G also forgets nothing.) This means that for each object A of **C**, there is an isomorphism $\eta_A : G \circ F(A) \to A$ such that for any arrow $f : A \to B$ in **C**, $\eta_B \circ G \circ F(f) = f \circ \eta_A$, and similarly for every object of **D**. When two categories are related by a functor that forgets nothing, we say the categories are *equivalent* and that the pair F, G realizes an *equivalence of categories*.

---

[9] The ideas in this section are developed in more detail by Barrett (2013).
[10] I will take for granted the basic definitions of category theory—that is, definitions of *category* and *functor*—but no more. For these notions, see Mac Lane (1998); Leinster (2014); and other papers in this volume.

Conversely, any functor that fails to be full, faithful, and essentially surjective forgets *something*. But functors can forget in different ways. A functor $F: \mathbf{C} \to \mathbf{D}$ forgets *structure* if it is not full; *properties* if it is not essentially surjective; and *stuff* if it is not faithful. Of course, "structure", "property", and "stuff" are technical terms in this context. But they are intended to capture our intuitive ideas about what it means for one kind of object to have more structure (resp., properties, stuff) than another. We can see this by considering some examples.

For instance, the functor $F : \mathbf{Top} \to \mathbf{Set}$ described earlier is faithful and essentially surjective, but not full, because not every function is continuous. So this functor forgets only structure—which is just the verdict we expected. Likewise, there is a functor $G : \mathbf{AbGrp} \to \mathbf{Grp}$ from the category $\mathbf{AbGrp}$ whose objects are Abelian groups and whose arrows are group homomorphisms to the categry $\mathbf{Grp}$ whose objects are (arbitrary) groups and whose arrows are group homomorphisms. This functor acts as the identity on the objects and arrows of $\mathbf{AbGrp}$. It is full and faithful, but not essentially surjective because not every group is Abelian. So this functor forgets only properties: namely, the property of being Abelian. Finally, consider the unique functor $H : \mathbf{Set} \to \mathbf{1}$, where $\mathbf{1}$ is the category with one object and one arrow. This functor is full and essentially surjective, but it is not faithful, so it forgets only stuff—namely all of the elements of the sets, since we may think of $\mathbf{1}$ as the category whose only object is the empty set, which has exactly one automorphism.

In what follows, we will say that one sort of object has more structure (resp. properties, stuff) than another if there is a functor from the first category to the second that forgets structure (resp. properties, stuff). It is important to note, however, that comparisons of this sort must be relativized to a choice of functor. In many cases, there is an obvious functor to choose—i.e. a functor that naturally captures the standard of comparison in question. But there may be other ways of comparing mathematical objects that yield different verdicts. For instance, there is a natural sense in which groups have more structure than sets, since any group may be thought of as a set of elements with some additional structure. This relationship is captured by a forgetful functor $F : \mathbf{Grp} \to \mathbf{Set}$ that takes groups to their underlying sets and group homomorphisms to their underlying functions. But any set also uniquely determines a group, known as the *free group* generated by that set; likewise, functions generate group homomorphisms between free groups. This relationship is captured by a different functor, $G : \mathbf{Set} \to \mathbf{Grp}$, that takes every set to the free group generated by it and every function to the corresponding group homomorphism. This functor forgets both structure (the generating set) and properties (the property of being a free group). So there is a sense in which sets may be construed to have more structure than groups.

13.3 Classical Space–Time Structure

To get a feel for how to apply these ideas to issues in philosophy of physics, I will begin by translating some well-understood examples into the present terms. In particular,

John Earman (1989b, ch. 2), building on Stein (1967) and others, characterizes several classical space-time structures that have been of interest historically. To focus on just two of the most important (the present discussion is easily extended), *Galilean space-time* consists of a quadruple (M, t_a, h^{ab}, ∇), where M is the manifold $\mathbb{R}^4$;[11] t_a is a non-vanishing one form on M; h^{ab} is a smooth, symmetric tensor field of signature $(0,1,1,1)$; and ∇ is a flat covariant derivative operator. We require that t_a and h^{ab} be compatible in the sense that $t_a h^{ab} = 0$ at every point, and that ∇ be compatible with both tensor fields, in the sense that $\nabla_a t_b = 0$ and $\nabla_a h^{bc} = 0$.

The points of M represent events in space and time. The field t_a is a "temporal metric", assigning a "temporal length" $|t_a \xi^a|$ to vectors ξ^a at a point $p \in M$. Since $\mathbb{R}^4$ is simply connected, $\nabla_a t_b = 0$ implies that there exists a smooth function $t : M \to \mathbb{R}$ such that $t_a = \nabla_a t$. We may thus define a foliation of M into constant-t hypersurfaces representing collections of simultaneous events—i.e. space at a time. We assume that each of these surfaces is diffeomorphic to $\mathbb{R}^3$ and that h^{ab} restricted these surfaces is (the inverse of) a flat, Euclidean, and complete metric. In this sense, h^{ab} may be thought of as a spatial metric, assigning lengths to spacelike vectors, all of which are tangent to some spatial hypersurface. We represent particles propagating through space over time by smooth curves whose tangent vector ξ^a, called the 4-*velocity* of the particle, satisfies $\xi^a t_a = 1$ along the curve. The derivative operator ∇ then provides a standard of acceleration for particles, which is given by $\xi^n \nabla_n \xi^a$. Thus, in Galilean space-time we have notions of objective duration between events; objective spatial distance between simultaneous events; and objective acceleration of particles moving through space over time.

However, Galilean space-time does *not* support an objective notion of the (spatial) velocity of a particle. To get this, we move to *Newtonian space-time*, which is a quintuple $(M, t_a, h^{ab}, \nabla, \eta^a)$. The first four elements are precisely as in Galilean space-time, with the same assumptions. The final element, η^a, is a smooth vector field satisfying $\eta^a t_a = 1$ and $\nabla_a \eta^b = 0$. This field represents a state of absolute rest at every point—i.e. it represents "absolute space". This field allows one to define absolute velocity: given a particle passing through a point p with 4-velocity ξ^a, the (absolute, spatial) velocity of the particle at p is $\xi^a - \eta^a$.

There is a natural sense in which Newtonian space-time has strictly more structure than Galilean space-time: after all, it consists of Galilean space-time plus an additional element. As Earman observes, this judgment may be made precise by observing that the automorphisms of Newtonian space-time—that is, its space-time symmetries—form a proper subgroup of the automorphisms of Galilean space-time. The intuition here is that if a structure has *more* symmetries, then there must be less structure

---

[11] More generally, the manifolds I consider throughout the chapter are smooth, Hausdorff, paracompact, and connected. All fields defined on these manifolds are likewise assumed to be smooth, unless I indicate otherwise. I work in the abstract index notation, as developed by Penrose and Rindler (1984). For further details on classical space-time structure, see Malament (2012, ch. 4).

that is preserved by the maps.[12] In the case of Newtonian space-time, these automorphisms are diffeomorphisms $\vartheta : M \to M$ that preserve t_a, h^{ab}, ∇, and η^a. These will consist in rigid spatial rotations, spatial translations, and temporal translations (and combinations of these). Automorphisms of Galilean space-time, meanwhile, will be diffeomorphisms that preserve only the metrics and derivative operator. These include all of the automorphisms of Newtonian space-time just described, plus Galilean boosts.

It is this notion of "more structure" that is captured by the forgetful functor approach described above. To recapitulate Earman, we define two categories, **Gal** and **New**, which have Galilean and Newtonian space-time as their (essentially unique) objects, respectively, and have automorphisms of these space-times as their arrows.[13] Then there is a functor $F: \mathbf{New} \to \mathbf{Gal}$ that takes arrows of **New** to arrows of **Gal** generated by the same automorphism of M. This functor is clearly essentially surjective and faithful, but it is not full for reasons already discussed, and so it forgets only structure. Thus, the criterion of structural comparison given earlier perfectly recapitulates Earman's condition—and indeed, may be seen as a generalization of the latter to cases where one is comparing collections of models of a theory, rather than individual space-times.

To see this last point more clearly, let us move to another well-trodden example. There are two approaches to classical gravitational theory: (ordinary) Newtonian gravitation (NG) and geometrized Newtonian gravitation (GNG), sometimes known as Newton–Cartan theory. Models of NG consist of Galilean space-time as described earlrier, plus a scalar field φ, representing a *gravitational potential*. This field is required to satisfy Poisson's equation, $\nabla^a \nabla_a \varphi = 4\pi\rho$, where ρ is a smooth scalar field representing the mass density on space-time. In the presence of a gravitational potential, massive test point particles will accelerate according to $\xi^n \nabla_n \xi^a = -\nabla^a \varphi$, where ξ^a is the 4-velocity of the particle. We write models as $(M, t_a, h^{ab}, \nabla, \varphi)$.[14]

The models of GNG, meanwhile, may be written as quadruples $(M, t_a, h^{ab}, \tilde{\nabla})$, where we assume for simplicity that M, t_a, and h^{ab} are all as described earlier, and where $\tilde{\nabla}$ is a covariant derivative operator compatible with t_a and h^{ab}. Now, however, we allow $\tilde{\nabla}$ to be curved, with Ricci curvature satisfying the geometrized Poisson equation, $R_{ab} = 4\pi\rho t_a t_b$, again for some smooth scalar field ρ representing the mass density.[15] In this theory, gravitation is not conceived as a force: even in the presence

---

[12] This way of comparing structure is also explored by Barrett (2015b), under the name SYM*. He goes on to argue that Minkowski space-time (see fn. 18) *also* has less structure than Newtonian space-time, but that Galilean and Minkowski space-times have incomparable amounts of structure.

[13] For simplicity, following Baez et al. (2004), I am restricting attention to groupoids, i.e. categories with only isomorphisms. There are several ways in which one could add arrows to the categories I discuss here and in what follows, but nothing turns on how or whether one includes them.

[14] Here and throughout, we suppress the source terms in differential equations when writing models of a theory, since such fields can generally be defined from the other fields—so, for instance, given a model $(M, t_a, h^{ab}, \nabla, \varphi)$ of NG, we define an associated mass density by Poisson's equation.

[15] The Ricci tensor associated with a covariant derivative operator ∇ is defined by $R_{bc} = R^a{}_{bca}$, where $R^a{}_{bcd}$ is the *Riemann tensor*, which is the unique tensor field such that given any point p and any smooth

of matter, massive test point particles traverse geodesics of $\tilde{\nabla}$—where now these geodesics depend on the distribution of matter, via the geometrized Poisson equation.

There is a sense in which NG and GNG are empirically equivalent: a pair of results due to Trautman (1965) guarantee that (1) given a model of NG, there always exists a model of GNG with the same mass distribution and the same allowed trajectories for massive test point particles, and (2), with some further assumptions, vice versa (see Malament, 2012, §4.2). But in a number of influential articles, Glymour (1970, 1977, 1980) has argued that these are nonetheless *inequivalent* theories, because of an asymmetry in the relationship just described. Given a model of NG, there is a unique corresponding model of GNG. But given a model of GNG, there are typically many corresponding models of NG. Thus, it appears that NG makes distinctions that GNG does not make (despite the empirical equivalence), which in turn suggests that NG has more structure than GNG.

This intuition, too, may be captured using a forget functor. Define a category **NG** whose objects are models of NG (for various mass densities) and whose arrows are automorphisms of M that preserve t_a, h^{ab}, ∇, and φ; and a category **GNG** whose objects are models of GNG and whose arrows are automorphisms of M that preserve t_a, h^{ab}, and $\tilde{\nabla}$. Then there is a functor $F : \textbf{NG} \to \textbf{GNG}$ that takes each model of NG to the corresponding model given by the Trautman results, and takes each arrow to an arrow generated by the same diffeomorphism.[16] Then results in Weatherall (2016a) imply the following.

Proposition 13.1 $F : \textbf{NG} \to \textbf{GNG}$ forgets only structure.

13.4 Excess Structure and "Gauge"

The relationship between NG and GNG captured by Proposition 13.1 is revealing. As noted, there is a sense in which NG and GNG are empirically equivalent—and more, there is a sense in which the theories are capable of representing precisely the same physical situations. And yet, NG has more structure than GNG. This suggests that NG has more structure than is strictly necessary to represent these situations, since after all, GNG can do the same representational work with less structure. Theories with this character are sometimes said to have "excess structure" and are often called *gauge theories* by physicists.[17]

The archetypal example of a gauge theory is classical electromagnetism. To be clear about the sense in which electromagnetism has excess structure, I will present two characterizations of its models. (For simplicity, I limit attention to electromagnetism

---

vector field defined on a neighborhood of that point, $R^a{}_{bcd}\xi^b = -2\nabla_{[c}\nabla_{d]}\xi^a$. The Riemann tensor vanishes iff the derivative operator is flat.

[16] That this is a functor is established in Weatherall (2016a).
[17] For more on the term "gauge theory", see Weatherall (2016e).

in a fixed background of Minkowski space–time, (M, η_{ab})).[18] On the first formulation, which I will call EM_1, the fundamental dynamical quantity is a two form F_{ab}, known as the *Faraday tensor*. This tensor field represents the electromagnetic field at each point.[19] It is required to satisfy Maxwell's equations, which may be written as $d_a F_{bc} = 0$ and $\nabla_a F^{ab} = J^b$, where d is the exterior derivative, ∇ is the Minkowski space–time derivative operator, and J^b is the *charge current density* associated with any charged matter present in space–time. On the second formulation, EM_2, the fundamental dynamical quantity is a one form, A_a, known as the *vector potential*. The vector potential is required to satisfy $\nabla_a \nabla^a A^b = \nabla^b \nabla_a A^a + J^b$. On the first formulation, a model of electromagnetism may be represented as an ordered triple, (M, η_{ab}, F_{ab}); on the second, a model would be a triple (M, η_{ab}, A_a).

As with NG and GNG in the previous section, there is a close relationship between EM_1 and EM_2. Given a model of EM_2, (M, η_{ab}, A_a), I can always construct a model of EM_1 that satisfies Maxwell's equations for the same charge-current density, by defining $F_{ab} = d_a A_b$. Conversely, given a model (M, η_{ab}, F_{ab}) of EM_1, there always exists a suitable vector potential A_a such that $F_{ab} = d_a A_b$. On both formulations, the empirical significance of the theory is taken to be captured by the associated Faraday tensor—directly for models of EM_1, and by the relationship $A_a \mapsto d_a A_b$ for EM_2. And so, we have a clear sense in which the two formulations are empirically equivalent.

These formulations of electromagnetism are analogous to NG and GNG, as follows. While every model of EM_2 gives rise to a unique model of EM_1, there are typically many models of EM_2 corresponding to any given model of EM_1. Once again, this appears to have the consequence that EM_2 makes distinctions that EM_1 does not—and that these distinctions between models of EM_2 add nothing to the empirical success of the theory. Again, this relationship is captured by a forgetful functor: define a category **EM₁** with models of EM_1 as objects and isometries of Minkowski space–time that preserve F_{ab} as arrows; and define a category **EM₂** with models of EM_2 as objects, and isometries of Minkowski space–time that preserve A_a as arrows. We may then define a functor $F : \mathbf{EM_2} \to \mathbf{EM_1}$ that acts on objects via $A_a \mapsto F_{ab} = d_a A_b$, and which takes arrows to arrows generated by the same isometry. Weatherall (2016e) then proves the following.

Proposition 13.2 $F : \mathbf{EM_2} \to \mathbf{EM_1}$ forgets only structure.

We thus have a sense in which EM_2 has more structure than EM_1—and indeed, since EM_1 and EM_2 are taken to have the same representational capacities, it would seem that EM_2 must have *excess* structure. This raises a question, though. If EM_2 has excess structure, why do physicists use it? The answer is purely pragmatic: vector potentials are often more convenient to work with than Faraday tensors. Moreover, the

---

[18] Minkowski space–time is a relativistic space–time consisting of the manifold $\mathbb{R}^4$ with a flat, complete Lorentzian metric η_{ab}. For more on relativistic space–times, see Section 13.6 and Malament (2012).

[19] For more on how to recover electric and magnetic fields from F_{ab}, see Malament (2012, §13.6).

excess structure does not cause any problems for the theory, because physicists are well aware that it is there, and it is easily controlled. In particular, if A_a is a vector potential, then $A'_a = A_a + \nabla_a \psi$, for any smooth scalar field ψ, will be such that $d_a A_a = d_a A'_b$; moreover, A'_a will satisfy the relevant differential equation for a given charge–current density just in case A_a does. The map that takes A^a to A'^a is an example of a *gauge transformation*. Vector potentials related by gauge transformations are taken to be physically equivalent—even though they are mathematically *inequivalent*, in the sense that they concern distinct fields on Minkowski space-time.

Once we take gauge transformations into account, the two formulations are usually taken to be equivalent ways of presenting electromagnetism. This relationship can also be captured in the language we have been developing. In effect, that the functor $F : \mathbf{EM}_2 \to \mathbf{EM}_1$ defined in Proposition 13.2 fails to be full reflects the fact that, on the natural way of relating models of the theories, there are arrows "missing" from $\mathbf{EM}_2$. The gauge transformations, meanwhile, may be understood as additional arrows— that is, they provide a second notion of isomorphism between models of $\mathbf{EM}_2$ that preserves the structure that we take to have representational significance in physics. These arrows can be added to $\mathbf{EM}_2$ to define a new category, $\mathbf{EM}'_2$, which has the same objects as $\mathbf{EM}_2$, but whose arrows $f : (M, \eta_{ab}, A_a) \to (M, \eta_{ab}, A'_a)$ are now pairs $f = (\chi, G_a)$, where $\chi : M \to M$ is an isometry of Minkowski space-time, $G_a = \nabla_a \psi$ for some smooth scalar field ψ, and $\chi^*(A'_a) = A_a + G_a$. We may then define a new functor $F' : \mathbf{EM}'_2 \to \mathbf{EM}_1$, which has the same action on objects as F, but which acts on arrows as $(\chi, G_a) \mapsto \chi$.[20] Weatherall (2016a) then proves the following.

Proposition 13.3 $F' : \mathbf{EM}'_2 \to \mathbf{EM}_1$ forgets nothing.

As Weatherall (2016a) also shows, one can identify an analogous class of gauge transformations between models of NG. These additional transformations, which reflect the fact that Newtonian physics cannot distinguish a state of inertial motion from one of uniform linear acceleration, have arguably been recognized as equivalences between models of Newtonian gravitation since Newton's *Principia*—indeed, one may interpret Corollary VI to the Laws of Motion as describing precisely these transformations.[21] More, one can define an alternative category with models of NG as objects, and with these Newtonian gauge transformations included among the arrows; this new category is then equivalent to **GNG**.

13.5 Yang–Mills Theory and Excess Structure

The discussion of gauge theories in the previous section makes precise one sense in which electromagnetism and Newtonian gravitation have excess structure: there

---

[20] That $\mathbf{EM}'_2$ is a category, and that F' is a functor, is shown in Weatherall (2016a).
[21] For more on the significance of Corollary VI, see DiSalle (2008), Saunders (2013), Knox (2014), and Weatherall (2016c).

are textbook formulations of these theories that have structure that apparently plays no role in how the theories are used, in the sense that there are other formulations of the theories that have the same representational capacities, but which have less structure in the sense described by Propositions 13.1 and 13.2. In many cases of interest, however, we do not have multiple formulations to work with, and we are confronted with the question of whether a particular formulation has excess structure without a comparison class.

It is here that the relationship between EM_2 and EM_2' becomes especially important. What we see in Proposition 13.3 is that the excess structure in EM_2, as captured by Proposition 13.2, may be "removed" by identifying an additional class of "gauge transformations" that relate non-isomorphic, but physically equivalent, models of the theories. This observation provides us with an alternative criterion for identifying when a formulation of a theory has excess structure: namely, when there are models of the theory that we believe have precisely the same representational capacities, but which are not isomorphic to one another.[22] With this criterion in mind, I will now turn to other theories that are often called gauge theories, to try to identify whether they have excess structure. In this section, I will consider (classical) Yang–Mills theory; in the next section, general relativity.

Classically, a model of Yang–Mills theory consists in a principal connection $\omega^{\mathfrak{A}}{}_\alpha$ on a principal bundle $G \to P \xrightarrow{\wp} M$ with structure group G, over a relativistic space-time (M, g_{ab}) (see Section 13.6).[23] We will write these models as $(P, g_{ab}, \omega^{\mathfrak{A}}{}_\alpha)$. The principal bundle P may be thought of as a bundle of frames for associated vector bundles $V \to P \times_G V \xrightarrow{\pi} M$, sections of which represent distributions of matter on space-time. The principal connection $\omega^{\mathfrak{A}}{}_\alpha$ determines a unique derivative operator on every such vector bundle; this derivative operator is the one appearing in matter dynamics, and in this way the connection affects the evolution of matter. Conversely, every matter field is associated with a horizontal and equivariant Lie algebra valued one form $J^{\mathfrak{A}}{}_\alpha$ on P, called the *charge-current density*. The principal connection is related to the total charge-current density on P by the Yang–Mills equation, $\star \overset{\omega}{D}_\alpha \star \Omega^{\mathfrak{A}}{}_{\beta\kappa} = J^{\mathfrak{A}}{}_\kappa$, where $\overset{\omega}{D}_\alpha$ is the exterior covariant derivative relative to $\omega^{\mathfrak{A}}{}_\alpha$, $\star$ is a Hodge star operator on horizontal and equivariant forms on P determined by the space–time metric g_{ab} on M, and $\Omega^{\mathfrak{A}}{}_{\alpha\beta}$ is the curvature associated with $\omega^{\mathfrak{A}}{}_\alpha$. Thus, Yang–Mills theory may be understood as a theory in which charged matter propagates in a curved space, the curvature of which is dynamically related to the distribution of charged matter.

---

[22] This criterion is also discussed in Weatherall (2016e).

[23] One also requires an inner product on the Lie algebra associated to G, but that will play no role in the following. For more on the principal bundle formalism for Yang–Mills theory, see Trautman (1980); Palais (1981); Bleecker (1981); Deligne and Freed (1999); and Weatherall (2016b). The notation used here, a variant of the abstract index notation, is explained in an appendix to Weatherall (2016b). Briefly, fraktur indices indicate valuation in a Lie algebra; lower case Greek indices label tangent vectors to the total space of a principal bundle; and lower case Latin indices label tangent vectors to space-time. So, for instance, $\omega^{\mathfrak{A}}{}_\alpha$ is a Lie algebra-valued one form on P, mapping tangent vectors at a point of P to the Lie algebra associated to G.

The connection, curvature, and charge–current density may all be represented as fields on M by choosing a local section $\sigma : U \subseteq M \to P$ and considering the pullbacks of these fields along the section. The resulting fields are generally dependent on the choice of section; changes of section are known as "gauge transformations". Note, however, that these gauge transformations are strongly disanalogous to those encountered in Section 13.4: they are not maps between distinct models of the theory, and so they do not indicate that there are non-isomorphic models of the theory that have the same representational capacities. Instead, they are most naturally construed as changes of local coordinate system—or frame field—relative to which one represents invariant fields on P as fields on M.[24]

In fact, it seems that Yang–Mills theory as just described does not have excess structure at all. One way of emphasizing this point is to observe that classical electromagnetism is a Yang–Mills theory in the present sense, which means we can compare EM_1 and EM_2 with a third formulation of the theory, EM_3. EM_3 is a Yang–Mills theory with structure group $U(1)$. The Lie algebra associated with this group is $\mathbb{R}$, and so Lie algebra valued forms are just real valued, which means we can drop the index $\mathfrak{A}$ from the curvature and connection forms. Continuing as above, we limit attention to Minkowski space–time. Thus, a model of the theory is a triple $(P, \eta_{ab}, \omega_\alpha)$, where ω_α is a principal connection on $U(1) \to P \xrightarrow{\wp} M$, the unique (necessarily trivial) $U(1)$ bundle over Minkowski space–time. Given such a model, we may generate models of EM_1 or EM_2 by choosing a (global) section $\sigma : M \to P$, and defining a Faraday tensor $F_{ab} = \sigma^*(\Omega_{\alpha\beta})$ or a vector potential $A_a = \sigma^*(\omega_\alpha)$, respectively. Since $U(1)$ is Abelian, F_{ab} is independent of the choice of section; A_a, however, depends on the section, with different sections producing different (gauge-related) vector potentials. If ω_α satisfies the Yang–Mills equation for a charge–current density J_α, then A_a and F_{ab} will satisfy their respective equations of motion for a field $J_a = \sigma^*(J_\alpha)$.

We now define a category $\mathbf{EM_3}$ whose objects are models of EM_3 and whose arrows are principal bundle isomorphisms that preserve both ω_α and the Minkowski metric η_{ab}. There is also a natural functor $F : \mathbf{EM_3} \to \mathbf{EM_1}$, which acts on objects as $(P, \eta_{ab}, \omega_\alpha) \mapsto (M, \eta_{ab}, d_a\sigma^*(\omega_\alpha))$ (for any global section σ) and on arrows as $(\Psi, \psi) \mapsto \psi$. Weatherall (2016e) then proves the following result.[25]

Proposition 13.4 $F : \mathbf{EM_3} \to \mathbf{EM_1}$ *does not forget structure.*

It follows that, by the criterion of comparison we have been considering, the principal bundle formalism has no more structure than EM_1. This is despite the fact

---

[24] One could recover a sense in which changes of section were gauge transformations in the other sense, discussed earlier, by stipulating that models come equipped with a preferred choice of section. But this is not natural for multiple reasons—the most important of which is that in general, there are no sections defined on all of M.

[25] In fact, Weatherall (2016e) erroneously claims that F forgets nothing. The argument there does, however, establish that F is full, and thus does not forget structure. It seems F does forget stuff, but I do not have space to discuss the interpretational significance of this fact.

that the dynamical variable of the theory is the *connection* on P, which is analogous—via a choice of section—to the vector potential. The reason the equivalence holds is that given any diffeomorphism that preserves the Faraday tensor, there is a corresponding principal bundle automorphism that also preserves ω_α, in effect by systematically relating the possible sections of P.

These variants of electromagnetism on Minkowski space–time are ultimately toy examples. That said, there is another issue in the neighborhood that has been a locus of recent debate. It concerns the relationship between the formalism for Yang–Mills theory just described—the so-called "fiber bundle formalism"—and a formalism known as the "holonomy formalism" or "loop formalism". Each of these is often associated with an "interpretation" of Yang–Mills theory, though I will not discuss those interpretations.[26] It is sufficient to note that one theme in these debates is the claim that the holonomy formalism posits (or requires) less structure than the fiber bundle formalism.

The idea behind the holonomy formalism for Yang–Mills theory is as follows.[27] Given a fiber bundle model of a Yang–Mills theory, $(P, g_{ab}, \omega^{\mathfrak{A}}{}_\alpha)$, and a fixed point $u \in P$, we may define a map $H : L_{\wp(u)} \to G$, where $L_{\wp(u)}$ is the collection of piecewise smooth closed curves $\gamma : [0, 1] \to M$ originating (and ending) at $\wp(u)$. This map assigns to each curve $\gamma \in L_{\wp(u)}$ the element $g \in G$ such that $\gamma_u(1) = \gamma_u(0)g$, where $\gamma_u : [0, 1] \to P$ is the (unique) horizontal lift of γ through u, relative to $\omega^{\mathfrak{A}}{}_\alpha$.[28] The group element $H(\gamma)$ is known as the *holonomy* of γ. One usually interprets $\gamma_u(1) \in P$ to be the *parallel transport* of u along γ, relative to the principal connection $\omega^{\mathfrak{A}}_\alpha$; thus, the holonomies associated with a connection encode information about the parallel transport properties of $\omega^{\mathfrak{A}}{}_\alpha$.

It is arguably the case that the holonomies of a principal connection contain all of the empirically significant data associated with a fiber bundle model of the theory.[29] The holonomy formalism attempts to characterize Yang–Mills theory directly with holonomy data. It does so via a *generalized holonomy map* $H : L_x \to G$, which simply assigns group elements to closed curves without ever mentioning a principal bundle. (Not any map will do; the properties required of a generalized holonomy map are described in Barrett (1991) and Rosenstock and Weatherall (2016).) Many commentators have had the intuition that this approach is more parsimonious than the fiber bundle formalism, since it posits just the structure needed to encode the possible predictions of the theory, without any geometrical superstructure.

---

[26] In particular, see Healey (2007) and Arntzenius (2012) for discussions of the interpretive options related to these formalisms; see also Rosenstock (2017).

[27] For details, see Rosenstock and Weatherall (2016); for further background, see Barrett (1991), Loll (1994), and Gambini and Pullin (1996).

[28] The horizontal lift of a curve $\gamma : [0, 1] \to M$ through $u \in \wp^{-1}[\gamma(0)]$ is the unique curve $\gamma_u : [0, 1] \to P$ such that $\pi \circ \gamma_u = \gamma$ and $\omega^{\mathfrak{A}}{}_\alpha \xi^\alpha = 0$ along the curve, where ξ^α is the tangent to γ_u. See Kobayashi and Nomizu (1963, ch. 2).

[29] See Healey (2007) for a defense of this claim. It is usually motivated by arguing that the empirical significance of a Yang–Mills theory concerns only interference effects exhibited by quantum particles.

But can this thesis of relative parsimony be made precise? It is not clear that it can be. In fact, the methods described here give a strikingly different answer. For any given Yang–Mills theory, circumscribed by some fixed choice of structure group G, we may define two categories of models, associated with each of these formalisms. The first, PC, is a generalization of EM_3: the objects are fiber bundle models $(P, g_{ab}, \omega^{\mathfrak{A}}{}_\alpha)$, with structure group G; and the arrows are principal bundle isomorphisms that preserve the connection and metric. The second, Hol, corresponds to the holonomy formalism. Here the objects are *holonomy models* $(M, g_{ab}, H : L_x \to G)$, where H is a generalized holonomy map, and the arrows are *holonomy isomorphisms*, which are maps that preserve the metric and holonomy structure. (These are somewhat subtle, and are described in detail in Rosenstock and Weatherall (2016).) There is a functor $F :$ Hol $\to$ PC that takes every holonomy model (M, g_{ab}, H) to a fiber bundle model that gives rise to the holonomies H, with appropriately compatible action on arrows. Rosenstock and Weatherall (2016) then prove the following result.

Proposition 13.5 $F :$ Hol $\to$ PC forgets nothing.

In other words—philosophers' intuitions notwithstanding—the formalisms underlying holonomy and fiber bundle interpretations have precisely the same amounts of structure, relative to a natural standard of comparison between models of the theories.[30]

13.6 General Relativity, Einstein Algebras, and the Hole Argument

Finally, I will return to space–time physics, to discuss the methods of the previous sections in the context of general relativity. I will briefly discuss two related topics. The first concerns the so-called "hole argument" of Earman and Norton (1987) and Stachel (1989); the second concerns a proposal for an alternative to the standard formalism of general relativity originally due to Geroch (1972), and later championed by Earman (1977a,b, 1979, 1986, 1989a,b).[31]

To begin, the models of relativity theory are *relativistic space–times*, which are pairs (M, g_{ab}) consisting of a 4-manifold M and a smooth, Lorentz-signature metric g_{ab}.[32] The metric represents geometrical facts about space–time, such as the spatiotemporal distance along a curve, the volume of regions of space–time, and the angles between vectors at a point. It also characterizes the motion of matter: the metric g_{ab} determines a unique torsion-free derivative operator ∇, which provides the standard of constancy

---

[30] Space constraints prevent further elaboration on this point; see Rosenstock (2017) for further details.
[31] For more on the hole argument, see Earman (1989b) and Pooley (2013); see also Weatherall (2016d), which extends the arguments given here.
[32] My notational conventions again follow Malament (2012). In particular, I work with a signature $(1, -1, -1, -1)$ metric.

in the equations of motion for matter. Meanwhile, geodesics of this derivative operator whose tangent vectors ξ^a satisfy $g_{ab}\xi^a\xi^b > 0$ are the possible trajectories for free massive test particles in the absence of external forces. The distribution of matter in space and time determines the geometry of space–time via Einstein's equation, $R_{ab} - \frac{1}{2}Rg_{ab} = 8\pi T_{ab}$, where T_{ab} is the *energy-momentum tensor* associated with any matter present, R_{ab} is the Ricci tensor, and $R = R^a{}_a$. Thus, as in Yang–Mills theory, matter propagates through a curved space, the curvature of which depends on the distribution of matter in space–time.

The most widely discussed topic in the philosophy of general relativity over the last thirty years has been the hole argument, which goes as follows. Fix some space–time (M, g_{ab}), and consider some open set $O \subseteq M$ with compact closure. For convenience, assume $T_{ab} = \mathbf{0}$ everywhere. Now pick some diffeomorphism $\psi: M \to M$ such that $\psi_{|M-O}$ acts as the identity, but $\psi_{|O}$ is not the identity. This is sufficient to guarantee that ψ is a non-trivial automorphism of M. In general, ψ will not be an isometry, but one can always define a new space–time $(M, \psi^*(g_{ab}))$ that is guaranteed to be isometric to (M, g_{ab}), with the isometry realized by ψ. This yields two relativistic space–times, both representing possible physical configurations that agree on the value of the metric at every point outside of O, but in general disagree at points within O. This means that the metric outside of O, including at all points in the past of O, cannot determine the metric at a point $p \in O$. Thus, Earman and Norton (1987) argue, general relativity, as standardly presented, faces a pernicious form of indeterminism. To avoid this indeterminism, one must become a *relationist* and accept that "Leibniz equivalent", i.e. isometric, space–times represent the same physical situations. The person who denies this latter view—and thus faces the indeterminism—is dubbed a *manifold substantivalist*.

One way of understanding the dialectical context of the hole argument is as a dispute concerning the correct notion of equivalence between relativistic space–times. The manifold substantivalist claims that isometric space–times are *not* equivalent, whereas the relationist claims that they are. In the present context, these views correspond to different choices of arrows for the categories of models of general relativity. The relationist would say that general relativity should be associated with the category $\mathbf{GR_1}$, whose objects are relativistic space–times and whose arrows are isometries. The manifold substantivalist, meanwhile, would claim that the right category is $\mathbf{GR_2}$, whose objects are again relativistic space–times, but which has only identity arrows. Clearly there is a functor $F: \mathbf{GR_2} \to \mathbf{GR_1}$ that acts as the identity on both objects and arrows and forgets only structure. Thus the manifold substantivalist posits more structure than the relationist.

Manifold substantivalism might seem puzzling—after all, we have said that a relativistic space–time is a Lorentzian manifold (M, g_{ab}), and the theory of pseudo-Riemannian manifolds provides a perfectly good standard of equivalence for Lorentzian manifolds *qua* mathematical objects: namely, isometry. Indeed, while one may stipulate that the objects of $\mathbf{GR_2}$ are relativistic space–times, the arrows of the category do not reflect that choice. One way of charitably interpreting the manifold substantivalist is to say that in order to provide an adequate representation of *all* the physical facts, one actually needs *more* than a Lorentzian manifold. This extra structure might be something like a fixed collection of labels for the points of the manifold, encoding which point in physical space–time is represented by a given point in the manifold.³³ Isomorphisms would then need to preserve these labels, so space–times would have no non-trivial automorphisms. On this view, one might *use* Lorentzian manifolds, without the extra labels, for various purposes, but when one does so, one does not represent all of the facts one might (sometimes) care about.

In the context of the hole argument, isometries are sometimes described as the "gauge transformations" of relativity theory; they are then taken as evidence that general relativity has excess structure. But as I argued in Section 13.5, one can expect to have excess structure in a formalism only if there are models of the theory that have the same representational capacities, but which are *not* isomorphic as mathematical objects. If we take models of GR to be Lorentzian manifolds, then that criterion is not met: isometries are precisely the isomorphisms of these mathematical objects, and so general relativity does *not* have excess structure.

This point may be made in another way. motivated in part by the idea that the standard formalism has excess structure, Earman has proposed moving to the alternative formalism of so-called Einstein algebras for general relativity, arguing that Einstein algebras have less structure than relativistic space–times.³⁴ In what follows, a *smooth n-algebra* A is an algebra isomorphic (as algebras) to the algebra $C^\infty(M)$ of smooth real-valued functions on some smooth n-manifold, M.³⁵ A *derivation* on A is an $\mathbb{R}$-linear map $\xi: A \to A$ satisfying the Leibniz rule, $\xi(ab) = a\xi(b) + b\xi(a)$. The space of derivations on A forms an A-module, $\Gamma(A)$, elements of which are analogous to smooth vector fields on M. Likewise, one may define a dual module, $\Gamma^*(A)$, of linear functionals on $\Gamma(A)$. A *metric*, then, is a module isomorphism $g: \Gamma(A) \to \Gamma^*(A)$ that is symmetric in the sense that for any $\xi, \eta \in \Gamma(A)$, $g(\xi)(\eta) = g(\eta)(\xi)$. With some further work, one can capture a notion of signature of such metrics, exactly analogously to metrics on a manifold. An *Einstein algebra*, then, is a pair (A, g), where A is a smooth 4-algebra and g is a Lorentz signature metric.

Einstein algebras arguably provide a "relationist" formalism for general relativity, since one specifies a model by characterizing (algebraic) relations between possible states of matter, represented by scalar fields. It turns out that one may then reconstruct a unique relativistic space–time, up to isometry, from these relations by representing an Einstein algebra as the algebra of functions on a smooth manifold. The question, though, is whether this formalism really eliminates structure. Let $\mathbf{GR_1}$ be as above, and define $\mathbf{EA}$ to be the category whose objects are Einstein algebras and whose

---

³³ A similar suggestion is made by Stachel (1993).
³⁴ For details, see Rosenstock et al. (2015). In many ways, the arguments here are reminiscent of those of Rynasiewicz (1992).
³⁵ These may also be characterized in purely algebraic terms (Nestruev, 2003).

arrows are algebra homomorphisms that preserve the metric g (in a way made precise by Rosenstock et al. (2015)). Define a *contravariant* functor $F : \text{GR}_1 \to \text{EA}$ that takes relativistic space–times (M, g_{ab}) to Einstein algebras $(C^\infty(M), g)$, where g is determined by the action of g_{ab} on smooth vector fields on M, and takes isometries $\psi : (M, g_{ab}) \to (M', g'_{ab})$ to algebra isomorphisms $\hat{\psi} : C^\infty(M') \to C^\infty(M)$, defined by $\hat{\psi}(a) = a \circ \psi$.[36] Rosenstock et al. (2015) prove the following.

Proposition 13.6 $F : \text{GR}_1 \to \text{EA}$ forgets nothing.

13.7 Conclusion

I have reviewed several cases in which representing a scientific theory as a category of models is useful for understanding the structure associated with a theory. In the context of classical space–time structure, the category theoretic machinery merely recovers relationships that have long been appreciated by philosophers of physics; these cases are perhaps best understood as litmus tests for the notion of "structure" described here. In the other cases, the new machinery appears to do useful work. It helps crystalize the sense in which EM_2 and NG have excess structure, in a way that clarifies an important distinction between these theories and other kinds of gauge theories, such as Yang–Mills theory and general relativity. It also clarifies the relationship between various formulations of physical theories that have been of interest to philosophers because of their alleged parsimony. These results seem to reflect real progress in our understanding of these theories—progress that apparently required the basic category theory used here. One hopes that these methods may be extended further—perhaps to issues concerning the relationships between algebras of observables and their representations in quantum field theory and the status of dualities in string theory.[37]

Acknowledgments

This material is based upon work supported by the National Science Foundation under Grant 1331126. Thank you to Thomas Barrett, Hans Halvorson, Ben Feintzeig, Sam Fletcher, David Malament, and Sarita Rosenstock for many discussions related to this material, and to Thomas Barrett, Ben Feintzeig, JB Manchak, Sarita Rosenstock, and two anonymous referees for comments on a previous draft.

---

[36] A contravariant functor is one that takes arrows $f : A \to B$ to arrows $F(f) : F(B) \to F(A)$. The classification described above carries over to contravariant functors, though two categories related by a full, faithful, essentially surjective contravariant functor are said to be *dual*, rather than *equivalent*.

[37] For more on the former, see Ruetsche (2011) and references therein.

References

Arntzenius, F. (2012). *Space, Time, and Stuff*. Oxford University Press, New York.

Baez, J., Bartel, T., and Dolan, J. (2004). Property, structure, and stuff. http://math.ucr.edu/home/baez/qg-spring2004/discussion.html

Baez, J., and Schreiber, U. (2007). Higher gauge theory, in A. Davydov (ed.), *Categories in Algebra, Geometry, and Mathematical Physics*, 7–30. American Mathematical Society, Providence, RI.

Baker, D. J. (2010). Symmetry and the metaphysics of physics. *Philosophy Compass* 5(12), 1157–66.

Barrett, J. W. (1991). Holonomy and path structures in general relativity and Yang-Mills theory. *International Journal of Theoretical Physics* 30(9), 1171–215.

Barrett, T. (2013). How to count structure. Unpublished.

Barrett, T. (2015a). On the structure of classical mechanics. *British Journal of Philosophy of Science* 66(4), 801–28.

Barrett, T. (2015b). Spacetime structure. *Studies in History and Philosophy of Modern Physics* 51, 37–43.

Belot, G. (1998). Understanding electromagnetism. *British Journal for Philosophy of Science* 49(4), 531–55.

Belot, G. (2003). Symmetry and gauge freedom. *Studies in History and Philosophy of Modern Physics* 34(2), 189–225.

Bleecker, D. (1981). *Gauge Theory and Variational Principles*. Addison-Wesley, Reading, MA, reprinted by Dover Publications in 2005.

Carnap, R. (1968 [1928]). *The Logical Structure of the World*. University of California Press, Berkeley, CA.

Curiel, E. (2013). Classical mechanics is Lagrangian; it is not Hamiltonian. *British Journal for Philosophy of Science* 65(2), 269–321.

Curiel, E. (2014). Measure, topology and probabilistic reasoning in cosmology. http://philsci-archive.pitt.edu/11071/

Dasgupta, S. (2016). Symmetry as an epistemic notion (twice over). *British Journal for the Philosophy of Science* 67(3), 837–78. https://doi.org/10.1093/bjps/axu049

Deligne, P., and Freed, D. S. (1999). Classical field theory, in P. Deligne, P. Etinghof, D. S. Freed, L. C. Jeffrey, D. Kazhdan, J. W. Morgan, D. R. Morrison, and E. Witten (eds), *Quantum Fields and Strings: A Course for Mathematicians*, 137–226. American Mathematical Society, Providence, RI.

DiSalle, R. (2008). *Understanding Space-Time*. Cambridge University Press, New York.

Earman, J. (1977a). Leibnizian space-times and Leibnizian algebras, in R. E. Butts, J. Hintikka, J. (eds), *Historical and Philosophical Dimensions of Logic, Methodology and Philosophy of Science*, 93–112. Reidel, Dordrecht.

Earman, J. (1977b). Perceptions and relations in the monadology. *Studia Leibnitiana* 9(2), 212–30.

Earman, J. (1979). Was Leibniz a relationist?, in P. French, T. Uehling, H. Wettstein (eds), *Midwest Studies in Philosophy*, Vol. 4, 263–76. University of Minnesota Press, Minneapolis, MN.

Earman, J. (1986). Why space is not a substance (at least not to first degree). *Pacific Philosophical Quarterly* 67(4), 225–44.

Earman, J. (1989a). Leibniz and the absolute vs. relational dispute, in N. Rescher (ed.), *Leibnizian Inquiries; A Group of Essays*, 9–22. University Press of America, Lanham, MD.

Earman, J. (1989b). *World Enough and Space-Time*. MIT Press, Boston.

Earman, J., and Norton, J. (1987). What price spacetime substantivalism? The hole story. *British Journal for the Philosophy of Science* 38(4), 515–25.

Fletcher, S. C. (2016). Similarity, topology, and physical significance in relativity theory. *British Journal for the Philosophy of Science* 61(2), 365–89.

French, S. (2014). *The Structure of the World: Metaphysics and Representation*. Oxford University Press, Oxford.

Gambini, R., and Pullin, J. (1996). *Loops, Knots, Gauge Theories, and Quantum Gravity*. Cambridge University Press, Cambridge, UK.

Geroch, R. (1972). Einstein algebras. *Communications in Mathematical Physics* 26, 271–5.

Glymour, C. (1970). Theoretical equivalence and theoretical realism. *PSA: Proceedings of the Biennial Meeting of the Philosophy of Science Association 1970*, 275–88.

Glymour, C. (1977). The epistemology of geometry. *Noûs* 11(3), 227–51.

Glymour, C. (1980). *Theory and Evidence*. Princeton University Press, Princeton, NJ.

Halvorson, H. (2012). What scientific theories could not be. *Philosophy of Science* 79(2), 183–206.

Halvorson, H. (2016). Scientific theories, in P. Humphreys (ed.), *The Oxford Handbook of the Philosophy of Science*, 585–608. Oxford University Press, Oxford.

Healey, R. (2001). On the reality of gauge potentials. *Philosophy of Science* 68(4), 432–55.

Healey, R. (2004). Gauge theories and holisms. *Studies in History and Philosophy of Modern Physics* 35(4), 643–66.

Healey, R. (2007). *Gauging What's Real: The Conceptual Foundations of Contemporary Gauge Theories*. Oxford University Press, New York.

Ismael, J., and van Fraassen, B. (2003). Symmetry as a guide to superfluous structure, in K. Brading, E. Castellani (eds), *Symmetries in Physics: Philosophical Reflections*, 371–92. Cambridge University Press, Cambridge, UK.

Knox, E. (2014). Newtonian space–time structure in light of the equivalence principle. *British Journal for the Philosophy of Science* 65(4), 863–88.

Kobayashi, S., and Nomizu, K. (1963). *Foundations of Differential Geometry*, Vol. 1. Interscience Publishers, New York.

Kock, A. (2006). *Synthetic Differential Geometry*, 2nd edn. Cambridge University Press, Cambridge, UK.

Ladyman, J., and Ross, D. (2009). *Every Thing Must Go: Metaphysics Naturalized*. Oxford University Press, Oxford.

Lawvere, W., and Schanuel, S. (1986). *Categories in Continuum Physics*. Springer-Verlag, Berlin.

Leinster, T. (2014). *Basic Category Theory*. Cambridge University Press, Cambridge, UK.

Loll, R. (1994). Gauge theory and gravity in the loop formulation, in J. Ehlers, H. Friedrich (eds), *Canonical Gravity: From Classical to Quantum*, 254–88. Springer, Berlin.

Lutz, S. (2015). What was the syntax-semantics debate in the philosophy of science about? *Philosophy and Phenomenological Research* 91(3). doi: 10.1111/phpr.12221. http://philsci-archive.pitt.edu/11346/1/lutz-syntax_semantics_debate.pdf

Mac Lane, S. (1968). *Geometrical mechanics: Lectures, Dept. of Mathematics, University of Chicago*. https://books.google.co.uk/books/about/Geometrical_mechanics.html?id=IMpUAAAAYAAJ

Mac Lane, S. (1998). *Categories for the Working Mathematician*, 2nd edn. Springer, New York.
Malament, D. (2012). *Topics in the Foundations of General Relativity and Newtonian Gravitation Theory.* University of Chicago Press, Chicago.
Nestruev, J. (2003). *Smooth Manifolds and Observables.* Springer, Berlin.
North, J. (2009). The "structure" of physics: a case study. *Journal of Philosophy* 106(2), 57–88.
Norton, J. D. (2011). The hole argument, in E. N. Zalta (ed.), *The Stanford Encyclopedia of Philosophy*, Fall 2011 Edition. http://plato.stanford.edu/archives/fall2011/entries/space-time-holearg/
Palais, R. S. (1981). *The Geometrization of Physics.* Institute of Mathematics, National Tsing Hua University, Hsinchu, Taiwan. http://vmm.math.uci.edu/
Penrose, R., and Rindler, W. (1984). *Spinors and Space-Time.* Cambridge University Press, New York.
Pooley, O. (2013). Relationist and substantivalist approaches to space–time, in R. Batterman (ed.), *The Oxford Handbook of Philosophy of Physics*, 522–86. Oxford University Press, New York.
Ramsey, F. P. (1931). *The Foundations of Mathematics*, 212–36. Routledge & Kegan Paul, London.
Reyes, G. (2011). A derivation of Einstein's vacuum field equations, in B. Hart, T. Kucera, A. Pillay, P. J. Scott, and R. A. G. Seely (eds), *Models, Logics, and Higher-Dimensional Categories*, 245–62. American Mathematical Society, Providence, RI.
Rosenstock, S. (2017). A categorical consideration of theoretical equivalence in Yang-Mills theories. Unpublished manuscript.
Rosenstock, S., Barrett, T., Weatherall, J. O. (2015). On Einstein algebras and relativistic space-times. *Studies in History and Philosophy of Modern Physics* 52B, 309–15.
Rosenstock, S., Weatherall, J. O. (2016). A categorical equivalence between generalized holonomy maps on a connected manifold and principal connections on bundles over that manifold. *Journal of Mathematical Physics* 57(10), 102902.
Ruetsche, L. (2011). *Interpreting Quantum Theories.* Oxford University Press, New York.
Russell, B. (2007 [1927]). *The Analysis of Matter.* Spokesman, Nottingham, England.
Rynasiewicz, R. (1992). Rings, holes and substantivalism: on the program of Leibniz algebras. *Philosophy of Science* 59(4), 572–89.
Saunders, S. (2013). Rethinking Newton's *Principia*. *Philosophy of Science* 80(1), 22–48.
Stachel, J. (1989). Einstein's search for general covariance, 1912–1915, in D. Howard, J. Stachel (eds), *Einstein and the History of General Relativity*, 62–100. Birkhauser, Boston.
Stachel, J. (1993). The meaning of general covariance, in J. Earman, A. I. Janis, G. J. Massey, N. Rescher (eds), *Philosophical Problems of the Internal and External Worlds: Essays on the Philosophy of Adolf Grünbaum*, 129–60. University of Pittsburgh Press, Pittsburgh, PA.
Stein, H. (1967). Newtonian space-time. *The Texas Quarterly* 10, 174–200.
Trautman, A. (1965). Foundations and current problem of general relativity, in S. Deser, K. W. Ford (eds), *Lectures on General Relativity*, 1–248. Prentice-Hall, Englewood Cliffs, NJ.
Trautman, A. (1980). Fiber bundles, gauge fields, and gravitation, in A. Held (ed.), *General Relativity and Gravitation*, 287–308. Plenum Press, New York.
Weatherall, J. O. (2016a). Are Newtonian gravitation and geometrized Newtonian gravitation theoretically equivalent? *Erkenntnis* 81(5), 1073–91.
Weatherall, J. O. (2016b). Fiber bundles, Yang-Mills theory, and general relativity. *Synthese* 193(8), 2389–425.

Weatherall, J. O. (2016c). Maxwell-Huygens, Newton-Cartan, and Saunders-Knox space–times. *Philosophy of Science* 83(1), 82–92.

Weatherall, J. O. (2016d). Regarding the "Hole Argument". *British Journal for Philosophy of Science*. doi: 10.1093/bjps/axw012. https://academic.oup.com/bjps/article-abstract/doi/10.1093/bjps/axw012/2669777/Regarding-the-Hole-Argument?redirectedFrom=fulltext

Weatherall, J. O. (2016e). Understanding "gauge". *Philosophy of Science* 83(5), 1039–49.

Worrall, J. (1989). Structural realism: the best of both worlds? *Dialectica* 43(1–2), 99–124.

14
Six-Dimensional Lorentz Category

*Joachim Lambek**

In memory of Basil Rattray, my friend, colleague, and collaborator.

14.1 Prologue

Pre-Socratic Greek philosophers were engaged in two intensive debates: are material objects continuous or discrete, and what is the nature of time? The claim that matter consists of infinitely divisible substances was first made by Thales, who postulated a single basic substance: water. In time, three other substances were added, notably by Empedocles, and even nowadays people accept four states of matter: liquid, solid, gas, and energy. The claim that all matter is made up of indivisible units seems to be implicit in the Pythagorean assertion that all things are numbers, but is ultimately replaced by the atomic theory of Democritus and Epicurus. The nature of time was debated by Heraclitus and Parmenides. The former emphasized the importance of time and change in his memorable slogans, while the latter insisted in his famous poem that time was not all that different from space. If I understand him correctly, he claimed that the one-dimensional flow of time is a human illusion not shared by the gods. His pupil Zeno seems to have pointed out that assuming time to be either discrete or continuous leads to contradictions.

---

* Joachim (Jim) Lambek (1922–2014) is well known as a mathematician, particularly for his work on ring theory and algebra, and for his work on category theory, especially for deductive systems and for categorical proof theory in general. He is also well known for his work in mathematical linguistics, particularly for his analysis of categorial grammar, the syntactic calculus, and pregroups. All this is referred to in other papers in this volume. What may be less well known is that he also had a professional interest in theoretical physics. He was particularly interested in the use of the language of quaternions as a tool to explain fundamental aspects of special relativity, for example. In fact the first version of his 1950 Ph.D. thesis included a section on this topic (though he removed that section upon learning that his results had been already proven by A. W. Conway in 1948). He visited this topic several times in his mathematical career, for instance, in a *Mathematical Intelligencer* article "If Hamilton Had Prevailed: Quaternions in Physics" (1995), and more recently in a series of articles, referenced in his chapter in this volume.

Jim's work in mathematics, logic, and linguistics may be inferred from references to it in other chapters in this volume—it seems fitting that this posthumous paper of his (which he dedicated to his former colleague Basil Rattray) should illustrate one of the many other sides of his scholarly interests. —R. A. G. Seely

Mathematicians too were wondering whether positive reals (which they called geometric quantities) or positive integers are more fundamental. The Pythagoreans at first assumed the latter and only reluctantly admitted the irrationality of the square root of 2. At Plato's Academy two ways of defining positive reals in terms of positive integers were proposed, which are now known as Dedekind reals and Cauchy reals, respectively. The former were introduced by Eudoxus and the latter by Theaetatus, who made use of continued fractions. Surprisingly, the ancient Greeks avoided zero and negative numbers, which were only introduced a thousand years later in India.

Modern physicists have definitely decided that the fundamental particles of nature are indivisible objects called fermions and bosons, but the matter of time is still being disputed. In Lambek (2000) I suggested that all fundamental particles of spin $\frac{1}{2}$ or 1 could be represented by four-vectors with entries $0, 1,$ and -1. More recently (Lambek, 2012), I observed that six-vectors with the same entries are more suitable if one wishes to distinguish between right-handed and left-handed particles. However, position in space–time is nowadays assumed to be subject to a probability distribution, best expressed as the norm of a quaternion, the Dirac spinor.

For reasons to be discussed later, I have also come to the conclusion that time has more than one dimension. Mathematical elegance would require three dimensions of time, but these may be reduced to two if one insists that Dirac's first-order equation is equivalent to the second-order Klein–Gordon equation. This may be proved as in Lambek (2011b), but better with the help of category theory as shown in the next section. In Lambek (2011a) I suggested that one should consider a finite additive category with three objects (called a ring with three objects by Barry Mitchell), whose arrows described four-vectors, six-vectors, and Dirac spinors of four-dimensional relativistic quantum mechanics. In the present chapter, the category is generalized to six-dimensional space–time and the six-dimensional classification of fundamental particles is exploited to present a proof of the probability density.

14.2 Six-Dimensional Lorentz category

Present-day theoretical physics relies on the representation of groups and Lie algebras. My personal preference is to make use of the regular representations of the algebra of quaternions instead.

The application of quaternions to special relativity has a long history and goes back to Conway (1912) and Silberstein (1912, 1914) a century ago. The original idea was to use *biquaternions*, i.e. quaternions with complex components. Thus, location in space–time was represented by the *Hermitian* biquaternion (one in which the quaternion and complex conjugates coincide):

$$\text{(i)} \quad x_0 + ii_1 x_1 + ii_2 x_2 + ii_3 x_3$$

or, equivalently, by

$$\text{(ii)} \quad ix_0 + i_1 x_1 + i_2 x_2 + i_3 x_3,$$

following Minkowski's suggestion that time be conceived as imaginary space.

When mathematicians turned their attention to Dirac's equation, they thought it convenient to replace i_1, i_2, and i_3 by their left regular matrix representations and i by the right regular matrix representation of one of them, say i_1. This idea was pursued by Lanczos (1929), Conway (1948), and Gürsey (1955). If we also admit the right regular representations of i_2 and i_3, we are led by (i) to think of a 10-dimensional space–time and by (ii) of a 6-dimensional one. It is the latter approach I will pursue here, thus admitting two additional dimensions of time rather than six additional ones of space.

The assumption that time has three dimensions was developed by me for special relativity in Lambek (2011b) and for general relativity by Gillen (2010). My original motivation for the extra dimensions was based on mathematical elegance. In retrospect, the additional dimensions of time make it easier to understand how Schrödinger's cat can be alive and dead "simultaneously", provided this adverb is interpreted to mean "at the same distance from the origin of temporal three-space".

With any quaternion x we associate two *regular* representations

$$L(x)[\psi] = [x\psi], \quad R(x)[\psi] = [\psi x],$$

where $[\psi]$ is the column vector consisting of the coefficients of the quaternion ψ. Evidently

$$L(xy) = L(x)L(y), \quad R(xy) = R(y)R(x), \quad L(x)R(y) = R(y)L(x).$$

The two representations are related by the diagonal matrix Γ with entries $(1, -1, -1, -1)$:

$$\Gamma L(x) \Gamma = -R(x), \quad \Gamma R(y) \Gamma = -L(y),$$

hence

$$\Gamma(L(x) + R(y))\Gamma = -(L(y) + R(x)).$$

Any quaternion may be written as $a_0 + \mathbf{a}$, where a_0 is a scalar and

$$\mathbf{a} = i_1 a_1 + i_2 a_2 + i_3 a_3$$

is called a *three-vector*. It is easily seen that

$$L(\mathbf{a}) + R(\mathbf{b})$$

is a skew-symmetric matrix and that every skew-symmetric 4×4 real matrix has this form. See Lambek (2011b).

It is our intention to represent space–time by the skew matrix

$$X = L(\mathbf{x}) + R(\mathbf{t}),$$

where the vector **t** now replaces the usual scalar t, and to treat other basic physical entities in the same manner.

Thus we have the *kinetic energy–momentum*

$$P = L(\mathbf{p}) + R(\mathbf{m}),$$

where **p** is the usual momentum vector and **m** is the three-dimensional analogue of the usual energy = matter = 4π frequency.

The skew matrix replacing the old *four-potential*

$$\Phi = L(\mathbf{A}) + R(\boldsymbol{\phi})$$

is composed of Maxwell's vector potential and the vector analogue of the usual scalar potential. This allows us to describe the *potential energy–momentum* $-e\Phi$ of the electron with charge $-e$, the minus sign being due to a choice made by Benjamin Franklin.

The *charge–current* density is described by

$$J = L(\mathbf{J}) + R(\rho),$$

where **J** is the usual current density and ρ replaces the usual charge density.

To these skew matrices we must add the *partial differentiation operator*

$$D = L(\nabla_x) - R(\nabla_t),$$

where

$$\nabla_x = i_1 \frac{\partial}{\partial x_1} + i_2 \frac{\partial}{\partial x_2} + i_3 \frac{\partial}{\partial x_3}, \quad \nabla_t = i_1 \frac{\partial}{\partial t_1} + i_2 \frac{\partial}{\partial t_2} + i_3 \frac{\partial}{\partial t_3},$$

the minus sign being due to the contravariance of differentiation.

In addition to the *basic* physical entities discussed so far, others may be represented by conjugation and composition of the above skew matrices, the *conjugate* of $A = L(a) + R(b)$ being $A^* = L(a) - R(b)$. Thus, every physical entity can be represented by a 4×4 matrix, but this should be accompanied by a *Lorentz transformation*, itself expressed with the help of a 4×4 matrix Q of determinant 1.

To start with, we have the basic entities transforming as follows:

$$\begin{aligned} X &\mapsto QXQ^T, & &\text{space-time,} \\ P &\mapsto QPQ^T, & &\text{kinetic energy–momentum,} \\ \Phi &\mapsto Q\Phi Q^T, & &\text{six-potential,} \\ J &\mapsto QJQ^T, & &\text{charge–current density,} \end{aligned}$$

where Q^T is the *transposed* matrix of Q. Note that the condition $\det Q = 1$ excludes $Q = \Gamma$, but it does not distinguish between transformations expressed by Q and $-Q$.

According to tradition, a Lorentz transformation is supposed to preserve the expression

$$X \odot X = \mathbf{x} \circ \mathbf{x} - \mathbf{t} \circ \mathbf{t} = -XX^*,$$

where $\mathbf{x} \circ \mathbf{x}$ is the usual Heaviside scalar product and $X \odot X$ is its extension to six dimensions.

The condition that $\det Q = 1$ ensures that $X \mapsto QXQ^T$ preserves the determinant of X, hence the square of $X \odot X$. But, if we also wish to preserve the sign of $X \odot X$, we can achieve this by postulating

$$\text{(iii)} \qquad X^* \mapsto Q^\# X^* Q^{-1},$$

where $Q^\#$ is the matrix of *cofactors* of Q, so that

$$Q^T = (Q^{-1})^\# = Q^{-\#}.$$

In fact, (iii) is a necessary and sufficient condition for $X \odot X$ to be Lorentz invariant. See Lambek (2011b).

We are now in a position to introduce the *Lorentz category* as an additive category (also called a *ring* by Barry Mitchell) with three objects 1, #, and 0, where

$$\#\# = 1, \ u\# = 0 = \#u, \ 1u = u = u1$$

for all objects u. The arrows $A: u \to v$ are matrices A such that

$$A \mapsto Q^u A Q^{-v},$$

where $Q^{-v} = (Q^v)^{-1}$. If $B: v \to w$ is another such arrow, so is the matrix product $AB: u \to w$, where we have reversed the conventional order of composition of arrows.

In particular, the basic entities $X, P, \Phi, J,$ and D all describe arrows $1 \to \#$, whereas $X^*, P^*,$ etc. are arrows $\# \to 1$. This had been done in Lambek (2011a) under the assumption that there was only one dimension of time, where the usual four-vectors, six-vectors, and Dirac spinors were represented by arrows $1 \to \#$, $\# \to \#$ and $1 \to 0$, respectively. The same category had been employed in Lambek (2011a), where the basic entities turned out to be Hermitian biquaternions, but here they are skew-symmetric 4×4 matrices.

Most (if not all) physical entities in pre-quantum physics live already in a ring with two objects, 1 and #, called a *Morita context*, but an understanding of the Dirac equation requires a third object 0 to admit the so-called *Dirac spinors* $1 \to 0$; see in the following.

To get an idea of how useful calculations are carried out in the Lorentz category, consider skew matrices $A, B, C: 1 \to \#$. Then $AB^*: 1 \to 1$ and $AB^*C: 1 \to \#$, but these arrows can be decomposed. Thus,

$$AB^* = \tfrac{1}{2}(AB^* + BA^*) + \tfrac{1}{2}(AB^* - BA^*),$$

where the first summand is the *trace* or scalar part of AB^*:

$$\tfrac{1}{2}(AB^* + BA^*) = -A \odot B : 1 \to 1.$$

Moreover,
$$AB^*C = \tfrac{1}{2}(AB^*C + CB^*A) + \tfrac{1}{2}(AB^*C - CB^*A),$$
where the first summand is the *skew* part of AB^*C, which may be calculated as follows:
$$\text{skew}(AB^*C) = -A(B \odot C) + B(C \odot A) - C(A \odot B).$$
This happens to be useful in discussing the Maxwell–Lorentz treatment of the electron, which may be summarized by the equation
$$d(P - e\Phi) = 0,$$
expressing the conservation of the total energy–momentum, where Φ may be subject to a gauge transformation. See Lambek (2011b).

Maxwell had defined the electromagnetic field F acting on the charged particle in four dimensions as the vector part of $D^*\Phi$. In six dimensions this becomes the symmetric part:
$$F = \tfrac{1}{2}(\overrightarrow{D^*}\Phi - \Phi\overleftarrow{D^*}).$$
It is supposed to be caused by J according to Maxwell's equation $DF = J$, where J satisfies the *equation of continuity* $D \odot J = 0$. On the other hand, the force of the field on an electron as described by Lorentz becomes
$$\frac{dP}{ds} = \text{skew}\left(-e\frac{dX}{ds}F\right) = \frac{d}{ds}(e\Phi),$$
where $(ds)^2 = dXdX^*$.

The relativistic treatment of quantum mechanics begins with the so-called *Klein–Gordon* equation, already known to Schrödinger,
$$DD^*[\psi] = -\mu^2[\psi],$$
where $\mu = PP^* : 1 \to 1$ is the rest-mass of a particle and $[\psi] : 1 \to 0$ is a *Dirac spinor*. If $\mu \neq 0$, this is equivalent to two first-order equations:
$$D^*[\psi_1] = \mu[\overline{\psi}_2], \quad D[\overline{\psi}_2] = -\mu[\psi_1].$$
Penrose (2004) might call $[\psi_1]$ the *zig* and $[\psi_2]$ the *zag*. However, Dirac would combine them into a single first-order equation.

Assuming that one time coordinate, say t_3, is redundant in a certain frame of reference, we may take $K = R(i_3)$ in this coordinate system and verify that
$$X^* = -K^{-1}XK^{-1}$$
for $K : 1 \to \#$, and similarly for P^*, D^*, etc. Now let
$$[\psi] = [\psi_1] - K[\overline{\psi}_2],$$

then we may calculate

$$\text{(iv)} \quad D^*[\psi] = -\mu K^*[\psi]$$

and take this to be the six-dimensional Dirac equation.

An explicit solution of (iv) is given by

$$[\psi] = \exp(\eta(X \odot P))[\psi_0],$$

where

$$\eta = \mu^{-1} K P^* = -\mu^{-1} P K^*$$

satisfies

$$\eta^2 = -1.$$

On the other hand, (iv) may be written in purely quaternionic form,

$$\vec{\nabla}_x \psi + \psi \overleftarrow{\nabla}_t + \mu(\mathbf{k}\psi - \psi\mathbf{k}') = 0,$$

provided

$$K = L(\mathbf{k}) + R(\mathbf{k}').$$

Multiplying (iv) by the row vector $[\psi]^T$ on the left, we obtain

$$[\psi]^T \vec{D^*}[\psi] = -\mu[\psi]^T K^*[\psi].$$

Here the right side is skew symmetric, hence so must be the left side, so that

$$[\psi]^T \overleftrightarrow{D^*}[\psi] = [\psi]^T \vec{D}[\psi] + [\psi]^T \overleftarrow{D}[\psi] = 0.$$

Multiplying this by a constant skew-symmetric matrix S on the left and assuming that $S: 0 \to 0$ is Lorentz invariant, we infer that

$$\text{(v)} \quad \text{trace}(\vec{D^*}[\psi]S[\psi]^T) = \text{trace}([\psi]^T \overleftrightarrow{D^*}[\psi]S) = 0.$$

Write $J_S = [\psi]S[\psi]^T$, check that $J_S : 1 \to 0 \to 0 \to \#$, and note that (v) then asserts that $D \odot J_S = 0$, which resembles Maxwell's equation of continuity and suggests a comparison of J_S with the electric charge–current density. A proof of this when time has only one dimensions is found in Sudbury (1986).

It remains to identify S. I will speculate that S is the quaternionic version of the six-vector characterization of fundamental particles of spin 1 and $\frac{1}{2}$ that I have discussed in Lambek (2012):

$$\begin{aligned} S = & \; L(s_1 i_1 + s_2 i_2 + s_1 i_3) + R(s'_1 i_1 + s'_2 i_2 + s'_3 i_3) \\ \sim & \; (S_1, S_2, S_3; s'_1, s'_2, s'_3), \end{aligned}$$

where the S_α and S'_β are all equal to 0, 1, or -1. Addition of these six-vectors or their skew-symmetric analogues helps to justify Feynman diagrams for fundamental particles. For example,

$$U = L(i_1 + i_3) + R(i_1) \sim (1, 0, 1; -1, 0, 0)$$

characterizes a first-generation left-handed blue up-quark and

$$W = L(-i_1 - i_2 - i_3) + R(-i_1 - i_2 - i_3) \sim (-1, -1, -1; -1, -1, -1)$$

characterizes a weak vector boson W^-. Adding these two expressions, we obtain

$$D = U + W = L(-i_2) + R(-i_2 - i_3) \sim (0, -1, 0; -0, -1, -1),$$

which characterizes a first-generation right-handed blue down-quark.

The equation $S_D = S_U + S_W$ serves to justify the Feynman diagram

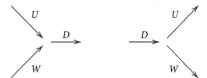

and allows us to infer the equation $J_D = J_U + J_W$. This seems reasonable, but fails to bring the coupling constant into the picture.

The two extra dimensions of time had been introduced for the sake of mathematical elegance and I have not settled on their physical meaning. For a while I had hoped that they might help to incorporate the direction of the spin axis, but did not succeed to make this idea work.

References

Conway, A. (1912). The quaternion form of relativity. *Philosophical Magazine* 24, 208.
Conway, A. (1948). Quaternions and quantum mechanics. *Pontificia Academia Scientiarum* 12, 204–77.
Dirac, P. (1945). Application of quaternions to Lorentz transformations, in *Proceedings of the Royal Irish Academy. Section A: Mathematical and Physical Sciences*, Vol. A50, 261–70.
Feynman, R. (1985). QED: *The Strange Theory of Light and Matter*. Princeton University Press, Princeton, NJ.
Feynman, R. P., Leighton, R. B., Sands, M., and Hafner, E. (1963/1977). *The Feynman Lectures on Physics*. Addison Wesley, Reading, MA.
Gillen, P. (2010). How general relativity contains the standard model. *arXiv* hep-th/0110296v3.
Gürsey, F. (1955). Contribution to the quaternion formalism in special relativity. *Rev. Fac. Sci. Univ. Istanbul. Sér. A* A20, 149–71.
Gürsey, F. (1956). Correspondence between quaternions and four-spinors. *Rev. Fac. Sci. Istanbul* A 21, 33–54.

Lambek, J. (2000). Four-vector representation of fundamental particles. *International Journal of Theoretical Physics* 39(9), 2253–8.

Lambek, J. (2011a). The Lorentz category in special relativity. *Models, Logics, and Higher-Dimensional Categories: A Tribute to the Work of Mihály Makkai* 53, 169–75 (ed. B. Hart, T. G. Kucera, A. Pillay, P. J. Scott, R. A. G. Seely, Centre de Recherches Mathématiques).

Lambek, J. (2011b). Quaternions and three temporal dimensions.

Lambek, J. (2012). A six-vector classification of fundamental particles.

Lambek, J. (2013). In praise of quaternions. *CR Math. Rep. Acad. Sci. Canada* 35, 1–16.

Lanczos, C. (1929). Three articles on Dirac's equation. *Zeitschrift für Physik* 57, 447–93.

Penrose, R. (2004). *The Road to Reality: A Complete Guide to the Laws of the Universe*. Jonathan Cape, London.

Silberstein, L. (1912). Quaternionic form of relativity. *The London, Edinburgh, and Dublin Philosophical Magazine and Journal of Science* 23(137), 790–809.

Silberstein, L. (1914). *The Theory of Relativity*. MacMillan, London.

Sudbury, A. (1986). *Quantum Mechanics and the Particles of Nature*. Cambridge University Press, Cambridge, UK.

Weiss, P. (1940). On some applications of quaternions to restricted relativity and classical radiation theory, in *Proceedings of the Royal Irish Academy. Section A: Mathematical and Physical Sciences*, 129–68.

15

Applications of Categories to Biology and Cognition

Andrée Ehresmann

Though the value of category theory is well recognized in different scientific domains, the situation is different in biology. The main mathematical models used in biology have been directly adapted from models developed for physics. They relate to specific local processes (e.g., protein folding, cellular model, ...), and cannot be extended to the global dynamic of a biological organism. The reason is that this dynamic has specific characteristics, namely cascades of structural and organizational changes with addition and loss of components and emergence of new properties. Models developed for physics are not well adapted in this case, so that "the biological is to be analyzed as a 'limit physical situation', an 'external limit' with regard to contemporary physical theories" (Bailly and Longo, 2006).

Realizing that "there is no record of a successful mathematical theory which would treat the integrated activities of an organism as a whole", Rashevsky (1954) proposed to develop a "Relational Biology". This program inspired his student Robert Rosen, who was the first to develop applications of categories to biology. In his 1958 seminal work, Rosen introduced a class of relational models of living organisms, called *(M, R)-systems* (M for metabolic and R for repair) to capture the minimal capabilities needed for a material system to be a functional organism (Rosen, 1958a,b). And he pursued this program in his subsequent works, in particular in his books on anticipatory systems (Rosen, 1985a) and on life (1991). He has had several followers, in particular Louie (2009). However, in these works, the aim is to give a model of the invariant structure and organization of the system, and not to analyze the characteristics of its multiscale dynamic in its becoming.

More recently, Simeonov (2010) emphasized the need for "general theoretical formalisms related to all aspects of emergence, self-assembly, self-organization and self-regulation of... living organisms", what he calls an *Integral Biomathics*. And the book *Integral Biomathics, Tracing the Road to Reality*, which he co-edited (2012), proposes different approaches in this direction. One of them is developed in this chapter; namely we show how a 'dynamic' category theory integrating time can help to realize this program. It will be done in the frame of the *Memory Evolutive Systems* (or MES) developed by A. Ehresmann and J.-P. Vanbremeersch since 1987; MES propose a methodology tailored to study evolutionary multiscale, multitemporality systems such as living systems. They model the dynamics of the system in the openness of its becoming, and analyze its changes over time both from an 'external' point of view and from the perspective of a net of internal agents acting as co-regulators. An application is given to MENS, an integrative model of the neural, mental, and cognitive systems, able to analyze the formation of higher cognitive processes.

Before beginning, it should be noted that our approach to modeling is more flexible than the modeling relation stressed by Rosen (1985a), because we use MES as a developing methodology (rather than a 'model') in two opposite ways:

(i) It gives a categorical 'translation' of the natural (or physical) properties of living systems (e.g. a transfer of information is modeled as a morphism in an adequate category), necessitating the introduction of new categorical notions such as a (Hierarchical) Evolutive System to model an evolutionary (multiscale) system, or the Multiplicity Principle to model the degeneracy property of biological systems.

(ii) Conversely having found some mathematical results in the categorical model, we search whether they can correspond to some known (or yet unknown) natural properties of living systems. For instance, the complexification process (Section 15.4) leads to the emergence of some particular morphisms called 'complex links', which model the "change in the condition of change" described by Popper (2002) in social systems; the introduction of the Archetypal Core (Ehresmann and Vanbremeersch (henceforth abbreviated as EV), 2002; cf. Section 15.6) has led to applications in creativity and anticipation (Ehresmann, 2012, 2013), and more recently design (Béjean and Ehresmann, 2015).

MES are progressively constructed: each section stresses a characteristic of living systems and gives a rough idea of how it is accounted for in MES. The appendix gives explicit definitions of the categorical notions followed by *. For more details on categories, we refer to the book *Categories for the Working Mathematician* (Mac Lane, 1971), and for MES, to the book *Memory Evolutive Systems* (EV, 2007).

15.1 The Primary Role of Time: Evolutive Systems

A first characteristic of living systems is that they are *evolutionary* systems, with their structure and organization varying with time. This section analyzes this evolution and introduces the notion of an Evolutive System to model it.

While time appears essentially as a parameter in physics, for living systems such as biological systems or social systems, time is related to changes (as in Augustine of Hippo), and "temporality has to be taken as primary with spatiality as emergent" (Kauffman and Gare, 2015). Indeed, living systems are 'open' evolutionary systems which have exchanges with their environment, are able to memorize their experiences for better adaptation, and progressively modify their structure, their organization, and their compartments. In particular their components change with time, as well as the links through which they can interact, with possible loss of components or links (e.g. 'death' of a cell, destruction of a synapse), formation of new ones (e.g. formation of a new organelle), and emergence of new properties. This essential variance makes it impossible to model a whole system using methods inspired by physics, which study the dynamic in specific phase spaces, while "In biological Evolution, the phase space itself changes persistently. More it does so in ways that cannot be prestated" (Longo et al., 2012).

How to account for the evolution of an evolutionary system (such as a living system) in its open-ended 'becoming', from its birth on, with all its successive changes? Knowing the time-life T of the system from its birth on, we give a snapshot of its configuration at each time t of T, and monitor the changes from t to a later time t' of T by measuring the differences between the two corresponding configurations.

(i) The time-life is represented by an interval of the real numbers beginning at the time t_0 of the birth of the system. The *configuration* at an instant t of T has for objects the states of the *components* C of the system existing at t, and they are connected by links representing directed channels through which they can interact, for instance via informational messages (e.g. presence of an antigen) or energy transfers (command of effectors). To account for the 'physical' constraints, a link has a *propagation* delay at t representing the delay necessary for transmitting the information, and the link is active or passive at time t depending on whether some information passes through it at t; for instance, a synapse from a neuron N to a neuron N' may exist and be passive at t if it does not transmit an influx from N to N' at t.

(ii) The change of configuration, or *transition*, from t to t' reflects the *structural changes* from t to t' (but does not describe the dynamic leading to them which will be considered later on). It associates to the state at t of a component C its new state at t', if C still exists at t', and the same for the links. In this way it also indicates which components and links existing at t have been 'suppressed' before t', and which components and links existing at t' did not exist at t.

Translating these notions in the categorical frame, an evolutionary system will be modeled by an *Evolutive System* (EV, 1987) defined as follows:

Definition 15.1 An Evolutive System K consists of:

(i) An interval T of the positive real numbers line, called the *timescale* of the system.
(ii) For each t in T, a *category** K_t called the *configuration* of K at t.
(iii) For each $t < t'$ in T a *functor** $k_{tt'}$ called *transition*, from a subcategory $K_{tt'}$ of K_t to $K_{t'}$. The transitions must respect a transitivity condition: if $t < t' < t''$ and if C_t is an object of $K_{tt'}$, then C_t is in $K_{tt''}$ if and only if $C_{t'} = k_{tt'}(C_t)$ is in $K_{t't''}$, and then $k_{t't''}(C_{t'}) = k_{tt''}(C_t) = C_{t''}$; and the same for the morphisms.

In the Evolutive System (ES) modeling a living system, the categories K_t are generally labeled as follows: a morphism f_t is equipped with a real number $d_t(f_t)$, called its *propagation delay*, and with an *index of activation* $i_t(f_t)$ equal to 1 or 0, so that

$d_t : K_t \to R_=$ defines a functor to the additive real numbers, and $i_t : K_t \to B$ a functor to the multiplicative group $\{0, 1\}$. The transitions may not respect them.

In an evolutionary system we have spoken of its components and the links between them. Definition 15.1 of an ES does not mention them, but they are reconstructed formally as follows: a *component* C of K is a maximal family of objects C_t of successive configuration categories related by transitions; we denote by T_C the timeline of C, i.e. the set of times $t \in T$ such that C has a state C_t at t. Links between components (resp. patterns of interlinked components) are similarly described as maximal families of morphisms (resp. patterns) in successive K_t related by transitions. For each interval $J = [t, t')$ of T, we define a category K_J having for objects the (restriction C_J to J of the) components C of K whose timeline T_C contains J, the morphisms being the restriction to J of the links between them. We often work in such a category.

Remark 15.1 An Evolutive System is a particular case of a *semi-sheaf of categories* (Ehresmann, 2008, 2009) over the category defining the order on T. Here T is an interval of R (to model biological systems) but it could be any totally ordered set. More generally a semi-sheaf of categories can be defined as a lax functor from a category to the 2-category *Cat* satisfying a pull-back condition which translates the transitivity condition of Definition 15.1.

Example 15.1 The *Evolutive System* of neurons NEUR is defined as follows. The configuration category NEUR$_t$ at an instant t is the category of paths of the graph of neurons and synapses between them existing at t. Formally, an object N_t is a triple $(N, t, n(t))$ which models the state of a neuron N existing at t with its activity $n(t)$ at t (where n is a real function of the instantaneous firing rate of N and of its threshold); a morphism $f_t = (f, t, s_t(f_t))$ from N_t to N'_t models a synaptic path f from N to N', with its strength $s_t(f_t)$ at t (which is the product of the strengths of its factor synapses); it is labeled by its propagation delay and its index of activity at t. The transition from t to $t' > t$ maps N_t on $N_{t'}$ if the neuron N still exists at t'. A component models a neuron by the temporal trajectory of its successive states; it is identified with the neuron if no confusion is possible.

15.2 The Tangled Hierarchy of Components: Hierarchical Evolutive Systems

A second characteristic of living systems is that they have a hierarchy of components of various complexities. In this section, we analyze the properties of this hierarchy and describe the notion of a Hierarchical Evolutive System to model it.

Living systems are multiscale systems, with components of various complexity levels. These properties have been emphasized for a long time, in biology (Jacob, 1970; von Bertalanffy, 1973), in neurosciences (Changeux, 1983; Laborit, 1983), or in evolution theory (Dobzhanski, 1970). For instance, Jacob (1970) writes: "Tout objet que considère la Biologie représente un système de systèmes; lui-même élément d'un système d'ordre supérieur, il obéit parfois à des règles qui ne peuvent être déduites de sa propre analyse". How to give a precise definition of a "system of systems"?

The idea is to classify the components of the system we want to study into a finite hierarchy of *complexity* levels numbered $0, 1, 2, \ldots$ so that the components of a given level be considered as (more or less) 'homogeneous' between them, but more 'complex' than those of lower levels; for instance, we can associate to a biological organism a hierarchy consisting of atoms, molecules, macromolecules, and subcellular structures such as organelles, cells, tissues and organs, and large systems. (The number of levels selected in the hierarchy depends on the problem we are interested in.) In this hierarchy, it seems natural to say that a molecule is 'more complex' than an atom because it admits a *decomposition* into the pattern formed by its atoms and their chemical bonds, and the molecule *binds* this decomposition. Similarly a cell binds its decomposition into a pattern consisting of its subcellular structures and their interactions in the cell. And so on at other levels.

First, we must give a precise definition of this dual notion of 'decomposition' and 'binding'. It relates to the "*binding problem*" (Aristotle): "a whole is something else than the sum of its parts"; here the whole can be the molecule, which is different from the sum of its parts (its atoms) because of the constraints imposed by the chemical bonds between the atoms. In a living system, a component C (e.g. a molecule) *binds* a pattern P of interacting objects (its atoms and their links) if it has, by itself, the same functional role, or output, as the components of P acting together synchronously in a way respecting their interactions in P. In the categorical frame, we model this notion of binding by the notion of a *colimit** of a pattern.

Definition 15.2

(i) A *pattern** (or diagram) P in a category K consists of a family of objects $(P_i)_{i \in I}$ of K and some distinguished morphisms between them. A *cone** from P to an object A is a family of morphisms s_i from P_i to A commuting with the distinguished morphisms f of P. A cone can be seen as modeling a collective action of the P_i on A, in accord with their distinguished morphisms in P.

(ii) A *colimit** of P, if it exists, is an object C such that there is a cone from P to C through which every other cone from P to an object A factors uniquely. (Cf. Appendix and Figure 15.1.)

Remark 15.2

(i) The colimit of a pattern reduced to a family $(P_i)_{i \in I}$ of objects is called its *sum*. If both the pattern P admits a colimit C and the family $(P_i)_{i \in I}$ admits a sum S in K, then there is a canonical morphism "comparison" from S to C which measures the difference between the 'whole' C and the 'sum of its parts' S.

(ii) If the morphisms f of K are labeled by propagation delays $d(f)$, a cone from P to A must satisfy $d(s_j) = d(f) + d(s_i)$ for each $f : P_i \to P_j$. Thus, if P is connected, it acts 'synchronously' on A, and P can have a colimit only if all zig-zags of

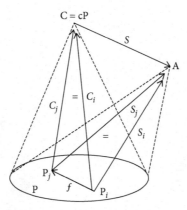

Figure 15.1 Colimit C of a pattern P.

morphisms between two of its objects have the same propagation delay. Such a pattern plays a main role in neuroscience where it is called *polychronous* (Izhikevich et al., 2004).

Let us come back to the hierarchy of a biological organism. Its characteristic is that a component C of complexity level $n+1$ binds at least one decomposition into a pattern P consisting of interacting components of levels $\leq n$ (shortly: lower level pattern). Depending on the time and the context, C may operate as the binding of different lower level patterns. While the binding operation is well defined, the 'complexity' of a component is contextual: a cell is 'complex' with respect to its organelles, but 'simple' when acting as an element of a tissue (forgetting its internal organization). Thus, a component C of a level $n+1$ has a double-face, leading Koestler (1965) to speak of a "Janus": C acts both as a simple "holon" (Koestler's terminology) with respect to a higher complexity level, and as a complex entity with respect to the elements P_i of one of its lower level decompositions P, which themselves are complex with respect to components of still lower levels, and the same down to atoms. Thus, C has at least one *ramification* down to level 0, representing a kind of tree-like structure, obtained by taking one of the lower level decompositions P of M, then one lower level decomposition of each P_i of P, and so on, down to patterns of interacting atoms. (Cf. Figure 15.2.)

Remark 15.3 The level of a component depends on the way we choose to construct the hierarchy; for instance, we can regroup molecules and macromolecules in the same level, and with respect to this new hierarchy the complexity level of a cell is reduced by 1 unit. We can also 'forget' the components of lower levels, for instance taking cells at level 0 (as in the MENS model, Section 15.6).

To account for the hierarchical organization of a biological organism, we enrich the notion of an Evolutive System into the notion of a Hierarchical Evolutive System.

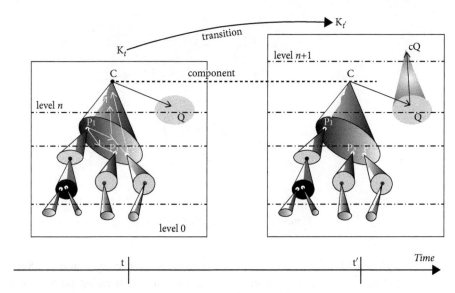

Figure 15.2 A HES with a ramification of the component C.

Definition 15.3

(i) A *hierarchical category* is a category H equipped with a map associating to each object of H a natural number $0, 1, \ldots, m$, called its *complexity level*, in such a way that an object A of level $n+1$ is the colimit of at least one pattern Q in H for which the Q_i are of complexity levels $\leq n$. We call such a pattern a *lower level decomposition* of A.

(ii) A *Hierarchical Evolutive System* (HES) is an Evolutive System **K** such that its configuration categories K_t are hierarchical and the transitions preserve the complexity levels, so that a component C of **K** keeps the same complexity level on its whole timeline T_C.

If C is a component of level $n + 1$ of the HES **K**, a *(lower level) decomposition* P of C on an interval J contained in T_C is a pattern P of the category K_J such that C_t, for each t in J, is the colimit of the state $P(t)$ of P in K_t (and the $P(t)_i$ are of level $\leq n$); in this case, we also say that C *is the colimit of* P on J. By definition of a HES, for each t in T_C, there is at least one such decomposition of C on $\{t\}$; C disappears if it has no more any lower level decomposition at t. It follows that C has also at least one *ramification* at t down to level 0, obtained by selecting one lower level decomposition P of C, then a lower level decomposition of each P_i of P, and so on down to patterns of level 0 which constitute the *base* of the ramification (cf. Figure 15.2). The smallest length of all the ramifications of C at different $t \in T_C$ is called the *complexity order* of C; it is less than, or equal to, the level of C. This order indicates the smallest number of successive binding processes necessary to reconstruct C from patterns of level 0 up.

In a biological organism, the components of a given level bind lower level patterns. Is there something analogous for the links? Only some links between components C and C' of a level $n+1$ bind a cluster of well-correlated interactions between components of lower level decompositions of C and C'. In the HES model K, we can characterize such links as follows: Let C and C' be components of K which have respectively lower level decompositions P and P' on an interval J. If there is a *cluster** of well-correlated links from P to P' in K_J, then the cluster binds into a link from C to C' in K_J (this comes from the 'universal property' of a colimit, cf. Appendix); this link is called a (P, P')-*simple link**, or just *n-simple link* if P and P' are of levels $< n+1$. Such a link sums up interactions between lower level components of C and C'. In K_J the composite of a (P, P')-simple link with a (P', P'')-simple link will be a (P, P'')-simple link. However, in the next section, we show how *composites of n-simple links* which are not *n-simple* can exist, and how such composites model properties emerging at the level $n+1$. (Cf. Figure 15.4.)

15.3 The Multiplicity Principle at the Root of Robustness and Flexibility

This section studies another characteristic of living systems: their flexibility which allows them to adapt to changing conditions. We show how it relies on a kind of 'functional redundancy' which will be modeled by the Multiplicity Principle.

We said earlier that, at each time t, a component C of a biological system has at least one lower level decomposition P. With time, this decomposition may progressively vary by renewal of its own components P_i through losses or additions, while C, as an individual (a holon), keeps its own identity: it "perseveres in existing" (Spinoza, 1677). For instance, the macromolecules of a cell are progressively renewed and its organization can change with time, while the cell as such 'perseveres'. This difference of comportment depending on the observed level is what Matsuno (2012) calls the individual/class distinction. The *stability span* of C at t is the longest period s_t during which C remains the colimit of a same lower level decomposition on the interval $(t, t + s_t)$.

The multiplicity, either simultaneously and/or over time, of lower level decompositions of a component C is possible in any HES, because in any category there may exist non-isomorphic patterns with the same colimit. However, biological systems have a much more specific property, namely the existence of patterns having the same binding, hence the same functional role (or output), but which moreover are *structurally non-connected** in a precise mathematical meaning (briefly that they are not wellconnected by a cluster*, cf. Appendix). This is a consequence of a ubiquitous property of biological systems, first introduced by Edelman for neural systems under the name of "degeneracy of the neural code" (Edelman, 1989), and later extended to a general setting: "Degeneracy, the ability of elements that are structurally different to perform the same function or yield the same output... a ubiquitous biological property... a feature of complexity..., both necessary for, and an inevitable outcome of, natural selection" (Edelman and Gally, 2001).

Degeneracy gives flexibility to the system by allowing that the components operate differently depending on the context; it is at the root of the complexity of living systems. In a HES it is formalized by the Multiplicity Principle (EV, 2007).

Definition 15.4 A component M of a HES is *n-multifaceted* at t if it has at least two lower level decompositions P and Q at t which are *structurally non-connected** in the category K_t; it is *n-multifaceted* on an interval J if it is *n*-multifaceted at each $t \in J$. The HES satisfies the *Multiplicity Principle* (MP) if, for each $n > 0$ and each $t \in T$, there are *n*-multifaceted components at t.

The HES modeling a biological organism satisfies MP. Let K be such a HES and M a component of K which is *n*-multifaceted on T_M. At a given time of its life, M can operate through any one of its lower level decompositions, and can *switch* between them depending on the context; a switch amounts to a 'random' lower level fluctuation with no effect for M at its own level. With respect to its different ramifications, the component acts as "un centre régulateur dont les différentes facettes représentent les déploiements qui caractérisent sa puissance" (Thom, 1988).

With time M may lose some lower level decomposition and/or acquire new ones while preserving its own identity. Thus, it gains its individuation (in the sense of Simondon, 2005), so that its successive instantaneous states are connected through the flow of time (cf. Figure 15.3). Such a component corresponds to what Longo et al. (2012) calls a "Kantian whole".

For the HES, a main consequence of MP is the possibility of emergence, at a level $n + 1$, of new properties, modeled by *n-complex links*. An *n-complex* link on J is the composite of *n*-simple links on J binding 'non-adjacent' clusters (i.e. separated by a 'switch' between structurally non-connected patterns, cf. Figure 15.4). More precisely,

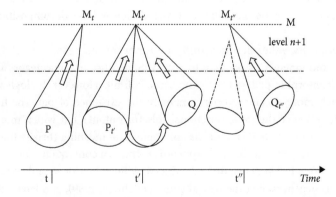

Figure 15.3 Successive states of a multifaceted component M.

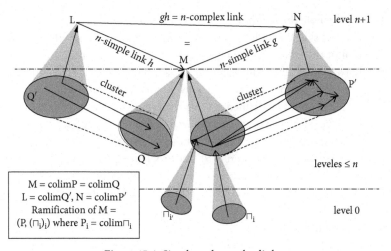

Figure 15.4 Simple and complex links.

let M be an n-multifaceted component having two structurally non-connected lower level decompositions P and Q on an interval J; then the (non-simple) composite gh in K_J of a (Q', Q)-simple link h from L to M with a (P, P')-simple link g from M to N is an n-complex link from L to N on J. In a living system, such a link represents a 'global' interaction between L and N, which does not combine local interactions between components of Q' and P' observable at levels $\leq n$. It really 'emerges' at the level $n + 1$ to reflect global properties of the lower levels (in particular, the fact that P and Q have M for their colimit). The complex links will play an important role in the emergence of higher order components and properties through complexifications, and they are necessary for going beyond a pure reductionism (cf. Section 15.4).

15.4 Structural Changes: Complexification Process

Here we analyze the different kinds of structural changes which a living system undergoes over time, and we model them by the complexification process. In particular, it explains the development of a robust but flexible memory.

In a physical system, initial conditions and physical laws entail the following states of the system. It is different in biology: knowing the state of the system up to t (included) leaves many possibilities for its future states. As said in Longo et al. (2012): "entailed causal relations must be replaced by 'enablement' relations, in evolutionary biology".

In a living system, the transitions between configurations are generated by the "standard changes" described by Thom (1988): birth, death, collision, scission. Birth corresponds to the addition of new components (endocytosis); death to the suppression of certain components (catabolism); collision to the formation of a new component C for binding or strengthening some pre-existing pattern P into a coherent and structured whole (biosynthesis of macromolecules; memorization of a new item; learning of a new skill through the synchronization of neural assemblies); scission to the reduction of the coherence of a pattern leading to the destruction of its binding (result of the action of alkylating agents or radiation on DNA, apoptosis of a cell).

On the configuration of the system at a given time, several procedures with objectives of these kinds could be defined, so that there is a large space of possible futures. Which one will be selected at time t and realized is not entailed by laws, but will depend on different factors, in particular the environmental context and the internal interplay among the co-regulators (cf. Section 15.5). The problem examined in this section is not the selection problem; on the contrary, we suppose that a procedure has been selected at t, and we want to describe what will be the new configuration which appears just after the 'physical' realization of the objectives of the procedure, which, physically, requires some time (say from t to t').

We approach this problem through the HES K modeling the system, using the categorical *complexification process** (EV, 1987, 2007). The procedure selected at t is modeled by a *pro-sketch** Pr on the configuration category K_t. To simplify here we suppose that $Pr = (S, \Pi, I)$ consists only of a set S of objects of the category K_t (modeling the components to be suppressed), a set Π of patterns P with no colimit in K_t (modeling the patterns P to bind), and a set I of cones in K_t (to become colimit cones). The *Complexification Theorem** (EV, 2007, ch. 4) ensures that there is a category $K_{t'}$, called the *complexification* of K_t for Pr, in which the objectives of the procedure modeled by the pro-sketch are realized 'in an optimal way'. (Cf. Figure 15.5, and Appendix for a more precise formulation.)

The complexification $K_{t'}$ models the configuration category of the living system at t', after realization of the procedure. Its mathematical construction has been given in EV (1987, 2007): its objects are the (new) states of the objects of K_t not in S and for each P in Π, there emerges a new object cP which becomes the colimit of (the new

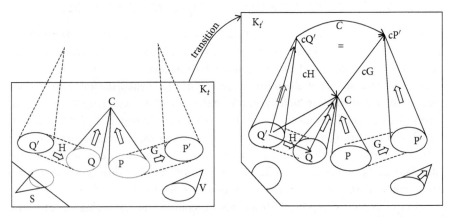

Figure 15.5 The category K_t and a complexification; c is an emergent complex link.

state of) P in $K_{t'}$. The cones in I become colimit-cones in $K_{t'}$. The morphisms are constructed stepwise, to 'force' each pattern P in Π to admit cP as its colimit in $K_{t'}$, each cone in I to become a colimit cone, and each path of morphisms to have a composite; this last process can add complex links to K. Being constructed by recurrence, the complexification may not be finitely computable.

The fact that the mathematical construction of the complexification effectively gives the model of a following configuration comes from the way the model has been tailored to adapt to the properties of living systems; in particular, the binding of a pattern corresponds to the formation of an attractor of the system's dynamic (cf. Ehresmann, 2009). We can go further and use mathematical properties of the complexification process and in particular the two following mathematical theorems, to deduce properties of a living system.

Let us recall (Section 15.2) that the *complexity order* of a component M of a HES is the smallest length of its different ramifications down to level 0 on T_C. In a pure reductionism, there would be no component of complexity order > 1. The following theorem shows that the HES model is not reductionist, but resorts to an "emergentist reductionism" (in the terms of Bunge, 1979).

Theorem 15.1 (Emergence Theorem (EV, 1996, 2007)) *The Multiplicity Principle is preserved through complexifications. In a HES, it allows for the existence of components of complexity order >1 and for the formation, through successive complexification processes, of multifaceted components of increasing complexity orders, with emergence of complex links between them.*

Applied to the HES K modeling a living system, this theorem explains how the system can develop a *memory*, containing more and more complex multifaceted components, called *records*. These records store information of any kind, such as features of the environment, innate or acquired processes at the basis of the functioning of the system, past events and experiences. They take their own individuation and can be recalled through different, possibly structurally non-connected, lower level decompositions and switch between them. Thus, the memory is both robust and flexible, and it has enough plasticity for adaptation to changes. It has a subsystem, the *Procedural Memory* whose records, called *procepts*, memorize automatic behaviors or selected procedures (e.g. associated to a complexification process), and send the corresponding commands to effectors (e.g. selecting the good muscular movement for seizing an object).

In a HES K, the memory is modeled by a sub-HES denoted by **Mem**, and still called the *Memory*, which develops over time with formation of multifaceted components of increasing complexity orders, connected by simple and complex links. **Mem** has a subsystem **Proc** whose records Pr are called *procepts*; such a Pr is the limit* of a pattern E(Pr) in K (modeling its effectors) which it 'commands' through the limit-cocone* from Pr to E(Pr). (Cf. Appendix for the notion of a limit, dual of a colimit.)

Now we give another property of the complexification process which will have important consequences for the formation of higher cognitive processes:

Theorem 15.2 (Iterated Complexification Theorem (EV, 2007)) *Let $K_{t'}$ be a complexification of K_t introducing complex links, and $K_{t''}$ a complexification of $K_{t'}$, where $t < t' < t''$. In presence of MP, the two successive complexifications of K_t are not always reducible to a unique one; that is: there can be no pro-sketch on K_t such that $K_{t''}$ is the complexification of K_t for this pro-sketch.*

Indeed, the first complexification leads to the formation of new components and to the 'acausal' emergence of complex links between them. These links represent new 'rules' or "changes in the conditions of change", and they allow for new "enabling constraints" (Kauffman and Gare, 2015) on the choice of a procedure Pr′ on $K_{t'}$. If some of the patterns that Pr′ must bind contain a complex link having emerged in the first complexification, the results of the second complexification are unpredictable from the data of K_t. (Cf. EV, 2007, ch. 4, sect. 6.1.)

Remark 15.4

(i) The Iterated Complexification Theorem permits justifying the distinction between "organisms" and "mechanisms" proposed by Rosen (1985b), based on their comportment with respect to Aristotle's causes (cf. EV, 2007).
(ii) Up to now, we have spoken of colimits and have just mentioned the dual notion of *limits** in the definition of a procept. In fact limits play some part in MES, especially to model procedural and semantic memories. The complexification process can be extended to also add limits to some patterns (cf. EV, 2007).

15.5 Self-Organization by a Network of Co-regulators: Memory Evolutive System

This section describes the multiagent, multitemporality self-organization of a living system; to model it we complement the notion of HES by adding a memory and a net of co-regulators, thus obtaining a *Memory Evolutive System*.

The dynamic of a living system is modulated by the cooperation/competition between a net of relatively autonomous internal agents (we call them *co-regulators*), operating with the help of the memory to "co-create their worlds with one another" (Longo et al., 2012). A co-regulator is a specialized sub-ES, with a more or less temporary existence, for instance: protein networks, cells, tissues, more or less specialized brain modules, co-regulators acting as effectors (e.g. replication complex, cell regulatory networks, motor modules, ...); it may or not bind into a higher level component. It develops a stepwise local dynamic at its own rhythm to select specific procedures which characterize its function and activate the corresponding effectors (e.g. start of the replication for the pre-replication complex, activation of muscles).

The global dynamic of the system results from the, possibly conflicting, interactions between the local dynamics of the co-regulators.

To model the dynamic of the system, we enrich the notion of HES as follows:

Definition 15.5 A *Memory Evolutive System* (MES) is a HES K satisfying MP and in which we distinguish:

(i) A sub HES, denoted by **Mem**, called its *Memory*, which develops with time (cf. Section 15.4); it has a sub-ES **Proc** whose records Pr, called *procepts*, are the limit of an 'effector pattern' E(Pr).

(ii) A family of sub-ES, called *co-regulators*, of various complexity levels. Each co-regulator CR has its timescale divided into successive steps, the (possibly varying) length of which defines its specific rhythm; it has also a differential access to **Mem**, in particular through links pr coming from particular procepts Pr, called its CR-*procepts*, satisfying the condition: the associated pattern E(Pr) takes its values in an effector coregulator and each command $p_i : \text{Pr} \to E(\text{Pr})_i$ factors through pr. (Cf. Figure 15.6.)

First, let us describe the local dynamic of a co-regulator CR during one of its steps; the duration of the step may depend on the context; for instance, a step of a cell coincides with the cell cycle and its length may vary. By definition, the step of CR from t to t' is divided in different more or less temporally overlapping phases:

(i) Formation of the landscape of CR on $J = [t, t')$: it consists of the partial information received by CR during the step; for instance, for a cell, a few biochemical events affecting its components and triggered by the binding to the membrane of a cellular mediator (e.g. hormonal), by changes in the concentration of some oligo-element, and so on. It represents a model of the system as 'seen' by CR during the step.

Definition 15.6 In a MES, the *landscape* of the co-regulator CR on $J = [t, t')$ is the Evolutive System L with timescale J whose components are links $b, c, \ldots$ from components of the system to CR whose index of activation on J is 1; the links between them in L are commutative squares $(g, f; b, c)$ in K_J such as $fb = cg$ (cf. Figure 15.6).

Thus, for each s in J, the configuration category L_s is a subcategory of the comma-category $K_s|CR_s$ and there is a *difference functor* from L_s to K_s mapping $(g, f; b, c)$ on g.

(ii) Selection via the landscape of a CR-procept (cf. Definition 15.6) Pr to respond, and command of its associated pattern E(Pr); for instance, replication of DNA for a cell. During the end of the step, the local dynamic leading to the 'physical' realization of the procedure modeled by Pr can be computed by classical mathematical models, e.g. based on non-linear differential equations. In the MES model, a CR-procept Pr is selected via a component pr of L, and its commands p_i to the associated pattern $E = E(\text{Pr})$ factorize into $e_i\, pr$, so that E can be 'commanded' via the e_i's. The update of K_t after realization of these commands should be a complexification K' of K_t for a specific pro-sketch (S, Π, I); let L_{ant} be the configuration at t' of the next landscape if K' is realized.

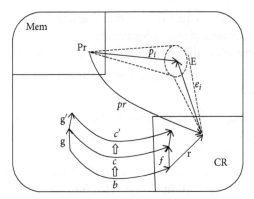

Figure 15.6 The landscape of CR on J: its components are the curved arrows b, c, c', pr. The effector pattern $E = E(Pr)$ of the selected CR-procept Pr is commanded via the e_i's.

(iii) At the beginning the next step defined on $J' = [t', t'')$, evaluation of the result, and its memorization. By construction, L_{ant} should be (isomorphic to) the configuration at t' of the new landscape $L_{J'}$ of CR; if it is not the case, we speak of a *fracture* for CR (e.g. DNA has not been replicated in time).

Earlier we described the one-step dynamic of a single co-regulator. However, there are several co-regulators CR_i, each operating at its own rhythm and level, with its own 'logic' (depending on its function), and they must work together for 'co-creating' their world. There is no problem as long as the commands sent to the effectors by each of them at a given time are more or less coordinated, and at least not conflicting, so that they can be synchronously realized.

Otherwise, an equilibrium process, called the *interplay among the co-regulators*, will be necessary to coordinate their actions while accounting for their different rhythms. It amounts to a kind of selection among the various commands sent to effectors at a given time; it has some flexibility since the commands can be realized through any of their lower level decompositions, with possibility of switches between them for selecting those that are most relevant to the context. The operative procedure resulting from the interplay may cause a fracture to some co-regulators.

Formally let us denote by (S_i, Π_i, I_i) the pro-sketch selected by the co-regulator CR_i at a given time t. The pro-sketch considered to update K_t via all the CR_i should be $(\cup_i S_i, \cup_i \Pi_i, \cup_i I_i)$. However, this pro-sketch may not exist, e.g. if a pattern in Π_i has some of its elements in S_j for another j. How to construct a pro-sketch which represents a 'best compromise' between the commands sent by the different co-regulators to effectors? This is an open problem (for some ideas on this, cf. EV, 2007, chs 7–8).

In particular, an important role is played by the structural temporal constraints or *synchronicity laws* that a co-regulator CR must respect; they relate the length of the step to the propagation delays of the components of the landscape and their stability span (cf. EV, 2007, ch. 7). Their non-respect is a cause of fractures. For instance, because of the propagation delays, small events for a lower co-regulator CR cannot be observed in real time by a higher co-regulator CR', but the accumulated change they produce to the system during the longer step of CR' may cause a fracture to CR'; the following repair can later backfire by also causing a fracture for CR. Whence we obtain a *dialectics* between co-regulators with different rhythms and complexities, which corresponds to loops of retroactions between levels. An example is given by the replication of DNA in a cell: specific self-repair mechanisms intervene to ensure a strict pairing of the bases; however, if too many errors occur, they are overrun, and a higher co-regulator, the SOS system for DNA repair (Radman, 1975), is activated and makes the replication complex tolerant to errors; later, these errors can lead to a mutation of the cell.

While the local dynamics of the co-regulators may resort to 'classical' computations, the real challenge is to deal with their interplay. Indeed, we cannot pre-state its result because of its flexibility and its several freedom degrees. It justifies to say that "the evolution of life marks the end of a Physics world view of law entailed dynamics" (Longo et al., 2012).

15.6 MENS, an Integrative Model Encompassing the Neural and Mental Systems

This section briefly evokes an application of MES to cognition in the frame of the integrative model MENS (for *Memory Evolutive Neural System*). Without falling into simple reductionism, MENS correlates the neural and mental systems, and analyzes the formation of mental states and cognitive processes of increasing complexity. Because of the limiting space, the notions and results are exposed in a more informal way; for a rigorous categorical treatment, cf. EV (2007, ch. 9).

How can cognitive functions arise from the brain dynamics? Let us recall some well-known approaches to this problem:

(i) *Computationalism* states that the mind functions as a symbolic operator, and that mental representations are symbolic representations; more complex mental states (thoughts) can be created, whence the compositionality and productivity of thought.

(ii) For *connectionism*, mental phenomena can be described by distributed interconnected networks of simple units (representing neural networks); it remains at the subsymbolic level.

(iii) *Neurophenomenology* intends "to marry modern cognitive science and a disciplined approach to human experience" (Varela, 1996); it takes into account both the first and third person perspectives and examines how the brain processes to support consciousness.

The three approaches have been criticized for being more or less reductionist.

MENS proposes another approach to human neurocognitive system, up to a theory of mind, which, as we are going to show, somewhat combines the three preceding approaches, while avoiding their pitfalls.

The construction of MENS relies on a common process in brain dynamics, already noted by Hebb in 1949, namely the representation of information (such as mental objects or processes) by distributed neuronal groups acting synchronously; these groups are self-reinforcing thanks to the *Hebb rule* for synaptic plasticity: "When an axon of cell A is near enough to excite B and repeatedly or persistently takes part in firing it... its efficiency, as one of the cells firing B, is increased" (Hebb, 1949). And the *degeneracy property* of the neural code shows that this representation is not 1-to-1: "More than one combination of neuronal groups can yield a particular output, and a given single group can participate in more than one kind of signaling function" (Edelman, 1989).

Formally, MENS is a MES having the Evolutive System of neurons NEUR (Example 15.1) as its sub-ES of level 0; its higher level components, called *category-neurons* (or cat-neurons) model mental objects (with their possible changes in time), and their hierarchy represents a dynamic "algebra of mental objects" (in the terms of Changeux, 1983). Cat-neurons and their links are obtained by iterations of the complexification process (Section 15.4) starting from NEUR.

The idea is to represent a mental object or process O by a cat-neuron colimit of each of the neuronal groups that O can synchronously activate at different times and in different contexts. A neuronal group is modeled by a pattern P in NEUR; generally P_t has no colimit in $NEUR_t$ (cf. EV, 2009). If P acts synchronously at t, the information it codes (e.g. a mental object O) will be modeled in MENS by a cat-neuron M which is constructed by a complexification process intended to bind P. This M, which becomes the colimit of P on an interval J, must also become the colimit of the other neuronal patterns coding the same information (i.e. activated by O) at different times or in different contexts. Thus, the cat-neuron M becomes a multifaceted component of MENS. More complex cat-neurons are obtained by iterative complexifications, binding more and more complex patterns of cat-neurons; they represent flexible mental objects and cognitive processes of increasing complexity. In EV (2009) we proved that the complexification process allows extending to MENS the notions of activity, propagation delays, and strengths of links, as well as Hebb rule.

A cat-neuron M is a component of MENS, hence a (dynamic) categorical object, not a 'physical' object; however, its ramifications have their base in NEUR which represents the 'physical' neural system. Thus, M has multiple physical realizabilities obtained by unfolding one of its ramifications down to its base. If a neuronal group is physically activated at t, the corresponding pattern P of NEUR is also activated (by construction of NEUR) and, if this pattern is a lower level decomposition of a cat-neuron M, its activity is reflected to M after an *activation delay* equal to the greatest propagation delay of its links to M. This activation delay extends to a multifaceted higher level cat-neuron M, through the unfolding of a ramification: M is activated at time t if the neuronal base of one of its ramifications is activated at $t - d$, where d is the activation delay; this delay increases with the complexity order of M.

Remark 15.5

(i) For the formal construction of cat-neurons and their links in MENS, cf. EV (2007, ch. 9).
(ii) The notions introduced in MENS can be compared to more recent notions introduced in neuroscience by Buzsàki (2010): a cat-neuron M acts as a mathematical specification of what he calls a "reader-actuator mechanism" since it detects neuronal groups with the same output (those of which M becomes the colimit), our hierarchy of cat-neurons gives a categorical translation of his "neural syntax", and his "synapsembles" correspond to patterns in MENS.

Since cat-neurons are activated through their neuronal bases, the dynamic of MENS depends on the dynamics of the neural system. NEUR has a modular self-organization, with more or less specialized brain modules developing their local dynamics. The interplay of their dynamics is facilitated by the existence of the structural core, introduced in Hagmann et al. (2008): "Our data provide evidence for the existence of a structural core in human cerebral cortex. This complex of densely connected regions in posterior medial cortex is both spatially and topologically central within the brain.... the core may be an important structural basis for shaping large-scale brain dynamics." The subsystem SC modeling it in NEUR will have an important role in MENS.

As in any MES, the global dynamic of MENS is directed by the interplay among the local dynamics of its different co-regulators. These co-regulators are sub-ES which are based on different brain modules (meaning that their components have ramifications with their base in these modules). The co-regulators operate with the help of a robust though flexible memory **Mem** whose records represent percepts, concepts, propcepts, past experiences, explicit or implicit knowledge of any kind, and non-conscious items such as innate or learnt automatic procedures and affects. Its records are plastic enough to adapt to changes. **Mem** develops through exchanges with the physical and social environment. (For the development of **Mem**, cf. EV, 2007.)

An important notion which MENS has led to introduce is that of the *Archetypal Core* (EV, 2002, 2007), which plays a main role in the formation of consciousness and higher mental and cognitive processes. It is a higher order sub-ES AC of **Mem** which develops over time by successive complexifications of the (later discovered) structural core SC (as a consequence of the Emergence Theorem, Section 15.4). Its components are higher order records with numerous non-connected ramifications based on SC; they model significant memories which integrate various modalities (sensorial, motor, affective,...) and are frequently recalled, so that their links are strengthened (via Hebb rule). They are densely connected by strong and fast complex links which form

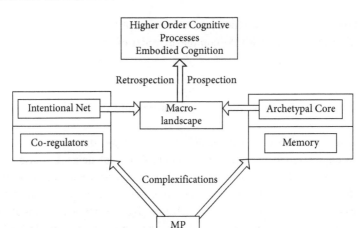

Figure 15.7 A synthetic representation of the interaction between the main notions.

archetypal loops; these loops propagate the activation of an archetypal record A to others, and then back to A, thus maintaining it for some time.

We have described (EV, 2007; Ehresmann, 2013) how AC acts as a driving force for developing higher cognition and embodying the 'Self'. For that, among the co-regulators of MENS we distinguish higher order co-regulators based on associative zones of the cortex and directly linked to AC; with their interconnections they form the *intentional network* (IN), which acts as a macro-regulator.

An unexpected event (possibly causing a fracture to IN) increases the attention, translated into the activation of a part of SC. This activation is transmitted (with a little delay) to components A of AC based on this part; it diffuses through self-maintained archetypal loops, and then propagates to lower levels through unfolding of ramifications and switches between them. Thus, a large domain of MENS remains activated for some time. The links transmitting this activation to IN are the components of the *macro-landscape* (ML) of IN, which hierarchically and temporally extends the landscapes of the co-regulators in IN; indeed, it contains information coming from lower levels and concerning the just passed (because of activation and propagation delays) as well as some opening to the near future since the activation of AC will be self-maintained for some time, whence a double integration of the time dimension (as in Husserl's retention-protention). ML can be seen as a model of the system for IN, its 'first person' perspective (to be compared to the "theater" of Baars, 1997). There is not enough space to explain how AC allows the development in ML of higher cognitive and mental processes, such as consciousness, embodied cognition, anticipation, or creativity, thanks to iteration of retrospection/prospection/complexification processes (cf. Figure 15.7). For that, we refer to Ehresmann (2012, 2013) and more recent papers and slides online: http://ehres.pagesperso-orange.fr (e.g. Ehresmann and Gomez-Ramirez, 2015).

To conclude, MENS (as well as MES) is a developing methodology rather than a model. Let us compare it with the models of cognition recalled earlier:

(i) As in *connectionism*, MENS basically considers the dynamic of neural networks. However, it does not remain at this subsymbolic level thanks to the formation of cat-neurons of level 1 which bind and classify such networks. These cat-neurons can themselves be taken as the units of new networks, and the operation can be reiterated. Whence MENS gives rise to a *hierarchy of connectionist models*, interconnected by the fact that each 'unit' has a Janus-face. As these units are themselves multifaceted, the hierarchical model avoids determinism.

(ii) As in *computationalism* MENS leads to a theory of mind with compositionality and productivity of thoughts (through the hierarchy of cat-neurons). If we forget the neuronal base of the cat-neurons, say of level 1, they could be taken as 'symbols' in a computational model; and higher level cat-neurons become higher level symbols. However, in this way, we lose the dynamic and flexibility given to cat-neurons by their multiple 'physical' realizations and the possibility of switches between them, necessary for avoiding pure reductionism and strict determinism.

(iii) It is also this multiplicity (due to MP) which avoids the "isomorphism between neural and mental" that Bayne (2004) sees lurking under *neurophenomenology*. MENS can be seen as a categorical "bridging strategy" (his terms) for closing the gap between neural and phenomenal data and accounting for intentionality and consciousness. It is done in the macro-landscapes, which reflect the "first person" perspective (inside the third person model MENS) and make possible the integration of different levels and times, at the root of embodied cognition and phenomenological experience.

Appendix: Mathematical Definitions

1. *Category, Functor*

    A (directed multi-)*graph* G consists of a set of vertices A, B, . . ., and a set of arrows $f: A \to B$ between them. A path of G is a finite sequence of consecutive arrows.

    A *category* K is a graph equipped with an internal *composition law* which maps a path (f, g) from A to C on an arrow $gf : A \to C$ (called its composite) and which satisfies the conditions: it is associative and each vertex A has an identity $id_A : A \to A$; a vertex of the graph is called an *object* of K, and an arrow a *morphism*.

    A *functor* F from K to a category K′ associates to an object A of K an object FA of K′, to a morphism $f : A \to B$ a morphism $F(f) : FA \to FB$ of K′, and it preserves the composition and the identities. A *partial functor* from K to K′ is a functor from a subcategory of K to K′.

    The *opposite category* of K, denoted K^{op}, has the same objects but the direction of the arrows is inverted.

2. *Colimit and Limit*

    A *pattern* (or diagram) P in a category K is a homomorphism of a graph sP to K; it is thus defined by the family of objects $(P_i)_{i \in I}$ of K (where I is the set of vertices of sP) and some distinguished morphisms between them (namely, the P(x) for $x : i \to j$ in sP).

A(n inductive) *cone* from P to an object A of K is a family $(s_i)_{i \in I}$ of morphisms $s_i : P_i \to A$ such that, for each distinguished link $f : P_i \to P_j$ of P, we have $s_i = s_j f$. (Cf. Figure 15.1.)

A *colimit* (or inductive limit; Kan, 1958) of a pattern P in K is an object C of K such that there is a cone $(c_i)_{i \in I}$ from P to C satisfying the "*universal condition*": for each cone $(s_i)_{i \in I}$ from P an object A of K, there is a unique morphism $s : C \to A$ such that $sc_i = s_i$ for each $i \in I$; we call s the binding of $(s_i)_{i \in I}$ (cf. Figure 15.1). A pattern P may have no colimit in K; if it has one, it is unique up to an isomorphism. Two patterns P and Q may have the same colimit C.

A *cocone* from an object A to a pattern P is a cone from P to A in the opposite category K^{op}. A *(projective) limit* of a pattern P is a colimit of P in K^{op}.

3. *Clusters between patterns. Structurally non-connected patterns* (cf. Figure 15.4)

If P and P' are two patterns in a category K, a *cluster* from P to P' is a maximal set G of morphisms between their objects satisfying the following conditions:

(i) For each P_i there is at least one morphism in G from P_i to some P'_j; and if there are several such morphisms, they are correlated by a zigzag of distinguished morphisms of P'.

(ii) The composite of a morphism in G with a distinguished morphism of P', and the composite of a distinguished morphism of P with a morphism in G belong to G.

If P and P' have colimits M and M', respectively, in K, it follows from the universal property of a colimit that a cluster G from P to P' binds into a unique morphism cG: $M \to M'$, called a *(P, P')-simple morphism*.

The (small) patterns and the clusters between them form a category Pat(K), which 'is' a free co-completion of K. The clusters from P to P' are in 1:1 correspondence with the elements of the set:

$\lim_i \mathrm{colim}_j (\mathrm{Hom}_K(P - , P'-) : sP^{op} \times sP' \to \mathbf{Sets})$,

so that Pat(K) has a full subcategory isomorphic to IndK.

Two patterns Q and P with the same colimit M are said *structurally non-connected* if they are not isomorphic in Pat(K), so that there is no cluster between them binding into the identity of M. Then there can exist *complex morphisms* gh : $L \to N$, where $g : L \to M$ is (Q', Q)-simple and h is (P, P')-simple.

4. *Complexification Process*

Let Pr = (S, Π, I) be a *pro-sketch* on a category K, where S is a set of objects of K, Π a set of patterns without a colimit in K, and I a set of cones in K. We denote by K° the largest subcategory of K not containing S. A functor F : $K^\circ \to K^*$ is said to be Pr-*compatible* if FP has a colimit cP in K^* for each P in Π and if F transforms a cone in I into a colimit cone.

Theorem 15.3 Complexification Theorem (EV, 2007, ch. 4, sect. 5.1). *There is a Pr-compatible functor $\varphi : K^\circ \to K'$ such that any other Pr-compatible functor $F : K^\circ \to K^*$ factors through φ. The category K' is called the complexification of K for Pr; it is explicitly constructed by recurrence.*

For the explicit construction of K', cf. (EV, 2007, ch. 4). It corresponds to the construction of the prototype of a sketch (Bastiani and Ehresmann, 1972), remarking that K' 'is' the prototype of the following sketch: the underlying graph is obtained by adding to K° a cone γP from P to cP for each P in Π, and the distinguished cones are those in I and the cones γP.

References

Baars, B. J. (1997). *In the Theatre of Consciousness: The Workspace of the Mind*. Oxford University Press, Oxford.

Bailly, F., and Longo, G. (2006). *Mathématiques et sciences de la nature. La singularit é physique du vivant*. Hermann, Paris.

Bastiani(-Ehresmann), A., and Ehresmann, C. (1972). Categories of sketched structures, *Cahiers Top. et Gom. Dif.* XIII-2. Reprinted in Charles Ehresmann: Oeuvres complètes et commentées, IV-2.

Bayne, T. (2004). Closing the gap? some questions for neurophenomenology. *Phenomenology and the Cognitive Sciences* 3(4), 349–64.

Béjean, M., and Ehresmann, A. (2015). D-MES: Conceptualizing the working designers. *International Journal of Design Management and Professional Practice* 9(4), 1–20.

Bunge, M. (1979). *Treatise on Basic Philosophy*, Vol. 4. Reidel, Dordrecht.

Buzsàki, G. (2010). Neural syntax: cell assemblies, synapsembles and readers. *Neuron* 68(3), 362–85.

Changeux, J.-P. (1983). *L'homme neuronal*. Fayard, Press.

Dobzhansky, T. (1970). *Genetics of the Evolutionary Process*. Columbia University Press, New York.

Edelman, G. (1989). *The Remembered Present*. Basic Books, New York.

Edelman, G. M., and Gally, J. A. (2001). Degeneracy and complexity in biological systems. *Proceedings of the National Academy of Science USA* 98(24), 13763–8.

Ehresmann, A. (2008). 'Semi-sheaves, distructures and Schwartz distributions'. Rapport 09, Universit de Picardie. http://ehres.pagesperso-orange.fr

Ehresmann, A. (2012). An info-computational model for (neuro-)cognitive systems up to creativity. *Entropy* 14, 1703–16.

Ehresmann, A. (2013). A theoretical frame for future studies. *On the Horizon* 21(1), 46–53.

Ehresmann, A., and Vanbremeersch, J.-P. (1987). Hierarchical evolutive systems: a mathematical model for complex systems. *Bulletin of Mathematical Biology* 49(1), 13–50.

Ehresmann, A., and Vanbremeersch, J.-P. (1996). Multiplicity principle and emergence in MES. *Journal of Systems Analysis, Modelling, Simulation* 26, 81–117.

Ehresmann, A., and Vanbremeersch, J.-P. (2002). Emergence processes up to consciousness using the multiplicity principle and quantum physics, in D. Dubois (ed.), *Conference Proceedings 627 (CASYS, 2001)'*, pp. 221–33.

Ehresmann, A., and Vanbremeersch, J.-P. (2007). *Memory Evolutive Systems: Hierarchy, Emergence, Cognition*. Elsevier, Amsterdam.

Ehresmann, A. and Vanbremeersch, J.-P. (2009). A propos des systèmes evolutifs à mmoire et du modèle MENS, in *Compte-rendu du SIC*, Paris. http://ehres.pagesperso-orange.fr

Hagmann, P., Cammoun, L., Gigandet, X., Meuli, R., Honey, C., Wedeen, V., and Sporns, O. (2008). Mapping the structural core of human cerebral cortex. *PLoS Biology* 6(7), 1479–93.

Hebb, D. O. (1949). *The Organization of Behaviour*. Wiley, New York.

Izhikevich, E. M., Gally, J. A., and Edelman, G. M. (2004). Spike-timing dynamics of neuronal group. *Cerebral Cortex* 14, 933–44.

Jacob, F. (1970). *La logique du vivant*. Gallimard, Paris.

Kan, D. M. (1958). Adjoint functors. *Transactions of the American Mathematical Society* 89, 294–329.

Kauffman, S., and Gare, A. (2015). Beyond Descartes and Newton: recovering life and humanity. *Progress in Biophysics and Molecular Biology* 119, 219–44.

Koestler, A. (1965). *Le cri d'Archimède*. Calmann-Lèvy, Paris.
Laborit, H. (1983). *La Colornbe assassinée*. Grasset, Paris.
Longo, G., Montévil, M., and Kauffman, S. (2012). No entailing laws, but enablement in the evolution of the biosphere, in *Proceedings of the 14th Annual Conference Companion on Genetic and Evolutionary Computation*, Vol. 1, pp. 1379–92.
Louie, A. H. (2009). *More Than Life Itself: A Synthetic Continuation in Relational Biology*. Ontos, Heusenstamm.
Mac Lane, S. (1971). *Categories for the Working Mathematician*. Springer, Berlin.
Matsuno, K. (2012). Time in biology as a marker of the class identity of molecules, *Integral Biomathics*. Springer, Berlin.
Popper, K. (2002). *The Poverty of Historicism*. Routledge Classics, London.
Radman, M. (1975). SOS repair hypothesis: phenomenology of an inducible DNA repair which is accompanied by mutagenesis, in *Molecular Mechanisms for Repair of DNA*, 355–67. Springer, Berlin.
Rashevsky, N. (1954). Topology and life: in search of general mathematical principles in biology and sociology. *Bulletin of Mathematical Biophysics* 16(4), 317–48.
Rosen, R. (1958a). A relational theory of biological systems. *Bulletin of Mathematical Biophysics* 20(3), 245–60.
Rosen, R. (1958b). The representation of biological systems from the standpoint of the theory of categories. *Bulletin of Mathematical Biophysics* 20(4), 317–41.
Rosen, R. (1985a). *Anticipatory Systems: Philosophical, Mathematical and Methodological Foundations*. Pergamon, Oxford.
Rosen, R. (1985b). Organisms as causal systems which are not mechanisms: an essay into the nature of complexity. *Theoretical Biology and Complexity* 165–203.
Rosen, R. (1991). *Life Itself: A Comprehensive Inquiry into the Nature, Origin, and Fabrication of Life*. Columbia University Press, New York.
Simeonov, P. L. (2010). Integral biomathics: a post-Newtonian view into the logos of bios. *Progress in Biophysics and Molecular Biology* 102(2), 85–121.
Simeonov, P. L., Smith, L. S., and Ehresmann, A. C. (eds) (2012). *Integral Biomathics*. Springer, Berlin.
Simondon, G. (2005). *L'individuation à la lumière des notions de forme et d'information*. Éditions Jérôme Millon, Grenoble.
Spinoza, B. (1677). *Ethics*, Livre II. Translated by R. H. Elwes.
Thom, R. (1988). *Esquisse d'une sémiophysique. physique aristotéliciene et théorie des catastrophes*. Dunod, Paris.
Varela, F. (1996). Neurophenomenology: a methodological remedy to the hard problem. *Journal of Consciousness Studies* 3(4), 330–49.

16
Categories as Mathematical Models

David I. Spivak

Mathematicians do not study objects, but relations between objects. Thus, they are free to replace some objects by others so long as the relations remain unchanged. Content to them is irrelevant: they are interested in form only.

– Henri Poincaré

Yet, I hope that I managed to convey the message: the mathematical language developed by the end of the 20th century by far exceeds in its expressive power anything, even imaginable, say, before 1960. Any *meaningful* idea coming from science can be fully developed in this language.

– Mikhail Gromov

16.1 Introduction

In the sciences, most of the prominent methods for incorporating mathematics involve setting up stochastic processes, dynamical systems, or statistical models that capture the relevant matters in the scientific subject. At bottom, these techniques all involve interplays of numbers. However, a biologist's sophisticated understanding of the complex aspects of life—heredity, reproduction, hierarchical nesting, symbiosis, metabolism, etc.—remains trapped in the realm of *ideas*.[1] Such ideas can often be reduced to numerical models, but the ideas themselves are confined to the background. This is suboptimal, because when an idea or theory is itself mathematically formalized, it gains clarity, systematicity, and falsifiability.

My hope for category theory is that it can be used to model many of the actual ideas and ways of thinking that exist within science. Modeling a phenomenon allows us to examine, and interact with, a simplified version of it, and involving mathematics generally provides an additional level of rigor and communicability. A category-theoretic model of the ideas, rather than of the mere quantities, may be able to formally capture a

---

[1] Here, we use biology only as a specific branch of science. Our intention is to consider *scientific ideas*, beyond their numerical shadows, as subject to mathematical formalization.

whole conceptual framework, say, ideas about the hierarchical nature of organizational systems. With such a formal description, one could apply rigorous conceptual—rather than numerical—tests to the ideas themselves. For example, one might check whether the ideas satisfy various internal consistencies, or one might ask about the nature of their integration with the other major conceptual frameworks that exist in the field.

There is a good deal of work on using category theory to model high-level conceptual aspects of scientific subjects. For example, categories have been used by John Baez to model signal flow and reaction networks, by Abramsky and Coecke to model aspects of quantum mechanics, and by Lambek to model computer programming languages. At the time of writing, I believe we are at the early stages of an effort to "categorify" science.

Category theory holds promise for such a conceptual integration of science because it has achieved such an integration of mathematics to a remarkable degree. It does so, not by finding a single syntax or format that can encode any structure, but by doing the opposite: leaving the encoding entirely absent. Rather than modeling a given object *in itself*, category theory models only the relationships between objects. For example, one may say that some object has an internal structure; however, in the categorical model this internal structure can only be seen by using other objects as probes. That is, category theory views an object's internal structure only in terms of its relationships with other objects.

My goal in this chapter is to give philosophers an intuitive idea about how category theory can be thought of as a universal modeling language, in which the relationships between objects are paramount. In particular, it is not my goal to present new interpretations of established philosophical concepts. At times I invoke philosophical ideas, e.g., those of Kant, but the reader should keep in mind that these are my attempt to forge a communication with the philosophical community, rather than to express substantial claims about Kant's work. It is also not my goal to explain exactly how category theory can be used to formalize the more complex ideas found throughout science. As mentioned already, this kind of work has certainly begun in earnest, but adequately addressing it is beyond the scope of this chapter. The discussion in this chapter is somewhat similar in spirit to that of Lawvere and Schanuel's excellent introductory book, *Conceptual Mathematics*, which I recommend for further reading.

16.1.1 Category theory as modeling language

In this chapter, I cast category theory (CT) as a universal modeling language. More precisely, I claim that CT is *a mathematical model of mathematical models*. Before I explain this assertion, I will ground the discussion with a prototypical example of a model to fix ideas, and then I will define what I mean by model.

The sonar system in a submarine models the distances between various objects in the physical environment of the submarine. It does so by plotting dots, which correspond to these objects, on a screen using polar coordinates (distance and angle). A representation of the information gathered by the sonar system is displayed to the submarine's pilot in a familiar "self-centric" way, i.e., with the submarine (and hence the pilot) shown at the center of the display. This allows the sonar system to become *transparent equipment*[2] for the pilot. In other words, she can seamlessly integrate the representation into her personal repertoire and thereby use familiar methods to cope appropriately with a situation as it unfolds, say to evade or pursue another object.

In this example, the sonar system of the submarine is the *model* and the pilot is its *user*. The physical environment of the submarine is the *subject* of the model; it is the thing being modeled by the sonar system. Because the locations of objects in the physical environment are emphasized, rather than suppressed, we say that these observable aspects of the subject are *foregrounded* by the model. The distances and angles between these locations are the *relationships* between foregrounded aspects. We refer to the translation, from sound-wave reflectance times to points in a polar coordinate system, as the *formalism* that founds the model.

Let us tie these ideas together by making some observations. The value of the sonar system (the model) is measured by the extent to which the pilot's (the user's) interaction with submarine's physical environment (the subject) is successfully mediated by the sonar system. This in turn depends on the extent to which the locations of the physical objects in the environment (the foregrounded aspects) are propitious for successful interaction with the subject and the extent to which the polar coordinate representation of these objects (the formalism) is faithful.

With this exemplar in mind, we make the following two philosophical postulates, which will help to organize the ideas in this chapter.

(1) Modeling a subject is foregrounding certain observable aspects of the subject, and then formalizing these aspects and certain observable relationships between them.
(2) The value of a model is measured by the extent to which the user's interactions with the subject are successfully mediated by the model. This in turn depends on the propitiousness of the foregrounded aspects and the faithfulness of the formalism.

Note that mathematical models seem to put more emphasis on, and care into, the formalism than do other types of models.

16.1.2 Using models is connecting models

In his *Critique of Pure Reason*, Immanuel Kant makes an important assertion:

[E]verything intuited or perceived in space and time, and therefore all objects of a possible experience, are nothing but phenomenal appearances, that is, mere representations, which in the way in which they are represented to us, as extended beings, or as series of changes, have no independent, self-subsistent existence apart from our thoughts.

---

[2] See Clark (2008).

In other words, our interactions with the subject, which is ostensibly *out there*, are actually interactions with our own familiar models (or as Kant says, representations) of it. Our mind is an economy of models and our thinking consists of negotiations within that economy. Thus, the value of a model is measured by the ease with which it negotiates or interfaces with the other models in our repertoire.

Using models is all about translating between models. To say it another way, the observable aspects of a model are known only by its relationships with other models. It follows that, in order to objectively understand our interactions with a model, it is useful to understand the more general question of how models relate to other models. We add to our list a third and final philosophical postulate, which will be clarified throughout the paper:

(3) A model is known only by its relationships with other models.

With our three modeling postulates in hand, we unpack the italicized statement from Section 16.1.1, that category theory is a mathematical model of mathematical modeling. We are claiming, then, that category theory mathematically foregrounds, and formalizes, certain observable aspects of the subject of mathematical modeling.

To make this claim, we must answer the question, *What observable aspect of mathematical models does category theory foreground?* The answer is, roughly, that CT foregrounds a sense in which each mathematical model is known by its relationship with other mathematical models. That is, CT foregrounds the third postulate as an observable aspect of modeling. It formalizes this postulate in terms of *morphisms*.

Clarifying these statements will be the subject of the present chapter. I will explain how notions in pure mathematics, such as vector spaces or groups, can be viewed as mathematical models, say of linearity or symmetry. I will show how models of linearity, symmetry, and action are all known by the interactions that exist between them. In other words, I am committing to the earlier description of, and postulates about, models—exemplified by sonar in submarines—and I will present several canonized mathematical concepts, e.g., vector spaces and groups, as mathematical models.

While many mathematicians would agree with the statement that category theory is valuable for understanding and working with mathematical subjects, this chapter does not attempt to prove it. However, our second postulate characterizes what such a statement would mean: the value of category theory should be measured by the extent to which it successfully mediates our interactions with mathematical models. If it is valuable, this would imply that category theory foregrounds and faithfully formalizes a propitious aspect of modeling: namely, that to gainfully use models, it is useful to be able to translate between them.

16.1.3 Plan of the chapter

Our first goal will be to gently introduce categories in Section 16.2 using what we hope is a familiar mathematical subject, that of matrix arithmetic. We will then discuss a model of linearity, or flat spaces, in Section 16.3, where we will emphasize the categorical perspective, i.e., how the mathematical model of flat spaces is reflected in (and determined by) the rules defining relationships between flat spaces. In Section 16.4, we define the sort of relationship between categories that captures their structure, namely the functorial relationship. This enables us to consider symmetry and action in Section 16.5. Finally, we give a few concluding remarks in Section 16.6.

Throughout the chapter we continually return to our three postulates about modeling. In this way, we will be able to view category theory as a mathematical model of mathematical modeling.

16.2 Matrices: From Groups to Enriched Categories

Our goal in this section is to introduce category theory by considering the case of matrix arithmetic. We will see that all the usual issues regarding dimension and invertibility are actually information about the structure of a category hidden behind the scenes.

Before we begin, it should be noted that matrices are among the most important tools in mathematical modeling. For example MATLAB, a highly popular technical computing program used by engineers of all kinds, is based primarily on matrix arithmetic. Thus, considering matrices is certainly fair game for thinking about mathematical modeling in the usual sense. Note that the problems usually considered in creating these tools, e.g., speed or accuracy issues, are not being addressed here. Instead, we are considering the abstract idea of matrices. In Section 16.2.1, we will foreground some observable aspects of matrix arithmetic, and in Sections 16.2.3–5 we will formalize them category-theoretically. The result will be a category $\text{Mat}_{\_\times\_}$ that models the subject of matrix arithmetic.

16.2.1 Matrices: a review of relevant aspects

A student who is new to linear algebra must learn a few things regarding when matrices can be added and multiplied, the properties of additive and multiplicative identity matrices, the issue of invertibility and non-invertibility, and so on. We now review these because they are precisely what is encoded in the single fact that *matrices form a group-enriched category*. That is, we will introduce category theory by explaining how a certain category models the subject of matrices.

For any natural numbers $m, n \in \mathbb{N}$, let $\text{Mat}_{m \times n}$ denote the set of $m \times n$ matrices—as usual, m is the number of rows and n is the number of columns—and let each entry be a real number. For a matrix $M \in \text{Mat}_{m \times n}$, we refer to (m, n) as the *dimension* of M and denote this fact by $\dim(M) = (m, n)$. It is well known that to add two matrices, say $M + P$, they must have the same dimension. However, to multiply two matrices, say MP, there is a different kind of restriction: the middle numbers must agree. More precisely, if $\dim(M) = (m, n)$ and $\dim(P) = (p, q)$, we require $n = p$ in order for the product MP to make sense. In this case, $\dim(MP) = (m, q)$.

There is a certain matrix $I_n \in \text{Mat}_{n \times n}$, called the $n \times n$ *identity matrix*, which looks like this:

$$I_n = \begin{bmatrix} 1 & 0 & 0 & \cdots & 0 \\ 0 & 1 & 0 & \cdots & 0 \\ 0 & 0 & 1 & \cdots & 0 \\ \vdots & \vdots & \vdots & \ddots & \vdots \\ 0 & 0 & 0 & \cdots & 1 \end{bmatrix}. \tag{16.1}$$

A matrix M is called *invertible* if there exists some matrix N such that $MN = I_n$ and $NM = I_n$. Not every matrix is invertible; for example a matrix of all 0's, as seen below on the left, is not invertible, but neither is the more average-looking matrix on the right:

$$\begin{bmatrix} 0 & 0 & 0 \\ 0 & 0 & 0 \\ 0 & 0 & 0 \end{bmatrix} \qquad \begin{bmatrix} 1 & 2 & 0 \\ 0 & 2 & -2 \\ -2 & -3 & -1 \end{bmatrix}.$$

Note that for any $m \times n$ matrix Q, we have $QI_n = Q$, and for any $n \times p$ matrix R we have $I_n R = R$.

16.2.2 The group of invertible $n \times n$ matrix multiplication

Many scientists and engineers, from physicists modeling the dynamics of elementary particles, to 3D-animators modeling the changes of camera-angle perspectives on a physical scene, use (either explicitly or implicitly) a *group* of invertible $n \times n$ matrices as part of their toolset. That is, group theory is used in mathematical modeling (in the usual sense). We will see later that group theory is a special case of category theory, and we will explain why groups are showing up in the theory of matrix multiplication. First, however, let us recall what a group is, using matrices as the working example.

Let InvMat_n denote the set of all invertible $n \times n$ matrices. Here are the rules that hold in InvMat_n, which make it a *group* in the sense of abstract algebra.

1. There is an established multiplication formula for InvMat_n. In other words, every two elements $M, N \in \text{InvMat}_n$ can be multiplied, and the result is again in the group, i.e., $MN \in \text{InvMat}_n$.
2. Multiplying is associative: for any $M, N, P \in \text{InvMat}_n$, we have $(MN)P = M(NP)$.
3. There is an established identity element in InvMat_n. In other words, there is an element $I_n \in \text{InvMat}_n$, such that $I_n M = M = MI_n$ for every $M \in \text{InvMat}_n$.
4. There is an established inverse operation in InvMat_n. In other words, for every element M, there is an established element N, often denoted $N = M^{-1}$, such that $MN = I_n = NM$.

These rules encode a notion of symmetry that we will return to later in Section 16.5.1.

16.2.3 The monoid of $n \times n$ matrix multiplication

Recall that $\mathsf{Mat}_{n \times n}$ denotes the set of all $n \times n$ matrices, including but not limited to the invertible ones. Rather than being a group, $\mathsf{Mat}_{n \times n}$ is a *monoid*. As such, three out of the four rules for groups, as enumerated earlier, are true of $\mathsf{Mat}_{n \times n}$. Namely,

1. There is an established multiplication formula for $\mathsf{Mat}_{n \times n}$. In other words, every two elements $M, N \in \mathsf{Mat}_{n \times n}$ can be multiplied, and the result is again in the monoid, i.e., $MN \in \mathsf{Mat}_{n \times n}$.
2. Multiplying is associative: for any $M, N, P \in \mathsf{Mat}_{n \times n}$, we have $(MN)P = M(NP)$.
3. There is an established identity element in $\mathsf{Mat}_{n \times n}$. In other words, there is an element $I_n \in \mathsf{Mat}_{n \times n}$, such that $I_n M = M = M I_n$ for every $M \in \mathsf{Mat}_{n \times n}$.

A monoid is like a group—elements can be multiplied, multiplication is associative, and there is an identity element—but there is no need for every element of a monoid to be invertible. Not every matrix is invertible, so if we want to think about $n \times n$ matrix multiplication in full generality, we need to use monoids.

A group is thus a special kind of monoid, one in which every element is invertible. In the same way, a monoid is a special kind of *category*, one in which every two elements can be multiplied. Just as we broadened our view from the set of invertible matrices (as a group) to the set of all $n \times n$ matrices (as a monoid), it is now time to further broaden our view to consider all $m \times n$ matrices, i.e., matrices that are not necessarily square. For this we will need a category.

16.2.4 The category of matrix multiplication

Let $\mathbb{N}$ denote the set of natural numbers. The set of all matrices forms neither a group nor a monoid, but an $\mathbb{N}$-category, which we will denote $\mathsf{Mat}_{\_ \times \_}$. Read the symbol $\mathsf{Mat}_{\_ \times \_}$ as "blank-by-blank matrices". In terms of sets, it is the union

$$\mathsf{Mat}_{\_ \times \_} := \bigcup_{m,n \in \mathbb{N}} \mathsf{Mat}_{m \times n}.$$

The three rules for $\mathbb{N}$-categories are like those for monoids, but with a slight relaxation in the multiplication rule:[3]

1. There is an established multiplication formula for $\mathsf{Mat}_{\_ \times \_}$, which is defined as long as the middle terms agree. In other words, matrices $M \in \mathsf{Mat}_{m \times n}$ and $N \in \mathsf{Mat}_{p \times q}$ can be multiplied if and only if $n = p$, and the result is again in the category, i.e., $MN \in \mathsf{Mat}_{m \times q}$.

---

[3] By $\mathbb{N}$-category, I mean a category with a position, or *object*, for every natural number $n \in \mathbb{N}$. For the definition of a general category, first add the following rule before the listed three:

0. There is an established set Ob, whose elements are called *objects*.

Then, replace every occurrence of $\mathbb{N}$ with an occurrence of Ob. In other words, an $\mathbb{N}$-category is a category in which Ob = $\mathbb{N}$. In the standard definition of a category, $\mathsf{Mat}_{m \times n}$ is playing the role of the set of *morphisms* $m \to n$.

2. Multiplying is associative: for any $M, N, P \in \mathsf{Mat}_{-\times-}$, if MN and NP can be multiplied, then $(MN)P = M(NP)$.
3. For each $n \in \mathbb{N}$ there is an established identity element in $\mathsf{Mat}_{n \times n}$. In other words, there is an element $I_n \in \mathsf{Mat}_{n \times n}$, such that $MI_n = M$ for every $M \in \mathsf{Mat}_{m \times n}$ and $I_n N = N$ for every $N \in \mathsf{Mat}_{n \times q}$.

Let's consider each dimension $n \in \mathbb{N}$ to be a kind of context. Then groups are about actions which do not change context and which are reversible; monoids are about actions which do not change context but which may be irreversible; and categories are about actions which may change context and which may be irreversible. While groups and monoids are said to have *elements*, the elements in a category (the elements of $\mathsf{Mat}_{-\times-}$ in the above case) are usually called *morphisms*.

Note that although our definition of category looks very much tuned to matrices, it is actually quite general. When someone speaks of a category, they mean nothing more than an establishment of the structures and rules shown in (0)–(3).

16.2.5 The group-enriched category of matrix arithmetic

We still have not grappled with the fact that matrices can be added. For every pair of natural numbers $m, n \in \mathbb{N}$ the set $\mathsf{Mat}_{m \times n}$ can be given the structure of a group, which encodes the addition of $m \times n$ matrices. It is a different group than the one discussed in Section 16.2.2—i.e., it encodes addition rather than multiplication—but it is a group nonetheless because it satisfies the same formal rules. That is:

1. There is an established addition formula for $\mathsf{Mat}_{m \times n}$. In other words, every two elements $M, N \in \mathsf{Mat}_{m \times n}$ can be added, and the result is again in the group, i.e., $M + N \in \mathsf{Mat}_{m \times n}$.
2. Adding is associative: for any $M, N, P \in \mathsf{Mat}_{m \times n}$, we have $(M + N) + P = M + (N + P)$.
3. There is an established (additive) identity element in $\mathsf{Mat}_{m \times n}$. In other words, there is an element $Z_{m,n} \in \mathsf{Mat}_{m \times n}$, such that $Z_{m,n} + M = M = M + Z_{m,n}$ for every $M \in \mathsf{Mat}_{m \times n}$.
4. There is an established (additive) inverse operation in $\mathsf{Mat}_{m \times n}$. In other words, for every element M, there is an established element N, often denoted $N = -M$, such that $M + N = Z_n = N + M$.

Of course, $Z_{m,n}$ is the $m \times n$ matrix of zeros, and $-M$ is the matrix obtained by multiplying each entry in M by -1. We saw in Section 16.2.4 that $\mathsf{Mat}_{-\times-}$ is a category; once we include the additive group structure on $\mathsf{Mat}_{m \times n}$, our structure becomes a *group-enriched category*.

Thus, we see two types of group structures arising in the story of matrix arithmetic: a group encoding matrix addition for $m \times n$ matrices, for any $m, n \in \mathbb{N}$, and a group encoding matrix multiplication for invertible $n \times n$ matrices, for any $n \in \mathbb{N}$. And there is further interaction between the additive and multiplicative operations in the category of matrices; namely, multiplication of matrices distributes over addition in the sense that $M(N + P) = MN + MP$.

The entire addition and multiplication story for matrices, discussed earlier and in any first course on linear algebra, is subsumed in a single category-theoretic statement: *matrices form a group-enriched category with objects* $Ob = \mathbb{N}$. This articulates:

- the dimensionality requirements for multiplication and addition,
- the roles of each identity and zero matrix,
- the associativity and distributivity laws for multiplication and addition,
- the existence of additive inverses, and
- the way invertible matrices fit into the picture.

Let us clarify the final point. From the perspective that matrices form a category, the notion of invertible matrices comes for free. That is, for every category $\mathcal{C}$, and for every object n in it, there is a group of invertible morphisms from n to itself, called the *automorphism group* of n. When $\mathcal{C}$ is the category $\mathsf{Mat}_{-\times-}$, in which each object is a natural number $n \in \mathbb{N}$, the automorphism group of n is InvMat_n. Invertibility is seen as an issue, not about matrices in particular, but about morphisms in any category, and hence the issue is placed in a much broader context.

16.3 Modeling Linearity

In Section 16.2, we showed how category theory models the subject of matrix arithmetic. But matrices themselves are one aspect of a highly valued mathematical modeling framework, namely that of vector spaces. Vector spaces are the mathematical model of linearity, as we will discuss in Section 16.3.1. The *category* of vector spaces, $\mathsf{Vect}_\mathbb{R}$, models this model by foregrounding the sense in which each vector space is known by its relationship with other vector spaces. We will discuss this in Section 16.3.2.

16.3.1 Vector spaces: the mathematical models of linearity

The notion of linearity shows up in our visual, linguistic, and cognitive interactions with the world. Indeed our visual system is hardwired to highlight straight lines. In our language, simplicity and goodness are often equated with flatness and straightness; English words such as *plain, straightforward, right, direct, correct,* and *true* invoke straightness. And linearity also appears to be inherent in our best scientific understanding of the physical universe. For example, general relativity postulates that the universe is locally linearizable (i.e., that close enough to any point, the curved space of the universe can be laid flat), and the predictions founded on that assumption match astoundingly with experiments. Even in mathematics we find that linearity often goes hand in hand with simplicity, where many of our most successful techniques work by reducing a difficult case to a linear one.

As may by now be clear, when I speak of linearity I am referring not only to lines, but to the general notion of flatness, e.g., to planes and higher-dimensional flat spaces. Unlike in curved spaces, such as soap bubbles, we find that in flat spaces a line can point in any direction without having to curve. In the cases mentioned above, the flat spaces also include a distinguished point, called *the origin*, and the straight line segment from the origin to any other point is called a *vector*. Vectors can be stretched by any scaling factor, and two vectors can be added together; in either case the result is another vector. In mathematics, this kind of abstract flat space is called a *vector space*.

But what are these vector spaces, and what are their relation with linearity in our visual perception, our language, and our cognition? The relationship here is that vector spaces are valuable *models* of our notions of linearity and flatness. That is,

1. Vector spaces foreground and formalize certain observed aspects of linearity and relationships between them.
2. Our visual, linguistic, and cognitive interactions with linearity are successfully mediated by the vector space model.
3. Vector spaces are known by the relationships between them.

Let's begin by considering the observable aspects of linearity. Newton's first law of motion is that the velocity of any object remains constant unless a force is applied to the object. In other words, time acts as a scalar multiplier for the motion of objects: doubling the time doubles the distance traveled but does not alter the direction. Scalar multiplication is at the heart of our notion of linearity: a line is the set of scalar multiples of a given vector.

But what about higher-dimensional linear spaces? In his thought experiments about motion, Galileo imagined a flat plane on which objects could move unobstructed. A plane can be imagined as a 2-dimensional analogue of a line; it is in some sense the simplest 2-dimensional space. A plane has enough structure to discuss not just distance but also direction. Both the angle between lines in a plane and the degree of inclination of a plane embedded in space were necessary for the laws of motion Galileo wished to discuss.

There are other aspects to planes and flat spaces that are important in modeling. Namely, in a flat space, the different directions do not interact. That is, moving forward in x does not cause any change to occur in y (compare with a parabola or sphere). An n-dimensional vector space is a space in which there are n degrees of freedom, which do not interact with one another. That is, there is a well-defined notion of *coordinate system*, whereby every point in the space is uniquely determined by n numbers. This does not hold on a sphere: while it is the case that every point is determined by its latitude and longitude (*lat, long*), this determination is not unique. For example, we have $(90°, long_1) = (90°, long_2)$ for every pair of longitudes $long_1, long_2$. In other words, the coordinates of latitude and longitude are not free from one another; they interact at the north and south poles. This issue may seem unimportant, but the point is that such caveats cannot be rectified on the sphere precisely because it is not flat.

As mentioned earlier, the established mathematical models of flat spaces are vector spaces. A (real) vector space is a collection of vectors, including a vector of length 0, such that each vector can be scaled by any real number, e.g., doubled or tripled in length, and such that any two vectors can be added together to form a new vector. When made precise, these ideas are sufficient to define a coordinate system, or *basis*, which is a minimal set of vectors that span the whole space. For example, if the basis consists of three vectors, x, y, z, then every vector v in the space can be uniquely obtained by adding together scalar multiples of the basis vectors, say $v = 4x + 3y - 1.5z$.

Thus, the observable aspects of linear spaces foregrounded by the mathematical model of vector spaces are:

- the existence of a zero vector,
- scalar multiplication of vectors, and
- addition of vectors.

These aspects and the relationships between them—e.g., commutativity of addition, distributivity of scalar multiplication, existence of additive inverses, etc.—are precisely the content of the formal definition of vector space.[4]

René Descartes (and simultaneously Pierre de Fermat) developed the notion of axes, whose coordinates specify any point in a plane (or 3D space). So coordinate systems were invented long before the abstract notion of vector spaces was. However, a major contribution of vector spaces is that they formalize the ability to change between coordinate systems, e.g., by scaling or adding axes to form new axes. Since one can change coordinates at will, perhaps one does not need them at all. In fact, the vector space concept formalizes the idea, due to Lagrange, that the flat space exists, and calculations can take place, a priori—without need for choosing coordinates.

Geometric considerations, such as how two intersecting lines span a plane and two intersecting planes (in space) form a line, are all completely captured by the vector space model. Moreover, various unexpected exceptions—such as the case where the two intersecting planes happen to be the same (and hence the intersection is not a line but the plane), or the case in which the two planes inhabit a space of more than three dimensions and hence intersect in a point rather than a line—are exposed in the mathematical model.

We have shown, then, that the vector space model of linearity successfully mediates our visual, linguistic, and cognitive interactions with flat spaces. We have also discussed the system of relationships (commutativity, associativity, distributivity) between various aspects of linearity (zero, scalars, sums). This justifies our first two postulates in this setting, so it remains to explain how vector spaces are known by the relationships between them. It is the job of category theory, as a model, to foreground

---

[4] Whenever we speak of vector spaces, we mean finite-dimensional vector spaces over the field $\mathbb{R}$ of real numbers.

this third postulate using morphisms. Thus we aim to show that vector spaces (the models of linearity) are known by the morphisms between vector spaces.

Section 16.3.2 will kill two birds with one stone: it will show that the vector space model of linearity satisfies the third postulate, and it will show that this satisfaction of the third postulate is itself foregrounded by the category-theoretic model $\text{Vect}_\mathbb{R}$ of the vector space model of linearity.

16.3.2 $\text{Vect}_\mathbb{R}$: *the categorical model of vector spaces*

Category theory formalizes relationships as morphisms, and the tried-and-true morphisms between vector spaces are called *linear transformations*, which we now recall. Let V and W be vector spaces. A linear transformation $f\colon V \to W$ is a method for converting vectors in V to vectors in W. But f cannot be an arbitrary function: in order to be called a linear transformation, it must preserve the structures we have formalized in our model, namely zero, scalars, and sums. That is, it must satisfy

- $f(0) = 0$,
- $f(r \cdot v) = r \cdot f(v)$, and
- $f(v + v') = f(v) + f(v')$

for any $v, v' \in V$ and $r \in \mathbb{R}$. In fact, there is a tight connection between the set of linear transformations $V \to W$ and the set $\text{Mat}_{|V| \times |W|}$ of $|V| \times |W|$ matrices, where $|V|, |W| \in \mathbb{N}$ are the dimensions of V and W. Once one chooses a basis for V and W, this connection becomes a one-to-one correspondence: each matrix represents a single linear transformation. There is an equivalence between the category $\text{Vect}_\mathbb{R}$ and the category $\text{Mat}_{\_ \times \_}$ described in Section 16.2.1; we will henceforth elide the difference.

The point so far is that the morphisms in the category $\text{Vect}_\mathbb{R}$ take seriously the structures (zero, scalars, sums) we have formalized in each individual model of linearity. In general, the morphisms in a category are designed to reflect (by preserving) the structures that define each object. Thus, an object in a category is not only (epistemologically) *known by* its relationships with other objects; it (ontologically) *is* what it is by virtue of its relationships with other objects. In a categorical model, the knowing of an object and the being of an object are essentially identical.

In the case of a vector space V, pure category-theoretic reasoning only allows one to consider those aspects of V that can be defined in terms of morphisms into or out of it. For example, the most important aspect of a vector space is the notion of vector. Any individual vector $v \in V$ defines a ruled line in V; that is, all scalar multiples of v lie on a single line, ruled by tick-marks at every integer multiple of v. But there is a vector space that represents ruled line-hood itself, namely the one-dimensional vector space $\mathbb{R}$ of all scalar multiples of 1. We can now say how the notion of vector is itself defined in terms of morphisms in $\text{Vect}_\mathbb{R}$: for any vector $v \in V$ there is a unique morphism $\mathbb{R} \to V$ for which v is the image of the vector $1 \in \mathbb{R}$. So if one wishes to think categorically about vectors in a vector space, one can think only in terms of morphisms in $\text{Vect}_\mathbb{R}$, i.e., linear transformations.

In fact, the four notions of vector, zero-vector, scalar multiplication, and addition of vectors can all be understood in terms of linear transformations, as we now explain. First, as mentioned above, a vector in V is the same thing as a linear transformation $\mathbb{R} \to V$, so we have interpreted single vectors in terms only of morphisms. Second, there is a unique morphism from the 0-dimensional vector space $\mathbb{R}^0$ to any vector space V. There is also a unique morphism $\mathbb{R} \to \mathbb{R}^0$, and the composite $\mathbb{R} \to \mathbb{R}_0 \to V$ picks out the zero-vector in V. Third, scaling a vector v by a real number $r \in \mathbb{R}$ can be understood using only composition of morphisms. It fits beautifully to realize that a scalar like r can be represented as a morphism, namely as the linear transformation $r: \mathbb{R} \to \mathbb{R}$ sending $x \mapsto rx$. We already know that v is represented by a morphism $\mathbb{R} \to V$. The scaled vector, $rv: \mathbb{R} \to V$, is then simply the composite of two morphisms $\mathbb{R} \xrightarrow{r} \mathbb{R} \xrightarrow{v} V$ in $\text{Vect}_\mathbb{R}$. Fourth, there is an similarly elegant way to understand addition of vectors in terms of morphisms.

The upshot is that the linear transformations between vector spaces can account for all the formal structure that makes a vector space a vector space. But what about the informal, cognitive aspects of linearity discussed in Section 16.3.1, namely ideas like lines, line segments, inclined planes, projections, intersections, and coordinate systems? We have seen that lines in V are captured by linear transformations $\mathbb{R} \to V$, and similarly the inclusion of an inclined plane in 3D space is a linear transformation $\mathbb{R}^2 \to \mathbb{R}^3$. The projection of 3D space onto a line or plane is given by a linear transformation $\mathbb{R}^3 \to \mathbb{R}$ or $\mathbb{R}^3 \to \mathbb{R}^2$. Any given coordinate system on a vector space of dimension n is given by a unique linear isomorphism $\mathbb{R}^n \to V$. The notion of intersecting various lines and planes are also beautifully and completely described by a kind of interplay between morphisms, known as the *pullback*.

It is something of a miracle that so many of our intuitive ideas about vector spaces are describable simply in terms of the morphisms in the category $\text{Vect}_\mathbb{R}$. However, to a category-theorist, this is routine. Every well-studied category has this property, because this is precisely the observable aspect of modeling that category theory foregrounds. In other words, in order for a subject to be model-able by category theory, its objects of study must be determined by their morphisms to and from other objects. There is a general theorem by Nobuo Yoneda (called Yoneda's lemma) to this effect: any object c in any category $\mathcal{C}$ is essentially determined by the morphisms into (or out of) c.

In other words, categories can only model "relationally-determined" subjects, subjects in which each object of study is ontologically determined by its relationships to the others. In this case, what is surprising is how extensive the reach of category theory is. The fact that category theory is consistently used to describe so many parts of mathematics, from topological spaces to groups to ordered sets to measure spaces, means that all of these subjects are relationally determined. This justifies our third postulate, at least for these specific cases.

In the next section, we define functors, which relate different categories like morphisms relate objects. This will prepare us to discuss symmetry and action in Section 16.5.

16.4 Relationships between High-Level Models

Throughout this chapter, we have roughly been interpreting models and relationships, in the context of category theory, as follows. Each object in a category is a model. For example, each vector space is a model of linearity: $\mathbb{R}$ is a model of line-hood, $\mathbb{R}^2$ is a model of plane-hood, etc. But each entire category is also a model, albeit one of a higher level. For example, Vect$_\mathbb{R}$ is our model of linearity itself. Each low-level model (object) is defined by its relationships (morphisms) to other low-level models, where these relationships are formalized in the higher-level model (the category). For example, we showed how the linearity—the vector spaceness, as formalized by zero, scalars, and sums—of each vector space $V \in$ Vect$_\mathbb{R}$ is defined in terms of morphisms between V and other vector spaces.

But if low-level models are known by the relationships between them, it should be that high-level models are as well. This is indeed the case. We now discuss the categorical model of higher-level models, i.e., the category of categories.

16.4.1 Cat: *the category of categories*

There is a larger-sized category, denoted Cat, which includes every normal-sized category as an object.[5] If $\mathcal{C}$ and $\mathcal{D}$ are categories, a morphism between them in Cat is called a *functor*, and can be denoted $F: \mathcal{C} \to \mathcal{D}$.

Just like any morphism, a functor is a relationship between two categories that preserves the defining structure of categories. Recall from Section 16.2.4 that the defining structure of a category $\mathcal{C}$ is a set of objects, a set of morphisms, an identity morphism for each object, and a formula for composing morphisms. These satisfy some laws, namely that composing any morphism g with an identity morphism gives back g and that composition is associative. To say that functors preserve the structures that define categories is shorthand for the following more formal statement. A functor $F: \mathcal{C} \to \mathcal{D}$ must

- assign to each object $c \in \mathrm{Ob}(\mathcal{C})$ an object $F(c) \in \mathrm{Ob}(\mathcal{D})$,
- assign to each morphism $g: c \to c'$ in $\mathcal{C}$ a morphism $F(g): F(c) \to F(c')$ in $\mathcal{D}$,
- ensure that the morphism assigned to each identity in $\mathcal{C}$ is an identity in $\mathcal{D}$, i.e., for all $c \in \mathrm{Ob}(\mathcal{C})$, a functor F must ensure that $F(\mathrm{id}_c) = \mathrm{id}_{F(c)}$, and
- ensure that the composition formula in $\mathcal{C}$ is compatible with the composition formula in $\mathcal{D}$, i.e., $F(g \circ_\mathcal{C} h) = F(g) \circ_\mathcal{D} F(h)$.

---

[5] I apologize for the vagueness with respect to the size issue, but it is not relevant to our discussion. For details, see Shulman (2008).

There is a very basic category, which represents object-hood in **Cat** the way that $\mathbb{R}$ represents ruled line-hood in **Vect**$_\mathbb{R}$. Let $\mathbb{O} \in \text{Ob}(\textbf{Cat})$ denote the category with one object, $\text{Ob}(\mathbb{O}) = \{\bullet\}$; only one morphism, $\text{id}_\bullet \colon \bullet \to \bullet$; and the composition formula says $\text{id}_\bullet \circ \text{id}_\bullet = \text{id}_\bullet$. This category $\mathbb{O}$ might be pictured

$$\boxed{\bullet} \tag{16.2}$$

For any category $\mathcal{C}$, the objects of $\mathcal{C}$ are the same as the functors $\mathbb{O} \to \mathcal{C}$. Similarly, there is a category $\mathbb{M}$, which might be pictured

$$\boxed{\bullet \longrightarrow \bullet} \tag{16.3}$$

that represents morphism-hood in **Cat**. That is, for any category $\mathcal{C}$, the morphisms of $\mathcal{C}$ are the same as functors $\mathbb{M} \to \mathcal{C}$. There is also a functor that represents identity morphisms as well as a functor that represents composition formulas.

In other words, the morphisms in **Cat** can account for all the formal structure that makes a category a category: objects, morphisms, identities, and compositions. This justifies our third postulate in the case of categories: the structure of any given category $\mathcal{C}$ is completely determined by the system of functors that map to it from other categories.

16.4.2 Two justifications for the third postulate in general

Our goal has been to show that category theory is a mathematical model of mathematical modeling. We discussed the rough meaning of this statement by introducing three postulates about modeling in Section 16.1, and we said that category theory foregrounds and formalizes the third: that each model is known by its relationships with other models. We have also justified this postulate in several cases; we now justify it in general.

In a single category, Yoneda's lemma offers one explication and formal justification of the third postulate, given the interpretation of models and relationships, laid out in the first paragraph of Section 16.4. However, any functor between two categories provides another way to relate these high-level models, namely by something reminiscent of analogy. That is, if $F \colon \mathcal{C} \to \mathcal{D}$ is a functor, it relates each object $c \in \text{Ob}(\mathcal{C})$ with an object $F(c) \in \text{Ob}(\mathcal{D})$. The morphisms between c and its neighbors in $\mathcal{C}$ are preserved by F, which makes F act as a sort of analogy, as observed by Brown and Porter.[6] Our third postulate was intended as a ordinary-language hybrid of the Yoneda concept and the functorial analogy concept.

We now return to a few of the most important areas of mathematical modeling in the usual sense: symmetry and dynamics. The Yoneda interpretation of our third postulate is always at work, but we will briefly explore the second interpretation: how high-level analogies bring out the features of a subject.

---

[6] See Brown and Porter (2006). Their aim is similar to, but whose scope is more ambitious than, the present chapter.

16.5 Symmetry and Action

Consider the reflective symmetry of the visible human body or the rotational symmetry of Escher's *Drawing Hands*. The question of symmetry is about reversible action-ability; e.g., the ability to reflect an object and the ability to rotate an object by 180° are both reversible. To say that an object is symmetrical with respect to an act is to say that it does not change when it undergoes this act; e.g., the visible human body is unchanged as it undergoes mirror reflection and *Drawing Hands* is unchanged as it undergoes 180° rotation.

We begin in Sections 16.5.1 and 16.5.2 by considering various questions of symmetry and action-abilities in the above sense, but this is mainly to set up our main point. Namely, a group G of symmetries (reversible action-abilities) exists independently of any thing that is so-symmetrical (unchanged under corresponding actions). But then what is the connection of the symmetry to the symmetrical? It is captured simply by a functorial connection between one category and another, namely the reversible action category G and the space category $\mathbf{Vect}_\mathbb{R}$. The functorial connection between a category like G[7] and another category $\mathcal{C}$ is called an *action* of G on an object of $\mathcal{C}$. We will study actions in Section 16.5.3. Finally in Section 16.5.4, we will consider dynamical systems, which are classical exemplars of mathematical modeling, and which again are nothing but functorial connections (roughly, between a time category and a space category).

16.5.1 Modeling reversible action-ability with groups

A square S, with unit-length sides, centered at the origin in $\mathbb{R}^2$ is symmetrical with respect to eight action-abilities: it can be rotated and reflected in any combination:

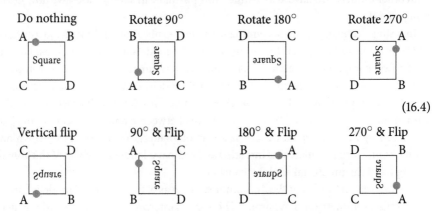

(16.4)

---

[7] By the phrase "a category like G", we mean a category with one object; see Section 16.5.1.

Neither the word *Square*, the gray dot, nor the labels A, B, C, D are symmetrical with respect to the same eight-element group that the square itself is. They are drawn in (16.4) to show that unlike the square, they undergo change when we act on them in these eight ways.

Every one of these eight actions on the square is reversible, and its reverse is another one of the eight actions. The actions are also serializable, in the sense that if each of $a_1, \ldots, a_n$ is one of the eight, then so is the process obtained by doing them in series, denoted $a_1 a_2 \cdots a_n$. In Section 16.2.2, we called this serialization *multiplication*, but in category-theoretic terms it is called *composition*. The eight-element set in (16.4), which has an identity element and is closed under inverses and compositions, is a group (see Section 16.2.2), called the *dihedral group of order 8*, and denoted D_8.

Note that the group D_8 acts on the square, but in fact D_8 exists independently of the square. That is, D_8 could just as well act on an octagon or on a single point at the origin. We refer to the elements of D_8 as *action-abilities* because each is able to act on a variety of things. The elements only become *actions* when they are applied to something, such as a square. We will discuss the notion of action itself in Section 16.5.3.

Every group can be modeled as a category $\mathcal{G}$ with many morphisms, each corresponding to an action-ability, but with only one object. The unique object of $\mathcal{G}$ stands for "the abstract thing that is unchanged under these actions". The identity morphism corresponds to the ability not to act, and the composition formula in $\mathcal{G}$ corresponds to the requirement that the serialization of action-abilities is an action-ability. Groups are categories that encode symmetries, which we define as action-abilities that can be undone. That is, for every morphism $a \in \mathcal{G}$, there is a morphism b for which the serializations a-then-b and b-then-a are equal to the identity (non-)action.

The group D_8 is one type of symmetry; there are many others. For example, consider the line that goes through the origin in $\mathbb{R}^2$ at an angle of $30°$. If you take each point on the line and multiply its distance from the origin by a non-zero number k, but do not change the angle, the total result is again the same $30°$ line: it is unchanged by non-zero scaling. Thus, there is a different group, $\mathbb{R}_{\neq 0}$, of non-zero scaling abilities. It is a different type of symmetry than D_8, but just as much able to encode a kind of ability to act reversibly.

Before we move on, note that we now know that each group is a category. A functor between two groups is called a *group homomorphism*. The category of all groups and group homomorphisms, denoted Grp, is the high-level model of symmetry itself, just like the category Vect$_\mathbb{R}$ is the high-level model of linearity.

16.5.2 Modeling action-ability with monoids

As mentioned in Section 16.2.3, a monoid is like a group, except that not all its elements need be invertible. For example, there is a monoid M consisting of four action-abilities, which we can depict using their action on a windowpane.

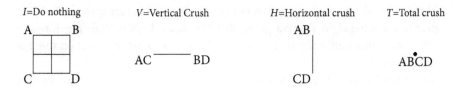

Any sequence of these action-abilities can be serialized, and the result is also one of these four action-abilities. Also, there is an identity, "do nothing" action-ability. Thus, we have a monoid, and it is not a group because three of the action-abilities are irreversible. Each element of M acts on the windowpane: it sends every point in the windowpane to another point in the windowpane.

A monoid is a category with one object, but with possibly many morphisms from that object to itself, as depicted here:

$$\tag{16.5}$$

The fact that there is only one object means that any two morphisms can be composed; this is the serializability.

In passing we note that the connection between groups and monoids (every group is a monoid but not vice versa) is captured by a high-level analogy, i.e., a functor Grp → Mon, where Mon is the category of all monoids. There are even functors going the other way, Mon → Grp, but we will not discuss any of them here.

16.5.3 Modeling action with outgoing functors

The four elements of the monoid M from the previous section (16.5.2) can be represented by the following matrices:

$$I \mapsto \begin{bmatrix} 1 & 0 \\ 0 & 1 \end{bmatrix} \quad V \mapsto \begin{bmatrix} 1 & 0 \\ 0 & 0 \end{bmatrix} \quad H \mapsto \begin{bmatrix} 0 & 0 \\ 0 & 1 \end{bmatrix} \quad T \mapsto \begin{bmatrix} 0 & 0 \\ 0 & 0 \end{bmatrix} \tag{16.6}$$

But how exactly can we enunciate this connection between elements of our monoid M and these matrices, i.e., morphisms in $\mathsf{Vect}_\mathbb{R}$?

The most concise and straightforward way I know is to use functors. A functor $F: M \to \mathsf{Vect}_\mathbb{R}$ assigns to each object in M an object in $\mathsf{Vect}_\mathbb{R}$. Since there is only one object in M, we get only one vector space; in the above case, it is $\mathbb{R}^2$. A functor F also assigns to each morphism in M a morphism in $\mathsf{Vect}_\mathbb{R}$; the four elements $\{I, V, H, T\}$ of the monoid are then sent to four linear transformations $\mathbb{R}^2 \to \mathbb{R}^2$. These are represented by the four matrices in (16.6). Moreover, the composition formula for M is preserved by F. This ensures that if an equation holds in M, e.g., $VH = T$, then the corresponding equation holds in $\mathsf{Vect}_\mathbb{R}$, e.g.,

$$\begin{bmatrix} 1 & 0 \\ 0 & 0 \end{bmatrix} \begin{bmatrix} 0 & 0 \\ 0 & 1 \end{bmatrix} = \begin{bmatrix} 0 & 0 \\ 0 & 0 \end{bmatrix}$$

In mathematics, if M is a monoid (or a group), we define an *M-action on a set* to be a functor $M \to $ **Set**; we define an *M-action on a vector space* to be a functor $M \to $ **Vect**$_\mathbb{R}$; and so on. Thus in (16.6) we have established an M-action on $\mathbb{R}^2$. Similarly, the group InvMat$_n$ of invertible $n \times n$ matrices, discussed in Section 16.2.2 acts on $\mathbb{R}^n$. The study of functors from a group to **Vect**$_\mathbb{R}$ is often called *representation theory* in mathematics.

There is a natural category structure on the class of functors between any two categories. These functor categories are often important, as they are in representation theory. In the next section, we briefly describe dynamical systems in these terms.

16.5.4 Dynamical systems

To model the behavior of a system that changes in time, one must decide whether to formalize its change as continuous or discrete. For example, the internal states of a computer may be best modeled with a discrete dynamical system, if it has an internal clock that dictates discrete times at which changes can occur. On the other hand, the concentrations of chemicals in a reaction process are changing continuously, so the reaction is often modeled by a continuous dynamical system. Here we will focus on autonomous dynamical systems.

A *discrete dynamical system* is defined to be a set S and a *where-to-go-next* function $f: S \to S$. A case where S has eleven elements may be depicted

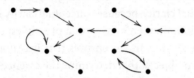

In fact a discrete dynamical system is the same thing as a functor $\mathbb{N} \to $ **Set**, where $\mathbb{N}$ is the monoid of natural numbers under addition. In other words, consider $\mathbb{N}$ as a one-object category with morphisms $0, 1, 2, \ldots$, where composing i with j is given by the formula $i + j$. A functor $\mathbb{N} \to $ **Set** consists of a set S and a function $f^n: S \to S$ for every natural number n. The fact that a functor must preserve the composition formula implies that $f^i(f^j(s)) = f^{i+j}(s)$ for every $s \in S$.

A *continuous dynamical system* (sometimes called a *flow*) is a topological space S and a continuous function $f: S \times \mathbb{R}_{\geq 0} \to S$, such that

$$f(s, 0) = s \quad \text{and} \quad f(f(s, t_1), t_2) = f(s, t_1 + t_2).$$

Here $f(s, 3.14)$ would tell us where the point s will be after 3.14 units of time. In fact, a continuous dynamical system can be modeled categorically as a topologically enriched functor $\mathbb{R}_{\geq 0} \to $ **Top**. The reader is not expected to understand this statement exactly, but the idea is that we can concisely capture the definition of dynamical systems using functors.

We have now shown that the class of dynamical systems (either discrete or continuous) is itself a *category of functors* from a time category to a space category. For discrete dynamical systems, time and space are modeled discretely (with time as $\mathbb{N}$ and space as a set). For continuous dynamical systems, they are modeled continuously (with time as $\mathbb{R}_{\geq 0}$ and space as a topological space). However, the dynamical system itself is a functor between these categories.

Time and space can be modeled as independent categories, but part of the human concept of time is that it acts on objects in space. That is, we understand our model of time and our model of space by connecting the two. This is an example of the analogical interpretation of the third postulate, and we have modeled it formally in this section using functors.

16.6 Conclusion

Karl Popper said, "A theory that explains everything, explains nothing." If category theory models algebra, geometry, logic, computer science, probability, and more, is it not trying to be a model of everything? The point is a bit subtle.

Category theory is not a theory of everything. It is more like, as topologist Jack Morava put it in Morava (2012), "a theory of theories of anything". In other words, it is a model of models. It leaves each subject alone to solve its own problems, to sharpen and refine its toolset in the ways it sees fit. That is, CT does not micromanage in the affairs of any discipline. However, describing any discipline categorically tends to bring increased conceptual clarity, because conceptual clarity is CT's main concern, its domain of expertise. And it does enforce certain principles; for example the theorem of Yoneda, discussed in Section 16.3.2, ensures that it is not the objects, but rather the relationships between objects, that determine the essence of any category. Finally, category theory allows one to compare different models, thus carrying knowledge from one domain to another, as long as one can construct the appropriate "analogy", i.e., functor.

Category theory has continually sharpened and refined its own toolset, i.e., its ability to articulate the various objects, relationships, properties, structures, and methods that show up throughout mathematics. It consistently shows itself as a powerful mode of mathematical thinking, and there is no a priori reason it cannot be similarly successful in science more broadly.

However, a question remains: what of the modeler? Who is the one that decides that a certain category adequately models a certain subject? Who is the one that finds value in the category-theoretic mode of thought? Perhaps category theory can aid in a mathematically rigorous form of phenomenological reduction (Cogan, 2016) in which the process of thinking is itself elucidated. Category theory could be considered a truly profound model of modeling if it could model the cognitive apparatus itself, i.e., if it could foreground and mathematically communicate the relationship between subject, model, and modeler.

Acknowledgments

I want to thank Allen Brown, Patrick Schultz, Tsung-Yun Tzeng, and Dmitry Vagner, as well as the referees, for their helpful comments on various drafts of this chapter. This project was supported by ONR Grant N000141310260 and AFOSR Grant FA9550-14-1-0031.

References

Brown, R., and Porter, T. (2006). Category theory: an abstract setting for analogy and comparison. *What is Category Theory* 3, 257–74.

Clark, A. (2008). *Supersizing the Mind: Embodiment, Action, and Cognitive Extension*. Oxford University Press, New York.

Cogan, J. (2016). The phenomenal reduction. *Internet Encyclopedia of Philosophy*. http://www.iep.utm.edu/phen-red/ Accessed July 19, 2016.

Morava, J. (2012). Theories of anything. arXiv:1202.0684.

Shulman, M. (2008). Set theory for category theory. arXiv:0810.1279.

17
Categories of Scientific Theories

Hans Halvorson and Dimitris Tsementzis

Our aim in this chapter is to recommend category theory to philosophers of science, in particular as a means to articulating the structure of scientific theories. We are not suggesting that we replace first-order logic, model theory, set theory, and similar formal tools with category theory—as if category theory were just one more competitor among various formal approaches to philosophy of science. Much less are we proposing to replace set theory with category theory as the foundation of mathematics. Rather, we suggest that category theory unifies various approaches to formal philosophy of science, and shows that some of the debates between various approaches have been misguided. But most importantly, our proposal is not ideological; i.e. we have no stake in the claim that category theory is the "one and only correct" approach to scientific theories, much less that "a scientific theory *is* a category". Rather, we are merely sketching a program of research in formal philosophy of science: we suggest that it might be interesting to think of the "universe" of scientific theories as a *category of categories*, or more precisely, as a 2-*category of categories*.

Our proposal includes the idea that familiar scientific theories (e.g. Hamiltonian mechanics, special and general relativity, quantum mechanics, quantum field theory) can themselves fruitfully be described as categories.[1] If we represent theories this way, then we can take philosophical questions—e.g. are Hamiltonian and Lagrangian mechanics equivalent theories?—make them precise, and then use mathematical tools to answer these questions. We can also suggest various explications of important notions, such as *equivalence* or *reducibility*, and can then try to prove general theorems about such notions. In short, we commend category theory to the attention of technically oriented philosophers of science.

17.1 Theories as Categories

Before we begin to discuss scientific theories, we review some basic notions of categorical logic, in order to frame the discussion of theories as categories. There are

---

[1] For more on this idea, including evidence of its fruitfulness, see Weatherall Chapter 13, this volume.

two salient ways in which a theory can be thought of as a category—a syntactic way, and a semantic way.

17.1.1 The syntactic category

In this entire chapter, when we speak of first-order logic, we mean first-order logic with possibly many sorts. Allowing the flexibility of many sorts doesn't truly add to the expressive power of first-order logic, but ignoring the possibility of many sorts can lead to needless confusions (see Barrett and Halvorson, 2015b).

What we mean by "many-sorted" logic is that a signature Σ comes with a (finite) list $\sigma_1, \sigma_2, \ldots$ of types, and variables, quantifiers, etc. are tagged by a particular type. For example, for each type σ, there is an equality symbol $=_\sigma$, which can be applied only to terms of type σ. Similarly, each predicate symbol $p \in \Sigma$ has an **arity** $\sigma_1 \times \ldots \times \sigma_n$, where $\sigma_1, \ldots, \sigma_n \in \Sigma$ are (not necessarily distinct) sort symbols. Likewise, each function symbol $f \in \Sigma$ has an **arity** $\sigma_1 \times \ldots \times \sigma_n \to \sigma$, where $\sigma_1, \ldots, \sigma_n, \sigma \in \Sigma$ are again (not necessarily distinct) sort symbols. Lastly, each constant symbol $c \in \Sigma$ is assigned a sort $\sigma \in \Sigma$. In addition to the elements of Σ we also have a stock of variables. We use the letters x, y, and z to denote these variables, adding subscripts when necessary. Each variable has a sort $\sigma \in \Sigma$.

Given a signature Σ, we define the terms and formulas of Σ in the normal way (see Barrett and Halvorson, 2015a). A *theory* T in Σ, is then a set of sentences (or sequents) of Σ. There are, of course, many well-known examples of such theories: e.g. the theory of partially ordered sets, the theory of groups, the theory of Boolean algebras, the theory of vector spaces over a field, the theory of categories.

The immediate goal is to associate a category C_T with a theory T. For several reasons, we will suppose that the theory T is formulated in the coherent fragment of first-order logic, whose only connectives are $\wedge$ and $\vee$, and whose only quantifier is $\exists$.[2] There is reason to think that the coherent fragment is adequate to formulate any theory that can be formulated in full-first order logic. In particular, via Morleyization, every first-order theory is Morita equivalent to a coherent theory (see Tsementzis, 2015). What's more, we agree that coherent logic is special: "there are good reasons why it is better to take $L_{\omega\omega}^g$ [the coherent fragment of (finitary) first-order logic] as basic rather than $L_{\omega\omega}$ [full (finitary) first-order logic]" (Makkai and Reyes, 1977, 121). In our case, two such reasons seem especially compelling. First, as emphasized by Makkai and Reyes, in coherent logic the distinction between intuitionistic and classical logic essentially disappears. Second, restricting ourselves to coherent logic does not

---

[2] Strictly speaking this means that a theory T is a set of sequents of the form $\phi \vdash \psi$ where ϕ and ψ are coherent formulas. Alternatively, one can understand such sequents as first-order sentences of the form $\forall \vec{x}(\phi \to \psi)$ (where $\vec{x}$ includes the unbound variables of both ϕ and ψ). In order not to deviate too much from the standard notation familiar to philosophers and logicians alike we will consider coherent theories to be sets of first-order sentences of the above-described form—this will allow us to speak of truth and satisfaction of a *sentence* rather than of a *sequent*, which is closer to the standard way of thinking about these matters.

practically diminish our expressive power at all: as Johnstone (2002) points out there are very few theories that are useful in mathematical practice that cannot be axiomatized by coherent sentences.

The standard way of building a "syntactic category" for T is described in many works on categorical logic—see Makkai and Reyes (1977, 241), Johnstone (2002, 841), van Oosten (2002, 39), and Mac Lane and Moerdijk (2012, 555). In outline: an object of the syntactic category C_T is a formula in context; i.e. if ϕ is a formula of Σ, and if $\vec{x}$ is a string of variables containing all those free in ϕ, then $\{\vec{x}.\phi\}$ is a formula in context. (Note that the objects of C_T depend only on the signature Σ, and not on the theory T.) Defining the arrows for C_T takes a bit more work. Let $\chi(\vec{x},\vec{y})$ be a formula of Σ, where $\vec{x}$ and $\vec{y}$ are mutually disjoint sequences of variables. We say that $\chi(\vec{x},\vec{y})$ is a T-provably functional relationship from $\{\vec{x}.\phi\}$ to $\{\vec{y}.\psi\}$ just in case T entails that "for any $\vec{x}$ such that $\phi(\vec{x})$, there is a unique $\vec{y}$ such that $\psi(\vec{y})$ and $\chi(\vec{x},\vec{y})$". (The precise definition can be found in the aforementioned works on categorical logic.) Then we define an arrow from $\{\vec{x}.\phi\}$ to $\{\vec{y}.\psi\}$ to be an equivalence class, relative to T-provable equivalence, of T-provably functional relations from $\{\vec{x}.\phi\}$ to $\{\vec{y}.\psi\}$.

The idea of a syntactic category might seem abstract and unfamiliar. But it is a direct generalization of the more familiar idea of a Lindenbaum algebra from propositional logic. Suppose that $\Sigma = \{p_0, p_1, \ldots\}$ is propositional signature, and let T be a theory in Σ. In this case, a formula in context simplifies to a sentence, and so the objects of C_T are just sentences. In this case, there is one provably functional relation (up to T-provable equivalence) between ϕ and ψ just in case $T, \phi \vdash \psi$; otherwise, there is no such provably functional relation. In other words, in C_T for a propositional theory T, there is an arrow from ϕ to ψ just in case $T, \phi \vdash \psi$.

Thus, from a theory T (considered as a set of sentences in Σ) we have constructed a category C_T. And we can immediately note that C_T partially eliminates the "language dependence" of T, which was so bemoaned by advocates of the semantic view of theories. Indeed, while T is bound to a particular signature Σ, the syntactic category C_T is independent of signature in the following sense: two theories T and T', in different signatures, can nonetheless have equivalent syntactic categories.

Furthermore, if our theory is coherent—as we are assuming—then this syntactic category C_T has the structure of a so-called coherent category. Coherent categories are categories that have just the right amount of structure to express models of coherent theories. In particular, they have a notion of conjunction (pullbacks), a notion of disjunction (pullback-stable disjoint coproducts), and a notion of existential closure (pullback-stable images).[3] To see that the syntactic category of a coherent theory is a coherent category is straightforward: the coherent structure of C_T emerges naturally out of the coherent formulas that are used to define its objects. (For example, the finite product of two objects $\{\vec{x}.\phi\}$ and $\{\vec{y}.\psi\}$ is given by the object $\{\vec{x},\vec{y}.\phi \wedge \psi\}$.) Moreover,

---

[3] One proper definition of a coherent category is a category that has finite limits, stable disjoint coproducts, and coequalizers of kernel pairs.

there are "as many" coherent categories as there are coherent theories: as we just saw, every coherent theory gives rise to a coherent category; conversely, every coherent category is equivalent to the syntactic category of a coherent theory.

But is the syntactic category C_T an adequate representative of the original theory T? Here we answer in the affirmative, following Makkai and Reyes:

> In Chapter 8 we will show that, in a sense made precise there, logical [i.e. coherent] categories are *the same* as theories in a finitary coherent logic $L^g_{\omega\omega}$. (Makkai and Reyes, 1977, 121)

And also:

> The content of 8.1.3 and 8.1.4 can be expressed by saying that for all practical purposes, T and C_T are the same. (Makkai and Reyes, 1977, 241)

In what sense are T and C_T the same? Roughly: replacing one with the other involves no essential loss of information. There are a couple of ways we can make this precise. First, the theory T can be reconstructed from C_T in the following sense: each coherent category C gives rise to a canonically specified (coherent) theory T_C. T_C is (essentially) the set of sentences satisfied by C when C is understood as a model of the theory of coherent categories, i.e. the "total" theory of C qua coherent category. More precisely, for any coherent category C we have its *canonical language* Σ_C whose sorts are the objects of C and function symbols the arrows of C (sorted in the obvious way). Over this language Σ_C we can then express in a straightforward way what it is for a diagram in C to commute, what it is for a diagram to be a product diagram, etc. T_C is then the collection of all those Σ_C-sentences expressing all those facts that are true of C as a coherent category. We then have that T is recoverable from C_T in the following sense:

Theorem 17.1 *Given a theory T, and its syntactic category C_T, the internal theory T_{C_T} of C_T is Morita equivalent to T.*

For the proof, see Tsementzis (2015, Corollary 4.6). We will further explain the notion of Morita equivalence in the following subsection as well as argue for its suitability as a good notion of equivalence between theories. For now, one may simply read the above result as "T is equivalent to T_{C_T}" and take it on faith that "Morita equivalence" is a sensible notion of equivalence between theories.

Second, the category Mod(T) of models of T can be reconstructed from its syntactic category C_T.

Theorem 17.2 *Let T be a coherent theory, and let C_T be its syntactic category. Let $\mathbf{Coh}(C_T, \mathcal{S})$ be the category whose objects are coherent functors from C_T into the category $\mathcal{S}$ of sets, and whose arrows are natural transformations. Let Mod(T) be the category whose objects are models of T, and whose arrows are homomorphisms between models. Then $\mathbf{Coh}(C_T, \mathcal{S})$ is equivalent to Mod(T).*

For the proof, see Makkai and Reyes (1977, 240).

The two preceeding results show the sense in which there is no loss of essential information in passing from a theory T to its syntactic category C_T. (In particular, the Morita equivalence class of a theory can be recovered from that theory's syntactic category.) It is tempting now to conjecture that if T and T' are Morita equivalent, then C_T and $C_{T'}$ are equivalent categories. But that conjecture fails, since the categories C_T and $C_{T'}$ might not be "conceptually complete" in the sense of Makkai and Reyes (1977). We discuss this issue further in the following subsection.

17.1.2 Equivalent theories

Before proceeding to discuss the semantic category $\text{Mod}(T)$ associated with a theory T, we will briefly discuss some ideas about when two theories are *equivalent* (for further discussion and technical results, see Barrett and Halvorson (2015a, 2016); Tsementzis (2015)). The first question to be asked here is what notion of equivalence are we intending to capture? Our answer here is that we have no intention of capturing any Platonic essence of "equivalence". Rather, just as a group theory gives us a fruitful notion of equivalence between groups (viz. isomorphism), and just as category theory gives us a fruitful notion of equivalence between categories (viz. categorical equivalence), so when theories are treated as mathematical objects, we hope to find a notion of equivalence that will be useful and illuminating.

The strictest notion of equivalence between theories is logical equivalence: two theories T and T' are said to be *logically equivalent* just in case they are formulated in the same signature Σ, and they have the same logical consequences among the sentences of Σ. Of course, logical equivalence is of no use for theories formulated in different signatures. For that case, we look to notions of how a theory can define new concepts that do not occur in the original signature Σ. Recall that a definitional extension T^+ in Σ^+ of the theory T in Σ is the result of adding new predicate symbols, function symbols, or constant symbols that can be defined by T in terms of formulas in the original signature Σ. There is every reason, moreover, to think of a definitional extension T^+ as equivalent to the original theory T. Thus, two theories T_1 (in Σ_1) and T_2 (in Σ_2) are said to be *definitionally equivalent* just in case there are definitional extensions T_i^+ of T_i in $\Sigma_1 \cup \Sigma_2$ (for $i = 1, 2$) such that T_1^+ is logically equivalent to T_2^+.

Nonetheless, there are reasons to think that definitional equivalence is not the most fruitful notion of equivalence between theories. One such reason is that it doesn't match well with the notions of equivalence between the corresponding syntactic categories (two theories can have equivalent syntactic categories without being definitionally equivalent). Another reason is that definitional equivalence cannot capture the sense in which, for example, the theory of categories can be equivalently formulated using objects and arrows, or just with arrows. In order to capture these intuitive verdicts of equivalence, the most plausible idea is *Morita equivalence*, which allows for equivalence of theories formulated not only in different signatures but also in different signatures with different sorts.

The notion of "Morita equivalence" of theories has two independent sources.[4] On the one hand, Morita equivalence is suggested by ideas from categorical logic, in particular from topos theory. To see this, we need to introduce the notion of the *pretopos completion* $P(C)$ of a coherent category C. This notion can be described in the following equivalent ways:

1. $P(C)$ is the result of freely adjoining finite coproducts and coequalizers of equivalence relations to C (see Johnstone (2002, A1.4)).
2. $P(C)$ is the subcategory of coherent objects in the topos $Sh(C)$ of sheaves on C, where the site C is equipped with the coherent Grothendieck topology.

Recall that the *classifying topos* $\mathcal{E}_T$ of the theory T is the unique (up to categorical equivalence) topos that contains a model of T, and such that any model of T in another topos $\mathcal{E}$ uniquely lifts to a geometric morphism from $\mathcal{E}_T$ into $\mathcal{E}$ (see Mac Lane and Moerdijk (2012, 561) and Makkai and Reyes (1977, 272)).

Now two coherent theories S and T are said to be *Morita equivalent* (in the categorical sense) just in case the following equivalent conditions hold Johnstone (see 2002):

1. The classifying toposes $\mathcal{E}_S$ and $\mathcal{E}_T$ are equivalent.
2. The pretoposes P_S and P_T are equivalent.[5]

The equivalence between these two statements follows from the fact that $\mathcal{E}_T \simeq Sh(C_T) \simeq Sh(P_T)$. Moreover, it follows from the second fact that if C_T is equivalent to C_S, then S and T are Morita equivalent. To see that the converse is not true, it suffices to display a theory T such that its syntactic category C_T is not a pretopos. Such theories are easy to find (see Example 17.1).

In a completely unrelated development, ideas related to Morita equivalence began to spring up in the works of logicians. As noted by Harnik (2011), Shelah's T^{eq} construction (also known as "elimination of imaginaries") is closely related to the pretopos completion construction. Indeed, the pretopos completion P_T of a theory C_T is the same thing as the syntactic category of T^{eq}, i.e. Shelah's construction applied to T. Similarly, Andréka et al. (2001) generalize the notion of a definitional extension so as to include the possibility of defining new sort symbols. We refer the reader to those works to see the original motivations for moving to a more expansive notion of equivalence between theories. For our purposes in this chapter, the following motivation suffices: the pretopos completion P_T of a syntactic category C_T represents a kind of "maximal mereological completion" of T, i.e. a theory in which not only everything implicitly defined by T receives its own name, but also every (suitable) combination of *types* of things in T receives its own name. Thus, for example, in P_T we have distinct names

---

[4] The name "Morita equivalence" originates in module theory, and was transmitted into category theory through the study of algebraic theories.

[5] Here we use P_T to abbreviate $P(C_T)$, the pretopos completion of the syntactic category of T.

both for constants and for pairs consisting just of those constants (the *type* of pairs is in some sense "reified").

But what exactly does this all mean at the level of syntax? As has been explained by Barrett and Halvorson (2015a) and Tsementzis (2015), the ideas about Morita equivalence coming from topos theory correspond to a completely natural generalization of the idea of having a common definitional extension. In particular, given a theory T in signature Σ, a *Morita extension* T^+ of T can be constructed either by defining new relation and/or function symbols or by defining new sorts from the sorts of Σ. The operation of defining new sorts via T corresponds roughly to taking the pretopos completion of C_T. Thus, intuitively speaking, two theories T and T' are Morita equivalent just in case T can define all the sorts, relation symbols, etc. of T', and vice versa, in a compatible fashion.[6] And in fact it can be shown that two theories are Morita equivalent in the syntactic sense just in case the pretopos completions of their syntactic categories are equivalent—see Tsementzis (2015, Theorem 4.7). This merely expresses the fact that the syntactic notion of Morita equivalence developed in Barrett and Halvorson (2015a) coincides with—and therefore characterizes—the topos-theoretic notion defined in Johnstone (2002). This justifies our free use of the same term "Morita equivalence" to refer to both notions.

Clearly if two syntactic categories C_T and $C_{T'}$ are equivalent, then T and T' are Morita equivalent. The converse, however, is not true.

Example 17.1 An easy way to see this is to take the an empty two-sorted theory T, i.e. the theory with no axioms whose signature Σ consists only of two sort symbols σ_1, σ_2. Then we can extend T to T' by adding a "coproduct sort" $\sigma_1 + \sigma_2$ together with function symbols $\rho_1 \colon \sigma_1 \to \sigma_1 + \sigma_2$ and $\rho_2 \colon \sigma_2 \to \sigma_1 + \sigma_2$ and axioms defining $\sigma_1 + \sigma_2$ as a "coproduct" with ρ_1 and ρ_2 as its coprojections. Indeed T' is exactly a Morita extension of T in the sense of Barrett and Halvorson (2015a), which means that T and T' are Morita equivalent. However, the syntactic categories C_T and $C_{T'}$ cannot be equivalent: the obvious embedding $C_T \hookrightarrow C_{T'}$ is full and faithful, but there can be no isomorphism from $\{z \colon \sigma_1 + \sigma_2.\top\}$ to any object of C_T (regarded as a full subcategory of $C_{T'}$). (Given the results of Tsementzis (2015), this is also an example of two theories which have equivalent pretopos completions but inequivalent syntactic categories.)

As such, we are left with two distinct notions of equivalence between theories T and T':

(SE) Equivalence of their syntactic categories, i.e. $C_T \simeq C_{T'}$

(ME) Morita Equivalence

---

[6] For an example of how this "definitional" understanding of Morita equivalence can be applied to issues of theoretical equivalence in physics, see the discussion on classical mechanics in Teh and Tsementzis (2016).

For reasons too detailed to go into here, we believe that (SE), although weaker than logical equivalence, is still too strong a notion. As noted earlier, (SE) is a sufficient condition for (ME) to hold, but not a necessary one. There is a strong sense in which (ME) captures exactly the right content of a theory as long as we care about that theory only up to the structure of its category of models. In order to clarify this remark, we must now go on to to say a few more words about this category of models.

17.1.3 The semantic category

We have already noted that there is a second category associated with a theory T, namely the category $\text{Mod}(T)$ of its models. We will call $\text{Mod}(T)$ the *semantic category* associated with T.

Before proceeding, let's be more precise about what we mean by the category of models of T. The objects of this category are simply set-valued models of T, in the sense of Tarski. But what are the arrows of the semantic category? There are two possible choices:

- Let T be a theory in signature Σ. We let $\text{Mod}(T)$ denote the category whose objects are Σ-structures that satisfy the axioms of T, and whose arrows are homomorphisms of Σ-structures. Recall that if M and N are Σ-structures, then a homomorphism $j : M \to N$ is a function that preserves the extensions of symbols in Σ. That is, for each relation symbol $r \in \Sigma$,

$$j(r^M) \subseteq r^N, \qquad (17.1)$$

and so on.

In contrast to the definition found in most model theory textbooks, we do *not* require the map j to be one-to-one, nor do we require equality in (17.1). The reason we don't impose these requirements is because they are unmotivated when the logic at hand doesn't have a negation symbol (as in the case of coherent logic).

For example, if $\Sigma = \{\circ, e\}$, and if T is the theory of groups (written in Σ), then the notion of a homomorphism of Σ-structures is simply the notion of a group homomorphism.

- We let $\text{Mod}_e(T)$ denote the category whose objects are (again) Σ-structures that satisfy T, and whose arrows are elementary embeddings of Σ-structures. Recall that if M and N are Σ-structures, then an elementary embedding $j : M \to N$ is a function that preserves extensions of all Σ-formulas. That is, for any formula $\phi(\vec{x})$ of Σ, and for any n-tuple $\vec{a}$ of elements of M,

$$M \models \phi(\vec{a}) \implies N \models \phi(j(\vec{a})). \qquad (17.2)$$

In particular, for any sentence ϕ of Σ,

$$M \models \phi \implies N \models \phi. \qquad (17.3)$$

We can think of Mod(T) as the "thick" category of models (more arrows) and Mod$_e$(T) as the "thin" category of models (fewer arrows) of T. Note that Mod$_e$(T) is a subcategory of Mod(T), and typically a proper subcategory.[7] For example, let Σ be a signature with one sort and no non-logical vocabulary, and let T be the empty theory in Σ, i.e. the theory whose models are bare sets. Let m_i be a model of T with i elements. Then Mod(T) has an arrow $j : m_1 \to m_2$, whereas Mod$_e$(T) has no such arrow (since elementary embeddings preserve the truth-value of numerical statements).

There are several questions one can ask of the relation between thin and thick categories. A well-known fact—alluded to earlier—is that the thin category of a full first-order theory is always equivalent to the thick category of a coherent theory, called its *Morleyization*—see Johnstone (2002, Lemma D1.5.13). But for now, the most important point of this subsection is that Mod(T), and a fortiori Mod$_e$(T), is *not* generally an adequate representative of the theory T. This means that even the thick categories of models are not "thick enough". To be more precise, the passage from T to Mod(T) loses information in the sense that neither T nor a theory T' that is Morita equivalent to T can be reconstructed from Mod(T). Examples from propositional logic makes this fact clear.

Example 17.2 For full first-order theories, there is an intuitive example. Let Σ be the propositional logic signature with symbols $p_0, p_1, \ldots$. Let T_1 be the empty theory in Σ, and let T_2 be the theory with axioms $p_0 \vdash p_i$ for all $i \in \mathbb{N}$. Clearly T_1 and T_2 are *not* Morita equivalent theories. And yet, the semantic categories Mod(T_1) and Mod(T_2) are equivalent—since both are discrete, and have $2^{\aleph_0}$ objects. Thus, T_1 and T_2 are inequivalent theories whose semantic categories are equivalent (see Halvorson (2012, 191)).

Example 17.3 For coherent theories, coming up with examples requires a bit more algebraic groundwork. Up to Morita equivalence, a coherent propositional theory is the same thing as the theory of prime filters of a (unique up to isomorphism) distributive lattice (see Johnstone (2002, Remark D1.4.14)). Given any such distributive lattice B the category of models of the corresponding theory can then be identified with the spectrum of B. And if B is Boolean (as a lattice) then its spectrum will be discrete. This means that up to equivalence of their categories of S-models we can only recover a coherent propositional theory up to the cardinality of its spectrum. However, there are many examples of non-isomorphic Boolean lattices whose spectra have equal cardinalities. Indeed a similar idea as our previous example works again here. Let B_1 be the Boolean algebra generated by a countably infinite number of elements $p_0, p_1, \ldots$ and let B_2 be the Boolean algebra generated by the same elements plus the relation $p_0 \leq p_i$ for all $i \in \mathbb{N}$. B_1 is atomless whereas B_2 has an atom and therefore B_1 and B_2 cannot be isomorphic. However, the cardinality of both their spectrums is equal to $2^{\aleph_0}$, a fact which can be seen by noting that homomorphisms $B_1 \to 2$ correspond exactly to homomorphisms $\phi : B_2 \to 2$ such that $\phi(p_0) = 0$.

---

[7] Indeed, Mod$_e$(T) = Mod(T) if and only if every first-order formula ϕ (over the signature Σ of T) is T-provably equivalent to a coherent formula (over classical logic)—see Johnstone (2002, Proposition 3.4.9).

The lesson here is that the semantic category of a theory—i.e. the models of that theory, and homomorphisms between models—generally contains less information than the theory itself does. A fortiori, the class of models of a theory contains less information than the theory itself does. (And this is what's wrong with the original semantic view of theories.)

It is completely natural to ask: if Mod(T) does *not* contain the same amount of information as T, then what information or structure must be added to Mod(T) in order to recover T? Think of the question this way: Mod(T) is a collection of models and arrows between models, including automorphisms (i.e. arrows from a model to itself). What other information about Mod(T) can be extracted from the theory T?

A classic answer to this question was given (for the propositional case) by Marshall Stone. Stone noted that T implicitly contains topological information about Mod(T). In particular, let's say that a sequence $m_1, m_2, m_3, \ldots$ of models in Mod(T) converges to a model m_0 just in case for any sentence ϕ of Σ, the truth value $m_i(\phi)$ is eventually equal to $m_0(\phi)$. This notion of convergence defines a topology on Mod(T). Letting Mod(T) denote the corresponding topological space, Stone's duality theorem establishes the following:

Theorem 17.3 (Stone duality) *The collection of compact open subsets of* Mod(T) *forms a Boolean lattice that is equivalent, as a category, to* C_T.

In other words, from Mod(T) we can reconstruct T up to its syntactic category, i.e. up to (SE). Thus, in the case of propositional theories, the *topological* semantic category Mod(T) contains as much information as T.[8]

But what now about the case of predicate logic? Here the situation is complicated by the fact that there are typically many non-trivial arrows between models. Can the category Mod(T) still be supplemented with topological information in order to recover T? The answer here is: Yes, sort of. Although there is still no result that perfectly generalizes Stone duality, some important partial results have been obtained by Makkai (1991) and Awodey and Forssell (2013).

Of course, philosophers of science should be eager to understand these duality results because of the important lesson they teach about the collection of models of a theory:

The category of models Mod(T) *of a theory T does not generally contain all the information that is contained in the original theory T. The content of T might include information, e.g., about topological relations between models.*

---

[8] Lest we confuse our physics-minded readers let us clarify that "dualities" such as Stone's are not to be thought of, from our point of view, as analogues of "dualities" in modern physics. The latter are relations *between* theories. The Stone-type dualities that we investigate here and in Section 17.3 are dualities that relate syntactic and semantic presentation of a *single* theory. The terminological coincidence is unfortunate (even more so since we are also interested in application of our framework to physical dualities) but, for historical reasons, unavoidable.

Let's rephrase that moral one more time, now trying to make it absolutely clear as a friendly amendment to the semantic view of theories:

The mathematical content of a scientific theory T is <u>not</u> exhausted by the class of models of T.

Obviously, this moral from first-order logic doesn't generalize directly to scientific theories in the wild (e.g. classical mechanics, general relativity, quantum mechanics). However, the results from first-order categorical logic strongly suggest that in the case of scientific theories, we would similarly go wrong if we identified the mathematical content of a theory T with its category $\mathrm{Mod}(T)$ of models[9]—for the theory might make use of further structures on $\mathrm{Mod}(T)$, perhaps topological (as in general relativity), or measure-theoretic (as in statistical mechanics), or perhaps some sort of monoidal or tensor structure (as in quantum mechanics and quantum field theory). For some evidence for this claim, with reference to specific scientific theories, see Curiel (2014) or Fletcher (2016), and for some related discussion see Lal and Teh (2017).[10]

That said, we do not want to fall into the trap—all too common in twentieth-century philosophy—of being blinded by the glow of a shiny new piece of formal apparatus. We are fully aware that, from a mathematical point of view, a syntactic category is regarded merely as a technical device useful for proving other results, e.g. completeness theorems.[11] Yet we believe that the conceptual value of the kind of formal apparatus best encapsulated by the syntactic category of a theory goes beyond its practical use *within* mathematics. And unlike W. v. O. Quine, we have no ideological commitment to regimenting theories in first-order logic. Rather, as just explained, we see first-order logic as providing a manageable testing ground for more general ideas about theoretical structure.

---

[9] An important clarification: when we say here that we would go wrong if we identified the mathematical content of a theory T with its semantic category $\mathrm{Mod}(T)$, recall that this means the semantic category *over the category of sets $\mathcal{S}$*. So what we are saying here is that a theory cannot be recovered up to (ME) from its category of models in $\mathcal{S}$. Nevertheless, conceptual completeness for coherent logic (see Johnstone (2002, Theorem D3.5.9)) says that this is very close to being true: equivalence of the semantic categories of T and T' does imply (ME) as long as this equivalence is induced by an interpretation I of T into T' at the level of syntax (i.e. a coherent functor $I : P_T \to P_{T'}$). Relatedly, there is another sense in which a theory T is actually recoverable up to (ME) from its semantic category: if we consider semantic categories over arbitrary Grothendieck toposes $\mathcal{E}$ and stipulate that the equivalence $\mathcal{E}-\mathrm{Mod}(T) \simeq \mathcal{E}-\mathrm{Mod}(T')$ is *natural* in $\mathcal{E}$, then from this alone we can conclude that T and T' are Morita equivalent. This means that the extra structure that we need to place on $\mathrm{Mod}(T)$ (understood as the category of $\mathcal{S}$-models) in order to recover T up to Morita equivalence corresponds exactly to the requirement of naturality in the class of Grothendieck toposes.

[10] We should also note that the question as to *how* these extra structures and topologies map on to actual "topologies" and "structures" between scientific theories remains to be explored. Clearly, the logical topologies on $\mathrm{Mod}(T)$ encode a notion of "nearness" between models of T—does this logical topology illuminate, or correspond in any way, to the notions of "nearness" between models used by practicing scientists, e.g. when mathematical physicists topologize the solution spaces of their differential equations? Are the logical topologies we consider here and in Section 17.3 good "frictionless" idealizations of these kinds of investigations? The question certainly deserves more attention and we intend to explore it in the future.

[11] And this extends to syntax very far removed from first-order logic; e.g. the metamathematics of simple type theories or of Martin-Löf type theories is also studied via structure-bearing syntactic categories.

Nevertheless, what we are claiming here clearly amounts to a methodology based on an analogy between the categorical metamathematics of first-order theories and the philosophy of scientific theories. Is there any reason to believe in the fruitfulness of this analogy? We believe so. Category theory brings to the table new constructions and concepts with which to study the metamathematics of first-order theories. And the metamathematics of first-order theories—if anything—is rich in concepts ("theory", "axioms", "interpretations", etc.) that are used heavily in the philosophy of scientific theories. So as long as one is not a complete skeptic with respect to the use of logical methods to come up with idealized versions of scientific theories, there is every reason—it seems to us—to take seriously the analogy on which our method relies.

17.2 The Category of Theories

We have talked about theories *as* categories. But as the title of this chapter makes clear, what we are interested in is categories *of* theories. Namely, we are interested in studying collections consisting of multiple scientific theories, and the relations between them. And since we've already argued that theories can be thought of as categories, the natural way to do so is to study the (2-)category of theories-as-categories. We can then apply the tools of category theory to understanding the structure of this larger category, how individual theories sit within it, and how theories are related to each other.

Why would we want to do that? To be sure, the category of "theories-as-categories" can be no more than a "frictionless" idealization of the actual zoo of scientific theories and the interactions between them that take place within it. But we are not interested in carrying out rigorous sociology—we are interested in a formal framework that is sufficiently expressive to act as such an idealization. In particular, a framework that formalizes the notions both of a scientific theory and of "structural similarities" *between* such scientific theories. Taking theories to be categories—as explained in Section 17.1—and then taking functors between such categories as expressions of such a "structural similarity" gives us just such a framework. Being the first to propose such an idea, we do of course realize that the choice of category-theoretic formalism might prove inadequate in the long run; perhaps it might appear as *too* idealized already. To this we say: may others come and find better frameworks—this is the one we, at this moment in time, consider the most fruitful idealization. (In this respect we believe we are making no more arbitrary a choice than, say, what Carnap did when he thought that the predicate calculus was an adequate idealization of the language of science.)

And why do we view it as fruitful? Because we want to view scientific theories not as contenders to the throne of absolute truth but rather as creatures in an active ecosystem whose interactions are essential if more advanced life is to evolve out of it. It is this view that we take of the practice of science today—and it is this picture that we want our formal framework to be an idealization of. And we think the state of physics today justifies this view. At the very least, because talking about the relationships between scientific theories captures a critical feature of modern physics, namely the existence of dualities—most prominently in string theory. Formal frameworks in the philosophy of science of the twentieth century (whether semantic or syntactic) are not, we believe, adequate to model the dualities of modern physics. Thus, our attempt to define a category of scientific theories can also be seen as our answer to the challenge of formally capturing such inter theoretical interactions.

With that said, let us now get down to the business of introducing a suitable category of theories. As before, our initial focus is on the case of theories in first-order logic. As described in the previous section, each first-order theory T corresponds to a syntactic category C_T (which we could also take to be P_T). We let the collection of all such C_T be the objects of a category Th, the category of all first-order theories.

Again, we have some fine-grained control over the definition of the category Th of theories. The main possibilities for Th are as follows:

- **Coh** the category of coherent categories (i.e. syntactic categories of coherent theories);
- **dCoh** the category of decidable coherent categories (cf. Awodey and Forssell, 2013);
- **BCoh** the category of Boolean coherent categories (i.e. syntactic categories of first-order theories over classical logic);
- **Pretop** the category of pretoposes.

These categories are arranged roughly as follows:

$$\text{BCoh} \subseteq \text{dCoh} \subseteq \text{Coh}$$

and

$$\text{Pretop} \subseteq \text{Coh},$$

where the subset symbol indicates a full inclusion of categories.

How do we choose among these (or other) options? An answer can be gotten by noting that such a choice amounts essentially to a choice of a "notion of equivalence" for our theories. This is a specific case of a very important (and far more general) observation about category-theoretic thinking, viz. that choosing the category where your objects of study live automatically determines what it means for two such objects to be "isomorphic". This important principle is worth stating:

(I) Choosing a notion of equivalence for our objects of study is the same thing as choosing a category of which they are the objects.

In our case, the objects of study are theories. Thus, as (I) makes clear, choosing a notion of equivalence for theories and choosing a "category of theories" are not two independent choices. Choosing one determines the other: there is only one degree of freedom here.

In our opinion, the two most natural choices for **Th** are **Coh** or **Pretop**. Given (I) and the discussion in Section 17.1, this shouldn't come as a surprise: we've already said that the two notions of equivalence that interest us the most are (SE) and (ME) and these correspond exactly to choosing **Coh** (for (SE)) and **Pretop** (for (ME)) as our preferred categories of theories. This is because (SE) identifies theories with their syntactic categories and every coherent category is the syntactic category of some coherent theory, whereas (ME) identifies theories with the pretopos completion of their syntactic categories and every pretopos can be seen to arise in this manner (although, of course, inequivalent coherent categories may have equivalent pretopos completions).

Furthermore there is a natural relation between these two categories: for each coherent category C, there is a unique pretopos $P(C)$ and functor $\eta_C : C \to P(C)$ satisfying a suitable universal property. In short: every coherent category has a unique pretopos completion—an operation corresponding roughly to taking a "maximal Morita extension" of the original theory. In fact, $P : \textbf{Coh} \to \textbf{Pretop}$ is a 2-functor (see Makkai, 1987). And even more is true: **Pretop** is "almost" a reflective sub-2-category of **Coh**. More precisely, **Pretop** is a full reflective sub-2-category of **FinSit**, the 2-category of "finitary" sites (i.e. categories equipped with finitely generated Grothendieck topologies)—this is a special case of a far more general result proven by Shulman (2012).

For the purposes of this paper, we needn't make a decision about the precise definition of **Th**. However, for concreteness, let us say that we are in favor of the identification **Th** = **Pretop**. (As explained earlier, this means that we are effectively choosing (ME) as our preferred notion of equivalence.) Although nothing we say here hinges on this choice, let us say a couple of things about why it is a natural choice to make (other than our faith in (ME)). First, there is a very precise sense in which—from a logical point of view—the pretopos completion of a coherent category adds only those concepts that are already *definable* from the coherent structure of the original category. Second, the pretopos completion is the *maximal* such extension; i.e. it contains *everything* that is definable from the coherent structure of a coherent category (for a precisification of this statement, see Harnik, 2011). As such, if one agrees with us that the initial syntactic presentation of a theory does not constitute its essential content, then moving from **Coh** to **Pretop** should seem a very reasonable move to make.

Now what are the arrows in **Pretop**? Since the objects of **Pretop** are categories, the arrows should be functors. Perhaps surprisingly, the arrows we care about in this particular case are obtained by considering a pretopos as a coherent category (recall that every pretopos is coherent). But do coherent categories have additional structure that ought to be preserved by our arrows? The answer, in short, is yes: a coherent category has limits and colimits that encode various syntactic structures—in particular, conjunction, disjunction, and existential quantification. Thus, we define an arrow between pretoposes P and P' to be a *coherent functor* in the sense of Johnstone

(2002, 34), also called a *logical functor* in Makkai and Reyes (1977, 121). In other words, we consider **Pretop** as a full subcategory of **Coh**.[12]

Choosing coherent functors as arrows has the nice consequence that arrows from P_T to P_S correspond to translations of the theory T into S (so do, incidentally, arrows from C_T to C_S). A way to see this is the following: each P_T contains the so-called *generic model* M_T of T (Johnstone, 2002, Proposition D1.4.12.(ii)). This is a model of T taken in the pretopos completion P_T (recall that P_T is coherent and therefore has the capacity to model any coherent theory T). It is *generic* in the sense that it satisfies exactly those sentences that are provable in T, i.e.

$$M_T \models \phi \iff T \vdash \phi.$$

Now, since coherent functors preserve the coherent structure and since models of coherent theories in coherent categories are built using (only) that coherent structure, we have that any coherent functor $F: P_T \to D$ into a coherent category D will give us a model $F(M_T)$ of T in D. In particular, when $D = P_{T'}$ for some other theory T' then $F(M_T)$ is a model of T in $P_{T'}$. But a ($\mathcal{S}$-)model of T' is simply a coherent functor from $P_{T'}$ into $\mathcal{S}$. Therefore, any model $G: P_{T'} \to \mathcal{S}$ of T' will also give rise to a model of T, viz. $GF: P_T \to \mathcal{S}$. In plain terms: any model of T' contains a model of T; this is just another way of saying that there is a translation of T into T'.[13] Finally, it is important to note that everything we've said in this paragraph can be said pretty much verbatim for syntactic categories themselves (rather than their pretopos completions). More on translations and definability at the level of syntactic categories can be found in Caramello (n.d.c).

It should immediately be pointed out that **Pretop** is most naturally thought of as a 2-category, rather than just a category. A 2-category **C** is (roughly speaking) a category such that for any two objects a, b of **C**, instead of $\mathbf{C}(a, b)$ being a *set* of arrows from a to b, it is a *category*; and the composition operation on arrows is functorial (see Borceux (1994), Lack (2010)). The arrows in the category $\mathbf{C}(a, b)$ are called *2-cells*. The paradigm example of a 2-category is **Cat**, the category of (small) categories, with functors as arrows, and natural transformations as 2-cells. That is, if C and D are categories, then $\mathbf{Cat}(C, D)$ is the category whose objects are functors $F : C \to D$, and whose arrows

---

[12] The fact that it is reasonable to do so essentially boils down to the above-mentioned fact that the pretopos completion of a coherent category is "definable" (in a precise sense) from the coherent structure of the category in question. In particular, even though coherent morphisms will not, in general, preserve arbitrary coproducts or coequalizers, they will preserve binary coproducts and coequalizers arising from equivalence relations. This means that "disjoint unions" and "quotients by equivalence relations" are concepts within the grasp of coherent logic and taking the pretopos completion of a coherent category amounts to a (maximal) "definitional extension" of the original coherent category by these definable concepts. Indeed, removing the scare quotes from the previous sentences and making this way of talking about pretopos completions fully precise was one of the motivations behind Barrett and Halvorson (2015a) and Tsementzis (2015).

[13] Indeed, in Pitts (1989) translations of a theory T into another theory T' are *defined* to be models of T in $P_{T'}$. This is the same thing as saying that translations are functors $C_T \to P_{T'}$ which in turn is the same thing as saying that translations are functors $P_T \to P_{T'}$.

are natural transformations between such functors. Similarly, we define **Pretop**(P, P') to be the category whose objects are coherent functors from P to P', and whose arrows (2-cells) are natural transformations.[14]

Basking in the full 2-categorical glory of **Pretop** is no mere pretension, nor is it a pointless exercise to prepare us for the altitude sickness that comes with the steep ascent towards higher category theory. It is, rather, a perspective that brings, among other things, new purely logical insights.[15] Whether these logical insights translate into insights about analogous concepts in the philosophy of science remains to be seen—we devote Section 17.4 to plausible lines of investigation in this direction. For now, we turn to an application of our point of view to a more concrete question in the philosophy of science: the debate between the semantic and the syntactic views of theories.

17.3 On the Duality of Syntax and Semantics

Recall that the logical positivists hoped to provide an explication for the notion of a *scientific theory*.[16] That is, they hoped to be able to say that a scientific theory is a certain sort of (rigorously defined) mathematical object. But what kind of object? According to the earliest proposals (by Carnap and others), a theory is a set of sentences in a formal language. This proposal and its later elaborations have come to be known as the *syntactic view of theories*.

As is well known, the syntactic view of theories was subjected to severe criticism in the later twentieth century. The consensus in the 1970s was the syntactic view couldn't be salvaged, and required a wholesale replacement. The proposed replacement was the so-called *semantic view of theories*, which claims that a scientific theory is a collection of models—perhaps the models of some first-order logical theory, or perhaps a collection of models of some more general sort.[17]

It has long been thought that the semantic view of theories has many advantages over the syntactic views—see e.g. the works of Suppe, van Fraassen, and Lloyd et al. Of course, that claim presupposes that there is a genuine dilemma of choice between the two points of view. Only a couple of isolated philosophers have suggested that this might be a false dilemma (see e.g. Friedman (1982)). In this section, we survey mathematical results that argue for a formal duality between the syntactic and semantic points of view. We also propose that this duality could be exploited in order to better understand the structure of scientific theories.

---

[14] A 2-category is a "strict" version of a bicategory in the sense of Bénabou (1967).

[15] To convince oneself of this one need look no further than Pitts' proof of conceptual completeness for coherent logic (resp. intuitionistic logic) using the 2-categorical structure of **Pretop** (resp. **HPretop**, the category of Heyting pretoposes)—see Pitts (1987, 1989).

[16] For further elaboration of this story, see Halvorson (2016).

[17] To be clear, van Fraassen (2014) has recently pointed out that for him, a theory is a class of models together with representational content. However, for this discussion, we are concerned only with the mathematically representable part of a theory.

Recall the previous discussion of Stone duality for theories in propositional logic. While it's not true that a propositional theory can be reconstructed from its category of models alone (i.e. the set of ultrafilters on the Lindenbaum algebra), it can be reconstructed from the category of models *plus* relevant topological information.

In recent years, logicians have attempted to generalize Stone duality to the case of full first-order logic. And while the results to date are only partial, they all point in a similar direction.[18]

First, Makkai (1991) makes use of an insight from Łos' theorem: if $\{m_i\}_{i\in I}$ are models of a theory T, then so is an ultraproduct $\prod_{i\in I} m_i/\mathcal{U}$, where $\mathcal{U}$ is an ultrafilter on I. What's more, in the case where T is a propositional theory, the ultraproduct $\prod_{i\in I} m_i/\mathcal{U}$ is simply the Stone topology limit of the sequence $\{m_i\}_{i\in I}$ along the ultrafilter $\mathcal{U}$. In other words, in the propositional case the Stone topology on Mod(T) can be alternatively described as "ultraproduct structure" on Mod(T); and the relevant functors $F : \text{Mod}(T) \to \text{Mod}(T')$ are those that preserve this ultraproduct structure (i.e. that are continuous in the Stone topology).

Now Makkai defines an *ultracategory* to be a category with a sort of ultraproduct structure (see Makkai, 1991). Of course, the motivating example of an ultracategory is the category Mod(T) of models of a first-order theory. Then the question arises:

Can a theory T be reconstructed from the corresponding ultracategory Mod(T)?

Makkai shows that the answer is yes. For any pretopos P, let $\Theta(P) = \textbf{Coh}(P, \mathcal{S})$ denote the category of coherent functors from P into the category $\mathcal{S}$ of sets (i.e. models of the theory corresponding to P). Makkai shows that there is another functor $\Gamma : \textbf{UCat} \to \textbf{Pretop}^{op}$, such that $(\Gamma \circ \Theta)(P)$ is equivalent to P. Stated more generally: there is a pair of adjoint functors as follows:

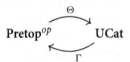

Unfortunately, this adjunction is *not* an equivalence of categories, as in the case of propositional theories (i.e. Stone duality for Boolean algebras). In particular, not every ultracategory is of the form Mod(T) $\simeq$ $\textbf{Coh}(P_T, \mathcal{S})$, for some first-order theory T. In slogan form: there are more ultracategories than there are coherent theories.

A more recent attempt to generalize Stone duality has been undertaken by Awodey and Forssell (2013). Here the insight comes not from model theory (as in the case of Makkai's ultraproducts), but from topos theory. Joyal and Tierney (1984) proved that for every Grothendieck topos $\mathcal{E}$, there is a localic groupoid G such that $\mathcal{E} \simeq \textbf{B}(G)$, where $\textbf{B}(G)$ is the topos of continuous actions of G. It was also shown by Butz and

---

[18] Such dualities as those we will outline in the following sometimes go under the name of "Isbell Duality" and many examples have been studied at a very high level of generality—see Porst and Tholen (1991) and Barr et al. (2008) for a sampling.

Moerdijk (1998) that when $\mathcal{E}$ has enough points—as is the case when $\mathcal{E} \simeq \mathcal{E}_T$ is the classifying topos of a coherent theory—then G may be taken to be a topological groupoid.[19]

The models of a (coherent) theory T naturally form a category $\text{Mod}(T)$. Now, if we eliminate all non-isomorphism arrows from $\text{Mod}(T)$ then the resulting category $\text{Mod}_i(T)$ is a groupoid, i.e. a category in which every arrow has a two-sided inverse. Intuitively speaking, $\text{Mod}_i(T)$ is the category of models of T and their symmetries (i.e. automorphisms).

Since $\text{Mod}(T)$ doesn't contain enough information to reconstruct T, a fortiori $\text{Mod}_i(T)$ doesn't contain enough information to reconstruct T. To reiterate, a theory's models and their automorphisms do not tell us everything about that theory! But now the insight of Awodey and Forssell was that if $\text{Mod}_i(T)$ is equipped with an appropriate topology, then the resulting topological groupoid G could be the very G that appears in the representation theorem of Butz and Moerdijk. To be more precise, if C_T is the syntactic category of T, and if $Sh(C_T)$ is the topos of sheaves on C_T, then

$$Sh(C_T) \simeq B(G_T),$$

where G_T is the topological groupoid of models of T, and $B(G_T)$ is the Grothendieck topos of continuous actions of G_T. Furthermore, since the pretopos completion of C_T can be recovered as the coherent objects in $Sh(C_T)$, it follows that the pretopos completion of C_T can be recovered from the topological groupoid G_T. In other words, T itself can be recovered from G_T up to Morita equivalence.

This result suggests that the syntactic and semantic categories of theories are dual to each other. On the one hand, we have **Pretop**, the category of theories (understood as (conceptually complete) syntactic categories). On the other hand, we have **TopGrpd**, the category of semantic categories, viz. topological groupoids. The "semantic functor" $\Theta : \textbf{Pretop} \to \textbf{TopGrpd}$ is defined by first taking $\textbf{Pretop}(P, \mathcal{S})$, the category of set-valued models of P, then restricting to the isomorphisms between models, and finally equipping the resulting groupoid with the "logical" topology. The "syntactic functor" $\Gamma : \textbf{TopGrpd} \to \textbf{Pretop}$ is defined by taking a topological groupoid G to the topos $B(G)$ of continuous G-sets, and then extracting the pretopos of coherent objects in $B(G)$.[20]

---

[19] Roughly, locales are topological spaces without a notion of a point, axiomatized instead with a primitive notion of a neighborhood and lattice operations on such neighborhoods (corresponding to unions and intersections). The advantage of locales is that they are amenable to a first-order axiomatization (albeit with a caveat: the category of locales is defined as the *dual* of the category of models of that axiomatization). The disadvantage is that they are strictly more general than topological spaces: every topological space is a locale but not every locale is a topological space. More precisely, it can be shown that the category of topological spaces is a coreflective subcategory of the category of locales. For this and more motivation on "pointless" topology see the introductory survey by Johnstone (1983).

[20] Technically, Awodey and Forssell work with the category **dCoh** rather than **Pretop**, and the results must be adjusted accordingly.

The two functors Θ and Γ are indeed adjoint to each other.

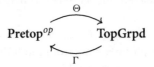

However, once again, this adjunction is *not* an equivalence of categories; i.e. **TopGrpd** is not exactly dual to **Pretop**. Thus, a natural question: can the semantic category **Pretop**^{op} be characterized independently of the functor Θ? That is, can we provide an independent characterization of the category of (semantically presented) theories? Doing so would lead to a so-called "perfect duality" but such an independent characterization has so far proved elusive.[21]

Why do mathematicians value duality results? One reason is that it enables them to transfer results and concepts from one category to its dual category. Thus, if we had a duality result for syntactic and semantic categories, then we could use information about theories as presented semantically in order to understand theories as presented syntactically, and vice versa.

For example, suppose that the theory T' results from adding a new predicate symbol (but no new axioms) to the theory T. There is then an obvious translation of T into T', namely the translation that takes each piece of non-logical vocabulary to itself. This translation corresponds to a functor $F : P_T \to P_{T'}$ that is faithful, full (since no new functional relations are created), but not essentially surjective (since no formula from the smaller language maps to the new predicate symbol). Recalling that F is an arrow in **Pretop**, there is a dual arrow $F^* : G_{T'} \to G_T$ in the semantic category **TopGrp**. It is natural to ask then: given that F has such and such features, what features does its dual arrow F^* have? In this case, the failure of essential surjectivity of F corresponds to the fact that F^* is not full, i.e. it forgets structure in the sense of Weatherall, Chapter 13, in this volume.[22]

Of course, the question can also be asked the other way around: given a functor $K : \mathbf{M} \to \mathbf{N}$ between categories of models, how do features of K correspond to features of its dual arrow K_* in the syntactic category? But here we pause, because in our opinion, the semantic category—whose objects are categories of models—has not yet

---

[21] One reason for this—mysterious as it may sound—is certainly the fact that there seems to be nothing inherently "category-theoretic" about ultraproducts. As Makkai's work proves and Los' theorem has long made obvious, taking ultraproducts is a fundamental operation when it comes to elementary classes: elementary classes are exactly those classes closed under elementary equivalence and the taking of ultraproducts. Since every pretopos corresponds to an elementary class (more precisely: to the category of models of a coherent theory) one would imagine that any such characterization of **Pretop** would amount to a characterization of "closure under ultraproducts". Absent any useful purely categorical description of ultraproducts (or even ultrafilters) this seems like a significant obstruction. Nevertheless the work of Leinster (2013) on ultrafilter monads as codensity monads might provide a way out, though this is still very far from being made precise.

[22] It ought to be remarked that the "functorial" explication of the idea of forgetting structure (or properties or stuff) is originally due to John Baez.

been adequately characterized. First, not every topological groupoid is the groupoid of models of a coherent theory; i.e. **Pretop**^{op} has fewer objects than **TopGrpd**. Second, **Pretop**^{op} is a 2-category; hence, if **Pretop**^{op} is to be seen as living inside **TopGrpd**, then we must understand the latter as itself a 2-category. But what is the appropriate 2-categorical structure and how should we understand it?[23]

So, for anyone who wishes to develop the semantic view of theories, the following is a pressing question:

Given two categories of models M *and* N, *what is a fruitful definition of an arrow* $K : M \to N$? *Furthermore, which arrows should be thought of as "reductions" of one theory to another, which as "equivalences" of theories, and which as other theoretical relations with which philosophers of science have been concerned?*

Recalling what we said earlier: to choose the arrows of a category is to choose a notion of equivalence, and hence to choose a notion of identity of the category's objects. (Here the objects of the semantic category are themselves categories, viz. *categories of models*.) Thus, until one proposes a notion of arrows between semantic categories, then one lacks a clear notion of equivalence of theories, and hence of how a theory can be identified semantically. For philosophers of science, this issue demands immediate attention.

But why do we care about these results and how do they relate to the framework we have been advocating? Essentially, because the earlier described duality between the category of syntactic and (suitably structured) semantic categories offers a large-scale dissolution of the syntax-semantics dichotomy. What this guarantees—and the reason it is important—is that intertheoretic relations get preserved regardless of whether we replace a theory with its syntactic category or its category of models. In other words, the way in which a syntactic category of a theory T "sits" within other syntactic categories is equivalent (in an appropriate sense) to the way in which Mod(T) "sits" within other categories of models. That this is the case is in no way trivial, or automatic. One could very well imagine it being true that moving back and forth between syntactic categories and categories of models changed the way in which these different presentations of a theory related to other theories. But it does not—this is what makes the duality results presented in this section essential, and what paves the way for several of the technical projects we will now go on to outline.

17.4 Future Directions

Given this elegant formal framework, numerous technical questions—of philosophical interest—suggest themselves. We will now list and discuss a few such questions.

---

[23] Moerdijk (1988, 1990) defines a bicategory **LocGrpd** of localic groupoids, with bimodules as arrows, and shows that $G \mapsto B(G)$ is an equivalence of categories between **LocGrpd** and the bicategory of Grothendieck toposes.

1. We saw that if the syntactic categories C_T and C_S are equivalent, then the theories T and S are Morita equivalent, i.e. that (SE) implies (ME). We also saw that the converse fails: two theories can be Morita equivalent even even though their syntactic categories are inequivalent.

    An interesting question we can now raise is this: What kind of conditions can we place on theories such that (SE) coincides with (ME)? (One well-known case in which they do coincide is that of algebraic theories.) Furthermore, what can we say about Morita equivalences over particular categories? For example, for any two first-order theories T and T' whose semantic categories are equivalent (over S, and not necessarily naturally) how can we characterize their relation from a purely syntactic standpoint? And what about their thin categories? Namely, if

    $$\mathcal{S}-\mathrm{Mod}_e(T) \simeq \mathcal{S}-\mathrm{Mod}_e(T')$$

    then—purely syntactically—how are T and T' related?

2. Let's consider some natural relations between theories. First, let T be a theory in signature Σ, and let T' be an extension of T by some additional axioms, also in the language Σ (such "extensions" are also called "quotients"). Then there will be a canonical functor $F : C_T \to C_{T'}$. What can we say about this functor? What features does it have? In the case of geometric theories a lot of work in this area has been done by O. Caramello starting with the "Duality Theorem" in her Phd thesis—for a big picture view see Caramello (n.d.b,n). See also Forssell (2013).

    Similarly, let T be a theory in signature Σ, and let T' be the same theory, but considered in a larger signature Σ'. Again there will be a canonical functor $F : C_T \to C_{T'}$. What features does this functor have?

    Furthermore, philosophers of science have been interested in questions about when one theory T' is *reducible* to another theory T. Can reduction be thought of as a functor $F : C_{T'} \to C_T$ Van Benthem and Pearce (see 1984)? How does this functorial account compare to the classical Nagelian account (see Nagel (1979, ch. 11))? How does this functorial account compare to semantic accounts (see Bickle (1998))?

3. If we were to think of **Coh** or **Pretop** as merely a category, then a natural technical question might be: does this category have limits? Or does this category have colimits? And if it does have limits or colimits, then do these have any sort of natural interpretation as operations on theories? For example, is there any sense in which a coproduct of two coherent categories represents a sort of amalgamation of the two theories? And if so, does the notion of an "amalgamation of theories" have a clear interpretation?

    However, it's more natural to think of either **Coh** or **Pretop** as a 2-category, in which case the better questions have to do with the existence of limits and colimits in the bicategorical sense. Does **Pretop** have 2-limits and 2-colimits?

And if it does, do these limits and colimits have a natural interpretation as operations on theories?[24]

Philosophers of science should be particularly interested in whether the categorical structure of **Pretop** can be used to explicate various relations between theories, such as limiting relations.

4. The relation between "thick" and "thin" semantic categories provides very fertile ground for investigation, as noted earlier. One obstruction here is that the thin semantic categories of first-order theories are almost invariably "too thin" in the sense that they rarely contain interesting categorical structure (e.g. limits, colimits)

On the other hand, it is perhaps worth investigating relations between pairs of theories (T_1, T_2) such that there are interesting functors

$$F: \text{Mod}_e(T_1) \to \text{Mod}(T_2),$$

where T_1 would be thought of as the "background space–time" theory and T_2 would be thought of as the "physical system theory". To be a little more specific, this formal situation seems to us an interesting generalization of the nowadays very common situation (e.g. with (quantum) field theories ((Q)FTs)) where we have a category of "state spaces" represented by algebraic objects (e.g. Hilbert spaces, C^*-algebras) and a category of "space–times" understood as categories of "physical space–times with embeddings as arrows" and where a theory of physics is defined as a functor relating those two.

For example, Fewster (2015) defines a locally covariant theory to be a functor

$$F: \textbf{Bkgnd} \to \textbf{Phys}.$$

In Fewster's set-up, **Bkgnd** seems to us to be best understood as essentially the thin category of models of some theory, since the morphisms are basically elementary embeddings (see Fewster (2015, 4)). On the other hand, the categories which he calls **Phys** seem to us to be best understood as the thick categories of models of some theory, since they are usually categories of algebraic structures. This suggests that it is worth looking at what kind of interesting things can be said about such functors, when T_1 and T_2 are first-order theories: what can we say about them, modulo some constraints on T_1, T_2, and F? Is this a fruitful general setup with which to study properties of (Q)FTs?

It might not be immediately clear how this is related to the category of theories perspective that we have been advocating. It very much does. Assuming for the time being that both $\text{Mod}_e(T_1)$ and $\text{Mod}(T_2)$ live in the same category **Th**, then what we are suggesting is to study certain objects in the category of arrows of **Th**; i.e. the category whose objects are the arrows of **Th** and morphisms

---

[24] Makkai (1995) proves a version of the Craig interpolation theorem using the 2-categorical version of a pushout of syntactic categories.

are commutative squares connecting these arrows. (Once again, since Th is a 2-category, its arrow category will also have a natural 2-categorical structure.) It appears to us to be a fruitful way of understanding field theories in general: as functors relating a "space–time" category (e.g. Minkowski space–times) to a "state-space" category (e.g. C*-algebras).[25]

5. The (2-)category **Pretop** might be used to explicate the notion of a *symmetry* of a theory.

In recent literature in philosophy of science, discussion of symmetries has proceeded in absence of an agreed-upon background framework. These discussions sometimes think of symmetries in terms of equations, and sometimes in terms of models. What's more, we find ambiguity about whether symmetries operate at the level of individual models of a theory, or at the level of the collection of all models of a theory.

A category-theoretic point of view makes it clear, however, that there are both syntactic and semantic notions of symmetries—as can be made precise with the category Th of theories (understood here as **Pretop**). Let T be a theory and let P_T be its syntactic category, which is an object of Th.[26] Then we propose:

A syntactic symmetry of T is an auto-equivalence of the syntactic category P_T.

Recall that a coherent functor $F : P_T \to P_S$ corresponds to a translation from T into S. Thus, an auto-equivalence $F : P_T \to P_T$ corresponds to a translation of T into itself (i.e. a sort of permutation of the vocabulary of T).

But more is true. A coherent functor $F : P_T \to P_S$ induces a functor $F^* : \mathrm{Mod}(S) \to \mathrm{Mod}(T)$ from models of S to models of T (see Makkai and Reyes (1977), Gajda et al. (1987)). In particular, an automorphism $F : P_T \to P_T$ induces a functor $F^* : \mathrm{Mod}(T) \to \mathrm{Mod}(T)$ on the category of models of T, which raises another technical question:

Is it true that any essentially invertible functor $G : \mathrm{Mod}(T) \to \mathrm{Mod}(T)$ has the feature that $G = F^*$ for some coherent functor $F : P_T \to P_T$?

The answer to this question is No, as can be seen by again looking at the example of the two propositional theories. Thus, philosophers of science should *not* necessarily suppose that any auto-equivalence $G : \mathrm{Mod}(T) \to \mathrm{Mod}(T)$ should count as a symmetry of T. But can we say something about further conditions on G so that it is indeed dual to some functor F on the syntactic category P_T?

---

[25] A complication arises here, however, with respect to how to actually view $\mathrm{Mod}_e(T_1)$ and $\mathrm{Mod}(T_2)$ as living in the same category. It is not immediately clear how this can be done in an interesting way when we've identified Th as either **Coh** or **Pretop**. This is an important question in itself to which we have no satisfying answer at present. As a result, we have decided to present the above question in terms of functors between "free-floating" categories of models. We thank an anonymous referee for pressing us on this point.

[26] A note of clarification on terminology: we are now using the term "syntactic category" to refer to an arbitrary object of the category of theories Th rather than the explicit construction as carried out in Section 17.1.1.

6. Since the demise of the syntactic view of theories, philosophers of science have been fond of pointing out that interesting scientific theories—even those in rigorous mathematical physics—typically fail to admit a first-order axiomatization. Thus, we might conclude that a typical scientific theory cannot be described by an object in **Coh** or **Pretop**. We already dealt with this point at the end of Section 17.1 but let us now add a few more words.

First, is this dismissal too fast? Note that some logics stronger than first-order logic—e.g. geometric logic, or even higher-order logics—will also give rise to syntactic categories that are coherent. In fact, if a logic is stronger than first-order logic, then the corresponding categories can be expected to have more structure than coherent categories.[27]

Relatedly, recent work in "cohesive" homotopy type theory (CoHoTT) uses syntactic methods (based on a logic much more "exotic" than first-order logic) to study higher gauge theories (see Schreiber (n.d.), as well as Corfield, Chapter 2 in this volume). Among other things, Schreiber envisions a certain class of ∞-toposes (the so-called "cohesive" ones) as the correct setting at which to study the foundations of higher gauge theories, *exactly because* these cohesive ∞-toposes are (conjectured to be) the syntactic categories of CoHoTT. Of course, Schreiber's use of syntactic methods is not motivated by considerations on theoretical equivalence and the structure of scientific theories—his mathematical work is carried out with the explicit goal of articulating a general foundation for higher gauge theories. Nevertheless, the way in which he blends syntactic and semantic methods is a great illustration, in our opinion, of how representing scientific theories as categories (in his case *higher* categories) is an illuminating perspective to take.

So, is being a coherent category a *minimal* necessary condition for representing a bona fide scientific theory? In any case, it would be natural to ask:

Given a formalized theory T of the empirical sciences, is there some category C_T that can be thought of as the "syntactic category" of T?

Consider a couple of examples. First, let T be Einstein's general theory of relativity (GTR). We do have some sense of what the *semantic* category—i.e. the category of models of T—ought to be, viz. the category of differentiable manifolds with Lorentzian metric and stress-energy tensor satisfying Einstein's field equations. But is there a category C_T that could be considered the syntactic category of GTR?[28] And could the failure to distinguish between semantic and

---

[27] For example, in the case of intuitionistic type theory (ITT), there is a syntactic category C_{ITT} (produced by a similar but different process than the one we outlined for first-order theories T) which bears the structure of an elementary topos—see Lambek and Scott (1988) for the classical account.

[28] One might hope for some help here from investigations in synthetic differential geometry. See, for example, Reyes (2009).

syntactic points of view be partially responsible for some of the difficulties that philosophers have had understanding the nature of symmetries in GTR?

As a second example, let T be quantum mechanics. In this case we also have a sense of what the semantic category of T ought to be—namely, the category of (finite-dimensional) Hilbert spaces and linear operators. (This example was one of the main motivators for the semantic view, at least in the mind of van Fraassen.) Now, what might the syntactic category C_T of T look like? Is there a way to present quantum mechanics syntactically? If so, what are its axioms, and what sorts of objects does it quantify over?

17.5 Conclusion

In this chapter, we have assumed that it can be useful, for a philosophical understanding of science, to represent theories as mathematical objects. But what is a good, fruitful mathematical framework for understanding theories? We have surveyed a number of concepts and results from category theory, which we believe provide strong evidence that it should be the locus of attention for formal philosophers of science.

First, we have shown that category theory provides the resources to get past philosophy of science's false dichotomy, viz. the dichotomy between syntactic versus semantic presentations of theories. On the one hand, a semantic presentation of a theory is nonetheless a presentation—written in a mathematical language. On the other hand, the syntactic category of a theory is a hybrid object, neither purely syntactic nor purely semantic. Enough of the original syntax of theory can be reconstructed from its associated syntactic category. Moreover, having equivalent syntactic categories guarantees having equivalent categories of models. As such, the syntactic category of a theory unites the semantic and the syntactic approach to (first-order) theories, via the notion of Morita equivalence: as long as we care about theories up to (a suitable notion of) definitional equivalence and as long as we care about the classes of models of a theory up to its categorical structure, then the tension between syntactic and semantic presentations disappears. With this formal groundwork in place our most urgent task now is to carry as much of this lesson as possible over to the philosophy of science.

Second, with this false dichotomy set aside, philosophers of science are now set free from the illusion of a direct access to the "theory in itself" (sometimes erroneously identified with a class of models), and to study how different representations of the world can be related one to another. The category Th is the first formal approximation to how this kind of project might be carried out. The kind of notions about theoretical relations that will emerge will no doubt be interesting and we hope the philosophy of science lends a keen ear. Aside from this however, the very notion of studying scientific theories in their totality (i.e. as a structured whole) should lead to a philosophy of science more directly attuned to the way theoretical physics is carried out today—where the interaction and interplay of (sometimes incompatible) theories is not seen merely as a temporary state of confusion as we approach some ultimate truth, but rather the necessary interactions that have to take place within any ecosystem before more advanced life can evolve out of it.

Acknowledgments

Thanks to Thomas Barrett for conversation and feedback and to two anonymous referees for numerous illuminating suggestions.

References

Andréka, H., Madarász, J., and Németi, I. (2001). Defining new universes in many-sorted logic. *Mathematical Institute of the Hungarian Academy of Sciences* 93.

Awodey, S. and Forssell, H. (2013). First-order logical duality. *Annals of Pure and Applied Logic* 164(3), 319–48.

Barr, M., Kennison, J. F., and Raphael, R. (2008). Isbell duality. *Theory and Applications of Categories* 20(15), 504–42.

Barrett, T., and Halvorson, H. (2015a). Morita equivalence. Unpublished manuscript. http://arxiv.org/abs/1506.04675

Barrett, T., and Halvorson, H. (2015b). Quine's conjecture on many sorted logic. Unpublished manuscript. http://philsci-archive.pitt.edu/11887/

Barrett, T. and Halvorson, H. (2016). Glymour and Quine on theoretical equivalence. *Journal of Philosophical Logic* 45(5), 467–83.

Bénabou, J. (1967). Introduction to bicategories, in *Reports of the Midwest Category Seminar* 1–77.

Bickle, J. (1998). *Psychoneural Reduction: The New Wave*. MIT Press, Cambridge, MA.

Borceux, F. (1994). *Handbook of Categorical Algebra*, Vol. 1. Cambridge University Press, Cambridge, UK.

Butz, C., and Moerdijk, I. (1998). Representing topoi by topological groupoids. *Journal of Pure and Applied Algebra* 130(3), 223–35.

Caramello, O. (n.d.a). Lattices of theories. arXiv:math.CT/0811.3547

Caramello, O. (n.d.b). The unification of mathematics via topos theory. arXiv:1006.3930

Caramello, O. (n.d.c). Universal models and definability. *Mathematical Proceedings of the Cambridge Philosophical Society* 152(2), 279–302.

Corfield, D. (2017). Reviving the philosophy of geometry, in E. Landry (ed.), *Categories for the working philosopher*, Oxford University Press, Oxford. This volume.

Curiel, E. (2014). Measure, topology and probabilistic reasoning in cosmology. http://philsci-archive.pitt.edu/11071/

Fewster, C. (2015). Locally covariant quantum field theory and the problem of formulating the same physics in all spacetimes. *Philosophical Transactions of the Royal Society A* 373(2047), 20140238. arXiv:1502.04642

Fletcher, S. (2016). Similarity, topology, and physical significance in relativity theory. *British Journal for Philosophy of Science* 67(2), 365–89.

Forssell, H. (2013). Subgroupoids and quotient theories. *Theory and Applications of Categories* 28(18), 541–51.

Friedman, M. (1982). The scientific image by Bas C. van Fraassen. *Journal of Philosophy* 79(5), 274–83.

Gajda, A., Krynicki, M., and Szczerba, L. (1987). A note on syntactical and semantical functions. *Studia Logica* 46(2), 177–85.

Halvorson, H. (2012). What scientific theories could not be. *Philosophy of Science* 79(2), 183–206.

Halvorson, H. (2016). Scientific theories, in P. Humphreys (ed.), *Oxford Handbook of the Philosophy of Science*, 585–608. Oxford University Press, Oxford, UK.

Halvorson, H. (2017). Categories of scientific theories, in E. Landry (ed.), *Categories for the Working Philosopher*. Oxford University Press, Oxford, UK, this volume.

Harnik, V. (2011). Model theory vs. categorical logic: two approaches to pretopos completion (aka t^{eq}), in B. Hartt, T. Kucera, A. Pillay, and P. Scott (eds), *Models, logics, and higher-dimensional categories*, American Mathematical Society, Providence, RI.

Johnstone, P. (1983). The point of pointless topology. *Bulletin of the American Mathematical Society* 8(1), 41–53.

Johnstone, P. (2002). *Sketches of an Elephant: A Topos Theory Compendium*. Oxford University Press, Oxford.

Joyal, A., and Tierney, M. (1984). An extension of the Galois theory of Grothendieck. *Memoires of the American Mathematical Society* 51(309).

Lack, S. (2010). A 2-categories companion, in J. C. Baez and J. P. May (eds), *Towards higher categories*, 105–91. Springer, Berlin.

Lal, R. and Teh, N. J. (2017). Categorical generalization and physical structuralism, *British Journal for the Philosophy of Science*. 68(1), 213–51. http://arxiv.org/abs/1404.3049

Lambek, J., and Scott, P. J. (1988). *Introduction to Higher-Order Categorical Logic*, number 7 in Cambridge Studies in Advanced Mathematics 7. Cambridge University Press, Cambridge, UK.

Leinster, T. (2013). Codensity and the ultrafilter monad. *Theory and Applications of Categories* 28, 332–70.

Mac Lane, S., and Moerdijk, I. (2012). *Sheaves in Geometry and Logic: A First Introduction to Topos Theory*. Springer, Berlin.

Makkai, M. (1987). Stone duality for first order logic. *Advances in Mathematics* 67(2), 97–170. doi:10.1016/0001-8708(87)90020-X

Makkai, M. (1991). *Duality and Definability in First Order Logic*, Vol. 503. American Mathematical Society, Providence, RI.

Makkai, M. (1995). On Gabbay's proof of the Craig interpolation theorem for intuitionistic predicate logic. *Notre Dame Journal of Formal Logic* 36(3), 364–81.

Makkai, M., and Reyes, G. (1977). *First Order Categorical Logic*. Springer-Verlag, Berlin.

Moerdijk, I. (1988). The classifying topos of a continuous groupoid. i. *Transactions of the American Mathematical Society* 310(2), 629–68.

Moerdijk, I. (1990). The classifying topos of a continuous groupoid. ii. *Cahiers de Topologie et Géométrie Différentielle Catégoriques* 31(2), 137–68.

Nagel, E. (1979). *The Structure of Science*. Hackett, Indianapolis, IN.

Pitts, A. (1987). Interpolation and conceptual completeness for pretoposes via category theory, in D. W. K. et al. (eds), *Mathematical logic and theoretical computer science*. M. Decker, New York.

Pitts, A. (1989). Conceptual completeness for intuitionistic first-order logic: an application of categorical logic. *Annals of Pure and Applied Logic* 41, 33–81.

Porst, H. E., and Tholen, W. (1991). Concrete dualities, in H. Heirlich and H. E. Porst (eds), *Category Theory at Work*, 11–136. Heldermann Verlag, Berlin.

Reyes, G. (2009). A derivation of Einstein's vacuum field equations, in B. Hartt, T. Kucera, A. Pillay and P. Scott (eds), *Models, logics, and higher-dimensional categories*, 245–62. American Mathematical Society, Providence, RI.

Schreiber, U. (n.d.). Differential cohomology in a cohesive ∞-topos. http://arxiv.org/abs/1310.7930

Shulman, M. (2012). Exact completions and small sheaves. *Theory and Applications of Categories* 27(7), 97–173.

Teh, N. J., and Tsementzis, D. (2016). Theoretical equivalence in classical mechanics and its relationship to duality. *Studies in the History and Philosophy of Modern Physics*. https://doi.org/10.1016/j.shpsb.2016.02.002

Tsementzis, D. (2015). A syntactic characterization of Morita equivalence. Manuscript. http://arxiv.org/abs/1507.02302

Van Benthem, J., and Pearce, D. (1984). A mathematical characterization of interpretation between theories. *Studia Logica* 43(3), 295–303.

Van Fraassen, B. (2014). One or two gentle remarks about hans halvorson's critique of the semantic view. *Philosophy of Science* 81(2), 276–83.

van Oosten, J. (2002). Basic category theory. http://www.staff.science.uu.nl/oostel10/syllabi/catsmoeder.pdf

18

Structural Realism and Category Mistakes

Elaine Landry

18.1 Introduction

Structural realism[1] is the view that, as a scientific realist, one should be ontologically committed to the *structure* of successful scientific theories. As proposed by Worrall (1989), it aims to provide the "best of both worlds", where these worlds are bound, on the one hand, by the realist's *no miracles argument* (NMA) and, on the other, by the anti-realist's *pessimistic meta-induction argument* (PMI). NMA is typically offered by the traditional, or metaphysical, scientific realist in favor of ontological commitment to the *objects* of successful scientific theories. Here I turn to Putnam (1975) to capture the conclusion of the argument: "[Metaphysical] realism is the only philosophy that does not make the success of science a miracle" (Putnam, 1975, 73). On the other side of the divide, the pessimistic meta-induction argument, as offered by Laudan (1981), seeks to undermine the position of the metaphysical realist by arguing that since past scientific theories, which were then held as successful, were since found to be false, we have no reason to believe the realist's claim that our currently successful theories are true, or approximately true, and, presuming that truth is explained by reference to objects, we have no reason to be ontologically committed to the objects of current successful scientific theories.

Aiming to cut a philosophical midpoint, and borrowing from Poincaré's structuralism,[2] Worrall shifts the focus of NMA from ontological continuity to *structural continuity*, arguing that we should be committed to the *structure* that is continuous between *successive* and successful scientific theories. For example, we should be epistemically committed to the structure of light, but not its nature, as expressed by the *shared structure* between Fresnel and Maxwell's equations. This *epistemic structural realist* position holds that we can explain the success of scientific theories

---

[1] For an extensive and excellent overview of structural realism, see Ladyman (2014).
[2] See especially Poincaré (1905).

in a non-miraculous manner, and so avoids the force of PMI. That is, the success of science is explained, not by reference to objects, but by reference to our *epistemic* grasp of the *continuity of the structure* of successful and successive scientific theories, all the while accepting the possibility of *ontological discontinuity* between successive theories. Thus, the slogan: *all we know is structure*.

Forgoing the presumption of the continuity of structure of *successive* scientific theories, structural realists, like French,[3] focus on the structure of *fundamental* and successful scientific theories. For example, we should be structural realists about the Lie-group structure of quantum mechanics. And further, as *metaphysical naturalists*, we should use this structure to carve out a conceptual space, for what is both *physically* and *metaphysically* possible. Accordingly, from a physical standpoint, when we consider the structure of quantum mechanical theories, since the symmetry permutations of the Lie-group are what carve reality into its *fundamental kinds*, e.g., bosons and fermions, we *physically* commit only to the structure of these kinds. And from a metaphysical standpoint, since quantum mechanics is a fundamental theory and since talk of quantum mechanical individuals is *metaphysically underdetermined*, we *ontologically* commit to the *ontic structural realist* claim that *all there is* to bosons, fermions, etc., is their Lie-group structure. French (2014), by presuming that quantum mechanics is the fundamental theory, goes even farther to argue that all there is to the *structure of the world* is Lie-group structure. Finally, he extends his ontic structural realism to underwrite his *metaphysical* commitment to a *structure-oriented metaphysics* used to analyse such notions as causality, modality, dispositions, laws, etc. As noted, Worrall and French's accounts offer the two standard interpretations of structural realism:[4] *epistemic* and the *ontic*. The first claims *all we know* is structure. The second claims *all there is* is structure. Yet another difference between the two is that Worrall uses a Ramsey-sentence approach as a meta-level framework for a *syntactic* account of the structure of scientific theories themselves, while French uses the da Costa and French (2003) partial-structures account as a meta-level framework for a *semantic* account of the structure of scientific theories. Against both interpretations, Psillos[5] has argued that it makes no sense to speak, either epistemically or ontically, of relations without relata or structures without objects.

Given this, albeit, brief characterization of structural realism, there are three points at which one can make appeal to the use of category theory. The first is as a *meta-level formal framework* for a structural realist account of the *structure of scientific theories*,

---

[3] See French (1999, 2000, 2014).

[4] For an account of the various varieties of, and standard objections to, structural realism, see Ladyman (2014).

[5] See Psillos (1995, 2001, 2006b). However, Psillos is not the first, nor the only, one to hold this view. As noted by Ladyman in his SEP entry, "this objection has been made by various philosophers including Dorato (2000), (Psillos, 2001, 2006), Busch (2003), Cao (2003), Morganti (2004). Even many of those sympathetic to the OSR of French and Ladyman have objected that they cannot make sense of the idea of relations without relata (see Lyre 2004; Stachel 2006; and Esfeld and Lam 2008)".

either syntactic or semantic. The second is to appeal to the category-theoretic structure of some successful, *successive* or *fundamental*, physical theory to argue that this is the structure we should be physically committed to, either epistemically or ontically. The third is to use category theory as a *conceptual tool* to argue, contra Psillos et al., that it does make *conceptual* sense to talk of relations without relata and structures without objects.

In what follows, I will consider how each appeal to the use of category theory stands up against the aims of the structural realist. In Section 18.2, I will argue that such *meta-level* formal frameworks, framed by logic, set theory, or, indeed, category theory, have to do with the *formal* structure of scientific theories, and not with the *physical* structure of some successive or fundamental theory, and, as a consequence, are of *no help* with structural realist *object-level* claims about the *physical structure of the world*. Thus, any structural realist move from the formal structure of scientific theories to the physical structure of the world will fall victim to a category mistake; that is, to a mistake the arises from conflating considerations of meta-level structure with those object-level structure. In Section 18.3, I will critically examine more recent claims by Bain (2013) and argue that he too falls victim to a similar category mistake. In so far as his claims for a *radical ontic structural realist* (ROSR) use of category theory are found at the *theoretical level*, and so are concerned with the *mathematical structure of a scientific theory* and not with the *object-level physical structure of the phenomena*, they are of little help to the structural realist who, again, needs to make claims about the *physical structure of the world*. However, along the way, and against the claims of both Psillos et al. and Lam and Wüthrich (2014), in Section 18.4, I will also argue, as I believe Bain (2013) intends, that at a *mathematical level* one *can* use category theory to answer the question of how we can *conceptually* speak of both relations *without* relata and structures *without* objects. Moreover, I will show that category theory so considered can be used to *conceptually collapse* the distinction between Bain's radical ROSR view and Wüthrich's (2014) "more balanced" BOSR view. Towards these ends, I will use an *in re* Hilbertian interpretation of category-theoretic mathematical structuralism to show that one can talk of *systems* that *have* a *type* of *structure*, *without* our having to concern ourselves with what these systems *are*, or *are made of*. Thus, I will conclude with the claim that category theory *can* be used as a *conceptual tool* to answer how we can *coherently speak* of relations *without* relata, and, more generally, structures *without* objects.

18.2 Category Theory as a Framework for the Structure of Scientific Theories

The syntactic view is typically taken as arising from the view of logical positivists and takes a theory to be framed by its logico-linguistic structure. That is, a theory is a set of sentences expressed, typically, in first-order logic, so the structure of scientific theories is formally framed by logic.[6] Modern advocates of the syntactic view, like Worrall and Zahar (2001), Maxwell (1971), and Lewis (1970), following Russell (1927), prefer to frame the structure of scientific theories by their Ramsey sentence,[7] whereby theoretical terms may be eliminated from a theory by means of an existentially quantified sentence containing only observational terms. The Ramsey sentence is thus held as retaining the empirical content of the theory without reference to theoretical terms. However, the use of the Ramsey sentence as a formal framework for the structure of scientific theories faces several problems. The first set of problems considers the set-theoretic notion of structure: if the Ramsey sentence frames the structure of scientific theories, then all that structure, expressed as set structure, can tell us is "the number of constituting objects".[8] The second set of problems concerns Worrall's supposition that the distinction between relations and relata and the structure and nature of objects does not appeal, as Russell (1927) does, to the "intrinsic character" of the relata.[9]

The semantic view of scientific theories was proposed by Suppes (1967b)[10] as an alternative to the syntactic view. The purpose was to shift the focus from the cumbersome philosophical task of expressing theories in logico-linguistic form to talk of models and the isomorphisms between them. According to Suppes, a theory, expressed by a set-theoretic predicate, consists of the various levels of models *qua* set structures[11] corresponding each to a level of theory: *theoretical* models, *experimental* models, and *data* models. The connection between each level is then made by an appeal to their *shared structure*, which is expressed by an *isomorphism*. Problematically, as Suppes himself notes, this raises the question of how, without a proper *theory of the data*, we make a connection to *the world*, and so make claims about *the truth* of a theory, i.e., claims that our data models apply to (or are isomorphic to) models of the world. Van Fraassen (1980), offering up a *constructive empiricist*, non-realist solution famously argued that, when the data models are taken as *embeddable* in the theoretical models, we can at best only get to the *empirical adequacy* of a theory. Again, problems remain.[12]

Da Costa and French (2003), attempting to carve out a realist alternative, yet one that is further amenable to both van Fraassen's empiricism and French's ontic structural realist position, propose a set-theoretic *partial structures* approach, wherein we need only appeal to *partial isomorphisms* between models as partial structures, and only those that capture the shared structure that is *actually used*. As argued in da Costa and French (2003), one can then appeal to the notion of *partial isomorphism*

---

[6] See, for example, Carnap (1966).   [7] See Ramsey (1929).

[8] See Newman (1928), Demopoulos and Friedman (1985).

[9] See especially Psillos (1995, 2000, 2001, 2006a,b). For Worrall's replies to some of these problems, see Worrall (2007).

[10] See also Suppes (1960, 1962, 1967a); Suppe (1989).

[11] To see the manner in which Suppes' semantic account is set-theoretically founded, in the Bourbaki sense of the term, see Landry (2007).

[12] See Monton and Mohler's (2014) SEP entry for a discussion of some of these problems.

to express *shared structure*, *horizontally*, between models of different theories and, *vertically*, between theoretical, experimental, and data models. And further, one can appeal to the notion of *partial homomorphism*, to express the shared structure between mathematical and physical theories in a way that respects the role of what Redhead (1996) has called "surplus structure".

More importantly, for the ontic structural realist, this meta-level framework can be used to make sense of both the epistemic and ontic *structural realisist* claims that all we know is structure, or all there is is structure. That is, one can express the structure that is shared (continuous) over theory change by looking to the *partial isomorphisms* that exist, *inter*-theoretically, between *successive* theories and claim that this is the structure that we are realist about. Similarly, as in French (2000), one can use *partial isomorphisms* to express the *shared structure* that exists, *intra*-theoretically, i.e., among theoretical, experimental, and data models, to capture the structure of successful *fundamental* scientific theories, and so run the ontic structural realist claim that this is the structure that carves the world into its *fundamental* kinds.

I have argued[13] that talk of such *meta-level* formal frameworks, either logical or set-theoretic, is beside the point of structural realism. Whether syntactically or semantically framed, the structure that does *structural realist work* in connecting a theory to the world, or more precisely theoretical models to experimental models to phenomena *qua* data models, is the *object-level mathematical structure* of a *particular theory* as determined in some physical context; it is *not* the formal framework of *the structure of scientific theories*. For example, in Landry (2007), I argued, against French, that what does structural realist work in the example of the role of group theory in quantum mechanics is the natural transformations that provide the connection between the Lie-group structure of the theory and the symmetry structure of the phenomena. That is, it is the structure of the permutation group, in *both* theoretical and phenomenal contexts, that imposes a group-theoretic structure, and so yields a division of kinds, e.g., boson and fermions, that share structure with the phenomena, e.g., with spin properties. Thus, to make the *object-level connection* between the shared structure of the theory and the phenomena, no meta-level appeal to the formally framed structure of the theory need be made.

French (2014) has recently acknowledged this point, but yet argues that there is still *philosophical work* to be done at the " 'meta'-level of the philosopher of science" (102), and that such work needs to consider the structure of scientific theories. I yet maintain that the only level that matters *for the structural realist* is, what French has termed, the " 'object'-level of scientific practice" (French, 2014, 102). Moreover, I have also argued[14] that even when we place our focus at the object level, we, as Suppes noted, reach an impasse. That is, without a proper *theory of the data*, which would allow us to connect the structure of *the phenomena* to the structure of *the world*, the best a structural realist can claim is that all that exists is expressed by the structure of *this* theory in *this* context

---

[13] See Landry (2007, 2012). [14] See Brading and Landry (2006); and Landry (2012).

given *this* methodological appeal to *this* mathematical structure. Thus, the position I have advanced is *methodological structural realism*.[15]

If, then, we appreciate that structural realists ought to focus on the object level of scientific practice and not on the meta-level of the philosopher of science, then we can better see where the mistakes may lie for those who simply want to replace talk of logical or set-theoretic meta-level formal frameworks with a category-theoretic one. While there might be good reasons, at the meta-level, to suppose that category theory can offer a better formal framework for expressing the *structure of scientific theories*,[16] this use is *distinct from* that needed by the structural realist, who must offer a framework, at the object level, for expressing the *structure of the world*. Thus, simply replacing either a logical formal framework, as in the syntactic approach, or a set-theoretic formal framework, as in the semantic approach, with a category-theoretic one is, for the structural realist, a category mistake.[17]

18.3 Category Theory as a Framework for the Structure of a Physical Structure

Bain (2013), seeking to "defend radical ontic structural realism (ROSR) against the charge that it rests on an incoherent claim; namely, that there can be relations devoid of *relata* in the physical world", argues that category theory provides a better formulation of the (ROSR) claim that "structure exists independently of objects that may instantiate it" (Bain, 2013, 1621). His argument is twofold: (1) category theory provides a more general notion of *mathematical structure* than does set theory, and (2) specific examples of *fundamental* physical theories show that this claim is "incoherent under a notion of *physical structure* informed by set theory, but not . . . by category theory" (Bain, 2013, 1621; italics added). Claim 1 has two components: (a) aims to show that *contra* those that hold that there is a *conceptual dependence* between relations and relata,[18] category theory, or more specifically, the category *Set* as framed topos-theoretically, *makes conceptually coherent* the claim that one can speak of relations without relata, and (b) aims to show, more generally, that one can use the "definition of structure as an object[19] in a category" to *make conceptually coherent* the claim that one can speak of structure existing independently of objects (Bain, 2013, 1634). Bain next seeks to bring these two components of Claim 1 together to underpin his own category-theoretic version of ROSR, and so to provide justification for Claim 2. He says:

---

[15] Again, see Brading and Landry (2006); and Landry (2012).
[16] See Weatherall, Chapter 13, and Halvorson and Tsementzis, Chapter 17, both this volume.
[17] See Landry (2007, 2011b).
[18] See French (1999, 2000, 2014); and French and Ladyman (2003).
[19] Note that by "object" Bain seems to mean an object in the category-theoretic sense. I will take-up the question of what he means in the final section. I will indicate such a category-theoretic use by 'object'.

[t]o guarantee that this mathematical fact [that category theoretic-objects need not be set-structured] has *physical significance*... [t]he category theoretic ROSRer [must] provide examples of "non-structured set" objects that have a role to play in articulating the relevant notions of structure in physics. (Bain, 2013, 1625; italics added)

Similar, then, to French's ontic structural realist conclusions from considerations of the role of group theory in quantum mechanics, Bain's claim is that category theory provides a better account of the *physical structure of the world*, as represented by the *mathematical structure* of a *fundamental* theory of physics. It is by use of *two examples* from fundamental physics that Bain seeks to underwrite the central components of Claim 2. The first example is the use of category-theoretic formulation of sheaves of Einstein algebras (EAs) of general relativity (GR), which "eliminates manifold points but retains global differential structure" and so shows that the *physically relevant* structure, e.g., global differentiable structure, cannot be predicated, on space–time point-correlates, i.e., manifold points, which serve as a domain of set-theoretic *relata* (as in the tensor formulation (TF) of GR), so we have a case of freestanding physical structure, without set-structured objects (Bain, 2013, 1626).

The second example considers the use, in topological quantum field theories (TQFT), of the categories *nCob*, with $(n-1)$-dimensional compact-oriented manifolds as 'objects' and n-dimensional oriented cobordisms as 'morphisms', and **Hilb**, with finite-dimensional Hilbert spaces as 'objects' and bounded linear operators as 'morphisms'. Both categories are, likewise, appealed to to underwrite the claim that "non-set-structured"[20] objects play a role in fundamental physics. And because this structure is "more general", as Bain (2013) claims, it can be used "as a method of reconciling the background independent nature of GR with quantum field theory" (Bain, 2013, 1631). Bain concludes, from these two examples, that ROSR is *physically* justified in its claim that structure exists independently of the (set-structured) objects that instantiate it, so that set-structured objects may be *eliminated*.

But, according to Bain, there yet remains work to be done, physically, mathematically, and representationally. *Physically*, "the category-theoretic ROSRer should also provide additional category theoretic reformulations of theories in physics that explicitly do not depend on structured sets" (Bain, 2013, 1632). *Mathematically*, we are left to face the following concern:

[o]ne might also argue that the generality afforded by the category-theoretic definition of structure is a moot point if it turns out that category theory presupposes set theoretic concepts
(Bain, 2013, 1631).

---

[20] For example, as Bain notes, their objects cannot be considered as set-structured since: (i) their morphisms are not functions, (ii) they are monoidal category so do not admit Cartesian products, and (iii) they are *-categories (Bain, 2013, 1630).

So the advocate of ROSR is left with yet another task: "[s]he should provide a rationale for the *fundamentality* of category theory over set theory" (1631; italics added). And finally, *representationally*, the ROSRer must note that

[t]he fundamentality of category theory would be a moot point if it turned out that the majority of structures in the physical world of physical relevance are better represented by set-theoretic constructions. (Bain, 2013, 1632)

I will take up the mathematical concern in the next section. In what remains of this section, I will show that Bain's physical examples are at the *wrong level* of analysis to be of any *representational* use by the structural realist. More pointedly,[21] as expressed by Esfeld and Lam (2008), the *representational concern* for the ontic structural realist is that

it is one thing to demonstrate that, *from a formal point of view*, category theoretic *representations of structure* are independent of *relata*, and another thing to argue that *structures in the physical world* are likewise independent of *relata*. (Esfeld and Lam, 2008, 1633; italics added)

To further appreciate the full impact of this concern, I turn to the recent work of Lam and Wüthrich (2014), who counter Bain (2013) with three possible objections: *logical* or *set-theoretic*, *physical*, and *metaphysical*. Again, I will consider the details of their logical/set-theoretic objections in the next section. In regards to their *physical objection*, Lam and Wüthrich claim that Bain has failed to show that a category-theoretic formulation of fundamental physical theories has the "physical significance"[22] needed by the ROSRer. I agree. As I will show, physical motivation at the *theoretical level* is *not* the same as physical significance at the *object level*.[23] Thus, in light of Esfeld and Lam's representational concern, Bain's arguments, in so far as they all concern physical motivation and *not* physical significance, are of little help to the structural realist.

Lam and Wüthrich next consider, on a suggestion of Pooley,[24] whether Bain's ROSR claims could be justified *metaphysically*. They hold, however, that Bain fails to consider such an argument. But, Bain *does* attempt just such a metaphysical argument and, indeed, does so with the aim of quelling both Esfeld and Lam's *representational* concern

---

[21] Bain considers two such representational concerns, one at the level of the theory of measurement and the other, as I take Esfeld and Lams to be, at the level of the theory of the phenomena.

[22] See Lam and Wüthrich (2014, 9, 17, 18, 21–2, 25).

[23] The intended difference between physical *motivation* and physical *significance* is that the former is at the *theoretical level*, that is, the level that considers the appropriateness of the *mathematical representation* of the *theoretical structure* of a physical theory. Such considerations may include the theoretical unity, simplicity, etc., of the structure of the physical theory. Physical significance, by contrast, is at the *object level*, that is, the level that considers the appropriateness of the *physical representation of the structure of the phenomena* of the world. Such considerations may include the (approximate) truth or empirical adequacy, or the predictive or explanatory value, of the physical theory. See Brading and Landry (2006) for more on the distinction between mathematical representation (which we there call mathematical *presentation*) and physical representation, and for the implications of this distinction on structural realism.

[24] See Lam and Wüthrich (2014, fn 30, p. 22).

and the *metaphysical* concern mentioned by Lam and Wüthrich.[25] Bain's ROSR argument, then, is not run either mathematically or physically; it is run *metaphysically*. As he clearly states, it is intended as an inference from "Jones underdetermination",[26] *plus* semantic realism *and* a naturalized metaphysical stance:

> [t]he current essay thus motivates ROSR by examples of different formulations of a single theory that disagree... over the ontological status of objects... The inference to ROSR from Jones underdetermination depends on some form of semantic realism with respect to theories in physics (i.e., it assumes that we should take the claims made by theories in physics at their face value). And it assumes a *naturalistic approach to metaphysics*; one in which metaphysical commitments are informed by contemporary theories in physics... [i]t suggests that the manner in which some contemporary theories in physics *represent* natural phenomena is in terms of structure devoid of relata; and it suggests that we should *take these representations at their face value* and accommodate them into what we take to be the ontology of the world.
> (Bain, 2013, 1634; italics added)

However, one important category mistake remains, and it's one that, I believe, results in Bain's failure to even address Esfield and Lam's representational concern and, as a result, entirely undermines his argument for ROSR. The mistake is this: Bain has conflated the *theoretical level* of the *mathematical representation* of the structure of a *physical theory*, with the *object level* of the *mathematical representation* of the structure of *physical phenomena*. Given his semantic realism and his naturalized metaphysical stance, his ROSR argument might yet be saved *if* it can be shown that the mathematical structure that does the *actual physical work*, i.e., does the object-level work of mathematically representing the *physical structure of the phenomena* of fundamental physical theories, is category-theoretic. But he has not shown this here.

While Bain's examples of the category-theoretic structure of theories of fundamental physics are, indeed, *physically motivated*, i.e., motivated at the *theoretical level* by considerations of the mathematical structure of the *physical theory*, it remains far from clear whether they are *physically significant*, i.e., motivated at the *object level* by considerations of the mathematical structure of the *phenomena*. The only motive Bain gives for preferring sheaf-theoretic EAs is that they "provides a more unifying description of phenomena of GR" (Bain, 2013, 1629), but ROSR is indifferent to *theoretical unification* and it's far from clear that "spacetimes of constant curvature and solutions involving certain types of curvature singularities" (Bain, 2013, 1627), which is what sheaf-theoretic EAs unify, count as "physical phenomena" (Bain, 2013, 1629).

---

[25] As Lam and Wüthrich (2014) note: "given [the naturalized metaphysicians] insistence that our metaphysics be scientifically informed, the radical [structural realist] can argue [against the metaphysical objection] that our naturalized metaphysics is subject to sometimes deep and counterintuitive revision, much like science" (6). See also Chakravartty (2003) for the claim that the "no relations without relata" presumption, whether presumed physically or metaphysically, simply begs the ROSR question (871).

[26] For the distinction between metaphysical underdetermination (which, as noted, French uses to reach his ROSR conclusion that all there is to quantum mechanical particles is Lie-group structure) and Jones underdetermination, see Bain (2013, 1633-4).

Thus, as noted by Lam and Wüthrich (2014), "such a generalization is only relevant if the space-times with the boundaries it incorporates have *physical significance*. The justification for this significance, unfortunately, is rather mute" (21; italics added).

Likewise with Bain's TQFT example; at least at this stage, TQFT is viewed "as a method of reconciling the background independent nature of GR with quantum field theory" (Bain, 2013, 1631), and so, as noted by Lam and Wüthrich (2014), the category-theoretic representation may be used to present TQFT as but a "tool or toy model" (19). That is, the use of ***nCob*** and ***Hilb*** are, again, only *physically motivated* at the *theoretical level*, and so any ROSR argument based *solely* at this level of analysis lacks the claim of being *physically significant* at the *object level*. Again, as noted by Lam and Wüthrich (2014), while the use of category theory in TQFT "does provide useful mathematical tools, in particular allowing us to build mathematical 'bridges'—the functors—between different space-times and quantum field theory . . . [this] does not entail that there are no physical objects" (19).

Thus, with respect to Bain's Claim 2, I conclude with an analogy. I have argued that French commits a category mistake, by conflating the meta-level of philosophy of science with the object level of scientific practice, so that there is no quick ROSR route from his use of a set-theoretic partial-structures account of the shared *mathematical structure* of *scientific theories*, to the *structure of the world*. Likewise, Bain commits a category mistake by conflating the theoretical level with the object level, so that there is no quick ROSR route from a category-theoretic *mathematical representation* of the shared (unified or simplified) *theoretical structure* of a fundamental physical theory to a *mathematical representation* of the *physical structure of the phenomena* of the world. So, in the end, it is unclear how Bain can use "Jones underdetermination" to assuage Esfeld and Lam's representational concern, that is, to do the *representational work* that would connect the mathematical structure of *the theory* to the mathematical structure *of the phenomena* and so license the inference to any ROSR claim about the structure of the world. Of course, this is *not* to say that category theory does no work in representing the *theoretical structure* of a physical theory,[27] but the ROSRer, in so far as he intends to make claims about the *physical structure* of the world, needs to additionally focus on connecting the theoretical structure of the theory to the physical structure of the phenomena.

18.4 Category Theory as a Conceptual Tool

I now turn to consider Bain's Claim 1. Recall that this claim has two components: (a) aims to show, *contra* those that hold that there is a *conceptual dependence* between relations and relata, that category theory, or more specifically the category *Set*, framed

---

[27] Indeed, it would seem that, at least at this point, the representational value of category theory is to be found at the theoretical level. See, for example, Lal and Teh's (2017) analysis of the role of category theory in physics and its relation to physical structuralism.

topos-theoretically, *makes conceptually coherent* the claim that one can speak of relations *without* relata, and (b) aims to show, more generally, that one can use category theory to *make conceptually coherent* the claim that one can speak of "a structure as an object in a category" (Bain, 2013, 1623) *without* set-structured objects (Bain, 2013, 1629).[28] Thus, together, these components are used to *conceptually* justify the ROSR claim that structures (and set-structure as defined in *Set*) exist independently of any objects that instantiate them, including set-objects *qua* elements as defined in *Set*.

As we will see, given his category-theoretic mistakes, it is far from clear whether Bain (2013) has succeeded in his arguments for Claim 1. As regards (b), he has confused and conflated the various category-theoretic levels, and so too the various types of category-theoretic structured systems. Moreover, as I will show, it simply makes no sense to say that a structure *is*, let alone is *defined* as, "an object in a category". Finally, with respect to both (a) and (b), *both* Bain, and Lam and Wüthrich have not fully appreciated the *in re* Hilbertian interpretation of the category axioms, which *does allow* one to *fully eliminate* any talk of objects or elements[29] that *make up* systems that are so structured. It is in this sense, then, that I claim, *contra* Psillos et al., and to quell both Bain and Lam and Wüthrich's *mathematical concern* about the *fundamentality* of set theory, that category theory *can* be used as a *conceptual tool* to organize what we say about systems that have a structure, wherein the 'objects' of such systems are to be taken as *nothing but* positions in a system that has a structure. Moreover, against Lam and Wüthrich, I will also show that this version of category-theoretic mathematical structuralism *can* be used to *conceptually collapse* the distinction between ROSR and their preferred BOSR. Simply, on an *in re* Hilbertian category-theoretic interpretation of mathematical structuralism, there are no *identity conditions* for 'objects'; 'objects' are nothing but positions in a system that has a structure.

Before we proceed, we must note that there are at least three levels for characterizing category-theoretic types of structure and these must be distinguished:

---

[28] Lam and Wüthrich (2014) conflate these components; they hold that Bain (2013) aims to show that "'the definition of a structure as an object in a category does not make ineliminable reference to relata in the set-theoretic sense' (2013, 1623)" (12). But this sentence does not exist in Bain. On this page Bain is using the category *Set* to talk about eliminating relata *qua* element; he is not talking about using the more general definition of a structure to eliminate talk of set objects *qua* elements. Lam and Wüthrich mistakenly assume that Bain's strategy is to use *Set* to argue for (b); this reading would collapse the distinction between components (a) and (b) of Bain's Claim 1. Moreover, it is this mistake that allows them to argue that Bain, in his attempt to eliminate relata, has thrown the baby (relations) out with the bathwater (relata). This is because in *Hilb* and *nCob*, which are not Cartesian closed, relations cannot be defined. But what Bain is trying to show by the use of *Hilb* and *nCob* is that one can speak of structures without objects as set elements.

[29] Throughout their paper, Lam and Wüthrich speak of the "elements of the objects of a category" (Lam and Wüthrich, 2014, 11). Indeed, such talk seems to arise from their own confusion with respect to the definition of a category. See, for example, their definition of a category, wherein 'morphisms' are claimed to be "between the elements of a collection", and where "A category, C, consists of a pair of *classes*" (Lam and Wüthrich, 2014, 10). But later they also claim that "if relational at all, the morphisms are relations between objects, not elements of objects" (Lam and Wüthrich, 2014, 15). Not only are these claims confusing but also, as I will show, any talk of elements of objects is problematic.

(1) The EM (Eilenberg–Mac Lane) axioms[30] implicitly define a category in terms of its 'objects' and 'arrows'. For example, *Set* characterizes set structure in terms of 'set' and 'function', *Grp* characterizes group structure in terms of 'group' and 'group homomorphisms', *Ded* characterizes proof structure in terms of 'formulas' and 'rules of inference', *Hilb* characterizes Hilbert-space structure in terms of 'finite-dimensional Hilbert spaces' and 'bounded linear operators', *nCob* characterizes manifold-structure in terms of '$(n-1)$ dimensional compact oriented manifolds' and 'n-dimensional oriented cobodisms', etc. Note, then, that the EM axioms *do not* tell us what a structure *is*; they characterize *systems that have a type of cat-structure* in terms of 'objects' and 'arrows'.

(2) The ETCS axioms, as given, say, by McLarty (1992), characterize the *internal set structure of Set*; that is, they logically characterize the structure of the 'objects' and 'arrows' *from within Set*, by offering a topos-theoretic account of set structure, in terms of the 'object' set and the 'arrow' function. As noted in both Bain and Lam and Wüthrich, unlike the usual set-theoretic formulations, where membership is primitive,[31] set-theoretic *global elements* are themselves characterized in *Set* as category-theoretic 'arrows'. That is, in a well-pointed topos, which is both Cartesian closed and so has terminal objects, a *global element* of an 'object' in *Set*, i.e., of a set in *Set*, is defined as an 'arrow' from the terminal object.

What is important, however, is that while element *qua* object, or bare-thing, might traditionally be appealed to in answering what sets are *made of*,[32] in *Set*, elements, as distinct from 'arrows' from the terminal objects, are *not essential* for characterizing what a set *is*; a set *is* an 'object', a global element, or set element, *is* an 'arrow'. Thus, in

---

[30] A cat-structured system C (a category) is an abstract system of two abstract kinds; objects $X, Y, \ldots$ and arrows $f, g, \ldots$ such that
  (a) Each arrow f has an object X as a *domain* and an object Y as a *codomain*, indicated by writing $f : X \to Y$.
  (b) If g is any arrow $g : Y \to Z$ with domain Y (the codomain of f) and codomain Z, there is an arrow $h = g \circ f$ called the *composition* of f and g.
  (c) For each object X there is an arrow $1_X : X \to X$ called the *identity* arrow of X.
  (d) These objects and arrows satisfy: (i) *Associativity*: $f \circ (g \circ h) = (f \circ g) \circ h$, (ii) *Identity*: for all X, the domain of $1_X x =$ codomain of 1_X and for all $f, f \circ 1_x = f, 1_y \circ f = f$.

[31] Note that while Zermelo–Fraenkel theory with choice (hereafter ZFC) takes membership as primitive, that this *does not* imply that sets are *made up* of elements, nor does it entail an "ineliminable reference to relata [*qua* element]" as Bain (2013, 1623) claims. Both of these claims result only if one is talking about a theory of sets formalized by ZFC *plus* urelements.

[32] Again note that this is *not* a presumption of set theory itself. For example, in ZFC, this is not, as Lam and Wüthrich (2014, 7) claim, a result of the axiom of extensionality (which, in ZFC, need only be taken as an axiom about sets). It *is* a result of a *urelement* account of set, which, while this might be consistently added to ZFC, need not be. Thus, while it might be the case that a set-theoretic formulation, e.g., a formulation in terms of ZFC, of set relations makes ineliminable reference to set relata, it is *not* the case, again as Lam and Wüthrich claim, that "the standard way of modeling the extensional understanding of the central ROSR notions of relation and structure in set theory makes ineliminable reference to the elements of the sets involved in the definitions" (Lam and Wüthrich, 2014, 6). This might be the way of Suppes (1972), who *did* presume an urelement account of set, and, as a result, this might be the account that has underpinned the way in which philosophers of science speak of relations and structure, but it need not be.

so far as, in a topos, the set-theoretic notion of a *relation* is given in terms of subsets of Cartesian products, then, as Bain rightly claims, we *do* have a characterization of the notion of *set relation* given in terms of *set relata*, i.e., in terms of global elements *qua* 'arrows', *without* requiring the notion of set relata *qua* element. Thus, and again in so far as the set-theoretic notion of a *structure* is given in terms of set relata and set relations, i.e., a set-structure $\langle D, R \rangle$ is characterized by an isomorphism class of sets structured by R, then, again, as Bain rightly claims, we *can* characterize the notion of a *set structure*, without requiring the notion of set-relata *qua* element. So bringing us back to the logical or set-theoretic objection of Lam and Wüthrich, Bain *has* shown that topos-theoretically it is *conceptually possible* to use **Set** to characterize *both* set relations and set structure *without* set-relata *qua* elements.

But neither the EM nor ETCS axioms get us to Bain's more general definition, "a structure is an object in a category". One thing he might have in mind is that a type of cat-structure can be itself be characterized as an 'object' in a category. This brings us, then, to our third level of characterizing category-theoretic types of structure.

(3) The category of categories as a foundation (CCAF) axioms, as given by, say, Lawvere (1966), characterize *types of cat-structure*, or *the structure of structured systems that are themselves categories*. That is, in **Cat** the 'objects' are categories and 'arrows' are functors. So, for example, **Set**, **Grp**, **Ded**, **Sh(X)**, **nCob**, and **Hilb** can be taken as 'objects' in **Cat**. But, again, this does not get us to "a structure is$_{df}$ an object in a category"; rather, it gets us to the claim that a category can be characterized as an 'object' in **Cat**. And, even if this is what Bain intends, as he notes, there are size problems with using **Cat** in this definitional way (Bain, 2013, 1632).

Before I move to my next point, I consider yet another possibility of what Bain might mean by "a structure is an object in a category". Lawvere (1963) showed that the structure of any algebraic (or equational) theory can be characterized category-theoretically, so that its models are functors. This result was then extended to any first-order theory.[33] Thus, we have a category-theoretic characterization of *the structure of a theory* in terms of its models, without having to invoke a problematic semantic/syntactic distinction and without requiring a logical or set-theoretic background or fundamental theory. Lam and Wüthrich seem to consider this interpretation (but they too conflate and confuse the various category-theoretic levels), and yet they note that on this reading:

the category as a whole articulates the theory, and its objects *represent the models of the theory*, not physical objects in their own right. (Lam and Wüthrich, 2014, 11; italics added)

I agree. As I claimed of French's set-theoretic partial structures approach, Bain must be careful not to confuse the *meta-level of philosophy of science*, which considers the *mathematical structure of scientific theories* and the object-level of scientific practice,

---

[33] See, for example, Makkai and Reyes (1977). For the historical details of this development, see also Marquis and Reyes (2012).

which considers the *physical structure of the world*. Thus, as Lam and Wüthrich rightly point out, the *physical claims* of ROSR must not be underwritten by mathematical structuralism (Lam and Wüthrich, 2014, 17).

But does mathematical structuralism have anything to offer ROSR at the *conceptual level*? That is, can it be used to make *conceptually coherent*, or conceptually frame, Bain's ROSR claim that "structure exist independently of objects that may instantiate it" (Bain, 2013, 1621), and so objects (*qua* sets, elements, things) are *entirely eliminated*, i.e., not eliminated in "name only" (Bain, 2013, 1625), as Bain considers, or "merely relabeled" as Lam and Wüthrich worry? The question before us, then, is this: Can category theory be used as a *conceptual tool* to show that one can talk of *systems* that *have a type of structure*, *without* our having to concern ourselves with what these systems *are*, or are *made of*, be these sets, elements, or bare-things? The answer is: Yes! Indeed, as I, and others[34] have repeatedly argued, it is precisely in this sense that an *in re* Hilbertian interpretation of category theory provides the best foundation[35] or mathematical structuralism. Moreover, and to display its conceptual bite, a simple Hilbert-inspired, category-theoretically framed, argument can be used to *conceptually collapse* the distinction between eliminating and relabeling, and so collapse the distinction between ROSR and Lam and Wüthrich's "more balanced" BOSR, *without* any foundational route through set theory or any other *fundamental* theory.

Lam and Wüthrich claim that what is at stake between ROSR and BOSR is that ROSR eliminates objects and BOSR retains them, but as "thin" (Lam and Wüthrich, 2014, 3), e.g., as perhaps "merely relabelled" (11). What, then, is this thin notion of object at the heart of BOSR? All we are told by Lam and Wüthrich is that

[t]he picture typically offered [by BOSR] is that of a balance between relations and their relata, coupled to an insistence that these relata do not possess their identity intrinsically [as Russell or Psillos would have it], but *only by virtue of occupying a relational position* in a structural complex.
(Lam and Wüthrich, 2014, 3; italics added)

But this does not distinguish BOSR from ROSR; indeed, it is just the familiar, and shared, structuralist slogan that objects are *positions* in a structure. What distinguishes them is that the ROSRer adds that objects so considered are *nothing but* positions in a structure.

---

[34] See Landry and Marquis (2005); Marquis (2006); and Landry (2011a, 2013).

[35] Indeed, Bain (2013, in fn 4, p. 1624), quotes Awodey (2004, 29), as expressing just this view, viz., "The idea is that objects and arrows are determined by the role they play in the category via their relations to other objects and arrows, that is, by their position in a structure and not by what they 'are' or 'are made of' in some absolute sense." The difference, however, between my *Hilbertian* interpretation and Awodey's *schematic* reading of the category axioms (see Awodey, 2004) is that Awodey limits the term 'foundation' to those *constitutive* accounts of "what things are or are made of", so *does not* accept category theory, schematically construed, as a foundation in this sense. I, however, use Hilbert to offer an *organizational* account of foundation and use this to rationally reconstruct an Hilbertian view of category theory as a foundation in the organization sense. That is, as I take category theory as a language used to organize what we say about *systems* that *have a type of structure*, *without* our having to concern ourselves with what these systems *are*, or are *made of*.

Shifting, then, from talk of relations and relata to talk of structures and objects, more generally, where does the difference lie between ROSR and BOSR? Lam and Wüthrich (2014) tell us that the ROSR, by *eliminating* objects, escapes having to confront "any of the known headaches concerning *the identity* of these objects" (3–4; italics added), whereas the BOSR does not. Let's now consider, from a philosophy of mathematics perspective, what it might mean to say that the BOSR still must face "known headaches concerning the identity of these objects". There are two standard replies: the *set-theoretic foundationalist* and the *philosophical mathematical structuralist*. The first holds that once we decide which set theory is *the* set theory and so *the* foundation, this would rule out such multiple interpretations and so provide *the* system or *the* "identity conditions for numbers as sets". Thus, when forced to face "known headaches concerning the identity" of numbers *as* sets, we are forced to claim that set theory is *fundamental* and so conclude that numbers *are* sets, and sets, or what they are *made of*, viz., elements, are ontologically prior.

The philosophical mathematical structuralist, in contrast, replies that numbers just *are positions* in a structured system. Yet, again feeling forced to face "known headaches concerning the identity of numbers *as positions*", they offer two ways of understanding 'position'. The *ante rem* structuralist, like Shapiro (1997),[36] says that positions are to be taken as instances of *places* in a *unique* system *qua* abstract platonist structure, and then goes on to provide structure-theoretic axioms for a *fundamental* background theory that founds the *actuality* of such structures. The *in re* structuralist, Hellman (1989)[37] for example, says that 'position' refers to a position in any or all *possible systems* that have the same structure. And then provides a modal nominalist *fundamental* background theory that founds our talk about the *possibility* of such systems, wherein possible systems are claimed to be *made up* of possible concreta.

The underlying worry, then, for the ontic structural realist, as expressed, for example, by French (2014), is whether all such *in re* talk of objects as *nothing but* positions in systems that have a structure commits us to the view that some *primitive system of things* must be presumed as *fundamental* and so as either *modally* or *ontologically prior*. Clearly, this is a problem for any ROSRer, like French and Bain, who wants to *eliminate* talk of fundamentally prior things. Interestingly, a similar objection was raised by Russell (1903) against Dedekind's (1888) structuralist account of natural numbers:

> it is impossible that the ordinals should be, as Dedekind suggests, *nothing but* the terms of such relations as constitute a progression. If they are to be anything at all, *they must be intrinsically something; they must differ from other entities* as points from instants, or colours from sounds. What Dedekind intended to indicate was probably a definition by means of the principle of abstraction ... But a definition so made *always indicates some class of entities having ... a genuine nature of their own*. (Russell, 1903, 249; italics added)

---

[36] See also Shapiro (1996). [37] See also Hellman (1996, 2001).

But what both French and Russell fail to realize is that neither Dedekind nor Hilbert, nor the *in re* Hilbertian interpretation of category-theoretic mathematical structuralist, have what Reck (2003) has aptly termed "metaphysical/semantic concerns".[38] At the heart of these concerns is the Fregean metaphysical/semantic presumption that structured systems must be constituted from or refer to some primitive system of things, which can be elements, sets, platonic places, or nominalist concreta, so that axioms, as truth about such primitive systems, must fix a domain which itself will provide *identity conditions* for the *positions* in structured systems.

However, what Russell, French, and both set-theoretic and philosophical mathematical structuralists ought to have learned from Dedekind (1888) and Hilbert (1899), respectively, is that we are to "entirely neglect the special character of the elements", so axioms are but *implicit definitions*, and, consequently "every theory is only a *scaffolding or as schema of concepts* together with their necessary relations and the basic elements *can be thought of in any way one likes*" (Hilbert, 1899, 41). Simply, if axioms are implicit definitions, no fundamental, ontologically or modally primitive, system of things is necessary. This shows why, to quell Bain's and Lam and Wüthrich's mathematical concern, an *in re* Hilbertian interpretation of the category axioms does not need either set theory or indeed any theory as a *fundamental* foundation. Thus, one need not be concerned that relata *qua* set elements are eliminated in *Set* "in name only" or "merely relabeled"; they are *nothing but* 'arrows'. And, more generally, 'objects' that are structured by the axioms are *nothing but* positions in systems that have a type of cat-structure, *full stop*.

Perhaps, then, this is what Bain intended (but ought to have explained with more detail) by "a structure is an object in a category". But note that, as Hilbert-inspired, category-theoretic *in re* structuralists, we talk of *systems* that *have a structure* and not of *structures*. Category theory cannot *explicitly define* what structures *are*. Bain's slogan, then, is more precisely expressed as "category theory *implicitly defines* (in the Hilbertian 'schematizes' sense) systems that have a type of cat-structure, in terms of 'objects' and 'arrows', where such objects and arrows are *nothing but* positions in a system that has that type of cat-structure, i.e., *without* our having to further consider what they *are*, or are *made of*". Thus, category theory *can* be used as a *conceptual tool* to answer how we can *coherently speak* of relations *without* relata *qua* set elements, and, more generally, structures *without* objects *qua* set elements.

Now, as promised, let's recast the supposed distinction between BOSR and ROSR in our Hilbert-inspired category-theoretic terms. ROSR and BOSR say we can speak of objects as *positions* in a system that has a structure. ROSR adds that in so doing we have thus *eliminated* reference to objects, and so can speak of objects as *nothing but* positions in a system that has a structure. BOSR adds that we have merely eliminated such objects in *name only*, or *merely relabeled* objects, and are thus left with a *thin* notion of object, so we still must face questions of their *identity conditions*. But both

---

[38] See Reck (2003) for a deep and thoughtful account of Dedekind's structuralism along these lines.

ROSR$_{ct}$ and BOSR$_{ct}$ say that, in characterizing the structure of a system by *in re* Hilbert-interpreted category-theoretic axioms, we can speak of an 'object' as *nothing but* a position in a system that has a structure. That is, by schematizing what we say about such objects by the category-theoretic axioms, we thereby *eliminate* reference to fixed domains and so to Fregean objects, and so need not face metaphysical/semantic questions of their identity conditions, i.e., questions of what they are *made of* or *refer* to.

18.5 Conclusion

I have shown, then, that at a mathematical level, a Hilbert-inspired category-theoretic *in re* version of mathematical structuralism *can* be used to *conceptually collapse* the distinction between ROSR and BOSR. It might be the case, however, that *physically* the distinction between ROSR and BOSR cannot be so collapsed. And if, like Bain and French, one adopts a *naturalized metaphysical* stance, wherein one infers ontological commitment from the *physical structure* of a fundamental theory, then it might turn out that physically, and so, *metaphysically*, ROSR comes out ahead of BOSR, as too it might well turn out, as Brading and Landry (2006) argue, that neither view is tenable. But these claims must be argued for *physically*, not mathematically or conceptually. As Chakravartty (2003) notes, it is not enough to simply *metaphysically* presume, as Russell and Psillos do, that nature is structured by either ontologically primitive relata or "intrinsic objects". Likewise too it would not be enough to simply *metaphysically* presume, as French seems to, that nature is structured by nothing but structure.[39] It is also not enough to *representationally* presume, as Bain seems to, that, at an object level, the phenomena of nature are structured by the theoretical-level mathematical structure of a physical theory. And, finally, it is not enough to simply *mathematically* presume that relations or "structures" (i.e., systems) are structured by set-theoretic relata or objects *qua* elements.[40] Under a category-theoretic conceptualization, we can show that one can coherently speak of *systems* that *have a type* of *structure*, without our

---

[39] I note here that in addition to naturalizing the metaphysics of physical structure, French (2014) also goes on to use this scientific structuralist approach to construct a more general structuralist account of metaphysics itself; that is, he adopts an ontic structural realist "Viking Approach" to metaphysical issues relating to notions of causality, modality, dispositions, laws, etc. The concern, however, is by so extending, French has opened himself to the objection that one's *scientific* metaphysics is itself but a choice of perspective. That is, French has indeed shown that the bottom-up Russellian approach, which "smacks of the pernicious influence of object-oriented metaphysics" (French, 2014, 205), is ill suited to capture the "core idea that the entities we are ... concerned with [in physics] have no identity or distinguishing features beyond those conveyed by the structure" (French, 2014, 20), *but* what he has not shown is that his ROSR top-down approach, which might equally be said to "smack of the pernicious influence of *structure-oriented metaphysics*", fares any better at explaining why we should presume that it is structure and not objects that provides the carving tool for our analyses of metaphysical notions like causality, modality, etc. And without such an explanation it is not clear that the structure-oriented realist and the object-oriented realist are not merely disagreeing about "a choice of perspective". I thank an anonymous referee for pushing me to make this point.

[40] This explains why the *logical* or set-theoretic objection, is, as Lam and Wüthrich (2014, 5) note, likewise, mute.

having to concern ourselves with what these systems *are*, or are *made of*, be these sets, elements, or bare-things. Thus, *at least conceptually*, the ROSR's dilemma, as presented by Lam and Wüthrich, of either "softening to a more moderate realist position [BOSR] or else [developing] the requisite alternative conceptualization of relations and of structures" (Lam and Wüthrich, 2014, 3), dissolves in favor of the second horn.

References

Awodey, S. (2004). An answer to hellman's question: "does category theory provide a framework for mathtematical structuralism?" *Philosophia Mathematica* 13(1), 54–64.

Baez, J. (2006). *Quantum Quanaries: A Category-Theoretic Perspective*. Oxford University Press, Oxford.

Bain, J. (2013). Category-theoretic structure and radical ontic structural realism. *Synthese* 190(9), 1621–35.

Brading, K., and Landry, E. (2006). Scientific structuralism: presentation and representation. *Philosophy of Science* 73(5), 571–81.

Busch, J. (2003). What structures could not be. *Philosophy of Science* 17(3), 211–25.

Cao, T. Y. (2003). Can we dissolve physical entities into mathematical structures? *Synthese* 136(1), 57–71.

Carnap, R. (1966). *Philosophical Foundations of Physics*. Basic Books, New York.

Chakravartty, A. (2003). The structuralist conception of objects. *Philosophy of Science* 70(5).

da Costa, N. C., and French, S. (2003). *Science and Partial Truth: A Unitary Approach to Models and Scientific Reasoning*. Oxford University Press, Oxford.

Dedekind, R. (1888). The nature and meaning of numbers.

Demopoulos, W., and Friedman, M. (1985). Bertrand Russell's the analysis of matter: its historical context and contemporary interest. *Philosophy of Science* 52(4), 621–39.

Dorato, M. (2000). Substantivalism, relationism, and structural space-time realism. *Foundations of Physics* 30(1), 1605–28.

Esfeld, M., and Lam, V. (2008). Moderate structural realism about space-time. *Synthese* 160(1), 27–46.

French, S. (1999). Models and mathematics in physics: the role of group theory, in J. butterfield and C. Pagonis (eds), *From Physics to Philosophy*, 187–207. Cambridge University Press, Cambridge, UK.

French, S. (2000). The reasonable effectiveness of mathematics: partial structures and the application of group theory to physics. *Synthese* 125, 103–20.

French, S. (2014). *The Structure of the World: Metaphysics and Representation*. Oxford University Press, Oxford.

French, S., and Ladyman, J. (2003). Remodelling structural realism. *Synthese* 136(1), 31–56.

Hellman, G. (1996). Structuralism without structures. *Philosophia Mathematica* 4(2), 100–23.

Hellman, G. (2001). Three varieties of mathematical structuralism. *Philosophia Mathematica* 9(2), 184–211.

Hilbert, D. (1899). *Grundlagen der Geometrie*. Open Court, Chicago.

Ladyman, J. (2014). Structural realism, in E. N. Zalta (ed.), *The Stanford Encyclopedia of Philosophy*, spring edn. http://plato.stanford.edu/archives/spr2014/entries/structural-realism/

Lal, R., and Teh, N. (2017). Categorical generalization and physical structuralism. *British Journal for the Philosophy of Science* 68(1), 213–51.
Lam, V., and Wüthrich, C. (2014). No categorial support for radical ontic structural realism. *British Journal for the Philosophy of Science*, 1–30.
Landry, E. (2007). Shared structure need not be shared set-structure. *Synthese* 158(1), 1–17.
Landry, E. (2011a). How to be a structuralist all the way down. *Synthese* 179, 435–54.
Landry, E. (2011b). A silly answer to a Psillos question, in *Vintage Enthusiasms: Essays in Honour of John Bell*, 361–81, Western Ontario Series for Philosophy of Science. Reidel, Dordrecht.
Landry, E. (2012). Methodological structural realism, in E. Landry and D. Rickles (eds), *Structure, Objects, and Causality*, 29–59, Western Ontario Series for Philosophy of Science. Reidel, Dordrecht.
Landry, E. (2013). The genetic versus the axiomatic method: responding to Feferman 1977. *Review of Symbolic Logic* 6(1), 24–50.
Landry, E., and Marquis, J.-P. (2005). Categories in context: historical, foundational, and philosophical. *Philosophia Mathematica* 13(1), 1–43.
Laudan, L. (1981). A confutation of convergent realism. *Philosophy of Science*, 19–49.
Lawvere, F. W. (1963). Functorial semantics of algebraic theories, PhD thesis. Columbia University.
Lawvere, F. W. (1966). The category of categories as a foundation for mathematics, in *Proceedings of the Conference on Categorical Alegebra*, 1–20. Springer, Berlin.
Lewis, D. (1970). How to define theoretical terms. *Journal of Philosophy* 67(13), 427–46.
Lyre, H. (2004). Horealism structuralism in u(1) gauge theory. *Studies in History and Philosophy of Modern Physics* 35, 643–70.
McLarty, C. (1992). *Elementary Categories, Elementary Toposes*. Clarendon Press, Oxford.
Makkai, M., and Reyes, G. E. (1977). *First Order Categorical Logic*, Vol. 611 of *Lecture Notes in Mathematics*. Springer, Berlin.
Marquis, J. (2006). Categories, sets, and the nature of mathematical entities, in J. van Benthem, G. Heinzmann, M. Rebuschi, and H. Visser (eds), *The Age of Alternative Logics: Assessing Philosophy of Logic and Mathematics Today*, 181–92. Kluwer, Dordrecht.
Marquis, J.-P., and Reyes, G. E. (2012). The history of categorical logic: 1963–1977, in *Handbook of the History of Logic: Sets and Extensions in The Twentieth Century* 6, 689–800. Elsevier, Amsterdam.
Maxwell, G. (1971). Structural realism and the meaning of theoretical terms, in S. Winokur and M. Radner (eds), *Analyses of Theories and Methods of Physics and Psychology*, Minnesota Studies in the Philosophy of Science. University of Minnesota Press, Minneapolis.
Monton, B., and Mohler, C. (2014). Constructive empericism, in E. N. Zalta (ed.), *The Stanford Encyclopedia of Philosophy*, spring edn. http://plato.stanford.edu/archives/spr2014/entries/constructive-empiricism/
Newman, M. H. (1928). Mr. russell's causal theory of perception. *Mind* 37(146), 137–48.
Poincaré (1905). *Science and Hypothesis*. Science Press.
Psillos, S. (1995). Is structural realism the best of both worlds? *Dialectica* 49, 15–46.
Psillos, S. (2000). Carnap, the ramsey-sentence and realistic empiricism. *Erkenntnis* 52(2), 253–79.
Psillos, S. (2001). Is structural realism possible? *Philosophy of Science* 68, S13–S24.

Psillos, S. (2006a). Ramsey's ramsey-sentences, *Cambridge and Vienna: Frank P. Ramsey and the Vienna Circle*. Springer, Berlin.
Psillos, S. (2006b). The structure, the whole structure and nothing but the structure. *Philosophy of Science* 73, 560-70.
Putnam, H. (1975). *Mathematics, Matter and Method*. Cambridge University Press, Cambridge, UK.
Reck, E. H. (2003). Dedekind's structuralism: an interpretation and partial defense. *Synthese* 137(3), 369-419.
Redhead, M. (1996). *From Physics to Metaphysics*. Cambridge University Press, Cambridge, UK.
Russell, B. (1903). *The Principles of Mathematics*. Cambridge University Press, Cambridge, UK.
Russell, B. (1927). *The Analysis of Matter*. Routledge Kegan Paul, London.
Shapiro, S. (1996). Space, number, and structure: a tale of two debates. *Philosophia Mathematica* 4(2), 148-73.
Shapiro, S. (1997). *Philosophy of Mathematics: Structure and Ontology*. Oxford University Press, Oxford.
Stachel, J. (2006). Structure, individuality, and quantum gravity, in D. Rickles, S. French, and J. T. Saatsi (eds), *The Structural Foundations of Quantum Gravity*, 53-82.
Suppe, F. (1989). *The Semantic Conception of Theories and Scientific Realism*. University of Illinois.
Suppes, P. (1960). A comparison of the meaning and uses of models in mathematics and the emperical sciences. *Synthese* 12, 287-301.
Suppes, P. (1962). *Logic Methodology and Philosophy of Science*. Stanford University Press, Palo Alto, CA.
Suppes, P. (1967a). *Set-Theoretical Structures in Science*. Stanford University Press, Palo Alto, CA.
Suppes, P. (1967b). What is a scientific theory? *Philosophy of Science Today*, 55-67. Basic Books, New York.
Suppes, P. (1972). *Axiomatic Set Theory*. Dover, London.
van Fraassen, B. C. (1980). *the Scientific Image*. Oxford University Press, Oxford.
Worrall, J. (1989). Structural realism: the best of both worlds? *Dialectica* 43, 99-124.
Worrall, J. (2007). Miracles and models: why reports of the death of structural realism may be exaggerated. *Royal Institute of Philosophy Supplement* 61, 125-54.
Worrall, J. and Zahar, E. (2001). Ramsification and structural realism, *Appendix IV in E. Zahar, Poincaré's Philosophy: From Conventionalism to Phenomenology*, 236-51. Open Court, Chicago.

Name Index

Please note that page references to Footnotes will be followed by the letter 'n' and number of the note.

Abramsky, S. 166n3, 180n21, 182n25, 270, 272, 275n4, 280, 281, 281–2, 287, 310–11, 382
Aczel, P. 46, 67
Adámek, S. 177n18
Aigner, M. 95
Andréka, H. 407
Arlo-Costa, H. 217n49
Arntzenius, F. 330, 340n26
Artin, M. 3
Aspect, A. 270
Atiyah, M. 23
Awodey, S. 20, 30, 37, 45n11, 48, 58, 60n1, 65, 66n5, 70, 87, 87n22, 90n2, 211n47, 217n49, 219, 411, 418, 419n20, 443n35

Baars, B. J. 376
Baez, J. 37, 43, 311, 330n8, 331, 334n13, 420n22, 436
Bailly, F. 358
Bain, J. 432, 435, 435n19, 436, 436n20, 437, 438, 438n26, 439, 440, 440n28, 441n31, 442, 443, 443n35, 444, 445, 446
Baker, D. J. 330n6
Balzer, W. 104n28
Barbosa, R. S. 280, 281, 282
Barnum, H. 286, 320
Barr, M. 115, 418n18
Barrett, J. 274
Barrett, T. 329n1, 331n9, 334n12, 340, 340n27, 344, 403, 406, 408, 416n12, 427
Bartosik, H. 262
Bastiani, A. 378
Bauer, A. 45n11, 51n19
Baumgarten, A. G. 102n20
Bayne, T. 377
Beeson, M. 46, 116
Béjean, M. 359
Bell, J. L. 90n2, 114n1, 262, 268n1, 270, 281
Belot, G. 330
Bénabou, J. 114, 146n27, 156, 298, 417n14
Benacerraf, P. 49n18
Bennett, C. H. 310
Ben-Zvi, D. 22
Bergfeld, J. M. 182n24
Bernays, P. 44n10
Bishop, E. 37, 38, 39, 40, 51, 51n20, 55
Blackburn, P. 175n14
Blass, A. 91n5

Bleecker, D. 338n22
Blute, R. 182n25, 227, 228n6, 234, 237, 241, 243, 245, 246, 249, 253, 257
Boileau, A. 120
Boolos, G. 84
Borceux, F. 98
Borger, J. 21
Bourbaki, N. 3, 5, 6, 100, 104n28, 107, 160n44
Brading, K. 434, 435n15, 437n23, 446
Brandenburger, A. 268n1, 272, 280, 281–2
Braüner, T. 210n46, 211n47, 213n48, 219
Bredon, G. E. 94, 95, 108
Brown, A. 401
Brown, R. 87n23, 395n6
Brunerie, G. 51n19, 52n23
Bunge, M. 369
Busch, J. 431n5
Butz, C. 418–19
Buzsàki, G. 375

Cabello, A. 280
Cantor, G. 4, 12n4, 13, 43, 85, 158
Cao, T. Y. 431n5
Carboni, A. 173n12
Carchedi, D. 29–30
Carnap, R. 67, 192, 329n1, 413, 417, 433n6
Carter, J. 86
Caru, G. 280–1
Cassirer, E. 19, 25, 26, 27, 33
Chagrov, A. 182n26
Chakravartty, A. 438n25, 446
Changeux, J.-P. 361, 374
Chellas, B. F. 178n20
Chiribella, G. 286–7, 325
Choi, M.-D. 310
Cisinski, D.-C. 109n36
Clark, A. 383
Cockett, J. R. B. 239, 253, 255, 256, 257, 258
Coecke, B. 287, 297, 310–11, 320, 322, 326, 382
Cogan, J. 400
Cohen, P. 3, 4, 115
Conradie, W. 172n11
Constable, R. L. 45n11
Constantin, C. 280, 281, 282
Conway, A. W. 349, 350, 351
Cook, R. T. 279, 280
Corfield, D. 21, 24, 27, 28, 37, 52n21
Cresswell 172n11

NAME INDEX

Curiel, E. 329n1, 412
Curry, H. B. 45

da Costa, N. C. 431, 433–4
Danos, V. 240–1
D'Ariano, G. M. 325
Dasgupta, S. 330n6
Dedekind, R. 19, 21, 350, 444, 445, 445n38
de Fermat, P. 391
Deligne, P. 338n22
Demopoulos, W. 433n8
Descartes, R. 9, 391
de Silva, N. 280, 282
Diaconescu, R. 115
Dickson, M. 19
Dirac, P. 309, 350, 351, 353, 354
DiSalle, R. 337n21
Dobzhanski, T. 361
Dolan, J. 311
Domski, M. 19
Dorato, M. 431n5
Drossos, C. A. 105
Dyckhoff, R. 172n9

Earman, J. 329, 330, 333, 341, 341n31, 342, 343
Eckmann, B. 100
Edelman, G. 365, 366
Ehresmann, A. 359, 369, 376, 378
Eilenberg, S. 12, 69, 75, 99, 108, 109, 113, 146, 154n35
Einstein, A. 18, 19, 24, 25, 38, 40, 248, 425
Ellerman, D. 102n19, 139n11
Erdös, P. 95
Ernst, M. 77
Escardó, M. 37, 47
Escher, M. C. 396
Esfeld, M. 431n5, 437–8, 439
Everett, H. 27

Fantechi, B. 5
Feferman, S. 76, 77, 81, 81n14, 82–3, 83n18, 84, 90n2, 91n5
Feintzeig, B. 344
Felix Klein, C. 19, 25, 67
Ferreirós, J. 106n31, 136n2
Fewster, C. 423
Feynman, R. P. 248
Fichte, J. G. 18, 24, 33
Fletcher, S. C. 329n1, 344
Forssell, H. 219, 411, 418, 419n20
Franklin, B. 352
Fréchet, M. 141n15
Freed, D. S. 338n22
Frege, G. 38, 48, 60
Freitag, E. 5

French, S. 330n6, 431, 433–4, 435n18, 442, 444, 445, 446n39
Freyd, P. 8, 100, 103n22, 145n25, 173n13
Friedman, H. 79
Friedman, M. 19, 25, 27, 30, 33, 80, 417, 433n8
Fritz, T. 326
Führmann, C. 253

Gabbay, D. M. 211n47, 217n49, 219
Gally, J. A. 366
Gambini, R. 340n27
Gare, A. 359, 370
Garner, R. 47
Gauss, C. 21
Gava, G. 90
Geroch, R. 341
Ghilardi, S. 210n46, 211n47, 213n48, 217n49, 219, 276, 326
Ghirardi, G.-C. 272
Gillen, P. 351
Giovanelli, M. 25
Girard, J.-Y. 224n3, 225, 240, 248, 253
Giustina, M. 263
Glymour, C. 335
Gödel, K. 3, 4
Goldblatt, R. 163, 211n47, 217n49, 219
Goré, R. 172n9
Gottlob, G. 281, 282
Greenberger, D. M. 277
Gromov, M. 381
Grothendieck, A. 3, 5, 21, 28, 32, 76n4, 99, 100, 102, 105n29, 108n34, 114, 115, 131, 132, 146n27
Gürsey, F. 351

Hagmann, P. 375
Hájek, P. 3
Halvorson, H. 329, 344, 403, 406, 408, 416n12, 417n16, 435n16
Hansen, H. H. 178n19
Hardy, L. 262, 270, 274–5, 281, 287
Harnik, V. 407, 415
Healey, R. 330, 340n26, 340n29
Hebb, D. O. 374
Hegel, G. W. F. 20, 26–7, 33
Heijltjes, W. 246
Heis, J. 24, 26, 27
Hellman, G. 81, 85, 90n2, 444, 444n37
Helmholtz, H. von 19, 24, 25
Hennessy, M. 260
Hensen, B. 263
Hermida, C. 158
Heunen, C. 287
Hilbert, D. 19, 20, 42n7, 44, 286, 423, 432, 443n35, 445
Hilton, P. J. 100

Hintikka, J. 185
Hofmann, M. 48n14
Houston, R. 246
Howard, M. 262
Howard, W. A. 45, 46n12
Hughes, D. J. 228n6
Hughes, G. E. 172n11
Husserl, E. 18, 24, 376
Hyttinen, T. 282

Isbell, J. 73, 80
Ismael, J. 330n6

Jacob, F. 361-2
Jacobi 94n10
Jacobs, B. 148n30, 173n12
Jamiołkowski, A. 310
Johnstone, P. T. 163n1, 164, 173n13, 404, 407, 408, 410, 412n9, 416, 419n19
Jónsson, B. 169
Joyal, A. 109n36, 114, 115, 120, 225, 248, 298, 418
Juhl, C. 180n21

Kamareddine, F. 148n30
Kamp, H. 282
Kan, D. 100, 146n27, 155, 378
Kanamori, A. 15
Kant, I. 23, 24, 26, 90, 102, 102n20, 103, 103n24, 104, 105, 382, 383, 384
Kapulkin, C. 48, 66n4
Kashiwara, M. 266
Kauffman, L. H. 310
Kauffman, S. 359, 370
Keller, B. 107
Kelley, J. 5
Kelly, G. M. 146n27, 311
Kelly, K. T. 180n21
Kiehl, R. 5
Kirchmair, G. 262
Kishida, K. 211n47, 217n49, 217n50, 219, 281, 282
Kisin, M. 13
Kissinger, A. 287, 298, 322, 326
Knox, E. 330n4, 337n21
Kobayashi, S. 340n28
Kochen, S. 262, 280, 281
Kock, A. 330n8
Koestler, A. 363
Koh, T. W. 228n6
Kolaitis, P. 281, 282
Kracht, M. 210n46, 211n47, 213n48
Kreisel, G. 1, 3, 90n1
Kripke, S. 115, 192, 199, 201, 202, 203, 205, 206, 209, 210, 212-18
Krömer, R. 102n21, 146n27

Kronecker, L. 95
Kunen, K. 15
Kutz, O. 210n46, 211n47, 213n48

Laborit, H. 361
Lack, S. 99n15
Ladyman, J. 37, 330n6, 431n4, 431n5, 435n18
Lakatos, I. 33
Lal, R. 277, 280, 281, 412, 439n27
Lam, V. 431n5, 432, 437-8, 437n24, 438, 438n25, 439, 440, 440n28, 440n29, 441n32, 442, 443, 444, 445, 446n40, 447
Lamarche, F. 228n6, 253
Lambek, J. 115, 115-16, 226, 228, 229, 246-7, 249, 349, 350, 351, 353, 354, 355, 382, 425n27
Lambert, K. 200
Lanczos, C. 351
Landry, E. 82, 82n15, 90n2, 433n11, 434, 434n13, 434n14, 435n15, 435n17, 437n23, 443n34, 446
Lang, S. 2, 8, 9, 10, 13
Langton, C. G. 282
Laudan, L. 430
Lawvere, F. W. 4, 8, 9, 10, 20, 30, 31, 33, 42, 43, 47, 69, 70-1, 72, 73, 80, 83, 85, 90n1, 113, 114, 114n1, 115, 117, 118, 119, 156n40, 196n39, 329n1, 330n8, 382, 442
LeCun, Y. 144n21
Leibniz, G. W. 47, 63, 105
Leinster, T. 2, 9, 10, 139n9, 147n29, 331n10, 420n21
Leitgeb, Hannes 67
Leng, M. 86
Levchenkov, V. 279
Lewis, D. 192, 199, 201-5, 208, 209, 210, 212, 213-18, 433
Liang, Y.-C. 280
Licata, D. R. 51n19, 52n22, 52n23
Lie, Sophus 24, 25
Linnebo, Ø. 2, 81-6, 85n21, 90n2
Logan, S. 84
Loll, R. 340n27
Longo, G. 358, 360, 366, 367, 370, 373
Louie, A. H. 358
Lumsdaine, P. L. 47, 48, 51n19
Lurie, J. 5, 21, 28, 29, 30, 100n16, 109n36, 156n41
Lutz, B. 329n1
Lyre, H. 431n5

MacIntyre, A. 27
McKinsey, J. C. C. 169
Mac Lane, J. O. 330n8

Mac Lane, S. 12, 69, 75–8, 79n11, 80, 87, 90n1, 99, 100, 113, 146, 154n35, 236, 237, 266, 298, 329n1, 331n10, 359, 407
McLarty, C. 2n1, 5, 6, 11, 13, 14, 20, 21, 43, 59, 61n1, 70, 73, 74, 77n6, 79, 79n12, 81, 83, 83n17, 84n19, 85, 86, 90n2, 91n6, 139n9, 441
Maddy, P. 78n8, 84, 86, 156n41
Makkai, M. 63n3, 90n2, 91n3, 91n6, 115, 136n1, 137, 139n10, 140n13, 141, 143n20, 148n30, 153, 154, 157, 158, 159, 160, 160n44, 211n47, 217n49, 403, 404, 405, 406, 407, 411, 416, 418, 420n21, 423n24, 424, 442n33
Malament, D. 333n11, 335, 336n19, 341n32, 344
Manchak, J. B. 344
Manchester, P. 102n20
Mancosu, P. 136n2
Manders, K. L. T. 91n6, 103, 106, 160n45
Mansfield, S. 277, 280, 281, 282
Marquis, J.-P. 20, 55, 82–3, 82n16, 83, 90n2, 91n3, 91n6, 102n21, 138n7, 139n11, 144n21, 146n27, 442n33, 443n34
Martin, D. 71
Martin-Löf, P. 43, 45, 46, 47, 48, 50, 58, 62, 116, 143n20, 148n30
Mathias, A. R. D. 75, 78, 78n10, 79, 80, 87
Matsuno, K. 365
Maxwell, G. 433
Mayberry, J. 137n3
Mazur, B. 95, 95n12
Melliès, P.-A. 253
Meloni, G. 211n47, 217n49
Mermin, N. 275n4
Meyer, J.-J. 185
Milne, J. 5
Milner, R. 260
Minkowski, H. 351
Mitchell, B. 350, 353
Mitchell, W. 114
Moerdijk, I. 266, 407, 418–19, 421n23
Mohler, C. 433n11
Montague, R. 168n6, 178
Monton, B. 433n11
Moore, G. H. 106n31, 136n2
Morava, J. 400
Morganti, M. 431n5
Moss, L. S. 176n16
Moulines, C. W. 104n28
Munkres, J. 2, 4–5, 7, 8, 9, 10, 13

Nagel, E. 23, 24, 422
Napier, J. 94n10
Nestruev, J. 343n35
Newman, M. H. 433n8
Nodelman, U. 138n7
Nomizu, K. 340n28
North, J. 329n1
Norton, J. 53–4, 330, 330n5, 341, 342

Ong, C. 228n6
Osius, G. 10, 79n12, 115
Otero, Daniel 94n10

Pacuit, E. 217n49
Palais, R. S. 338n22
Palmigiano, A. 172n11
Paquette, É. O. 287, 297
Parikh, R. 180n21
Pavlovic, D. 287
Pearce, D. 422
Pedicchio, M. C. 106n32
Pedroso, M. 90n2
Pelayo, A. 37
Penrose, R. 248, 263, 298, 310, 333n11, 354
Perinotti, P. 325
Pettigrew, R. 2, 81, 83, 84, 85–6, 85n21, 90n2
Pitts, A. M. 196n39, 416n13, 417n15
Planck, M. 320
Plotkin, G. 248n9
Poincaré, H. 19, 381, 430n2
Pollard, S. 140n14
Pooley, O. 330n5, 341n31, 437
Popescu, S. 272, 276
Popper, K. 359, 400
Porst, H. E. 418n18
Porter, T. 395n6
Pratt, V. R. 166n3
Prawitz, D. 228, 233
Priest, G. 262
Psillos, S. 431, 431n5, 432, 433n9, 440, 446
Pudlák, P. 3
Pullin, J. 340n27
Putnam, H. 430
Pym, C. 253

Quine, W. V. O. 11, 412

Radman, M. 373
Ramsey, F. P. 329n1, 433n7
Ranta, A. 33–4
Rashevsky, N. 358
Rathjen, M. 46
Rattray, B. 349
Raussendorf, R. 262, 282
Reck, E. H. 136n2, 445, 445n38
Redhead, M. 434
Regnier, L. 240–1
Reichenbach, H. 18, 19, 24, 27
Resnik, M. 60n1
Reyes, G. E. 115, 211n47, 217n49, 330n8, 403–7, 416, 424, 425n28, 442n33
Reyle, U. 282
Richardson, A. 19, 27
Riehl, E. 37
Riemann, B. 19, 21, 23–4
Rindler, W. 333n11

NAME INDEX 455

Robinson, A. 115, 280
Rodin, A. 37, 44
Rohrlich, D. 272, 276
Rosen, R. 358, 359, 370
Rosenstock, S. 330n7, 340, 340n26, 340n27, 341, 341n30, 343n34, 344
Ross, D. 330n6
Rudin, M. E. 5
Ruetsche, L. 344n36
Russell, B. 4, 24, 66, 329n1, 433, 444, 445, 446
Ryckman, T. 19, 24–5, 27
Rynasiewicz, R. 343n34

Sachs, R. K. 38
Sadrzadeh, M. 172n9, 281, 282
Sambin, G. 116
Samuel, P. 3n2
Saunders, S. 330n4, 337n21
Scedrov, A. 173n13
Schanuel, S. 329n1, 330n8, 382
Schapira, P. 266
Schiemer, G. 136n2
Schlick, M. 18, 24
Scholz, E. 24, 25
Schreiber, U. 20, 29–33, 37, 52n21, 330n8
Schrödinger, E. 299, 300, 351, 354
Schulte, O. 180n21
Schultz, P. 401
Scott, D. S. 178, 200
Scott, P. J. 115, 116, 182n25, 425n27
Seely, R. A. G. 228n6, 239, 253, 255–8, 349
Segerberg, K. 178
Selinger, P. 227, 309, 310, 311, 320
Serre, J.-P. 100
Shalm, L. K. 263
Shapiro, S. 60n1, 90n2, 444, 444n36
Shoenfield, J. 106n31
Shulman, M. 20, 30, 52n21, 52n22, 77, 394
Silberstein, L. 350
Simeonov, P. L. 358
Simondon, G. 366
Simpson, S. 3, 14
Smith, J. M. 116
Specker, E. P. 262, 280, 281
Spinoza, B. 365
Spivak, M. 2, 6
Stachel, J. 330, 341, 343n33, 431n5
Steenrod, N. 108
Stein, H. 329, 333
Steiner, J. 19
Straßburger, L. 228n6, 253
Street, R. 109n35, 225, 248, 298
Streicher, T. 48n14
Sudbury, A. 355
Sullivan, D. 94
Suppe, F. 433n10

Suppes, P. 433, 433n10, 433n11, 434, 441n32
Szabo, M. E. 115

Tamme, G. 5
Tarski, A. 67, 113, 169
Teh, N. J. 408, 412, 439n27
Tholen, W. 418n18
Thom, R. 366, 367
Tierney, M. 114, 114n1, 115, 119, 418
Tiu, A. 172n9
Toën, B. 29, 30, 32, 33
Torretti, R. 18, 19
Trautman, A. 335, 338n22
Trimble, T. 234
Tsementzis, D. 37, 403, 406, 408, 435n16
Tzeng, T.-Y. 401

Urban, C. 228n6

Vagner, D. 401
van Bentham, J. 422
Van Benthem, J. 180n22, 185n30, 190n32
Vanbremeersch, J.-P. 359
Van Dalen, D. 181
van den Berg, B. 47
van Fraassen, B. 330n6, 417n17, 433
van Oosten, J. 404
Varela, F. 373
Venema, Y. 163, 171, 177n17
Verdier, J. 28
Vicary, J. 287
Vickers, S. 180n21
Voevodsky, V. vii, 45n11, 48, 49n17, 58, 65
von Karger, B. 172n9
von Neumann, J. 91–2, 92n7, 311, 322

Wadler, P. 45
Waldhausen, F. 108
Waldrop, M. M. 282
Walicki, M. 279, 280
Walters, R. F. C. 173n12
Warren, M. A. 37, 48n14, 51n19, 58
Weatherall, J. 10, 12n4, 330n4, 330n7, 335, 337, 338n22, 339, 340, 340n27, 341, 402n1, 435n16
Weber, H. M. 21
Weil, A. 6
Wells, C. 3–4
Wen, L. 279
Wester, L. 281
Weyl, H. 18, 19, 22, 24–5, 27, 33, 140n14
Whitehead, A. N. 288
Wiles, A. 13
Williamson, R. 37
Wittgenstein, L. 66

Worrall, J. 330n6, 430, 431, 433, 433n9
Wu, H. 38
Wüthrich, C. 432, 437, 437n24, 438, 438n25, 439, 440, 440n29, 441n32, 442, 443, 444, 445, 446n40, 447

Xu, C. 47

Yanofsky, N. 268n1
Ying, S. 281

Yoneda, N. 393, 395

Zahar, E. 433
Zakharyaschev, M. 182n26
Zalta, E. N. 138n7
Zangwill, J. 120
Zermelo, E. 13, 76
Ziegler, G. M. 95
Zu, C. 262
Zucker, J. 228n6

Subject Index

Please note that page references to Footnotes will contain the letter 'n' followed by the number of the note. Diagrams are represented by the italicized letter 'f' following the page number.

Abelian Categories (Freyd) 100
Abelian group 224n4, 332
abstraction dissipation 250
abstract mathematical concepts 136–62
  being abstract 140–2
  and conceptual holism 142
  contrast and compare 138–42
  development 143
  epistemological distinction between abstract and concrete 138n7
  FOLDS (formal system) 148–55
  functions, abstract sets connected by 145
  generalities 138–48
  instances 141–2
  levels of abstraction 144n21
  natural organization of 142
  presentation mode 144
  pure sets 138–9
  set-theoretic approach 138, 140, 143, 146, 158
  syntax and logic 142–4
  universe of 144–8
accessibility and mathematical thought 86–7
action
  dynamical systems 399–400
  modeling action-ability with monoids 397–8
  modeling action with outgoing functors 398–9
  modeling reversible action-ability with groups 396–7
  and symmetry 396–400
activation delay 374
Adjoint Functor Theorem 77
adjoints
  and connectedness 307–8
  modalities 31
  and unitarity 304–6
algebraic geometry 3, 41, 113
  see also algebras; geometry
  canonical maps 99, 100, 107, 108
  derived 21, 28, 29, 32
algebraic theory 113, 117
algebraic topology 43, 50, 86, 113
  canonical maps 103, 108
algebras
  algebraic semantics 163

  Boolean algebras *see* Boolean algebras
  Einstein algebras 343, 344
  first-order modal logic 210–11
  Frobenius algebras 326
  Heyting algebras 114, 164, 183
  Hopf-algebras 326
  Lie algebras 338, 339, 350
  Lindenbaum–Tarski 165
  "spatial" semantics 163
  syntax as 163–5
analytical philosophy tradition 142
analytic geometry 41–2
applied mathematics 103
a priori 103n24, 108
archetypal loops 375–6
architectonic, canonical 102–7
  see also canonical maps
  defining "architectonic" 102
  difference from other systems 105–7
  underlying principles 102–5
  why philosophically relevant 107–9
Aristotle 102n20, 362, 370
arithmetic
  and geometry 21, 22
  matrix, group-enriched category of 388–9
  set theories close to 3
Aspects of Topoi (Freyd) 100
assertiveness, autonomy 81
Association for Symbolic Logic 6
atomism, logical 142
Augustine of Hippo 359
automorphisms 48, 55, 335
  Galilean space–time 333, 334
  non-trivial 343
autonomy
  see also categorical foundations
  assertiveness 81
  categorical foundations 81–6
  definition and examples 189–92
  justification 84–6
  lauter Einsen view 85, 86
  prior theory 82–4
  results on 86
  substructure 185–92
axiomatic cohesion 52n21, 54n24
Axiom of Foundation, ZFC 4, 13

SUBJECT INDEX

axioms
  see also ZFC (Zermelo-Fraenkel set theory with Axiom of Choice)
  assertiveness, autonomy 81
  axiomatic systems 23–4, 42
  Axiom Scheme of Replacement 4, 6n3
  Bounded Separation Axiom Scheme 6n3
  categorical foundations 71, 73, 74, 84
  compared to rules 44
  defining 44
  Elementary Theory of the Category of Sets (ETCS) 2, 10, 71, 74
  extensionality 139
  first-order axiom scheme 4
  genetic versus axiomatic methods 44
  non-logical 229, 230
  set theories 3, 4
  univalence 48, 49, 55, 65, 66
  ZFC see ZFC (Zermelo-Fraenkel set theory with Axiom of Choice)
Axiom Scheme of Replacement 4, 6n3, 13

balanced ontic structural realism (BOSR) 432, 440, 443–7
behavioral equivalence 177n17
Bell inequality, logical 269, 270, 281–2
Bell's theorem 263, 270
Bell table, logical analysis 269–70
Benacerraf's problem, philosophy of mathematics 159
bicartesian closed categories 116
bicategories 147
bijectiveness 38, 39, 49
binding problem 362
biology, application of categories to 358–80
biquaternions 350
bisimulation 177n17
Boolean algebras 164, 225, 331, 403
  categorical logic 114, 118, 119
  categories and modalities 163, 164, 165, 167, 168, 170, 173, 178, 197, 210
  equations 279
  scientific theories 410, 418
Boolean toposes 114, 115, 131
bounded quantifiers 78
Bounded Separation Axiom Scheme 6n3
box topologies 5
broadcasting
  cloning compared 318
  defining 319
  no-broadcasting theorem 286, 320

calculus see sequent calculus
canonical correspondence 100
canonical isomorphisms 108
canonical maps 90–112
  algebraic topology 103, 108
  architectonic, canonical 102–7
  canonical morphisms 101, 104, 105, 107, 109, 362
  examples 96–8
  first-order logic (FOL) 91, 103
  folklore 93–6, 99
  "God-given" 94, 95
  historical perspective 98–100
  natural transformation 94, 95, 99
  role in mathematics 91
  taking stock 101–2
canonical matrix 98n14
canonical morphisms 101, 104, 105, 106, 107, 109, 362
canonical projections 97
cardinality/cardinalities 4, 38, 49, 50
cardinals/cardinal numbers 38
  see also cardinality
  inaccessible 80
  large 1, 3, 80
    categorical foundations 79, 80
  measurable 15
Cartesian products 7, 9, 44, 77, 206, 436n20, 442
categorical foundations
  accessibility and mathematical thought 86–7
  autonomy 81–6
  categorical replacement 79, 85
  Category of Categories as a Foundation (CCAF) 69, 72–5
  Elementary Theory of the Category of Sets (ETCS) 69, 70–2, 78
  motivations for introducing 69–70
  operation and collection 82, 83
  psychological priority 82, 84
  technical adequacy 69, 78–80, 87
  unlimited categories 75, 77, 80
categorical framework 90–1, 92, 96
categorical logic
  deductive systems and categories 116–17
  and model theory 113–35
categorical proof theory 226–7
categorical quantum mechanics (CQM) 286–328
  see also quantum mechanics
  biproducts, limitations of 287
  cap-effect 300, 311
  category-theoretic counterparts 296–8, 308–10, 320, 325
  causality postulate 287
  comparison to previous versions 287
  diagrams
    benefits of use 286, 287
    boxes, drawing of 286, 288, 289, 290
    circuits 292–6
    description 289
    empty 292, 293

SUBJECT INDEX 459

formation 289
processes as 288–92
string diagrams 299–311
wires 286, 289, 290
generalized Born rule 292, 313
interchange law 294
no-broadcasting theorem 286, 320
normalization issues 287, 314
overviews, surveys and tutorials on 287
parallel composition operation 287, 292
process ontology 287
process–state duality 301, 304, 317, 318
process theories 288–98
   category-theoretic counterpart 296–8
   circuit diagrams 292–6
   definition of "process" 288
   definition of "process theory" 290–2
   effects 292
   non-signalling 324
   processes as diagrams 288–92
   states 291
quantum types 286, 311–14
sequential composition operation 287, 293–4
spider processes 287
standard category-theoretic structures, limitations of 287
swaps 294
system-types 288, 290, 321
wires 286, 289, 290
   directed path/cycle 295
categorical replacement 79, 85
categories
see also categorical foundations; category theory (CT)
application to biology and cognition 358–80
basic notions 226
bicartesian closed 116
category of interpretations 128
compact closed categories 308
concrete 103
dagger 253
and deductive systems 116–17
first-order languages in, model theory 127–30
functors, connected by 145
geometric 130
hierarchical 364
higher-dimensional categories (HDC) 156–8
of Kripke frames 174–7, 187
linearly distributive 243–8
as mathematical models 381–401
and modalities 163–222
monoidal *see* monoidal categories
multicategories 228–37, 238
semantic 409–13
semi-sheaf of 361
six-dimensional Lorentz category 350–6

syntactic *see* syntactic categories
and theories *see* theories and categories
thin and thick 410
Categories for the Working Mathematician (Mac Lane) 359
category mistakes 430–49
category-neurons (cat-neurons) 374–5, 377
Category of Categories (CAT) 394–5
Category of Categories as a Foundation (CCAF) 72–5, 442
see also categorical foundations
   goals 73
   introduction of 69, 72
   and ZFC 69
category theory (CT)
see also categorical foundations
American school 99, 100
applied mathematics 103
and classical space–time theories 332–5
coequalizers 98
collections (types) 37
as a conceptual tool 439–46
constructions 96–7, 101
contributions of 219
development 55
foundational status 90
as framework for the structure of a physical structure 435–9
as framework for the structure of scientific theories 432–5
functions 92, 93
and HoTT/UF 42
invention 113
language and concepts 92
and mathematical structure 60n1
and philosophy 33
and set theories 138n6
structure 60n1, 331–2
unification 91n4
as a universal modeling language 382–3
Cauchy reals, sequences of 37n2, 58, 59
causality 320–6
   category-theoretic counterpart 325
   causal processes 321–2
   doubling 312, 322
   evolution 322–3
   no-signalling 323–5
   postulate 286–7
causal processes 321–2
CH *see* continuum hypothesis (CH)
charge-current density 352
Chern–Weil theory 31
Church–Rosser theorem 235
Church–Turing Thesis 96
circuits 225, 226, 227, 248–53
   ⊗-circuits 251–3
   abstraction dissipation 250

circuits (*cont.*)
   categorical quantum mechanics (CQM) 292–6
   circuit diagrams 292–6
   coherence conditions 237, 238f, 244–5
   components 248
   covariables 249
   diagrams 292–6
   input variables 248
   for monoidal categories 232–3
   non-planar 249, 250–1
   normal forms 235–6
   planar juxtaposition 249
   poly-circuits 239
   primitive expression 248
   rewriting 233–6
   swaps 294
   switching 232–3, 240
   typed 248
   unit rewirings 234, 235f, 247, 250, 252
cloning process, broadcasting compared 318
coequalizers 98
cognition, application of categories to 358–80
coherence conditions 237, 238f, 244–5, 297
cohesion 30, 115
   axiomatic 52n21, 54n24
cohesive homotopy type theory (CoHoTT) 33, 425
cohomology 28
   differential 20, 31
colimits 101, 362, 363f, 364, 365, 367, 374, 375
   colimit-cones 368, 369, 378
   defining 377–8
collections (types) 36, 37, 55
   see also type theory
   and sets 38–9, 41
compact closed categories 308
complex function field 21, 22
complexification process 367–70, 374, 378
computationalism 373
computer programs 288, 291
   MATLAB 385
concrete categories 103
conjugate 305, 306
connectedness, and adjoints 307–8
connecting models 383–4
connectionism 373, 377
consistency
   see also contextuality
   global 265, 266, 272
      global inconsistency 262, 263
   local 262, 265, 266
   logical 262
   paraconsistent logics 262
constructibility 3, 14
constructions 45, 46
constructive type theory 58, 62

contextuality 262–85
   see also consistency
   Alice–Bob scenario 263, 264, 266, 267, 268, 272, 276–7
   basic formalization 264–6
   Bell inequality, logical 269, 270, 281–2
   Bell's theorem 263, 270
   data 274
   empirical models 273–4
   example 266–70
   global assignment 278
   Hardy paradox 262, 274–5, 276, 282
   Kochen–Specker paradox 262, 282
   Liar cycles 278–9
   logical consistency 262
   logic and paradoxes 262, 278–80
   measurement 263, 273, 275
   no-signalling principle 272, 276–7
   observables 264, 275, 277
   perspective 280–2
   possibilistic contextuality/collapse 273, 274
   quantum advantage 262, 264
   quantum information 274, 281
   quantum mechanics 272, 275n4
   Robinson Joint Consistency Theorem 280
   semantics, contextual (in classical computation) 282
   strong 276–7
   univocal path 278
   visualizing 277–8
continuous dynamical system 399
continuum hypothesis (CH) 1, 2, 3, 4–5, 115
contrast and compare
   being abstract 140–2
   pure sets 138–9
converse Barcan formula 215–17
co-regulators 370, 371, 372
counterpart theory (Lewis) 202
covariance 40
   general 53–5
CQM *see* categorical quantum mechanics (CQM)
Critique of Pure Reason (Kant) 102, 104, 383
Critique of the Power of Judgment (Kant) 90
CT *see* category theory (CT)
Curry–Howard correspondence 45, 46, 62
Curry–Howard isomorphism 226
cut rule, proof theory 228–30
cyclicity 302

dagger categories 253
decimal expansions 37, 38, 39
decomposition 362
deductive systems and categories 116–17
degeneracy, neural codes 365–6
Democritus 349

SUBJECT INDEX   461

diagrams
  categorical quantum mechanics (CQM)
    benefits of use 286, 287
    boxes, drawing of 286, 288, 289, 290
    circuits 292–6
    description 289
    empty 292, 293
    formation 289
    processes as 288–92
    string diagrams 299–311
    wires 286, 289, 290
  circuits 292–6
  processes as 288–92
  string 299–311
diffeomorphisms 38, 334, 342
differential cohomology 20, 31
Dirac spinors 353
discrete dynamical system 399
distributive lattices, theory of 225
DNA 373
doubling 312, 322
Drawing Hands (Escher) 396
duality 93, 106–7
  higher duality of relations and modalities 172–4
  Lindenbaum–Tarski 168
  perfect 420
  process–state 301, 304, 317, 318
  Stone 183–5, 418
  symmetric self-duality 308
  of syntax and semantics 167–8, 181, 184, 417–21
dynamical systems 399–400

Eilenberg–Mac Lane (EM) axioms 441
Einstein algebras 343, 344
electrical devices, theory of 290
electromagnetism 18
  excess structure 337
  gauge theory 335
  set theories 10, 11
elementary existential doctrine 118
"An Elementary Theory of The Category of Sets" (Lawvere) 70
Elementary Theory of the Category of Sets (ETCS) 70–2, 78
  see also categorical foundations
  axioms 2, 10, 71, 74
  goals 70–1
  and HoTT/UF 43, 44
  introduction of 69
  justification of 84
  limitations 72, 78
  local membership 11
  and roles of set theories in mathematics 2, 3, 6

structural realism 441, 442
subsets of a set 71–2
Elements (Euclid) 39, 42n7
elliptical geometry 24, 25
empirical models 273–4
energy–momentum tensor 342
entanglement 299
Epicurus 349
epistemic structural realist position 430–1
equivalence
  behavioral 177n17
  categorical proof theory 226–7
  category theory (CT) 331–2
  circuit rewrites 234
  equivalence classes 38, 40, 119, 236
  equivalence relation 38–9, 40, 41, 52
  equivalence set 96
  equivalent theories 406–9
  in FOLDS 154–5
  homotopy 65, 67, 155
  and identifications 47–50
  and identity 65
  logical 406
  Morita equivalence 405–6, 407, 408, 410
Erlanger Programme 25, 27, 33, 54, 55
ETCS *see* Elementary Theory of the Category of Sets (ETCS)
Euclidean geometry 24, 25, 42n7
Euler characteristic 28
evolution/evolutionary systems 359–61
  causality 322–3
  Evolutive System (ES) 360–1
  Hierarchical Evolutive Systems (HES) 361–5, 366
  Memory Evolutive Neural System (MENS) 373–7
  Memory Evolutive System (MES) 358–9, 370–3
exchange rule 230
expansion rule 236, 251, 256f
extensionality principles 66

Faraday tensors 336, 340
Fermat's Last Theorem 13
fiber bundle formalism 340, 341
field equations 24, 425
first-order axiom scheme 4
first-order logic (FOL) 148, 201–9
  ascendancy of 46
  and axioms 44
  canonical maps 91, 103
  categorical approach to first order 192–201
  development 136
  first-order model logic 210–19
  full first-order theories 410
  functorial semantics 118
  and HoTT/UF 45, 46

462 SUBJECT INDEX

first-order logic (FOL) (*cont.*)
   quantification, interpretations 206–9
   satisfaction 201–3
   scientific theories 410, 412
   semantic structures 203–6
First-Order Logic with Dependent Sorts *see* FOLDS (First-Order Logic with Dependent Sorts)
first-order modal logic 210–19
   algebras, family of 210–11
   categorical perspective 218–19
   converse Barcan formula 215–17
   Kripke semantics *see* Kripke semantics
   Lewis's semantics 213–15
flat modality 31
flexibility 365
FOLDS (First-Order Logic with Dependent Sorts) 148–55
   equivalences in 154–5
   first-order logic (FOL) as extension of 148
   FOLDS-signatures 148–54
      arbitrary, fixing 150–1
      as categories 153
      defining 150, 155
      and equivalences 154
      formulas 150, 151, 152
      language of 148, 152
      one-way categories 149–50
      semantics 153
      variables 151, 153
   higher-dimensional categories (HDC) 156–8
   language of 149, 152
   motivations for introducing 147
   significance for philosophers 159, 160
forcing, notion of 115
forgetful functor concept 331, 332
formal class terms 129
formalism
   alternative, of Einstein algebra 343
   excess structure 343
   fiber bundle formalism 340, 341
   general relativity 341
   holonomy formalism 340, 341
   loop formalism 340
   mathematical models 383
   standard 341, 343
   theoretical 358, 413
   Yang-Mills theory 340
formalization 271–4
   measurement scenarios and empirical models 273–4
foundations for mathematics 59
   see also categorical foundations
   categorical foundations 69, 71
   category theory (CT) 114
   and HoTT/UF 36, 37, 41–3

science 136–7, 159
set-theoretic approach 288
frames
   homomorphisms of 183, 184
   Kripke frames, categories of 174–7, 187
   neighborhood 178
free logic 199–201
functor boxes 253–8, 259f
functorial semantics 113, 117–19
functors
   categories connected by 145
   dagger 310
   forgetful 331, 332
   linear 226, 256–60
   multifunctors 230
   outgoing, modeling action with 398–9
   syntax as 193–6

Galilean space–time 333, 334
Galois theory 31
gauge theory/gauge field theory
   and excess structure 335–7
   gauge invariance 2, 10–13, 39
   and geometry 18, 29
   higher gauge theory 29, 31, 33, 330
   meaning of gauges 38
   transformations 338
   use of gauge 13–15
generalized Born rule 292, 313
generalized holonomy map 340
general theory of relativity (GTR) 11, 38, 40, 425
   see also relativity theory
geometric implication 130
geometric site 130
geometric theory 115
geometrized Newtonian gravitation (GNG) 334, 335, 336, 337
geometry 2, 18–35
   algebraic *see* algebraic geometry
   analytic 41–2
   and arithmetic 21, 22
   basic forms (k-cells) 158
   capturing 28–32
   circles 22, 26, 28
   current 21–3, 28–32
   demise of philosophy 23
   elliptical 24, 25
   Erlanger Programme and everyday perception 25
   Euclidean 24, 25, 42n7
   geometric theories and classification of toposes 130–2
   Grothendieck-inspired 5
   hyperbolic 24, 25
   intersection, points of 26, 28
   intuition, going beyond 26

isomorphisms 24, 29
mathematical versus physical 18
and physics 18, 19, 25, 29, 33
projective 23
revival of philosophy 20, 23–7
Riemannian 23–4
'rigorous' 23
rings 21, 28
scheme theory 21, 28, 29, 32n1
space, essence of 24–5
structuralist approach to 55
synthetic 42
theories 115, 130–2
unity of 21
geometry, algebraic 114
global inconsistency 262, 263
global membership relation 4, 11, 12
GNG *see* geometrized Newtonian gravitation (GNG)
Gödel–Henkin completeness theorem 115
Gödel translation 181–3
graph theory 87n23, 92, 116
gravitational field 38, 40
gravitational potential 334
Grothendieck toposes 114
Grothendieck universes 1, 5, 76
group homomorphisms 332, 397
group isomorphisms 140, 143
∞-groupoids 29, 31, 36, 37–41
  see also Homotopy Type Theory and Univalent Foundations (HoTT/UF)
  basic objects of mathematics as 43, 55
  bijectiveness 38, 39
  decimal expansions 38, 39
  freely generated 52
  in HoTT/UF versus in set theory 48
  real numbers 37, 38
  as substitutes of sets 39, 41
  synthetic theory 41, 42, 43, 48
group-theoretical properties 140

Hardy paradox 262, 274–5, 276, 282
Hebb rule 374, 375
Helmholtz–Lie theorem 24
Heraclitus 349
Hermitian biquaternion 350, 353
HES *see* Hierarchical Evolutive Systems (HES)
Heyting algebras 114, 164, 183
Heyting category 129
hierarchical categories 364
Hierarchical Evolutive Systems (HES) 361–5, 366
  co-regulators 370, 371, 372
  Memory Evolutive System (MES) 370–3
  structural changes 367–70
hierarchy of connectionist models 377
Higher Algebra (Lurie) 100n16

higher-category theory 20
higher-dimensional categories (HDC) 109, 156–8, 160
higher gauge theory 29, 31, 33, 330
higher inductive types (HITS) 50–3
higher-order logic 119, 120
  higher-order intuitionistic logic 115
higher-topos theory 20
high-level models, relationships between 394–5
Hilbert spaces 181n23, 298, 311
hole argument 40, 53, 341–3
holism, conceptual 142
holonomy formalism 340, 341
holonomy isomorphisms 341
holonomy models 341
homeomorphism 143, 145, 219
homomorphisms 103, 140, 175, 411
  complete 198
  group 332, 397
  interpretations as homomorphisms perspective 184
  semantics as 165–7
homotopical circle 52n21
homotopical perturbation 29
homotopy coequalizer 52
homotopy equivalence 65, 67, 155
homotopy theory 20
  see also Homotopy Type Theory and Univalent Foundations (HoTT/UF)
  abstract homotopy theory 42, 58
  abstract mathematical concepts 137n4
  cohesive homotopy type theory (CoHoTT) 20, 33, 425
  extensions of 33
  and geometry 20, 29, 30
  homotopy *n*-types 20, 49
  and HoTT/UF 36, 52n22
  modalities 30, 32
  propositional truncation 219
Homotopy Type Theory and Univalent Foundations (HoTT/UF) 36–57
  see also ∞-groupoids
  collections (types) 36, 37, 38
  covariance 53–5
  encoding into mathematics 43, 46
  equivalences 47–50
  higher inductive types (HITS) 50–3
  identifications 36, 47–50, 55
    and equivalences 47–50
    general covariance 53
    identification-constructors 51, 52
  isomorphism 40, 49, 53, 55
  manifolds 38, 39, 40–1, 53
  mathematics, foundations for 41–3
  points 44
  rules 44, 47, 48, 50, 53

SUBJECT INDEX

Homotopy Type Theory and Univalent
    Foundations (HoTT/UF) (*cont.*)
  sameness 37, 39–40, 55
  and set theories 36
  as a synthetic theory of higher
      equalities 41n6
  types 44
  ZFC compared 37, 43n9, 44, 49
HoTT/UF *see* Homotopy Type Theory and
    Univalent Foundations (HoTT/UF)
hyperbolic geometry 24, 25
hyperdoctrines 196, 198, 210

identifications 36, 55
  and equivalences 47–50
  general covariance 53
  identification-constructors 51, 52
identity
  criterion of and abstract mathematical
      concepts 143
  identity identification 47
  identity of types 63–4
  Leibniz's principle of 105
  rules of 65
impredicative type theory 63
index of activation, Evolutive Systems 360–1
indiscernibility of identicals 47
induced maps 100, 154, 273, 331
inductive types
  higher inductive types (HITS) 50–3
  ordinary 50, 51
inference, rules of 122, 126, 441
infinity, points at 26
integers 7, 12, 31, 52, 58, 132, 291
  'integer-as-function' 21–2
  positive 148, 350
  ring of 21
Integral Biomathics (Simeonov) 358
intentional network (IN), macro-landscape 376
interchange law, CQM 294
intersection, points of 26, 28
intuition 26, 32, 43
  geometric 25
  topological 28
intuitionistic logic 177, 181, 183
intuitionistic set theory 46, 115
intuitionistic type theory (ITT) 115, 123,
    425n27
invariance 24, 39, 61
  gauge invariance 2, 10–13, 39
  Principle of Invariance 64, 65
isometry 38, 40
isomorphism/isomorphisms 55, 64
  see also morphism
  canonical 108
  concept 60
  Curry–Howard 226

and geometry 24, 29
group isomorphisms 140, 143
holonomy isomorphisms 341
and HoTT/UF 38, 40, 49, 53, 55
isomorphic copies 59
of manifolds 38
natural 99, 231, 297
objects 58, 59, 61–3, 66
partial 433, 434
and set theories 2, 9, 12
structure 60
Iterated Complexification Theorem 370

Kantian philosophy 24, 26
kinetic energy-momentum 352
Klein–Gordon equation 350, 354
Kochen–Specker paradox 262, 282
Kripke–Joyal semantics 115
Kripke semantics 169–77, 185
  basics 169–70
  categories of Kripke frames 174–7, 187
  first-order model logic 212–13
  higher duality of relations and
      modalities 172–4
  modalities 171–2, 185

lauter Einsen view 85, 86
Leibniz's Law 63
Liar cycles 278–9
Lie algebras 338, 339, 350
Lie group 331
Lie-group structure 431
Lindenbaum algebra/Lindenbaum–Tarski
    algebra 165, 418
Lindenbaum doctrine 119
Lindenbaum–Tarski duality 168
linearity
  basic components of linear logic 238–48
  circuits 225, 226, 227
  higher-dimensional linear spaces 390
  linear distribution maps 243
  linear functors 256–60
  linearly distributive categories 243–8
  linear transformations 392, 393
  modeling 389–94
  multiplicative fragment of linear logic 224,
      228n6, 245
  par connective 224, 239
  poly-logic 239–41
  tensor connective 223, 239
  thinning links 241–3
locales 419n19
local membership 11
local set theories 119–27
logarithmetic tables 94n10
logic
  see also set theories

abstract mathematical concepts 142–4
Bell inequality, logical 269, 270, 281–2
categorical *see* categorical logic
doxastic 185
dynamic 171
epistemic 185
first-order logic *see* first-order logic (FOL)
free 199–201
graphical representation of logical derivations 227
higher-order 119, 120
intuitionistic 177, 181, 183
linear *see* linearity
multi-logic 223, 228, 230, 231, 236
natural 46
poly-logic 239–41
propositional *see* propositional logic
set theories 3
temporal 171, 172
and type theory 43–7
logical atomism 142
logical notion, concept 55, 67
loop formalism 340
loop space 50
Lorentzian distance function 54
Lorentzian manifolds 342

Mac Lane set theory (MAC) 78
magnitudes 39
manifolds 38–41, 53, 333n11
  Lorentzian 342
  pseudo-Riemannian 342
  symplectic 94n9
manifold substantivism 342
marginalization 271
material objects, whether continuous or discrete 349
mathematical models
  categories as 381–401
  category theory (CT) as a universal modeling language 382–3
  connecting, using as 383–4
  high-level models, relationships between 394–5
  linearity 389–94
  minimal models 178n20
  modeling action-ability with monoids 397–8
  modeling action with outgoing functors 398–9
  modeling reversible action-ability with groups 396–7
  symmetry and action 396–400
  third postulate, justifications for 395
mathematical objects *see* objects
mathematical physics 19, 20, 23, 54, 330, 425
  see also mathematics; physics
mathematical structuralism 54

mathematics
  see also mathematical models
  abstract concepts 136–62
  accessibility and mathematical thought 86–7
  applied 103
  architecture 104n28
  basic acts 44
  basic objects, as ∞-groupoids 43, 55
  categories as mathematical models 381–401
  construction aspect 45
  dualism in 106–7
  encoding into HoTT/UF 43, 46
  encoding into ZFC 45
  epistemological aspects 91
  foundations for 36, 37, 41–3, 59, 69, 71
  proof in 13
  set theories, roles in 1–17
MATLAB (computing program) 385
matrices 385–9
  category of matrix multiplication 387–8
  invertible $n \times n$ matrix multiplication 386
  matrix arithmetic, group-enriched category of 388–9
  monoid of $n \times n$ matrix multiplication 387
  relevant aspects 385–6
membership tree structure 13–14, 15, 43
  see also global membership relation; local membership
Memory Evolutive System (MES) 358–9, 363, 370–3
Memory Evolutive Neural System (MENS) 373–7
MES *see* Memory Evolutive System (MES)
metamathematics 91
metaphysics 330n6
metascience 19, 32, 33
methodological structural realism 435
metric structure 38
minimal models 178n20
Minkowski space-times 53, 54, 55, 336, 337
mix rule 238
modalities
  and categories 163–222
  higher duality of relations and modalities 172–4
  Kripke modalities 171–2, 185
  topology and modality 177–85
  universal 171
model theory
  see also mathematical models
  and categorical logic 113–35
  of first-order languages in categories 127–30
  toposes, models in 130–2
modulo equations 226, 233
monoidal categories 223, 227, 230, 231
  circuits for 232–3, 251–3
  strict 298, 309, 310

monoidal categories (*cont.*)
    symmetric monoidal categories (SMCs) 286, 297, 298, 310
    traced 253
    traditional presentation, limitations of 286
monoids
    elements 388
    modeling action-ability with 397–8
    $n \times n$ matrix multiplication 387
monomorphisms 71, 72, 97
Morita context 353
Morita equivalence 405–6, 407, 408, 410
morphism/morphisms 55
    see also isomorphism
    1-morphisms 147
    2-morphisms 147
    and abstract mathematical concepts 145
    abstract mathematical concepts 146, 150
    bounded 175, 176, 187, 213, 214, 215, 218
    canonical 101, 104, 106, 107, 109, 362
    categorical quantum mechanics (CQM) 298, 308, 310, 320
    categories, application to biology and cognition 359, 360, 361, 377, 378
    circuits 226
    composition 98
    compound 247
    concept 60, 388
    geometric 131, 219
    identity 149, 229, 296, 394, 397
    of interpretation 128
    mathematical models 393, 394, 395
    morphisms, between 147
    of multicategories 230
    null morphism 98
    and objects 91, 93
    objects, between 147
    proof theory 233, 254
    proper morphism 149, 150
    simple 378
    swap 297
    unique 101, 325, 378, 393
    universal 99n15, 102n19
Mostowski embedding theorem 14
multicategories 228–37, 236, 238
    Lambek's 229
    morphism 230
multifunctors 230
multi-logic 223, 228, 230, 231, 236
multiplicative fragment of linear logic 224, 228n6, 245
multiplicative units 224
Multiplicity Principle 365–7

natural numbers 95, 444
    categorical logic and model theory 117, 132
    category theory and foundations 76, 78
    functorial semantics 117
    and HoTT/UF 49n18, 50, 51
    and mathematical models 385, 387, 388, 399
    set theories, roles in mathematics 7, 10, 11
natural transformations
    canonical maps 94, 95, 99
    FOLDS (First-Order Logic with Dependent Sorts) 157
    semantics as 196–9
n-complex links 366
negation 224, 244–8
neighborhood semantics 178–9, 185
neurophenomenology 373
New Foundations with Urelements (NFU) 77
Newtonian gravitation (NG) 330, 334–8
    excess structure 337
    geometrized Newtonian gravitation (GNG) 334, 335, 336, 337
Newtonian space–time 333, 334
NG *see* Newtonian gravitation (NG)
no-broadcasting theorem 286, 320
no miracles argument (NMA) 430
non-deterministic polynomial time (NP) algorithm 240
non-planar circuits 249, 250–1
non-separability
    existence of non-separable states, in quantum mechanics 300
    string diagrams 299–302
no-signalling principle 272, 276–7
    causality 323–5
numbers 3, 61, 291, 349, 444
    see also cardinality; cardinals/cardinal numbers
    and analytic geometry 41–2
    categorical quantum mechanics (CQM) 313, 314
    complex 7
    large 281
    and mathematical models 381, 390
    natural *see* natural numbers
    negative 350
    prime 22
    rational 11
    real *see* real numbers
    whole 12

objects
    Category of Categories as a Foundation (CCAF) 73
    isomorphic 58, 59, 61–3, 66
    material, whether continuous or discrete 349
    mathematical 61, 73, 156
    canonical maps 91, 95
    comparing 332
    equivalence 65

FOLDS (First-Order Logic with Dependent Sorts) 137, 140, 148
   and HoTT/UF 37n2
   identical 59, 67
   identity 63
   scientific theories 406, 417, 426
   space–time theories, classical 331, 332, 342, 343
   and morphisms 91, 147
   terminal 325
observables 264, 275
   compatible 277
ontological discontinuity 431

paraconsistent logics 262
parallel composition operation 287, 292
par connective 224, 228n6, 239, 243
Parmenides 349
partial isomorphisms 433, 434
partial structures 433, 442
Peano–Lawvere axiom 114
Penrose tribar 263
pessimistic meta-induction argument (PMI) 430
The Phenomenology of Knowledge (Hegel) 26–7
philosophy of geometry, reviving 20
Philosophy of Geometry from Riemann to Poincare (Torretti) 18
physical geometry 18
physics 282, 337, 360
   classical 138n6, 264
   and geometry 18, 19, 25, 29, 33
   and HoTT/UF 38, 41
   mathematical 19, 20, 23, 54, 330, 425
   metaphysics 330n6, 431, 438n25, 446n39
   models in 358
   modern 20, 37, 38, 350, 411n8, 414
   Newtonian 138n6, 337
   and philosophy 332, 344
   pre-quantum 353
   quantum 138n6, 262, 321
   theoretical 349, 350, 426
   and time/space–time 341, 359
planar abstraction 250
plane curves 28
plasticity, synaptic 374
Poincaré group 54
point-constructors 51, 52
points 26, 28, 44
polycategories 223
polychromous patterns 363
poly-circuits 239
Polykeitos (Greek sculptor) 95n11
poly-logic 223, 239–41
   correctness condition 233, 239, 240

polynomials 21
possibilistic contextuality/collapse 273, 274
possible-world semantics 178, 179
postulates 44, 307, 391, 395
   causality 286–7
   and mathematical models 383, 384, 385, 389
predicates 39, 47, 63, 132, 203n45
   calculus 413
   categorical approach to first order 193, 196
   predicate logic 127–8, 163, 411
   set-theoretic 433
   symbols 403, 406, 420
   truth 278
presentations 37, 38, 39
presheaf 265, 266, 274
pre-Socratic Greek philosophers 349
Principia Mathematica 66
Principle of Invariance 64, 65
Principle of Structuralism 58–9, 61, 64, 65
prior theory, autonomy 82–4
probabilistic empirical models 274
probability theory 269, 274
Proceedings of the National Academy of Sciences of the USA (Lawvere) 113
Process and Reality (Whitehead) 288
processes, as diagrams 288–92
process–state duality 301, 304, 317, 318
process theories 288–98
   see also categorical quantum mechanics (CQM)
   category-theoretic counterpart 296–8
   circuit diagrams 292–6
   definition of "process" 288
   definition of "process theory" 290–2
   effects 292
   non-signalling 324
   processes as diagrams 288–92
   states 291
pro-étale morphisms 32n1
projections 97
proof nets 225
proof theory 223–61
   basic notions of categories 226
   categorical 226–7
   construction and proof 45, 46
   cut rule 228–30
   encoding of proofs 45
   logical derivations, graphical representation 227
   multicategories 228–37, 238
   multi-logic 228, 230
   negation 224, 224–8
   prerequisites 226–8
   representability 233
   sequent calculus 226
propagation delay, Evolutive Systems 360

468　SUBJECT INDEX

propositional logic 163, 219, 227
　categorical approach to first order 192, 193, 196
　intuitionistic 228n6
　scientific theories 404, 410, 418
propositional theories 418, 424
propositions 44, 45, 46, 64, 66
pseudo-Riemannian manifolds 342
psychological priority, categorical foundations 82, 84

quantification, interpretations 206–9
quantifiers 45, 113, 171, 195
　bounded 78, 148
　classical laws 200
　interpretation of 199–200, 208, 209
　universal 118
quantum advantage 262, 264
quantum information 274, 281
quantum mechanics 272, 275n4
　categorical quantum mechanics (CQM)
　　see categorical quantum mechanics (CQM)
　characteristic trait 300
　Klein–Gordon equation 350, 354
quantum processes 311–20
　category-theoretic counterpart 320
　causality 311
　cloning process 318
　discarding effect 315
　doubling 312, 313, 322
　impure/mixed 315
　normalization issues 315
　pure processes and discarding 314–18
quantum teleportation 299, 304, 310
quantum theory/quantum gauge theories 18, 19, 20, 299, 326
　see also gauge theory/gauge field theory
quantum types 286, 311–14
quaternions 350, 351
quotient, of equivalence relation 38
quotient map 96

radical ontic structural realism (ROSR) 432, 435–8, 443
Ramsey sentence 433
rationality, retrospective 27
real numbers 31, 78, 132, 288, 391n4
　categories, application to biology and cognition 360, 361
　Cauchy reals, sequences of 37n2, 58, 59
　decimal expansions 38, 39
　∞-groupoids 37, 38
　and HoTT/UF 37, 37n2, 38, 39, 51, 54
　set theories, roles in mathematics 4, 7, 11, 12
reflexivity witness 47
The Reign of Relativity (Ryckman) 19

relativity theory 18, 20, 286, 341–2
　see also general theory of relativity (GTR)
representability 233
representation theory 399
restriction map 265
Riemannian geometry 23–4
Robinson Joint Consistency Theorem 280
Rosetta Stone (Weil) 21, 22
rules 117, 122
　box formation 258
　cut rule 228–30
　deduction 226, 227, 228
　expansion 236, 251, 256f
　formal 388
　generalized Born rule 292, 313
　Hebb rule 374, 375
　and HoTT/UF 44, 47, 48, 50, 53
　of identity 65
　of inference 122, 126
　intuitionistic 115
　Kripke modalities 172, 185
　of local set theories 120
　of logic 44, 108, 115, 123, 239
　mathematical models 385, 386, 387, 388
　of quantifiers 195
　recursive 137
　rewriting 228
　semantic 166
　sequent calculus 232, 233
　structural 223, 224, 228, 231
　of surgery 250, 251
　twist 258f

sameness 37, 39–40, 55
satisfaction, first-order logic (FOL) 201–3
scheme theory 21, 28, 29, 32n1
science/scientific theories
　Boolean algebras 410, 418
　categories see theories and categories
　category theory as a framework for the structure of 432–5
　foundations for mathematics 136–7, 159
　future directions 421–6
　mathematical structure 442
　structure 434
self-organization 370–3
semantics
　algebraic 163
　category theory (CT) 113
　contextual, in classical computation 282
　duality of syntax and semantics 181, 184, 417–21
　FOLDS (First-Order Logic with Dependent Sorts) 153
　functorial 113, 117–19
　as homomorphisms 165–7
　Kripke 169–77, 185, 212–13

semantics (*cont.*)
   Kripke–Joyal 115
   Lewis's 213–15
   as natural transformations 196–9
   neighborhood 178–9, 185
   possible-world 178, 179
   rules 166
   semantic categories 409–13
   semantic structures 203–6
   sheaf semantics 115
   spatial 163
   syntax-semantics duality 167–8
   theories, semantic view 329, 404, 411, 412, 417, 421, 426, 433
   topological 179–81
   topos semantics 115
sequent calculus 226, 232–3
sequential composition operation 287, 293–4
set-quotient 51
set-theoretic approach 40
   see also set theories
   and abstract mathematical concepts 138, 140, 143, 146, 158
   canonical maps 91, 92, 96
   categorical foundations 79
   categorical logic and model theory 115, 120, 123
   categorical quantum mechanics (CQM) 286, 288, 297
   equivalence classes 40
   foundations for mathematics 288
   and geometry, reviving philosophy of 19, 20, 24
   higher-dimensional categories (HDC) 158
   predicates 433
   properties 59
   and role of set theories in mathematics 1–2, 3, 5, 7, 11, 19, 20
   structural realism 433, 434, 435, 436, 437, 439, 440n28, 441, 442, 445, 446
   and univalence 66
set theories
   see also set-theoretic approach
   alternative set theory (Lawvere) 43
   base sets 62
   canonical maps 91
   categorical 8
   and category theories 138n6
   collections and sets 38–9, 41
   conception of "set" 37
   constructibility 3, 14
   Continuum Hypothesis (CH) 4–5
   continuum hypothesis (CH) 1, 2, 3
   development 136
   effective knowledge 1
   ETCS *see* Elementary Theory of the Category of Sets (ETCS)
   explicit, spread of 6
   extensional 91
   foundational ascendancy of sets 36
   foundations for mathematics 36, 37, 41–3, 59
   functions, sets connected by 145
   gauge invariant 10–13
   Grothendieck universes 1, 5, 76
   and HoTT/UF 36
   and isomorphism 2, 9
   local 119–27
   membership tree structure 13–14, 15, 43
   and numbers 3
   ontological unification 91
   orthodox 2, 3–5
   overview 1–3
   pure sets 138–9
   roles in mathematics 1–17
   sets for working mathematicians 6–10
   subsets of a set, in ETCS 71–2
   variable sets 114
   variation of role with mathematics type 1–2
   ZFC not synonymous with orthodox set theory 2, 3–5
shape modality 31
sharp modality 31
sheaf semantics 115
six-dimensional Lorentz category 350–6
SMCs *see* symmetric monoidal categories (SMCs)
sonar system, submarines 383
space
   see also space–time theories, classical
   essence of (Weyl) 24–5
   extension of notion 26
   model spaces 32
   n-dimensional topological 24
space–time theories, classical 330
   foundations 329–48
   Galilean 333
   and isometric space–times 342
   Minkowski 53, 54, 55, 336, 337
   Newtonian 333
   relativistic 336n18, 341
   structure 332–5, 344
Specker triangle 280
Stinespring dilation 323, 326
Stone duality 183–5, 418
string diagrams 299–311
   see also categorical quantum mechanics (CQM)
   adjoints and connectedness 307–8
   adjoints and unitarity 304–6
   Aleks and Bob scenario 303–6, 311, 324–5
   category-theoretic counterpart 308–10
   compact closed categories 308
   conjugate 305
   defining 299

string diagrams (*cont.*)
  non-separability 299–302
  process–state duality 301, 304
  quantum processes 316
  traces and transposes 302–4
string theory 20, 344, 414
structural continuity 430
structuralism 48, 54, 55, 138n7
  Principle of Structuralism 58–9, 61, 64, 65
structural realism 430–49
  balanced ontic structural realism
    (BOSR) 432, 440, 443–7
  methodological 435
  no miracles argument (NMA) 430
  pessimistic meta-induction argument
    (PMI) 430
  radical ontic structural realism (ROSR) 432,
    435–8, 443
structure-preserving maps 60, 179
  see also morphism
structure/structures
  canonical maps 98
  category theory (CT) 60, 331–2
  causal 323–4
  excess structure, and gauge theory 335–7
  and FOLDs 148
  and geometry 24
  isomorphism 60
  logico-linguistic 432
  partial structures 433, 442
  semantic 203–6
  shared structure 433, 434
  space–time theories, classical 332–5, 344
  structural changes 367–70
  structural rules 223, 224, 228, 231
  synthetic theory of structures, HoTT/UF
    as 48
  ultraproduct structure 418
submarines, sonar system 383
substantivism 329
substructure, autonomy of 185–92
  closure and irrelevance 187–9
symmetric monoidal categories (SMCs) 286,
  297, 298, 310
synchronicity laws 372
syntactic categories 116, 414, 415, 416, 419, 421,
  422, 423n24, 425, 426
  theories as categories 403, 404, 406,
    408, 412n11
syntax
  abstract mathematical concepts 142–4
  as algebras 163–5
  category theory (CT) 113
  duality of syntax and semantics 181, 184,
    417–21
  as functors 193–6
  syntax-semantics duality 167–8

synthetic geometry 42
synthetic theories 41, 42, 43, 48
system-types 288, 290, 321

technical adequacy
  categorical foundations 69, 75, 78–80
  results 80
  ZFC (Zermelo-Fraenkel set theory with
    Axiom of Choice) 75–8
teleportation 299, 304, 310
temporal logic 171, 172
tensor connective 223, 224, 228n6, 239,
  243, 298
tensorial strength 224
tensor unit 223
terminal objects 325
Thales 349
theorems
  Bézout's theorem 28
  Church–Rosser theorem 235
  completeness 126–7
  Fermat's Last Theorem 13
  Helmholtz–Lie theorem 24
  justification of 4
  Mostowski embedding theorem 14
  no-broadcasting theorem 286, 320
  soundness 126
theories
  see also category theory (CT); *specific
    theories, such as Chern–Weil theory*
  algebraic 113, 117
  and categories *see* theories and categories
  distributive lattices 225
  duality of syntax and semantics 417–21
  electrical devices 290
  electromagnetic 10, 11
  elementary 113
  equivalent 406–9
  first-order 115
    as foundations 82n16
  geometric 115, 130–2
  graph theory 87n23
  inequivalent 335
  prior theory, autonomy 82–4
  scheme theory 21, 28, 29
  scientific 402–29
  semantic view 329, 404, 411, 412, 417, 421,
    426, 433
  set theories *see* set theories
  synthetic 41, 42, 43, 48
  "theory in itself" 426
  type theory *see* type theory
theories and categories 402–17
  the category of theories 413–17
  equivalent theories 406–9
  semantic categories 409–13
  syntactic categories *see* syntactic categories

theories as categories 402–13
thinning links 233, 241–3
time
   see also space–time theories, classical
   dimensions of 351
   as imaginary space 351
   nature of 349
   primary role of 359–61
   three dimensions 351
topological quantum field theories (TQFT) 436, 439
topologies
   algebraic 43, 50, 86, 103, 108, 113
   box 5
   Gödel translation 181–3
   and modalities 177–85
   topological semantics 179–81
   topological spaces 179, 183, 185, 419n19
toposes/topos theory 28
   Boolean toposes 114, 115, 131
   classifying 115
   elementary toposes 114, 119
   geometric theories and classification of toposes 130–2
   Grothendieck toposes 114, 131
   higher-topos theory 20
   internal language 114–15
   internal logic 114
   and local set theories 120
   models in 130–2
   scientific theories 418, 419
topos semantics 115
traces and transposes, string diagrams 302–4
transposes, and traces 302–4
triple-dual diagram 246, 247f
truncation 45, 46, 47, 49, 50
truth tables 163
truth values 45, 46, 49
   categorical logic 113, 114, 119
type theory
   see also Homotopy Type Theory and Univalent Foundations (HoTT/UF); theories
   categories 116
   computational character 47
   constructive 58, 62
   identity of types 63–4
   impredicative 63
   intuitionistic type theory (ITT) 115, 123, 425n27
   and logic 43–7
   of Martin-Löf 48
   properties 67
   propositions as types 45
   quantum types 311–14
   rules 44
   types and truth values 49

ultraproduct structure 418
unitarity, and adjoints 304–6
univalence/Univalence Axiom 48, 49, 55, 58, 65
   as extensionality principle 66
   historical perspective 66–7
Univalent Foundations Program 43
   see also Homotopy Type Theory and Univalent Foundations (HoTT/UF)
universe
   of abstract mathematical concepts 144–8
   Grothendieck universes 1, 5, 76
   higher-dimensional categories (HDC) 156–8, 160
unlimited categories 75, 77, 80

vectors/vector spaces 390–2
   categorical model of vector spaces 392–4
Vienna Circle 18, 19
Von Neumann–Bernays–Gödel (NBC) 75
von Neumann ordinals 13

Yang-Mills theory 330, 338–9, 340, 342, 344
Yoneda's lemma 393, 395

ZBQC (Zermelo set theory with bounded quantifiers and choice) 78
Zermelo-Fraenkel set theory with Axiom of Choice see ZFC (Zermelo-Fraenkel set theory with Axiom of Choice)
Zermelo set theory with bounded quantifiers and choice (ZBQC) 78
ZFC (Zermelo-Fraenkel set theory with Axiom of Choice) 2
   ascendancy of 46
   Axiom of Foundation 4, 13
   axioms 3, 4, 7–8, 139
   and categorical foundations 69
   circuits for 66n4
   cumulative hierarchy, interpreted in 106
   encoding mathematics into 45
   and ETCS 6, 72
   function evaluation, formalizing 8
   global membership relation 11, 12, 71
   HoTT/UF compared 37, 43n9, 44, 49
   iterated hierarchy for ZFC sets 3, 14
   justification of 84
   not synonymous with orthodox set theory 2, 3–5
   proof issues 3, 4
   standard approach 7
   supplementation of 80
   technical adequacy 69, 75–8, 87